LIQUID MATTER

Editor

Meinard Reiter

scitus academics

Liquid Matter

Edited by **Meinard Reiter**

This Book is Printed on Acid-Free Paper. ♾

ISBN: 9781681178554
Library of Congress Control Number: 2017948882

SCITUS Academics LLC,
616, Corporate Way, Suite 2, 4766,
Valley Cottage, NY 10989

www.scitusacademics.com

Notice

Printed in United States of America.

Preface

The liquid state of matter is an intermediate phase between solid and gas. A liquid is a fluid. Unlike a solid, the molecules in a liquid have a much greater freedom to move. The forces that bind the molecules together in a solid are only temporary in a liquid, allowing a liquid to flow while a solid remains rigid. Like the particles of a solid, particles in a liquid are subject to intermolecular attraction; however, liquid particles have more space between them, so they are not fixed in position. The attraction between the particles in a liquid keeps the volume of the liquid constant. The movement of the particles causes the liquid to be variable in shape. Liquids will flow and fill the lowest portion of a container, taking on the shape of the container but not changing in volume. Liquid is one of the four primary states of matter, with the others being solid, gas and plasma. The limited amount of space between particles means that liquids have only very limited compressibility.

This volume addresses modern problems in the fields of liquids, solutions and confined systems, critical phenomena, as well as colloidal and biological systems. The book focuses on state-of-the-art developments in contemporary physics of liquid matter structure of liquids in confined systems, phase transitions, supercritical liquids and glasses, and covers the most recent developments in the broader field of liquid state. Recent trends and progress in the field of liquid matter are explored in this volume by a wide spectrum of contributions from liquid state physicists, chemists and chemical engineers.

Table of Contents

CHAPTER 1

A Review of Ionic Liquids, Their Limits and Applications

Khashayar Ghandi

Department of Chemistry, Mount Allison University, Sackville, Canada

ABSTRACT

Since environmental pollution caused by chemical and energy industries has increased for several decades, there is a social expectation that scientists and engineers try to design sustainable chemical processes, to generate less hazardous materials and more environmentally friendly sources of energy production. In this review the roles of Ionic Liquids (ILs) and IL based solvent systems as proposed alternative for conventional organic solvents are described. Since there are already many reviews on benefits of ILs, after a very brief review of ILs we focus mostly on aspects that are not covered in other reviews, in particular the known limits of these solvents. In addition, different methods to measure the physicochemical properties relevant to their use in energy storage applications such as fuel cells and batteries are introduced. The physicochemical properties that are reviewed are thermal properties, conductivity and chemical reactivity. The focus of the review is on the literature after 2008, with the exception of some important historic articles on ILs.

Keywords: Technology; Preference for Quality; Volume of Trade; Vertical Intra-Industry Trade

1. INTRODUCTION

One of the major sources of waste is solvent losses that end up in the atmosphere or in ground water [1-6]. Solvent use has been reported to account for about 60% of the overall energy in pharmaceutical production, and it has been responsible for 50% of post-treatment greenhouse gas emissions [7,8]. Therefore, solvent selection should be considered systematically to improve synthesis conditions within the framework of green chemistry principles.

Various methods and tools have been developed for the identification and selection of appropriate solvents for synthesis. Consequently, there are a number of solvent selection guides available in the literature [1-8]. We have categorized the common solvents in three different classes of preferred, usable, and undesirable (Table 1) [1-8]. The first three items in the preferred column are the most desired solvents and ILs on their own are a large class of solvents. However not all ILs are green solvents.

The focus of this review is on the literature after 2008, with the exception of some important historic articles. We will first have a very brief review of ILs and then we will discuss the limits of these solvents. Most of the limits in the literature are reported for non protic ILs therefore in our review of protic ionic liquids instead of discussion of their limits we focused mostly on their applications in energy industry. Consequently several methods to measure the physicochemical properties relevant to their use in energy industry applications are reviewed.

2. IONIC LIQUIDS

Room-temperature ILs, organic salts that are liquid below 100°C, have received considerable attention as substitutes for volatile organic solvents. Since they are nonflammable, non-volatile and recyclable, they are classified as green solvents. Due to their remarkable properties, such as outstanding solvating potential [9], thermal stability [10] and their tunable propertiesby suitable choices of cations and anions [11], they are consideredfavourable medium candidates for chemical syntheses.

ILsare usually categorized into four types based on their cation segment: 1) alkylammonium-, 2) dialkylimidazolium-, 3) phosphoniumand 4) N-alkylpyridiniumbased ILs (Figure 1).

Table 1. Guide for solvent selection.

Preferred	Usable	Undesirable
Sc-CO_2	Cyclohexane	**Pentane**
ILs	Heptane	**Hexane(s)**
Water	Toluene	**Di isopropyl ether**
Acetone	Methyl cyclohexane	**Diethyl ether**
Ethanol	Isooctane	**Dichloromethane**
2-propanol	Acetonitrile	**Dichloroethane**
1-propanol	2-MeTHF	**Chloroform**
Ethylacetate	THF	**Pyridine**
Isopropylacetate	Xylenes	**Dioxane**
Methanol	DMSO	**Dimethoxyethane**
1-Butanol	Acetic acid	**Benzene**
***tert*-Butyl alcohol**	Ethylene glycol	**Carbon tetrachloride**
	Methyl Ethyl Ketone	

Note: Sc-CO_2 is referred to Supercritical-CO_2.

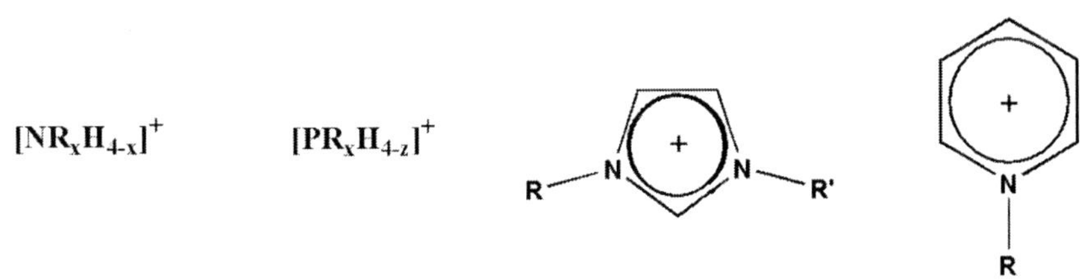

Figure 1. Alkylammonium, phosphonium, dialkylimidazolium and N-alkylpyridinium cations.

Although these ILs are used successfully as solvents and catalysts in many reactions, there are some limitations in their use. In the following, we will describe the known advantages and disadvantages of each class of ILs.

The first synthesized IL was an ammonium-based one (ethanolammonium nitrate, EOAN), which was reported by Gabriel in 1888 [12]. Ammonium-based ILs have been used widely as electrolytes in high-energy electrochemical devices owing to their good electrochemical cathodic stabilities, low melting points and low viscosities [13-15].

Popular imidazolium-based ILs are among the most studied ILs. Selection of the imidazolium ring as a cation (Figure 2) is often due to its stability within oxidative and reductive conditions [16], low viscosity of imidazolium ILs and their ease of synthesis [17]. There are also several reports regarding the application of imidazolium-based ILs as catalysts for the improvement of reaction time, yield and chemoselectivity of many organic reactions [18-21].

However, Olofson et al., in 1964 [22], reported on a kinetics studydemonstrating that the proton sandwiched between the two nitrogen atoms (H2)in the imidazolium cation undergoes deuterium exchange in deuterated solvent because of its acidic nature. Two later studiesreported that deprotonation of the imidazolium cation to the highly reactive carbene and hence showedthe noninnocent nature of immidazolium-based ILs under basic conditions [23,24].

In another investigation [25], the low yield of a basecatalyzed Baylis-Hillman reaction in the presence of imidazolium-based ILs was attributed to a side reaction involving the imidazolium-based IL (Scheme 1). This observation also confirmed that using this type of IL under basic conditions needs to be considered with more precaution to avoid unexpected side reactions (Scheme 2).

Pyridinium-based ILs are more novel in comparison with their imidazolium-based counterparts, and research on their stability, reactivity and catalytic role in organic synthesis is still in progress. Although they show poor regioselectivity in palladium-catalyzed telomerization of butadiene with methanol [26], and they have a negative effect on the rate of some Diels-Alder reactions [27], applications of this type of ILs are quite successful in reactions such as Friedel-Crafts [28] and Grignard [29]. The catalytic role of pyridinium-based ILs has been shown to be remarkablein the synthesis of some pharmaceutical agents such as 1,4-dihydropyridine [30], dihydropyrimidinones [31] and 3,5-bis(dodecyloxycarbonyl)- 1,4-dihydropyridine derivatives [32].

Phosphonium-based ILs aremore novel than the imidazoliumand pyridinium-based ILs. They are more thermally stable (in some cases up to nearly 400°C!) [33] in comparison with ammonium and imidazolium salts, and this remarkable property makes them suitable for reactions that are carried out at greater than 100°C. Phosphonium-based ILs are used as the catalyst and solvent for hydroformylation [34], palladium-catalyzed Heck re actions [35] and palladium-mediated Suzuki crosscoupling reactions [35]. In addition, they arealso powerful phase-transfer catalysts for the Halex reaction [36].

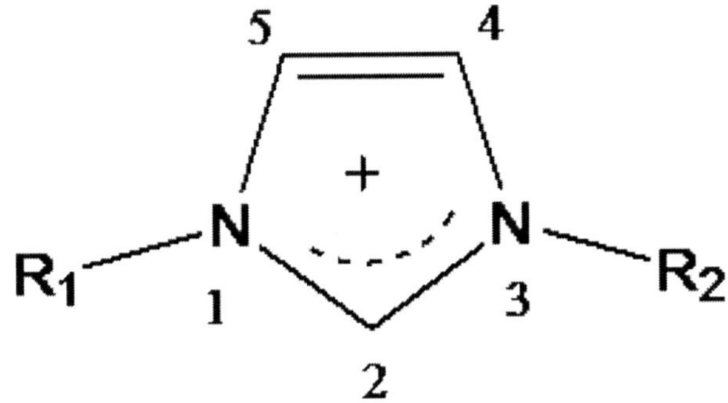

Figure 2. The imidazolium cation.

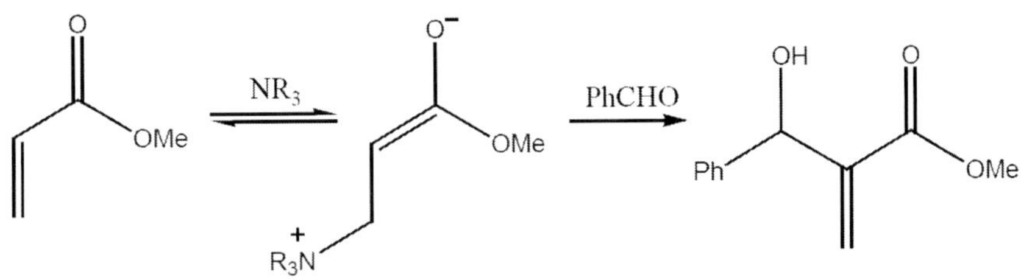

Scheme 1. Base-catalyzed Baylis-Hillman reaction.

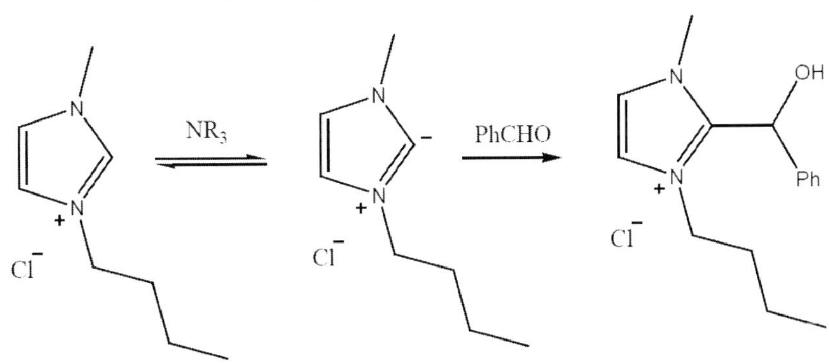

Scheme 2. The side reaction of an imidazolium-based ionic liquid with benzaldehyde in the Baylis-Hillman reaction.

Recently, phosphonium-based ILs have been used for CO_2 capture [37]. Along with their application in the synthesis of a novel polystyrene-based material [38], the styrenic derivatives of phosphonium-based ILs are used as monomers in the synthesis of phosphonium-containing random copolymers [39]. Thecyclohexadieneylradical in trihexyl (tetradecyl) phosphonium chloride (IL101) [40] has been studied, and the effects of temperature and solvent on the reaction have been investigated using muon techniques at the TRIUMF National Laboratory of Canada [41]. These studies showed [40,41] reactive free radicals do not react with phosphonium ionic liquids.

Although phosphonium-based ILs showed good stability in the presence of bases (even in reactions involving strong bases such as Grignard reagents [42]), they are still susceptible to reaction with small bases [43] (Scheme 3).

Unlike ammonium-based ILs, which undergo Hoffman or -elimination in the presence of a base at high temperature, phosphonium-based ILs tend to decompose to tertiary phosphine oxides and alkanes under alkaline conditions[44] (equation (1)).

$$R_3P\text{-}CH_2\text{-}R' + OH^- \rightarrow R_3P{=}O + CH_3\text{-}R' \qquad (1)$$

3. PROTIC ILS

ILs can be divided into two broad categories: protic ILs (PILs) and aprotic ILs (APILs). PILs are produced through proton transfer from a Brønsted acid to a Brønsted base.

Historically, the first PIL, EOAN was reported in 1888 by Gabriel [12]. There are a large number of reports on the properties of APILs and their applications in different fields [9,45-51]; however, there are few reviews on PILs [52,53].

In comparison with APILs, PILs often have higher conductivity and fluidity as well as lower melting points [54]. They are also cheaper and more convenient to prepare as their synthesis does not involve the formation of byproducts [52]. However, there are several reports of the ability of PILs to form a hydrogen bond network, which can limit the ionicity of PILs in comparison with APILs [55-57]. The hydrogen bonding of PILs has been identified by NMR [55], X-ray diffraction [58] and neutron diffraction [58]. The most studied example of the ability of a PIL to form supramolecular networks through hydrogen bonding is related to ethylammonium nitrate (EAN) [59-61].

Normally, PILs are prepared through the neutralization of a base by an acid [62,63] or the mixture of equimolar amounts of acid and base [54,55]. Ideally, the proton transfer is completed from Brønsted acid to Brønsted base, but, in most cases, a neutral species is formed owing to incomplete proton transfer. Aggregation or the formation of ion complexes also can happen to prevent complete proton transfer, which limits the ionicity of the

PILs [51]. Although there is still no standard method to measure the ionicity of PILs, some qualitative techniques such as NMR spectroscopy [56,64], changes in thermal properties as a function of stoichiometry [56,64], IR spectroscopy [64], Raman spectroscopy [64] and ionic conductivity by using Walden plots [65,66] have been used to provide information about the ionicity of PILs. Obviously, proton transfer improves by using stronger acids and bases.

PILs have wide applications in biological systems [52] and chromatography [52,67]. In addition, they have been applied as proton-conducting electrolytes for polymer membrane fuel cells [68-72], because of the advantage of having a defined proton activity as well as high proton conductivity, allowing the fuel cell to operate under nonhumidified and high-temperature conditions. PILs also have been widely used as a Brønsted acid or base [72] in many acid-base-catalyzed organic reactions such as Knoevenagelcondensation [73], the Diels-Alder reaction [74], aldol condensation[75], Fischer esterification [76] and pinacol rearrangement [77] owing to their non-corrosive, non-volatile and recyclable nature in comparison with mineral acids.

Since ILs, including PILs, are great microwave absorbents, they are good candidates for application as a medium or catalyst in many microwave-assisted reactions. Henderson and Byrne [77] used several ammoniumbased PILs as potential mediators for pinacol rearrangements under microwave irradiation; complete conversion was observed in optimized conditions in a pinacol rearrangement of hydrobenzoin (Scheme 4).

Esterification of benzoic acid with a variety of alcohols and a variety of acids with benzoic alcohols are also reported to be efficient when used with some types of imidazoliumor pyridinium-based PILs under microwave irradiation[78] (Scheme 5).

The solvent-free synthesis of coumarins [79], the estrification of salicylic acid [80] and the dehydration of dfructose and glucose [81] are other examples of using PILs as acidic catalysts under microwave irradiation.

Compared to the disadvantages of ordinary ILs (mostly their reactivity under certain reaction conditions where ILs are used as medium for reaction) there are limited disadvantages of PILs including side reactions during the planned reactions. On the other hand PILs have been considered mostly for their applications in energy industry not as medium for chemistry. For such applications knowledge of their physicochemical properties is important.

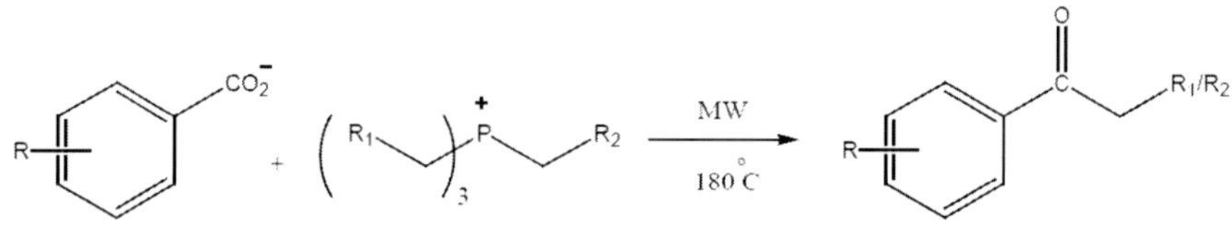

Scheme 3.Reactions of benzoate salts with trihexyl (tetradecyl) phosphonium chloride(IL101)under microwave irradiation.

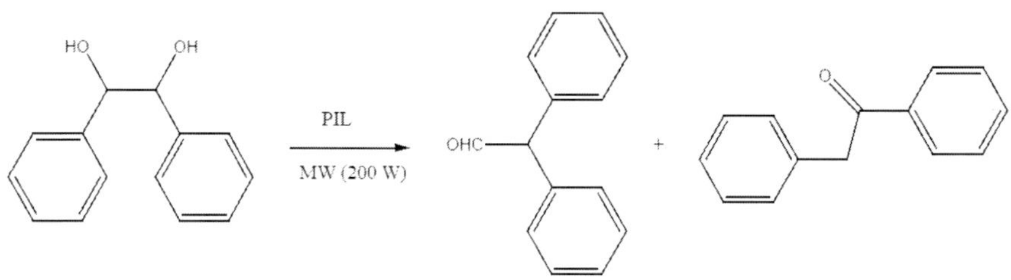

Scheme 4.Microwave-assisted hydrobenzoin pinacol rearrangements mediated by protic ionic liquids (PILs) [78].

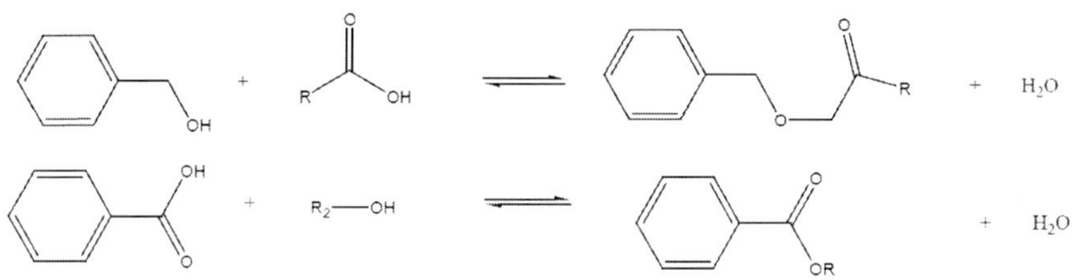

Scheme 5.Fisher esterification reaction in Brønsted acidic ionic liquids under microwave irradiation [79].

3.1. Physicochemical Properties of PILs

Different potential applications of PILs rely on their physicochemical properties, which vary based on the structures of the cation and anion used in the system. In the following sections, different methods for the investigation of thermal and ionic properties of PILs are described briefly.

3.1.1. Thermal-Phase Behaviour

Differential scanning calorimetry (DSC) is one of the thermoanalytical techniques used to study phase transitionsof materials (Figure 3). In DSC, both the sample and the reference are maintained at the same temperature ($\Delta T = T_s - T_r = 0$) and any heat transfer between the sample and reference materials is recorded against the temperature [82,83]. The reference in the DSC method is a material that does not show phase change over a wide temperature range, Alumina (Al_2O_3) and silicon carbide (SiC) are mostly used as the reference materials in DSC. The DSC trace usually is plotted as the heat flow versus temperature; deviation from the baseline of the DSC trace is representative of a phase change such as melting of the sample. Steps in the baseline position of the DSC curves usually refer to the glass-transition temperature (T_g) of the materials, which is a transition that happens for amorphous and semi-crystalline materials including some ILs. For ILs, the T_g indicates the cohesive energy within the salt which is decreased by repulsive Pauli forces and increased through attractive Coulomb and van der Waals interactions [52]. Therefore, it is feasible to reach a lower T_g by decreasing the cohesive energy of the ILs through the modification of the their cations and anions.

Usually the dependence of logarithm of viscosity on inverse temperature, i.e. $\log(\eta)$ vs. T_g/T are used to show the fragility of materials. More deviation from a linear trend indicates more fragility, which means that as the temperature goes up, the viscosities will decrease at a faster rate than the Arrhenius relationship. Almost all the data on room temperature PILs provide evidence that they show fragile behavior [52,84-88].

3.1.2. Thermal Stability

Thermogravimetry analysis (TGA) is a type of thermal analysis that examines the mass loss of the sample as a function of temperature in a controlled atmosphere [83] (Figure 4).

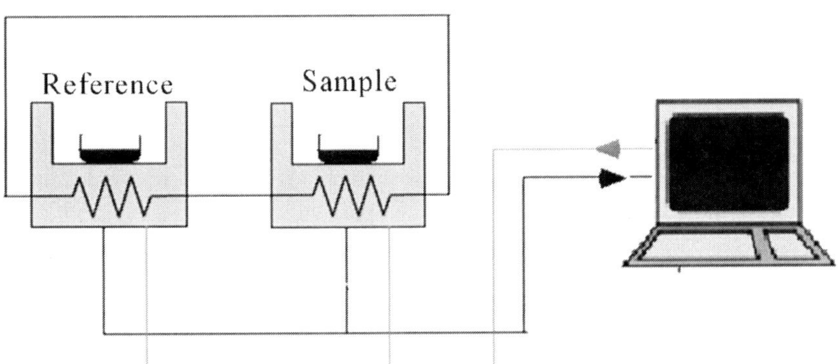

Figure 3. Simple schematic diagram of differential scanning calorimetry (DSC).

It has been established that PILs with large protontransfer energies decompose before reaching their boiling points [52,84,89]. The decomposition temperature varies between 100°C and 360°C [52,68,85,90]. PILs with a bis (trifluoromethane) sulfonamide anion (TFSI) with alkylammonium cations, the imidazolium cation, and a variety of heterocyclic cations are known as the most stable PILs with decomposition temperatures of >200°C [52, 90].

3.1.3. Viscosity

Viscosity is an important property of ILs for different applications. Normally, materials with greater van der Waals interactions and hydrogen bonding have higher viscosities [52,86,87,90].

Although it has been observed that the size of the PIL components has little effect on viscosity, the structure of the anion has a large effect on the viscosity and usually more than the structure of the cation [88]. This is an interesting observation that needs to be further substantiated by theoretical modeling. Specific structural features can affect viscosity of ILs. E.g. in substituted imidazolium PILs, stacking of the aromatic rings leads to higher viscosity [52,90]. Increasing the cation size by increasing the ring number in lactam-based PILs increases the viscosity by enhancing the cation-anion interactions [68]. Analysis of the temperature dependence of viscosity for PILs with different glass transitions, T_g, suggests that PILs in general are among fragile material and the PILs with higher fragility have lower viscosity (Figure 5) [90].

3.1.4. Conductivity

The ionic conductivity, which depends on the available charge carriers and their mobility (which depends on viscosity), varies with the molecular weight, and size of the ion. The conductivity of PILs is limited usually by their ion mobility resulting from aggregation [52,88]. Therefore, less ionic interaction and more delocalized charge lead to higher conductivity; therefore, high ionic conductivity values will be expected for the stronger Brønsted acids and bases [69,89]. Ion conductivity decreases by increasing the size of the cation (less mobility); consequently, the conductivity of PILs with longer alkyl chains decreases [52]. Therefore, the higher ionic conductivity of the 1-methyl-2-methyl imidazolium-based PILs compared with the 1-benzyl-2 methyl imidazoliumbased PILs [81], and the higher conductivity value for methyl formate over butyl ammonium formate [84], should be due to the increase in the size of the cation. The ionic conductivity of heterocyclic PILs increases with less symmetrical cation structure and smaller molecular weight [52,68]. No obvious trend can be found for the

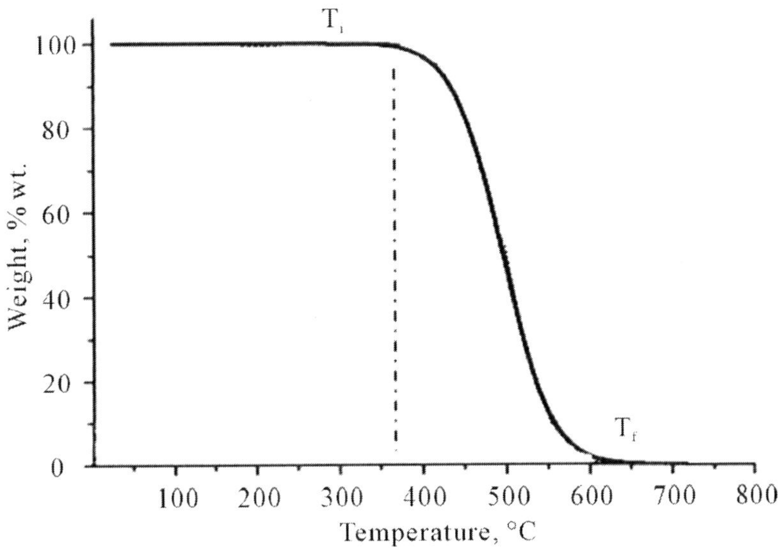

Figure 4. Thermogravimetry analysis (TGA) sample curve for a one-stage process. Ti, intial temperature; Tf, final temperature.

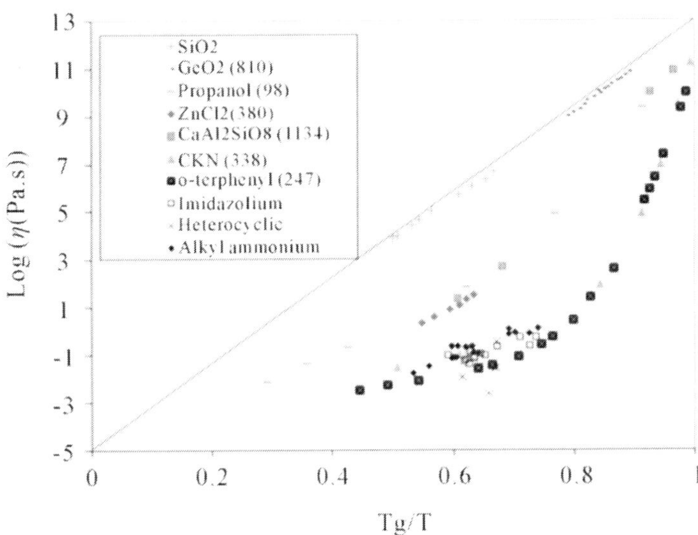

Figure 5. Logarithm of viscosity vs. Tg/T (in Kelvin) for several materials with different glass temperatures (in Kelvin, presented in the legend) as well as for several known PILs [84]. The ellipse shows the range of data for PILs.

anions used in the system: as an example, nitrate has the highest ionic conductivity in ethyl ammonium-based PILs in comparison with formate, acetate, but rate and lactate anions, but in the series of ethanol ammonium-based PILs, nitrate has the lowest ionic conductivity compared with the same anions [84]. There are a few alkylammonium-based PILs, such as methylammoniumformate (MAF), EAN, ethylammonium acetate (EAA), and ethylammonium formate (EAF), that have a high ion conductivity over 10 $mS\cdot cm^{-1}$ at 25°C [52].

A Waldon plot, which is of the equivalent conductivity against the log of the fluidity (inverse viscosity), is a good indication of the ionicity of the ILs. The Walden rule is shown in equation (2), where Λ is the molar conductivity and η is the viscosity.

$$\Lambda\eta = \text{Constant} \tag{2}$$

Negative deviation from the straight line can be representative of an incomplete proton transfer or aggregation of PILs [65] (Figure 6). The solid ideal line corresponds to a dilute aqueous KCl solution in which the system is known to be fully dissociated and to have ions of equal mobility. The presence of parent acid and base molecules in the system can be indicated by vertical deviation from the Walden line as well. The PILs are usually categorized as poor ILs, except for those with a BF_4 anion, which have good ionicity [52].

4. CONCLUSIONS

Over the past five years, ILs have continued to be used significantly as a medium and catalyst for many reactions. Despite the significant potential of ILs in this regard, there are reports that show in some cases ILs react with reactants and therefore they cannot be considered as inert solvents. Therefore synthetic chemists should be cautious when designing reactions in ILs depending on reactions they want to do. Overall PILs are less studied compared to aprotic ILs, however among PILs that have been studied so far there is no report of reactivity of PILs with reagents used for chemical reactions. Moreover PILs have an easily tunable and interesting range of physicochemical properties that make them potential candidates for applications in alkaline and alkaline earth ion batteries and fuel cells.

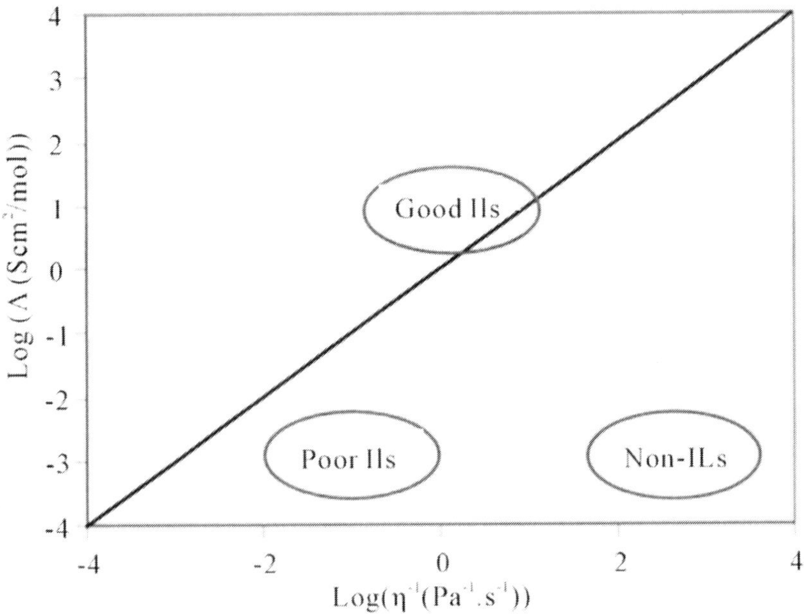

Figure 6. Walden plot of log (equivalent conductivity) against log (fluidity).

There is a significant need for modelling IL and PIL chemical and physicochemical properties to guide the applications of these materials more efficiently. On the experimental side, the least understood reactions in ILs and PILs are the free radical reactions. There is a need for a microscopic understanding of free radical chemistry in ILs and PILs.

REFERENCES

1. D. J. C. C. Concepción Jiménez-González, "Green Chemistry and Engineering—A Practical Design Approach," John Wiley & Sons Inc., Hoboken, 2011, pp. 3-39.
2. V. K. Ahluwalia, "Green Chemistry, Environmentally Benign Reaction," CRC Press & Francis Group, Boca Raton, 2009.
3. R. A. Sheldon, "Fundamentals of Green Chemistry: Efficiency in Reaction Design," Chemical Society Reviews, Vol. 41, No. 4, 2012, pp. 1437-1451. http://dx.doi.org/10.1039/c1cs15219j
4. P. J. Dunn, "The Importance of Green Chemistry in Process Research and Development," Chemical Society Reviews, Vol. 41, No. 4, 2012, pp. 1452-1461. http://dx.doi.org/10.1039/c1cs15041c
5. D. Ghernaout, B. Ghernaout and M. W. Naceur, "Embodying the Chemical Water Treatment in the Green Chemistry—A Review," Desalination, Vol. 271, No. 1-3, 2011, pp. 1-10. http://dx.doi.org/10.1016/j.desal.2011.01.032

6. C. Capello, U. Fischer and K. Hungerbuhler, "What Is a Green Solvent? A Comprehensive Framework for the Environmental Assessment of Solvents," Green Chemistry, Vol. 9, No. 9, 2007, pp. 927-934. http://dx.doi.org/10.1039/b617536h
7. C. S. Slater and M. Savelski, "A Method to Characterize the Greenness of Solvents Used in Pharmaceutical Manufacture," Journal of Environmental Science and Health, Part A, Vol. 42, No. 11, 2007, pp. 1595-1605.
8. E. M. Rundquist, C. J. Pink and A. G. Livingston, "Organic Solvent Nanofiltration: A Potential Alternative to Distillation for Solvent Recovery from Crystallisation Mother Liquors," Green Chemistry, Vol. 14, No. 8, 2012, pp. 2197-2205. http://dx.doi.org/10.1039/c2gc35216h
9. T. Welton, "Room-Temperature Ionic Liquids. Solvents for Synthesis and Catalysis," Chemical Reviews, Vol. 99, No. 8, 1999, pp. 2071-2084. http://dx.doi.org/10.1021/cr980032t
10. N. Meine, F. Benedito and R. Rinaldi, "Thermal Stability of Ionic Liquids Assessed by Potentiometric Titration," Green Chemistry, Vol. 12, No. 10, 2010, pp. 1711-1714.http://dx.doi.org/10.1039/c0gc00091d
11. S. Ahrens, A. Peritz and T. Strassner, "Tunable Aryl Alkyl Ionic Liquids (TAAILs): The Next Generation of Ionic Liquids. Angewandte Chemie International Edition, Vol. 48, No. 42, 2009, pp. 7908-7910. http://dx.doi.org/10.1002/anie.200903399
12. S. Gabriel and J. Weiner, "Ueber Einige Abkömmlinge des Propylamins," Berichte der Deutschen Chemischen Gesellschaft, Vol. 21, No. 2, 1888, pp. 2669-2679. http://dx.doi.org/10.1002/cber.18880210288
13. M. Armand, F. Endres, D. R. MacFarlane, H. Ohno and B. Scrosati, "Ionic-Liquid Materials for the Electrochemical Challenges of the Future," NatureMaterials, Vol. 8, No. 8, 2009, 8, 621-629. http://dx.doi.org/10.1038/nmat2448
14. Guerfi, M. Dontigny, P. Charest, M. Petitclerc, M. Lagacé, A. Vijh and K. Zaghib, "Improved Electrolytes for Li-Ion Batteries: Mixtures of Ionic Liquid and Organic Electrolyte with Enhanced Safety and Electrochemical Performance," Journal of Power Sources, Vol. 195, No. 3, 2010, pp. 845-852. http://dx.doi.org/10.1016/j.jpowsour.2009.08.056
15. Lewandowski and A. Świderska-Mocek, "Ionic Liquids as Electrolytes for Li-Ion Batteries—An Overview of Electrochemical Studies," Journal of Power Sources, Vol. 194, No. 2, 2009, pp. 601-609. http://dx.doi.org/10.1016/j.jpowsour.2009.06.089
16. D. Weingarth, I. Czekaj, Z. Fei, A. Foelske-Schmitz, P. J. Dyson, A. Wokaun and R. Kötz, "Electrochemical Stability of Imidazolium Based Ionic Liquids Containing Cyano Groups in the Anion: A Cyclic Voltammetry, XPS and DFT Study," Journal of The Electrochemical Society, Vol. 159, No. 7, 2012, pp. H611-H615. http://dx.doi.org/10.1149/2.001207jes
17. Wang, H. Luo, X. Luo, H. Li and S. Dai, "Equimolar CO_2 Capture by Imidazolium-Based Ionic Liquids and Superbase Systems," Green Chemistry, Vol. 12, No. 11, 2010, pp. 2019-2023. http://dx.doi.org/10.1039/c0gc00070a
18. J. Safaei-Ghomi, M. Emaeili, et al., "Mild and Efficient Method for Oxidation of Alcohols in Ionic Liquid Media," Digest Journal of Nanomaterials and Biostructures, Vol. 5, No. 4, 2010, pp. 865-871.
19. J.-I. Yu, H.-Y. Ju, K.-H. Kim and D.-W. Park, "Cycloaddition of Carbon Dioxide to Butyl Glycidyl Ether Using Imidazolium Salt Ionic Liquid as a Catalyst," Korean Journal of Chemical Engineering, Vol. 27, No. 2, 2010, pp. 446-451. http://dx.doi.org/10.1007/s11814-010-0074-1
20. Sarkar, S. R. Roy, N. Parikh and A. K. Chakraborti, "Nonsolvent Application of Ionic Liquids: Organo-Catalysis by 1-Alkyl-3-methylimidazolium Cation Based Room-Temperature Ionic Liquids for Chemoselective N-tert-Butyloxycarbonylation of Amines and the Influence of the C-2 Hydrogen on Catalytic Efficiency," The Journal of Organic Chemistry, Vol. 76, No. 17, 2011, pp. 7132-7140. http://dx.doi.org/10.1021/jo201102q

21. Sarkar, R. Bhattarai, D. A. Headley and B. Ni, "A Novel Recyclable Organocatalytic System for the Highly Asymmetric Michael Addition of Aldehydes to Nitroolefins in Water," Synthesis, Vol. 2011, 2011, pp. 1993-1997.
22. S. Sowmiah, V. Srinivasadesikan, M.-C. Tseng and Y.-H. Chu, "On the Chemical Stabilities of Ionic Liquids," Molecules, Vol. 14, No. 9, 2009, pp. 3780-3813. http://dx.doi.org/10.3390/molecules14093780
23. J. P. Canal, T. Ramnial, D. A. Dickie and J. A. C. Clyburne, "From the Reactivity of N-Heterocyclic Carbenes to New Chemistry in Ionic Liquids," Chemical Communications, Vol. 2006, No. 17, 2006, pp. 1809-1818. http://dx.doi.org/10.1039/b512462j
24. V. K. Aggarwal, I. Emme and A. Mereu, "Unexpected Side Reactions of Imidazolium-Based Ionic Liquids in the Base-Catalysed Baylis-Hillman Reaction," Chemical Communications, Vol. 2002, No. 15, 2002, pp. 1612-1613. http://dx.doi.org/10.1039/b203079a
25. L. Magna, Y. Chauvin, G. P. Niccolai and J.-M. Basset, "The Importance of Imidazolium Substituents in the Use of Imidazolium-Based Room-Temperature Ionic Liquids as Solvents for Palladium-Catalyzed Telomerization of Butadiene with Methanol," Organometallics, Vol. 22, No. 22, 2003, pp. 4418-4425. http://dx.doi.org/10.1021/om021057s
26. N. D. Khupse and A. Kumar, "The Cosolvent-Directed Diels-Alder Reaction in Ionic Liquids," The Journal of Physical Chemistry A, Vol. 115, No. 36, 2011, pp. 10211-10217. http://dx.doi.org/10.1021/jp205181e
27. J. M. Snelders and P. J. Dyson, "Efficient Synthesis of β-Chlorovinylketones from Acetylene in Chloroaluminate Ionic Liquids," Organic Letters, Vol. 13, No. 15, 2011, pp. 4048-4051. http://dx.doi.org/10.1021/ol201182t
28. L. Ford, F. Atefi, R. D. Singer and P. J. Scammells, "Grignard Reactions in Pyridinium and Phosphonium Ionic Liquids," European Journal of Organic Chemistry, Vol. 2011, 2011, pp. 942-950.
29. X. Y. Wu, "Facile and Green Synthesis of 1,4-Dihydropyridine Derivatives in n-Butyl Pyridinium Tetrafluoroborate," Synthetic Communications, Vol. 42, No. 3, 2011, pp. 454-459. http://dx.doi.org/10.1080/00397911.2010.525773
30. R. Hajipour and M. Seddighi, "Pyridinium-Based Brønsted Acidic Ionic Liquid as a Highly Efficient Catalyst for One-Pot Synthesis of Dihydropyrimidinones," Synthetic Communications, Vol. 42, No. 2, 2011, pp. 227- 235. http://dx.doi.org/10.1080/00397911.2010.523488
31. K. Pajuste, A. Plotniece, K. Kore, L. Intenberga, B. Cekavicus, D. Kaldre, G. Duburs and A. Sobolev, "Use of Pyridinium Ionic Liquids as Catalysts for the Synthesis of 3,5-Bis(dodecyloxycarbonyl)-1,4-dihydropyridine Derivative," Central European Journal of Chemistry, Vol. 9, No. 1, 2011, pp. 143-148. http://dx.doi.org/10.2478/s11532-010-0132-x
32. K. Tsunashima, A. Kawabata, M. Matsumiya, S. Kodama, R. Enomoto, M. Sugiya and Y. Kunugi, "Low Viscous and Highly Conductive Phosphonium Ionic Liquids Based on Bis(fluorosulfonyl)amide Anion as Potential Electrolytes," Electrochemistry Communications, Vol. 13, No. 2, 2011, pp. 178-181.http://dx.doi. org/10.1016/j.elecom.2010.12.007
33. S. A. Dake, R. S. Kulkarni, V. N. Kadam, S. S. Modani, J. J. Bhale, S. B. Tathe and R. P. Pawar, "Phosphonium Ionic Liquid: A Novel Catalyst for Benzyl Halide Oxidation," Synthetic Communications, Vol. 39, No. 21, 2009, pp. 3898-3904. http://dx.doi.org/10.1080/00397910902840835
34. H. Cao and H. Alper, "Palladium-Catalyzed Double Carbonylation Reactions of o-Dihaloarenes with Amines in Phosphonium Salt Ionic Liquids," Organic Letters, Vol. 12, No. 18, 2010, pp. 4126-4129. http://dx.doi.org/10.1021/ol101714p
35. K. L. Luska, K. Z. Demmans, S. A. Stratton and A. Moores, "Rhodium Complexes Stabilized by PhosphineFunctionalized Phosphonium Ionic Liquids Used as Higher Alkene Hydroformylation Catalysts: Influence of the Phosphonium Headgroup on

Catalytic Activity," Dalton Transactions, Vol. 41, No. 43, 2012, pp. 13533-13540. http://dx.doi.org/10.1039/c2dt31797d

36. Fan, G.-K. Chuah and S. Jaenicke, "Phosphonium Ionic Liquids as Highly Thermal Stable and Efficient Phase Transfer Catalysts for Solid-Liquid Halex Reactions," Catalysis Today, Vol. 198, No. 1, 2012, pp. 300-304. http://dx. doi.org/ .1016/j.cattod.2012.02.063

37. N. D. Harper, N. D. Nizio, A. D. Hendsbee, J. D. Masuda, K. N. Robertson, L. J. Murphy, M. B. Johonson, C. C. Pye and J. A. C. Clyburne, "Survey of Carbon Dioxide Capture in Phosphonium-Based Ionic Liquids and EndCapped Polyethylene Glycol Using DETA (DETA = Diethylenetriamine) as a Model Absorbent," Industrial & Engineering Chemistry Research, Vol. 50, No. 5, 2011, pp. 2822-2830. http://dx.doi.org/10.1021/ie101734h

38. K. Ghandi, "Process for the Production of Polystyrene and Novel Polymers in Phosphonium Ionic Liquids," US Patent: 20,120,049,101, 2012.

39. S. Cheng, M. Zhang, T. Wu, S. T. Hemp, B. D. Mather, R. B. Moore and T. E. Long, "Ionic Aggregation in Random Copolymers Containing Phosphonium Ionic Liquid Monomers," Journal of Polymer Science Part A: Polymer Chemistry, Vol. 50, No. 1, 2012, pp. 166-173. http://dx.doi.org/10.1002/pola.25022

40. J. M. Lauzon, D. J. Arseneau, J. C. Brodovitch, J. A. C. Clyburne, P. Cormier, B. McCollum and K. Ghandi, "Generation and Detection of the Cyclohexadienyl Radical in Phosphonium Ionic Liquids," Physical Chemistry Chemical Physics, Vol. 10, No. 39, 2008, pp. 5957-5962. http://dx.doi.org/10.1039/b804800b

41. B. Taylor, P. J. Cormier, J. M. Lauzon and K. Ghandi, "Investigating the Solvent and Temperature Effects on the Cyclohexadienyl Radical in an Ionic Liquid," Physica B: Condensed Matter, Vol. 404, No. 5-7, 2009, pp. 936-939. http://dx.doi.org/10.1016/j.physb.2008.11.224

42. C. J. Bradaric, A. Downard, C. Kennedy, A. J. Robertson and Y. Zhou, "Industrial Preparation of Phosphonium Ionic Liquids," Green Chemistry, Vol. 5, No. 2, 2003, pp. 143-152. http://dx.doi.org/10.1039/b209734f

43. M.-C. Tseng, H.-C. Kan and Y.-H. Chu, "Reactivity of Trihexyl (tetradecyl)phosphonium Chloride, a Room-Temperature Phosphonium Ionic Liquid," Tetrahedron Letters, Vol. 48, No. 52, 2007, pp. 9085-9089. http://dx.doi.org/10.1016/j.tetlet.2007.10.131

44. K. J. Fraser and D. R. MacFarlane, "Phosphonium-Based Ionic Liquids: An Overview," Australian Journal of Chemistry, Vol. 62, No. 4, 2009, pp. 309-321. http://dx.doi.org/10.1071/CH08558

45. N. V. Plechkova, R. D. Rogers and K. R. Seddon, Eds., "Ionic Liquids: From Knowledge to Application," American Chemical Society, Vol. 1030, 2009, p. 472.

46. J. S. Wilkes, P. Wasserscheid and T. Welton, "Introduction," In: P. Wasserscheid and T. Welton, Eds., Ionic Liquids in Synthesis, Wiley-VCH Verlag GmbH & Co. KGaA, Hoboken, 2007, pp. 1-6.

47. J. D. Holbrey, R. D. Rogers, R. A. Mantz, P. C. Trulove, V. A. Cocalia, A. E. Visser, J. L. Anderson, J. L. Anthony, J. F. Brennecke, E. J. Maginn, T. Welton and R. A. Mantz, "Physicochemical Properties," In: P. Wasserscheid and T. Welton, Eds., Ionic Liquids in Synthesis, Wiley-VCH Verlag GmbH & Co. KGaA, Hoboken, 2007, pp. 57-174.

48. S. Sowmiah, C. I. Cheng and Y.-H. Chu, "Ionic Liquids for Green Organic Synthesis," Current Organic Synthesis, Vol. 9, No. 1, 2012, pp. 74-95. http://dx.doi .org/10.2174/157017912798889116

49. M. J. A. Shiddiky and A. A. J. Torriero, "Application of Ionic Liquids in Electrochemical Sensing Systems," Biosensors and Bioelectronics, Vol. 26, No. 5, 2011, pp. 1775-1787.http://dx.doi.org/10.1016/j.bios.2010.08.064

50. C. M. Gordon, "New Developments in Catalysis Using Ionic Liquids," Applied Catalysis A: General, Vol. 222, No. 1-2, 2001, pp. 101-117. http://dx.doi. org/10.1016/S0926-860X(01)00834-1

51. Karadas, M. Atilhan and S. Aparicio, "Review on the Use of Ionic Liquids (ILs) as Alternative Fluids for CO_2 Capture and Natural Gas Sweetening," Energy & Fuels, Vol. 24, No. 11, 2010, pp. 5817-5828. http://dx.doi.org/10.1021/ef1011337
52. T. L. Greaves and C. J. Drummond, "Protic Ionic Liquids: Properties and Applications," Chemical Reviews, Vol. 108, No. 1, 2007, pp. 206-237. http://dx.doi.org/10.1021/cr068040u
53. K. E. Johnson, R. M. Pagni and J. Bartmess, "Brønsted Acids in Ionic Liquids: Fundamentals, Organic Reactions, and Comparisons," Monatshefte für Chemie, Vol. 138, No. 11, 2007, pp. 1077-1101. http://dx.doi.org/10.1007/s00706-007-0755-6
54. H. Markusson, J.-P. Belières, P. Johansson, C. A. Angell and P. Jacobsson, "Prediction of Macroscopic Properties of Protic Ionic Liquids by ab Initio Calculations," The Journal of Physical Chemistry A, Vol. 111, No. 35, 2007, pp. 8717-8723. http://dx.doi.org/10.1021/jp072036k
55. M. S. Miran, H. Kinoshita, T. Yasuda, M. A. B. H. Susan and M. Watanabe, "Hydrogen Bonds in Protic Ionic Liquids and Their Correlation with Physicochemical Properties," Chemical Communications, Vol. 47, No. 47, 2011, pp. 12676-12678. http://dx.doi.org/10.1039/c1cc14817f
56. Noda, M. A. B. H. Susan, K. Kudo, S. Mitsushima, K. Hayamizu and M. Watanabe, "Brønsted Acid-Base Ionic Liquids as Proton-Conducting Nonaqueous Electrolytes," The Journal of Physical Chemistry B, Vol. 107, No. 17, 2003, pp. 4024-4033.http://dx.doi.org/10.1021/jp022347p
57. K. Fumino, A. Wulf and R. Ludwig, "Hydrogen Bonding in Protic Ionic Liquids: Reminiscent of Water," Angewandte Chemie International Edition, Vol. 48, No. 17, 2009, pp. 3184-3186. http://dx.doi.org/10.1002/anie.200806224
58. D. Wakeham, A. Nelson, G. G. Warr and R. Atkin, "Probing the Protic Ionic Liquid Surface Using X-Ray Reflectivity," Physical Chemistry Chemical Physics, Vol. 13, No. 46, 2011, pp. 20828-20835. http://dx.doi.org/10.1039/c1cp22351h
59. D. F. Evans, S.-H. Chen, G. W. Schriver and E. M. Arnett, "Thermodynamics of Solution of Nonpolar Gases in a Fused Salt. Hydrophobic Bonding Behavior in a Nonaqueous System," Journal of the American Chemical Society, Vol. 103, No. 2, 1981, pp. 481-482.http://dx.doi.org/10.1021/ja00392a049
60. D. F. Evans, A. Yamauchi, G. J. Wei and V. A. Bloomfield, "Micelle Size in Ethylammonium Nitrate as Determined by Classical and Quasi-Elastic Light Scattering," The Journal of Physical Chemistry, Vol. 87, No. 18, 1983, pp. 3537-3541. http://dx.doi.org/10.1021/j100241a035
61. H. Beesley, D. F. Evans and R. G. Laughlin, "Evidence for the Essential Role of Hydrogen Bonding in Promoting Amphiphilic Self-Assembly: Measurements in 3-Methylsydnone," The Journal of Physical Chemistry, Vol. 92, No. 3, 1988, pp. 791-793.http://dx.doi.org/10.1021/j100314a039
62. L. Timperman, P. Skowron, A. Boisset, H. Galiano, D. Lemordant, E. Frackowiak, F. Beguin and M. Anouti, "Triethylammonium Bis(tetrafluoromethylsulfonyl)amide protic Ionic Liquid as an Electrolyte for Electrical Double-Layer Capacitors," Physical Chemistry Chemical Physics, Vol. 14, No. 22, 2012, pp. 8199-8207. http: //dx.doi.org/10.1039/c2cp40315c
63. Z. Du, Z. Li, S. Guo, J. Zhang, L. Zhu and Y. Deng, "Investigation of Physicochemical Properties of LactamBased Brønsted Acidic Ionic Liquids," The Journal of Physical Chemistry B, Vol. 109, No. 41, 2005, pp. 19542- 19546. http: //dx.doi. org/10.1021/jp0529669
64. B. Schwenzer, N. S. Kerisit and M. Vijayakumar, "Anion Pairs in Room Temperature Ionic Liquids Predicted by Molecular Dynamics Simulation, Verified by Spectroscopic Characterization," RSC Advances, Vol. 4, No. 11, 2014, pp. 5457-5464.
65. B. Nuthakki, T. L. Greaves, I. Krodkiewska, A. Weerawardena, M. I. Burgar, R. J. Mulder and C. J. Drummond, "Protic Ionic Liquids and Iconicity," Australian Journal of Chemistry, Vol. 60, No. 1, 2007, pp. 21-28. http://dx.doi.org/10.1071/CH06363

66. J. Stoimenovski, E. I. Izgorodina and D. R. MacFarlane, "Ionicity and Proton Transfer in Protic Ionic Liquids," Physical Chemistry Chemical Physics, Vol. 12, No. 35, 2010, pp. 10341-10347. http://dx.doi.org/10.1039/c0cp00239a
67. M. Yoshizawa, W. Xu and C. A. Angell, "Ionic Liquids by Proton Transfer: Vapor Pressure, Conductivity, and the Relevance of ΔpKa from Aqueous Solutions," Journal of the American Chemical Society, Vol. 125, No. 50, 2003, pp. 15411-15419.http://dx.doi.org/10.1021/ja035783d
68. C. F. Poole, "Chromatographic and Spectroscopic Methods for the Determination of Solvent Properties of Room Temperature Ionic Liquids," Journal of Chromatography A, Vol. 1037, No. 1-2, 2004, pp. 49-82. http://dx. doi.org/10. 1016 /j.chroma.2003.10.127
69. W. Wang, L. Shao, W. Cheng, J. Yang and M. He, "Brønsted Acidic Ionic Liquids as Novel Catalysts for Prins Reaction," Catalysis Communications, Vol. 9, No. 3, 2008, pp. 337-341. http://dx.doi.org/10.1016/j.catcom.2007.07.006
70. S.-Y. Lee, A. Ogawa, M. Kanno, H. Nakamoto, T. Yasuda and M. Watanabe, "Nonhumidified Intermediate Temperature Fuel Cells Using Protic Ionic Liquids," Journal of the American Chemical Society, Vol. 132, No. 28, 2010, pp. 9764-9773. http://dx.doi.org/10.1021/ja102367x
71. Fernicola, S. Panero and B. Scrosati, "Proton-Conducting Membranes Based on Protic Ionic Liquids," Journal of Power Sources, Vol. 178, No. 2, 2008, pp. 591-595.http://dx.doi.org/10.1016/j.jpowsour.2007.08.079
72. H. Ye, J. Huang, J. J. Xu, N. K. A. C. Kodiweera, J. R. P. Jayakody and S. G. Greenbaum, "New Membranes Based on Ionic Liquids for PEM Fuel Cells at Elevated Temperatures," Journal of Power Sources, Vol. 178, No. 2, 2008, pp. 651-660. http://dx.doi.org/10.1016/j.jpowsour.2007.07.074
73. H. Nakamoto and M. Watanabe, "Bronsted Acid-Base Ionic Liquids for Fuel Cell Electrolytes," Chemical Communications, No. 24, 2007, pp. 2539-2541. http://dx.doi.org/10.1039/b618953a
74. N. B. Darvatkar, A. R. Deorukhkar, S. V. Bhilare and M. M. Salunkhe, "Ionic Liquid-Mediated Knoevenagel Condensation of Meldrum's Acid and Aldehydes," Synthetic Communications, Vol. 36, No. 20, 2006, pp. 3043-3051. http://dx.doi.org/10.1080/00397910600775218
75. Janus, I. Goc-Maciejewska, M. Łożyński and J. Pernak, "Diels-Alder Reaction in Protic Ionic Liquids," Tetrahedron Letters, Vol. 47, No. 24, 2006, pp. 4079-4083.http://dx.doi.org/10.1016/j.tetlet.2006.03.172
76. Zhu, T. Jiang, D. Wang, B. Han, L. Liu, J. Huang, J. Zhang and D. Sun, "Direct Aldol Reactions Catalyzed by 1,1,3,3-Tetramethylguanidine Lactate without Solvent," Green Chemistry, Vol. 7, No. 7, 2005, pp. 514-517. http://dx.doi.org/10.1039/b501925g
77. Zhou, J. Yang, L. Ye, H. Lin and Y. Yuan, "Effects of Acidity and Immiscibility of Lactam-Based BronstedAcidic Ionic Liquids on Their Catalytic Performance for Esterification," Green Chemistry, Vol. 12, No. 4, 2010, pp. 661-665. http://dx.doi.org/10.1039/b921081d
78. L. C. Henderson and N. Byrne, "Rapid and Efficient Protic Ionic Liquid-Mediated Pinacol Rearrangements under Microwave Irradiation," Green Chemistry, Vol. 13, No. 4, 2011, pp. 813-816. http://dx.doi.org/10.1039/c0gc00916d
79. X. Li, W. Eli and G. Li, "Solvent-Free Synthesis of Benzoic Esters and Benzyl Esters in Novel Brønsted Acidic Ionic Liquids under Microwave Irradiation," Catalysis Communications, Vol. 9, No. 13, 2008, pp. 2264-2268. http://dx.doi.org/10.1016/j.catcom.2008.05.015
80. Shi, W. Zhu, H. Li, H. Liu, M. Zhang, Y. Yan and Z. Wang, "Microwave-Accelerated Esterification of Salicylic Acid Using Brönsted Acidic Ionic Liquids as Catalysts," Catalysis Communications, Vol. 11, No. 7, 2010, pp. 588-591. http://dx.doi.org/10.1016/j.catcom.2009.12.025

81. X. Tong and Y. Li, "Efficient and Selective Dehydration of Fructose to 5-Hydroxymethylfurfural Catalyzed by Brønsted-Acidic Ionic Liquids," ChemSusChem, Vol. 3, No. 3, 2010, pp. 350-355. http://dx.doi.org/10.1002/cssc.200900224
82. P. Gabbott, "Principles and Applications of Thermal Analysis," Wiley, Hoboken, 2008. http://dx.doi.org/10.1002/9780470697702
83. M. Sorai and N. N. Gakkai, "Comprehensive Handbook of Calorimetry and Thermal Analysis," Wiley, Hoboken, 2004.
84. J.-P. Belieres and C. A. Angell, "Protic Ionic Liquids: Preparation, Characterization, and Proton Free Energy Level Representation," The Journal of Physical Chemistry B, Vol. 111, No. 18, 2007, pp. 4926-4937. http://dx.doi.org/10.1021/jp067589u
85. M. A. B. H. Susan, A. Noda, S. Mitsushima and M. Watanabe, "Bronsted Acid-Base Ionic Liquids and Their Use as New Materials for Anhydrous Proton Conductors," Chemical Communications, No. 8, 2003, pp. 938-939. http://dx.doi.org/10.1039/b300959a
86. S. R. Varma, "Solvent-Free Organic Syntheses Using Supported Reagents and Microwave Irradiation," Green Chemistry, Vol. 1, No. 1, 1999, pp. 43-55. http://dx.doi.org/10.1039/a808223e
87. Davoodnia, M. M. Heravi, Z. Safavi-Rad and N. Tavakoli-Hoseini, "Green, One-Pot, Solvent-Free Synthesis of 1,2,4,5-Tetrasubstituted Imidazoles Using a Brønsted Acidic Ionic Liquid as Novel and Reusable Catalyst," Synthetic Communications, Vol. 40, No. 17, 2010, pp. 2588-2597. http://dx.doi.org/10.1080/00397910903289271
88. T. L. Greaves, A. Weerawardena, C. Fong, I. Krodkiewska and C. J. Drummond, "Protic Ionic Liquids: Solvents with Tunable Phase Behavior and Physicochemical Properties," The Journal of Physical Chemistry B, Vol. 110, No. 45, 2006, pp. 22479-22487. http://dx.doi.org/10.1021/jp0634048
89. Davoodnia, M. Bakavoli, R. Moloudi, N. Tavakoli-Hoseini and M. Khashi, "Highly Efficient, One-Pot, SolventFree Synthesis of 2,4,6-Triarylpyridines Using a Brønsted Acidic Ionic Liquid as Reusable Catalyst," Monatshefte für Chemie, Vol. 141, No. 8, 2010, pp. 867-870. http://dx.doi.org/10.1007/s00706-010-0329-x
90. S. Nazari, K. Ghandi, S. B. Cameron and M. B. Johonson, "Physicochemical Properties of Imidazo Pyridine Protic Ionic Liquids," Journal of Materials Chemistry A, Vol. 1, No. 38, 2013, pp. 11570-11579.

CHAPTER 2

Viscosity of Liquid Co–Sn Alloys: Thermodynamic Evaluation and Experiment

Andriy Yakymovych[a*], Yuriy Plevachuk[b], Stepan Mudry[b], Jürgen Brillo[c], Hidekazu Kobatake[c] and Herbert Ipser[a]

[a] *Department of Inorganic Chemistry (Materials Chemistry), University of Vienna, Währinger Str. 42, 1090 Vienna, Austria;*

[b] *Department of Metal Physics, Ivan Franko National University, Kyrylo and Mephodiy str. 8, 79005 Lviv, Ukraine;*

[c] *Deutsches Zentrum für Luft- und Raumfahrt (DLR), Institut für Materialphysik im Weltraum, 51170 Köln, Germany*

ABSTRACT

Shear viscosity measurements were performed for liquid Co–Sn alloys over a wide temperature range above the respective liquidus temperatures. A high temperature oscillating-cup viscometer was used. It was found experimentally that viscosity as a function of temperature obeys an Arrhenius law. The data were compared with calculated values, obtained from different thermodynamic approaches. A good agreement was found between experimental results and calculated ones by the Budai– Benkö-Kaptay model.

Keywords: Co–Sn; liquid binary alloys; thermodynamic models; viscosity.

INTRODUCTION

Viscosity is one of the most important transport properties of materials in the liquid state. Due to its high sensitivity to structure and phase transformations, the temperature and concentration changes of viscosity can provide exhaustive information about different structural changes, especially those occurring above the liquidus temperature [1,2]. From the practical point of view, viscosity is one of the key parameters for design and optimisation of metallurgical processes.

Sn-based alloys are widely applied in various branches of industry. Due to their ability to form the lithium rich phase $Li_{4.4}Sn$, tin containing alloys are prospective anode materials for rechargeable lithium batteries [3–5]. Tin, as the active component of the electrode, reacts with lithium, while the inactive component (Co, Cu, Fe, etc.) provides a matrix to reduce the massive volume change during the lithiation/delithiation process. The reaction of lithium with Co–Sn alloys leads to the formation of disordered $Li_{4.4}Sn$ alloys and nanoscopic grains of cobalt. The reason of using inter metallic Co–Sn compounds is the reversible mechanism, in which nanosized cobalt forms. Such type materials act as a spectator upon the lithiation/delithiation reaction and supports also electronic contact in the material for better cycling performance. The Co–Sn alloys can also be used in specific high-tech applications for devices, chip interconnections and packaging [6]. The replacement of the chromium electroplating by deposited Co–Sn alloys could help solving the problem with the toxicity of chromium.

Co–Sn belongs to the compound-forming systems with the existence of several intermetallic phases in the diagram [7]. Numerous studies of structural [8–10] and thermodynamic properties [11–14] of this system have been performed in both, the liquid and solid state. The preferable interaction between unlike kind atoms for the Co–Sn system is displayed in negative values of the enthalpy of mixing, structure studies in the solid as well as liquid state. Taking into account this type of interatomic interaction, Ivanov [15] calculated the enthalpy of mixing for liquid Co–Sn alloys based on the association theory.

To our knowledge, the viscosity, in contrast to the above-mentioned properties, has not yet been studied for liquid Co–Sn alloys. In general, most transport properties, such as the diffusion coefficient or the electrical conductivity have been studied for Co–Sn either in the solid state [16,17] or using various models [18,19]. The changes of the electrical conductivity of liquid tin upon adding small amounts of cobalt were experimentally investigated in Reference [20] and the diffusion coefficient was studied in Reference [21].

The main goal of the present study is the investigation of the temperature and concentration dependencies of the viscosity over the entire concentration range. The existence of an inhomogeneous short-range order in the liquid should also be displayed in the concentration dependence of the viscosity. The obtained experimental data are compared with thermodynamic models, calculated from different approaches [22–25] at a constant temperature of 1773 K, as well as with literature data.

2. EXPERIMENTAL

The viscosity measurements were carried out using a high temperature oscillating-cup viscometer [26]. In this method, a cylindrical alumina crucible containing the sample is placed in a graphite container which is attached to the torsion wire inside the high temperature furnace. The experiments were performed in an atmosphere of argon (400 mbar) after initially evacuating the working volume of the furnace down to a pressure of approximately 10–8 mbar. The temperature was estimated by a pyrometer looking at the bottom of the graphite crucible container which has an emissivity close to unity. In addition, the pyrometer was calibrated at the melting temperatures of several pure materials (Al, NaCl, Ag, Cu, Ni and Fe). Samples were prepared by initially melting accurately weighed (within ±0.1 mg) amounts of the pure components Co (99.9+ % metallic purity, Alfa Aesar, USA) and Sn (99.98% metallic purity, Alfa Aesar, USA) in the furnace, which was heated up to 1873 K. This temperature was kept for at least one hour allowing the samples to homogenise. The viscosity was measured during cooling from highest temperature to the liquidus point with a constant cooling rate of 2 K min^{-1}.

To study the viscosity of Sn-rich samples with liquidus temperatures, lower than the values, which could be measured by the pyrometer, another oscillating-cup viscometer was used [27]. The temperature was determined by WRe-5/20 thermocouples. All other experimental conditions were almost the same in both experimental sets, and the viscosity was determined with an accuracy of about 5%.

After the measurement, the weight of each sample was checked. In all cases, the loss of material by vaporisation was lower than 0.2% of the ingot.

Using the modified Roscoe equation, the dynamic viscosity, η, was calculated from the logarithmic decrement and the period of oscillations [28]. The required density values were calculated for the melts from the densities of the constituents, *ρi*, under the assumption that the excess volume is zero [29].

3. RESULTS AND DISCUSSION

Measurements were performed for binary alloys with compositions $Co_{80}Sn_{20}$, $Co_{70}Sn_{30}$, $Co_{60}Sn_{40}$, $Co_{50}Sn_{50}$, $Co_{20}Sn_{80}$, $Co_{15}Sn_{85}$, $Co_{10}Sn_{90}$, Co_5Sn_{95} and Co_3Sn_{97} (in at. %). We choose these compositions because they are close to the compositions of the chemical compounds and/or the eutectic point. The temperature dependencies of the viscosities of liquid Co–Sn alloys together with viscosity data of the pure components [30,31] are presented in Figure 1.

The viscosities increase when the temperature is lowered according to the Arrhenius type equation:

$$\eta = \eta_0 \exp\left(\frac{E_\eta}{RT}\right) \tag{1}$$

where $\eta 0$ is the constant, E_η is the energy of activation of viscous flow, T is the absolute temperature and R is the ideal gas constant. Such behaviour is illustrated in Figure 1, where the viscosity of the samples as a function of temperature is given.

A temperature dependence of this type is often found for the viscosity in melts of simple metals, if the properties of the melt can be well described by a hard-sphere model [32]. The lack of any deviation of the experimental viscosity from the exponential law in Co–Sn melts indicates that no significant structure transformations occur in the liquid state when approaching the solidification temperature. Values of the activation energy of viscous flow for all investigated samples were determined by least square regressional fits of the experimental data to Equation (1). These values are presented in Table 1 together with η0 data.

Figrure 2 shows the variation of the activation energy with concentration superimposed on the phase diagram [7]. Clearly, a pronounced similarity between the activation energy and the liquidus curve is obvious, with the exception of the near eutectic region (80 at. % diagram.

Co). According to X-ray diffraction data [33], short-range order in the liquid$Co_{79.5}Sn_{20.5}$ alloy consists of βCo_3Sn_2 type clusters and βCo clusters, each accounting for about 50% in the melt. Such a structure leads to nucleation difficulties and is the reason of the high value of the activation energy of viscous flow. Therefore, the obtained positive deviation of the activation energy at the eutectic point can be related to the existence of an inhomogeneous short-range order in the liquid state. As seen in Figure 2, the activation energy values obtained using two different viscometers are also in good agreement.

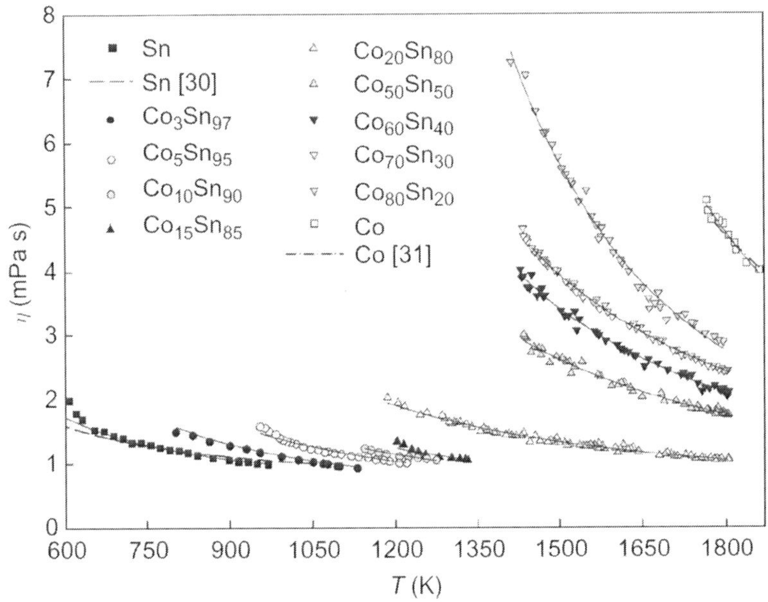

Figure 1. The temperature dependence of the viscosity for liquid Co–Sn alloys.

Table 1. Fitting numerical parameters of the Arrhenius-type Equation (1).

Sample, at. %	η_0 (mPa s)	E_η (kJ mol^{-1})
Sn	0.407	7.2
Co_3Sn_{97}	0.273	11.7
Co_5Sn_{95}	0.256	13.9
$Co_{10}Sn_{90}$	0.246	15.3
$Co_{15}Sn_{85}$	0.254	15.8
$Co_{20}Sn_{80}$	0.318	17.8
$Co_{50}Sn_{50}$	0.245	29.5
$Co_{60}Sn_{40}$	0.186	36.3
$Co_{70}Sn_{30}$	0.214	36.4
$Co_{80}Sn_{20}$	0.072	54.4
Co	0.048	68.1

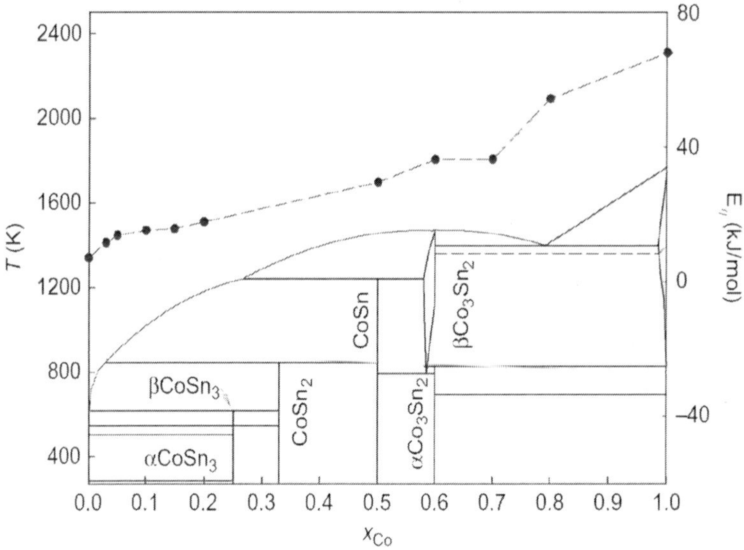

Figure 2. The variation of the activation energy with concentration superimposed on the phase

Several thermodynamic models and semi-empirical equations were developed to evaluate the viscosity of liquid alloys:

According to the Budai–Benkö–Kaptay equation, the viscosity of a multi-component alloy can be written as [22]

$$\eta = A\frac{\left(\sum_i x_i M_i\right)^{1/2}}{V^{2/3}} T^{1/2} \exp\left[\frac{B}{T}\left(\sum_i x_i T^*_{m,i} - \frac{\Delta H}{qR}\right)\right] \quad (2)$$

where M_i and x_i are the atomic mass and concentration of the given component *i*, respectively; *q* is a semi-empirical parameter equal to $q \cong 25.4 \pm 2$ [34]; *V* is the molar volume of the alloy; ΔH is the enthalpy of mixing; *A* and *B* are fitting parameters equal to $(1.80 \pm 0.39) \times 10^{-8}$ $(J \cdot K^{-1} \cdot mol^{-1/3})^{1/2}$ and (2.34 ± 0.20), respectively; $T^*_{m;i}$ is the effective melting temperature of the component *i:*

$$T^*_{m,i} = \frac{T}{B}\ln\left(\frac{\eta_i V_i^{2/3}}{A M_i^{1/2} T^{1/2}}\right) \quad (3)$$

where η_i and V_i are the viscosity and atomic volume of the component *i*.

Singh and Sommer [35] described the viscosity of liquid metals and metal alloys using the Andrade equation [23]:

$$\eta_{m,i} = \frac{\alpha_i k_B}{h} \frac{M_i \theta_i^{l}}{a_i^{l}}$$

,(4)

$$\eta_i = \eta_{m,i} \exp\left[\frac{E_\eta^i}{RT}\left(1 - \frac{T}{T_{m,i}}\right)\right],$$

(5)

$$\eta_m = \frac{\left(\frac{\left(\sum x_i \alpha_i\right) k_B}{h}\right)\left(\sum x_i M_i\right)\left(\sum x_i \theta_i^{l}\right)}{\sum x_i a_i^{l}} \frac{\left(1 - \frac{0.0341 \cdot \Delta H}{R \sum x_i \theta_i^{l}}\right)}{\left(1 + \frac{\left(\Delta V / N_A\right)^{1/3}}{\sum x_i a_i^{l}}\right)},$$

(6)

$$\eta = \eta_m \exp\left[\frac{E_\eta}{RT}\left(1 - \frac{T}{T_m}\right)\right],$$

(7)

where η_m and $\eta_{m,i}$ are the viscosity of the alloy and component *i* at the melting temperature T_m and $T_{m,i}$, respectively, α^i is equal to 1.75 for normal metals and 1.35 for semimetals. The term $\theta^1{}_i$ can be determined through the relation

$$\theta_i^{l} = \theta_i \exp\left(\frac{-\Delta H_{m,i}}{3RT_{m,i}} + \frac{S_{\text{conf}}}{3R}\right),$$

(8)

where θ_i and $\Delta H_{m,i}$ are the Debye temperature of the solid metal and the heat of fusion, k_B and h are the Boltzmann and Planck constant, respectively. S_{conf} is the configurational entropy arising from disorder in the liquid state.

The Hirai equation is a semi-empirical extrapolation of the Andrade's equation, estimating the viscosity in the following way [24]:

$$\eta = 1.7 \cdot 10^{-7}\frac{\rho^{2/3}T_m^{1/2}}{M^{1/6}}\exp\left[\frac{2.65 \cdot T_m^{1.27}}{R}\left(\frac{1}{T}-\frac{1}{T_m}\right)\right], \quad (9)$$

where M is the molar mass of the alloy calculated by the addition rule.

Recently, the following equations were proposed for the viscosity of the binary system Al-Cu by Schick et al. [25]:

$$\ln(\eta_0) = \sum_i x_i \ln\left(\eta_{0,i}\right), \quad (10)$$

$$E_\eta = \sum_i x_i E_\eta^i + \Delta H, \quad (11)$$

$$\ln(\eta) = \sum_i x_i \ln\left(\eta_{0,i}\right) + \frac{\sum_i x_i E_\eta^i + \Delta H}{RT}, \quad (12)$$

where $\eta_{0,i}$ and E^i_η are the pre-exponential factor and the activation energy of the viscous flow of the Arrhenius type equation for component i, respectively.

In our calculation, it was assumed that the excess volume of the alloys equals zero. The enthalpy of mixing data was taken from [13]; the density, atomic volume and configurational entropy of components were taken from [29].

The experimental and calculated dynamic viscosities as functions of concentration for binary liquid Co–Sn alloys are presented in Figure 3. The four evaluated models, Equations (2), (7), (9) and (12), are within a band of ±0.6 mPa s around the experimental data. Hence, the agreement between the models and the data as well as among the four models is reasonable.

The semi-theoretical Andrade equations (Equations (4–7)) systematically underestimate the data. According to this approach, atoms in the liquid state at the melting temperature may be regarded as performing vibrations around their equilibrium positions with random periods and direction, like in the solid state. In other words, the melting is assumed to occur when the

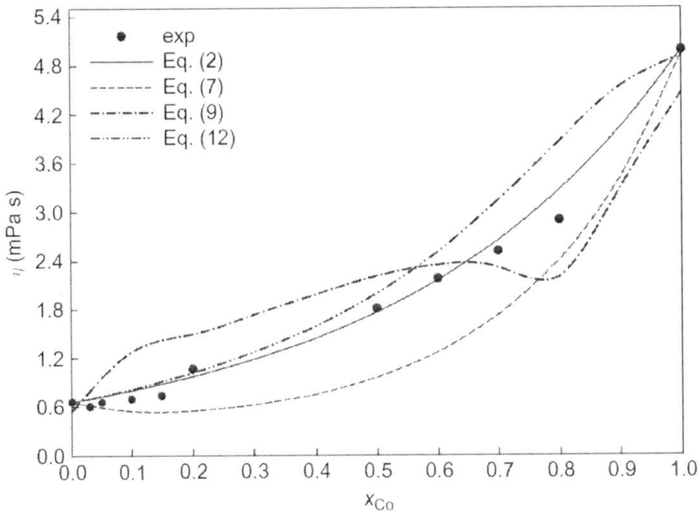

Figure 3. The concentration dependence of the viscosity of liquid Co–Sn at 1773 K compared with different models.

amplitude of atomic vibrations exceeds a certain limit, and the frequency can be calculated by Lindemann's equation [36]:

$$T_m \sim \theta^2 V^{2/3} M \quad , \tag{13}$$

if a Debye model for the solid is assumed. In that case liquids at their melting temperatures behave like simple monoatomic liquids what is in disagreement with present-day theories of the structure of metal alloys at the melting temperature. Nevertheless, we obtained a qualitatively good correlation between experimental and calculated viscosity using Andrade`s equation.

The Hirai model (Equation (9)) underestimates the viscosity for $x_{Co} > 0.6$. It overestimates it for all other concentrations. Figure 3 shows that the concentration dependence of the calculated viscosity by Hirai's model is strikingly similar to the phase diagram. This fact is connected with the idea to relate the term corresponding to the activation energy with the melting temperature ($2.65T_m$ 1.27). This correlation is more evident in the region close to the melting temperature of the liquid.

The model proposed by Schick et al. (Equations (10–12)) is found to be in very good agreement with the data for cobalt mole fractions lower than 0.5. For $x_{Co} > 0.5$, however, it deviates from the experimental data, and this deviation increases with the increase of cobalt concentration. The authors of Reference [25] assumed for the calculation of the activation energy of the liquid alloys that it would be related to the free enthalpy of mixing, which requires knowledge of both, the enthalpy and the entropy of mixing. While the first can be taken from thermodynamic databases, the latter was assumed to be ideal in order to guarantee that EA, as required for the Arrhenius law, is independent of the temperature. This assumption may not always be fulfilled.

The best agreement is found for the Budai–Benkö–Kaptay approach, see Figure 3. Their model is focused on the cohesion energy in the pure liquid constituents and the enthalpy of mixing of the investigated liquid alloy using a semi-empirical factor. This factor was estimated by testing many alloys. Therefore, this model has been very successful for the description of the concentration dependence of experimental viscosity data [37].

It should be noted that because of the absence of molar volume data for liquid Co–Sn alloys we assumed in our calculations that the excess molar volume is equal to zero. It is also assumed that the entropy of mixing is ideal. Possibly, the calculations with experimental entropy of mixing and molar volume data would give a more realistic viscosity behaviour and would improve the agreement between experimental viscosity values and calculated ones using certain models

CONCLUSION

The viscosity of the binary $Co_{80}Sn_{20}$, $Co_{70}Sn_{30}$, $Co_{60}Sn_{40}$, $Co_{50}Sn_{50}$ and $Co_{20}Sn_{80}$ alloys (at. %), which correspond to the chemical compounds as well as to the eutectic and peritectic points has been studied in a wide temperature range. The correlation between the activation energy curve and the liquidus curve of the Co–Sn phase diagram suggests a correlation between interatomic interactions in the liquid and solid states.

ACKNOWLEDGEMENTS

Part of this work has been carried out at the Institute of Materials Physics in Space at the German Aerospace Center in Cologne. Financial support from the Austrian Science Fund (FWF) under Grant No. M 1324 and from the Austrian Agency for International Mobility and Cooperation in Education, Science and Research (OeAD), under Research Project No. UA 03/2013 is gratefully acknowledged. In addition, we acknowledge financial support from the German Science Foundation DFG under contract number BR 3665/3-2.

REFERENCES

1. Frenkel YaI. Kinetic theory of liquids. Leningrad: Nauka; 1975. Russian
2. Plevachuk Y, Sklyarchuk V, Yakymovych A, Gerbeth G. Microsegregation in liquid Pb-based eutectics. J Non-Cryst Solids. 2008;354:4443–4447
3. Yoshio M, Brodd RJ, Kozawa A. Lithium-ion batteries: science and technologies. New York: Springer; 2009
4. Kamali AR, Fray DJ. Tin-based materials as advanced anode materials for lithium ion batteries: a review. Rev Adv Mater Sci. 2011;27:14–24
5. Chen JJ. Recent progress in advanced materials for lithium ion batteries. Materials. 2013;6:156–183
6. Brusic VA, Marino JR, O'Sullivan EJ, Sambucetti CJ, Schrott AG, Uzoh CE. International Business Machines Corporation, assignee. Cobalt-tin alloys and their applications for devices, chip interconnections and packaging. United States patent US 5,755,589. 1998 May 26
7. Jiang M, Sato J, Ohnuma I, Kainuma R, Ishida K. A thermodynamic assessment of the Co-Sn system. CALPHAD. 2014;28:213–220
8. Singh M, Singh NP. Effect of alloying additions on the structure of Co75Sn25XMX (M = As, In, and Sb) alloys. Phys Stat Sol A. 1987;100:K111–K116
9. Komarnytsky MS, Mudry SI, Halchak VP. The structure of liquid Co-Sn alloys. J Phys Stud. 1998;2:54–61. Ukrainian
10. Kazimirov VP, Roik AS, Sokol'ski VE. Structural features of Co-Si, Co-Ge, and Co-Sn binary melts. Metally (Russian Metallurgy). 2010;2:77–83
11. Cömert H, Pratt JN. The thermodynamic properties of solid cobalt-tin alloys. Thermochim Acta. 1985;84:273–286
12. Torgersen AN, Bros H, Castanet R, Kjekshus A. Enthalpy of formation for CoGe, CoSn, Ni3.14Sn4, Ni3.50Sn4, AuCo1.66Sn4, AuNi2Sn4 and Au1.17Pt1.82Sn4. J Alloys Comp. 2000;307:167–173
13. Flandorfer H, Yakymovych A, Fürtauer S, Elmahfoudi A, Ipser H. Enthalpy of mixing of liquid Co-Sn alloys. J Chem Thermodyn. Forthcoming 2013
14. Vassilev GP, Lilova KI. Contribution to the thermodynamics of the Co-Sn system. Arch Metallurg Mater. 2006;51:365–375
15. Ivanov M. Thermodynamics of self-associated liquid alloys. Z Metallkde. 1991;82:53–58
16. Xu J, Wang N, Wei B. Microstructural characteristics and electrical resistivity of rapidly solidified Co-Sn alloys. Chin Scie Bull. 2004;49:2242–2247
17. Edwards JB, Hucke EE, Martin JJ. Diffusion in binary liquid-metal systems. Met Rev 1968;120:1–28
18. Roy AK, Chhabra RP. Prediction of solute diffusion coefficients in liquid metals. Met Trans A. 1988;19A:273–279
19. Hitara K, Waseda Y, Jain A, Srivastava R. Resistivity of liquid transition metals and their alloys using the t matrix. J Phys F: Metal Phys. 1977;7:419–425
20. Tamaki S. Localized impurity states in liquid metals; electrical resistivity of dilute 3d-transition metals in liquid tin. J Phys Soc Japan. 1968;25:1596–1601
21. Kuzmenko PP, Kharkov EI, Lozovoy VI. Transmission by electric current of silver in liquid lead and of cobalt in liquid tin. Ukr Phys J. 1964;9:881–889. Ukrainian
22. Budai I, Benkö MZ, Kaptay G. Comparison of different theoretical models to experimental data on viscosity of binary liquid alloys. Mater Sci Forum. 2007;537–538:489–496

23. Andrade ENC. A theory of the viscosity of liquids – Part I. Phil Mag. 1934;17:497–511
24. Hirai M. Estimation of viscosities of liquid alloys. ISIJ Int. 1993;33:251–258
25. Schick M, Brillo J, Egry I, Hallstedt B. Viscosity of Al-Cu liquid alloys – measurement and thermodynamic description. J Mater Sci. 2012;47:8145–8152
26. Kehr M, Hoyer W, Egry I. A new high-temperature oscillating cup viscometer. Int J Thermophys. 2007;28:1017–1025
27. Mudry S, Sklyarchuk V, Yakymovych A. Influence of doping with Ni on viscosity of liquid Al. J Phys Stud. 2008;12:1601–1605
28. Roscoe R. Viscosity determination by the oscillating vessel method I: theoretical considerationsProc Phys Soc. 1958;72:576–584
29. Iida T, Guthrie RIL. The physical properties of liquid metals. Oxford: Clarendon Press, Oxford University Press; 1988
30. Plevachuk Y, Sklyarchuk V, Hoyer W, Kaban I. Electrical conductivity, thermoelectric power and viscosity of liquid Sn-based alloys. J Mater Sci. 2006;41:4632–4635
31. Assael MJ, Armyra IJ, Brillo J, Stankus SV, Wu JT, Wakeham WA. Reference data for the density and viscosity of liquid cadmium, cobalt, gallium, indium, mercury, silicon, thallium, and zinc. J Phys Chem Ref Data. 2012;41:033101 (16 pages)
32. van Loef JJ. Hard-sphere results and atomic transport properties of liquid-metals. J Phys F: Metal Phys. 1974;4:L190–L195
33. Sun JJ, Zheng HL, Shi XY, Tian XL. Structure correlation of Co79.5Sn20.5 eutectic alloy between liquid and solid states. Sci China Ser E: Tech Sci. 2009;52:3674–3677
34. Kaptay G. Modelling interfacial energies in metallic systems. Mater Sci, Test Info. 2005; II:473–474, 1–10
35. Singh RN, Sommer F. Viscosity of liquid alloys: generalization of Andrade's equation. Monatsh Chem. 2012;143:1235–1242
36. Lindemann FA. The calculation of molecular vibration frequencies. Physik Z. 1910;11:609- 612
37. Brillo J, Brooks R, Egry I, Quersted P. Viscosity measurement of liquid ternary Cu–Ni–Fealloys by an oscillating cup viscometer and comparison with models. Int J Mater Res. 2007;6:457–462

CHAPTER 3

Liquid Crystals in Micron-Scale Droplets, Shells and Fibers

Martin Urbanski[1], Catherine G Reyes[1], JungHyun Noh, Anshul Sharma, Yong Geng, Venkata Subba Rao Jampani and Jan P F Lagerwall

University of Luxembourg, 162 A Avenue de la Faïencerie, 1511 Luxembourg, Luxembourg

ABSTRACT

The extraordinary responsiveness and large diversity of self-assembled structures of liquid crystals are well documented and they have been extensively used in devices like displays. For long, this application route strongly influenced academic research, which frequently focused on the performance of liquid crystals in display-like geometries, typically between flat, rigid substrates of glass or similar solids. Today a new trend is clearly visible, where liquid crystals confined within curved, often soft and flexible, interfaces are in focus. Innovation in microfluidic technology has opened for high-throughput production of liquid crystal droplets or shells with exquisite monodispersity, and modern characterization methods allow detailed analysis of complex director arrangements. The introduction of electrospinning in liquid crystal research has enabled encapsulation in optically transparent polymeric cylinders with very small radius, allowing studies of confinement effects that were not easily accessible before. It also opened the prospect of functionalizing textile fibers with liquid crystals in the core, triggering activities that target wearable devices with true textile form factor for seamless integration in clothing. Together, these developments have brought issues center stage that might previously have been considered esoteric, like the interaction of topological defects on spherical surfaces, saddle-splay curvature-induced spontaneous chiral symmetry breaking, or the non-trivial shape changes of curved liquid crystal elastomers with non-uniform director fields that undergo a phase transition to an isotropic state. The new research thrusts are motivated equally by the intriguing soft matter physics showcased by liquid crystals in these unconventional geometries, and by the many novel application opportunities that arise when we can reproducibly manufacture these systems on a commercial scale. This review attempts to summarize the current understanding of liquid crystals in spherical and cylindrical geometry, the state of the art of producing such samples, as well as the perspectives for innovative applications that have been put forward.

Keywords: liquid crystal, functional fiber, actuator, sensor, microfluidics, electrospinning, photonics

1. INTRODUCTION

Liquid crystals present phenomena that no other materials can reproduce. They uniquely combine the long-range order that we know from solid crystals with the mobility and flexibility of fluid phases. The fact that they are liquid allows them to easily change shape, accommodating very different boundaries from the standard flat and rigid confining substrates typically used when studying or applying liquid crystals (LCs) in display-like geometries. Curved confinement can trigger fascinating effects, since curvature of the sample boundary, in case of sufficiently strong anchoring of the LC alignment direction at the boundary, is transmitted into the bulk in form of deformation of the orientational field. In many cases topological defects and sometimes also dislocations form, often with beautiful as well as informative consequences. In fact, the nanoscopic response of a smectic LC to the macroscopic curvature of a droplet (the droplet becomes 'stepped') was instrumental to Grandjean's experiments leading up to the identification of the molecular organization in smectics, and Mauguin's experiments of liquid crystal flow through capillary tubes demonstrated the surprising result that the viscosity could be lower in the nematic than in the isotropic phase [1].

As these classic experiments took place at the dawn of the 20th century, one might think that the mysteries of droplets, cylinders and other curved geometries for studying LCs would by now be elucidated and the research field exhausted. This is very far from the case. With the considerable advances in understanding of LCs that have developed during their first 128 years in human awareness (the first report of liquid crystals, by Reinitzer, dates from 1888 [2]) the enormous complexity and beauty of these states of matter has become apparent, continuously leading to new stimulating mysteries and providing many good reasons to revisit the classic experiments. Modern theoretical knowledge allow a deeper analysis of the behavior of LCs with curved boundaries, and the experimental equipment has developed to such an extent that today investigations that Mauguin and Grandjean could not dream of are possible.

Beyond superior imaging equipment, we now also have methods for producing LC samples with curved geometries with unprecedented control. The key development in this respect is the advent of microfluidics, which not only allows rapid production of thousands of identical LC droplets with extraordinary monodispersity, but even the 'excavation' of the droplet, by including a droplet of isotropic liquid inside it, thus creating a thin shell of LC. With shells we have obtained a truly fascinating new configuration for studying LCs, with two closely spaced curved interfaces, with positive curvature on one side and negative on the other. Another powerful technique is electrospinning, which allows us to prepare LC samples in a geometry that fundamentally is cylindrical with very small radius (on the order of a micron), but which can also be modulated into a beads-on-a-string type morphology, if desired.

This review is motivated by the many new opportunities that are now available for studying LCs in the shape of drops, shells or fibers, and the intriguing behavior that different LC phases grant us with when subjected to these sample geometries. These new configurations are intriguing not only from a fundamental science point of view, but they can be very interesting also in an applied perspective. Recent research has demonstrated application potential in sensors, actuators, photonics and even security. We believe that such cross-fertilization opportunities render the systems that we discuss interesting also outside the LC community. Even without a focus on applications, researchers working with microfluidics of isotropic liquids may find the new opportunities arising by incorporating LCs stimulating. The electrospinning and fiber technology research community will find that the responsiveness that LCs can add to textile and other fibers open many new opportunities.

Therefore, we begin the review with a brief introduction to the liquid crystalline states of matter and the terminology used to describe them in section 2, directed at researchers curious in the topic but without the LC background. Section 3 is devoted to shells and droplets of LCs, where we also provide a brief introduction to microfluidics in section 3.1, limiting ourselves to the key issues relevant for LC work. In the rest of section 3 we discuss the technicalities of producing and tuning LC droplets and shells in section 3.2, moving to the intriguing physics related to the topological defects that arise when preparing droplets or shells of LC in section 3.3, and then to two application opportunities, in LC elastomer actuators (section 3.4) and in photonics (section 3.5), respectively. The section ends with a brief discussion in section 3.6 of the very recent addition of LC droplet swimmers and active LCs in shell geometry.

Cylindrical confinement of LCs is the topic of section 4. We first give a general overview, summarizing seminal work on LCs filled into glass capillaries or in narrow Anopore or Nucleopore membranes. We then give a general introduction to the electrospinning technique in section 4.2, focusing on the specific challenge of incorporating LCs within polymer fibers in section 4.3. Finally, we move on to application opportunities of LC fibers in section 4.4. We end the article with an outlook in section 5.

2. THE LIQUID CRYSTALLINE STATES OF MATTER

The long-range order in arrangement of molecules, molecule aggregates or particles that make up the LC phase gives it anisotropic macroscopic properties, like birefringence, direction-dependent viscosity and the ability to rearrange its internal structure in response to electromagnetic fields. Reflecting the unique combination of anisotropy with fluidity, LCs are sometimes referred to as anisotropic liquids. Many LCs are colloids, the phase building blocks then being nanoparticles or aggregates of amphiphilic molecules in an isotropic solvent (frequently water). This group constitutes the broad class of *lyotropic* LCs. The phase diagram of lyotropic LCs depends primarily on the volume fraction of the disperse phase.

Because the vast majority of studies of LC drops, shells and fibers were not done with lyotropics but rather with *thermotropic* LCs, we will focus mainly on the latter class in this review. Here no solvent is needed, and the phase is homogeneous down to the molecular scale, rather than colloidal. The molecules making up a thermotropic LC are called mesogens, as they generate *mesophases* (from Greek μέσος = intermediate), an alternative name for liquid crystal phase. An LC-forming compound is sometimes called mesogenic, but most often it is simply referred to as a liquid crystal. This somewhat careless terminology can be confusing for the newcomer, who may have difficulties to tell LC *phases* apart from LC *materials*, in particular since the latter exhibit crystalline solid states at low temperatures and isotropic liquid states at high temperatures.

Indeed, the phase diagram of thermotropics depends primarily on temperature (pressure is also relevant, but for practical reasons most experiments are done at atmospheric pressure), a fact that is reflected in the name of this LC class. While ordinary materials melt directly from the crystalline solid into the isotropic liquid state, LCs exhibit at least two transitions separating these states: first the crystal melts into a liquid crystalline state, then there may be transitions between different LC phases, before the highest-temperature LC phase is replaced by the isotropic liquid state at the *clearing point*. The latter name reflects the fact that a bulk LC without control of its alignment scatters light, in contrast to the behavior in the isotropic phase of the same material, where it becomes transparent and clear. In case of mixtures between multiple mesogens (sometimes also involving non-mesogenic constituents), also the mixture composition becomes a key parameter in determining the phase diagram.

A large number of thermotropic LC phases have been identified over the years. In this review we will only deal with a small selection that we consider the most important for the field that we cover: the *nematic*, in its non-chiral and chiral versions (the latter often called *cholesteric*), and the *smectic*phases SmA and SmC. In the following subsections each phase is introduced in some detail, and thereafter we will say a few words about the peculiar behavior of LC rubbers, or elastomers (LCEs).

2.1. Thermotropic nematics

The least ordered thermotropic LC phase is called *nematic* (abbreviated N), and it is characterized by long-range order in orientation of the molecules (which evidently must be anisometric, e.g. rod- or disc-shaped) but no long-range positional order, see figure 1. In terms of positional order the N phase is similar to an ordinary liquid, with only short-range positional order and no periodic lattice, no grains and no grain boundaries. Indeed, nematics typically flow easily, making them an ideal candidate for exploring liquid crystalline drops, shells or fibers. To describe the long-range orientational order and analyze how it gives rise to the anisotropic physical properties, one introduces the *director*, abbreviated $\boldsymbol{n}$. It indicates the direction along which the molecules tend to align, thus the principal symmetry axis of the phase. Because there are equally many molecules 'pointing upwards' as are 'pointing downwards' in an averaging ensemble, $+\boldsymbol{n}$ and $-n$ are indistinguishable. This is often expressed by the *director sign invariance*, formalized as $\boldsymbol{n} = -\boldsymbol{n}$. More specifically, the director can be said to be a signless pseudovector.

2.1.1. The nematic order parameter. The long-range orientational order of LCs is quantified by an orientational order parameter (also called nematic order parameter) that in 3D takes the form:

$$S = \frac{1}{2}\langle 3\cos^2\beta - 1\rangle \tag{1}$$

The angle β describes the deviation in orientation away from $\boldsymbol{n}$ of a particular molecule in the averaging ensemble, see figure 1. The pointed brackets here express averaging over 3D space, thus integrating over the angles β and φ, defined in figure 1.

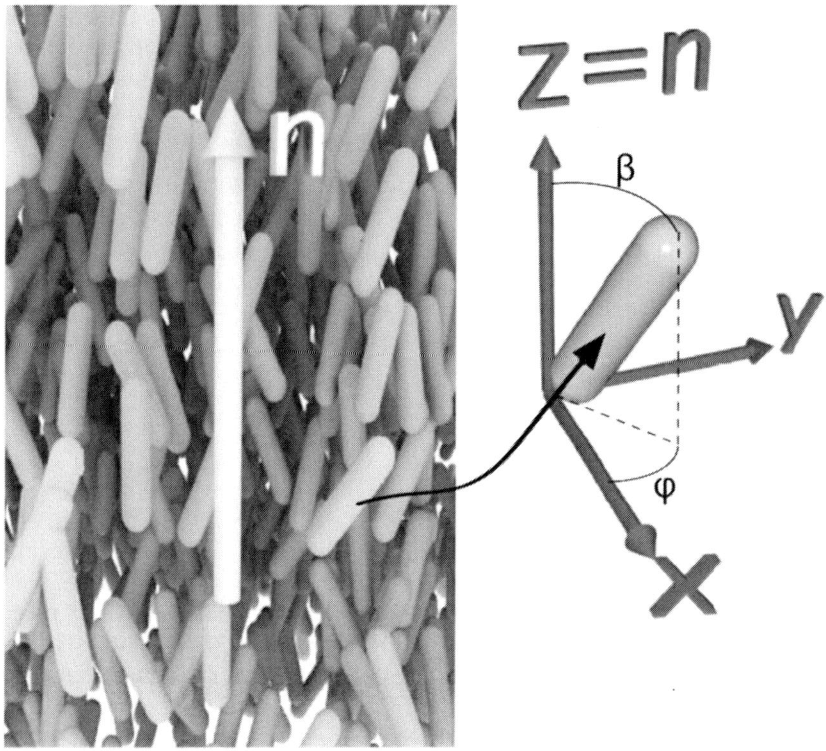

FIGURE 1. Cartoon of the order in a nematic phase of rod-like mesogens (left), aligning on the average along the director ***n***, without any long-range positional order. The right drawing illustrates the parameters required for calculating the orientational order parameter according to (1). The polar angle β expresses the deviation of a certain molecule from ***n*** and the azimuthal angle φ defines the deviation direction. The variable φ does not appear explicitly in (1) but it is implicit in the pointed brackets, which express averaging over all three space dimensions.

In 3D, the order parameter (1) is zero for total disorder, whereas for complete orientational order, $S = 1$. The isotropic-nematic transition in thermotropics is first-order, i.e. S jumps discretely at the phase transition from zero to a non-zero value, typically around $S = 0.4$. On further cooling, S increases, initially fast but rapidly approaching saturation around $S = 0.6$–0.7.

2.1.2. Optical anisotropy and the terminology for describing LC orientation. The mesogen anisometry together with the long-range orientational order generally render LCs optically anisotropic, i.e. their refractive index $n_{\|}$ parallel to **n** is different from that perpendicular to the director, $\mathbf{n}_{\perp}$. The director defines the optic axis. The magnitude of optical anisotropy, the *birefringence*, is defined as $\Delta n = n_{\|} - n_{\perp}$. Consequently, if $n_{\|}$ is the maximum refractive index, $\Delta > 0$ and we call the LC positive uniaxial (normal case for rod-shaped mesogens). Light passing through the sample will experience $n_{\perp}$ as the ordinary refractive index n_o, while the extraordinary index n_e varies depending on the direction of light propagation, $n_{\|}$ being the limiting value.

Like any birefringent material, LCs can affect the polarization ***E*** of light, changing it continuously between linear, elliptical and circular as well as changing the main oscillation direction or reversing rotation handedness. The effect disappears, however, if the light propagates along ***n***, because a birefringent material appears isotropic to light traveling along the optic axis: since light is a transverse wave, ***E*** is then by necessity perpendicular to the optic axis, hence the light experiences only n_o. An LC sample that is prepared such that the director is perpendicular to its boundary is called *homeotropic*, a name that reflects the 'isotropic-like' appearance for light entering the sample at normal incidence, along the director. If a flat homeotropic LC sample is placed between crossed polarizers during observation and if the alignment persists throughout the sample (also the other boundary must thus be homeotropic), it appears black. When working with curved samples, like those considered in this review, the meaning of the term 'homeotropic' as director perpendicular to the boundary is carried over. Note, however, that the curvature then means that light will generally *not* enter the sample along the director, and the curved sample will thus not (except in certain points) appear black.

The geometry orthogonal to homeotropic, where ***n*** is in the plane of the sample boundary, is called *planar*. If the sample is flat and if also the opposite boundary promotes the same planar alignment, such that ***n*** is uniform throughout the sample, the full birefringence ***Δn*** is experienced by light entering at normal incidence. Inside the LC the light is split up into two eigenmodes with polarization perpendicular and parallel to ***n***, respectively, experiencing the refractive indices $n_o = n_{\perp}$ and $n_e = n_{\|}$. Homeotropic and planar anchoring are the two limiting cases of anchoring, but the anchoring conditions can often promote an arbitrary tilt of ***n*** out of the interfacial plane. A tilt of 0° corresponds to planar alignment, a tilt of 90° is homeotropic, whereas we here refer to anchoring with any other inclination simply as *tilted*.

For curved samples the situation is again more complex. As for the term 'homeotropic', the terms 'planar' and 'tilted' are carried over in the sense of how ***n*** is oriented with respect to the sample boundary. A planar interface thus has the director in the boundary plane, regardless of how this curves. This means that, across the surface of a planar-aligned curved sample, $\boldsymbol{n}$ and thus the optic axis change direction with respect to the light propagation. The resulting optical behavior is very different compared to a planar flat sample. We refer the reader to an optics textbook for a deeper analysis, here settling by pointing out that the effective birefringence in most points is lower than for a flat planar sample, the difference depending on the inclination of $\boldsymbol{n}$ in each point. The reduction in $\Delta\boldsymbol{n}$ is partially compensated by an increased length of light passage through the curving LC. For a shell, moreover, we frequently have top and bottom surfaces and—as discussed in section 3.3—both contribute to the birefringence, even if only one of them is in focus.

If light propagates perpendicular to a uniformly planar-aligned LC, the orientation that $\boldsymbol{n}$ has within the sample plane—which is then equal to the light's polarization plane—is also very important for the optical effect of the LC. For tilted alignment the same holds for the orientation of the *projection* of $\boldsymbol{n}$ onto this plane, which we here call $\boldsymbol{n_p}$. Whenever the light polarization $\boldsymbol{E}$ is along or perpendicular to $\boldsymbol{n}$ (or $\boldsymbol{n^p}$), one of the two eigenmodes has zero amplitude, and the effect of the birefringence is lost. Therefore, when a sample is observed between crossed polarizers and $\boldsymbol{E}\|\boldsymbol{n}$ or $\boldsymbol{E} \perp \boldsymbol{n}$ (or $\boldsymbol{E}\|\boldsymbol{n}_p$ or $\boldsymbol{E} \perp \boldsymbol{n}_p$), it appears black. For all other polarizations both components exist, the birefringence changes the polarization state, and the sample typically appears bright between crossed polarizers, see figure 2.

Planar alignment means that the director is in the sample boundary plane, and tilted alignment that it has a component $\boldsymbol{n_p}$ in that plane, but the terms say nothing about *which* orientation in this plane $\boldsymbol{n}$ or $\boldsymbol{n_p}$ should adopt. With a solid substrate a preferred orientation can often be induced, for instance by rubbing the substrate in a specific direction, but many planar- or tilted-aligning interfaces induce what is called 'degenerate planar/tilted alignment', giving no preference for any particular direction within the interface plane. In a nematic LC subject to such boundary conditions, the direction of $\boldsymbol{n}/\boldsymbol{n_p}$ in the interface plane can vary smoothly as a function of location. Since $\boldsymbol{n}$ defines the optic axis, this means that such a spatially varying director field $\boldsymbol{n}(\boldsymbol{r})$ is recognized in the polarizing microscope through a characteristic *schlieren* texture, named after the German word for streak. An example is shown in figure 2. Softly curving black 'brushes' appear, often emanating from crosses or cusps that are defects in the director field (see section 3.3). Along a black brush, $\boldsymbol{n}$ is parallel to one of the polarizers.

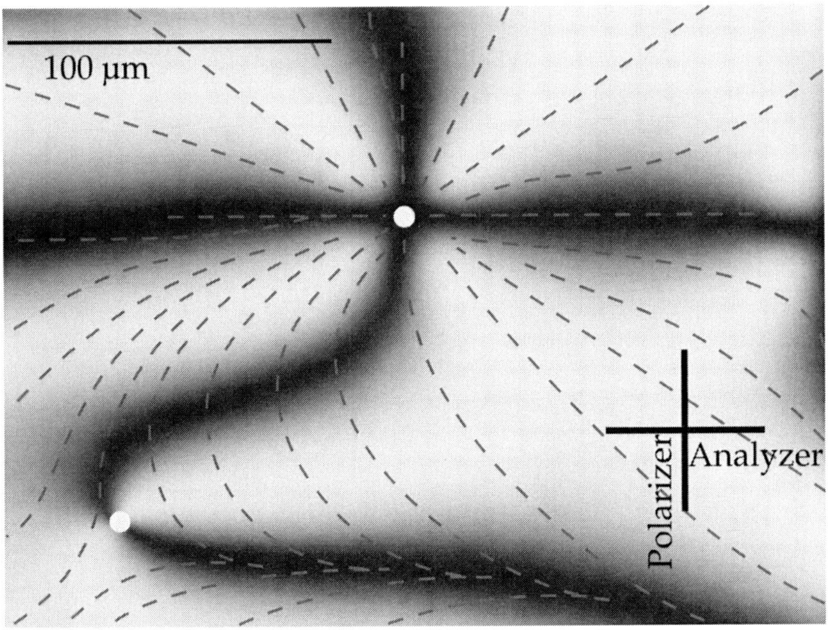

FIGURE 2. The *schlieren* texture of a flat nematic sample with degenerate planar alignment prevailing uniformly from one substrate to the other, observed between crossed polarizers. The dashed lines are sketches of the director field variations within the sample. Two topological defects are highlighted with grey spots, the upper one with strength $s = 1$ and the lower with strength $s = 1/2$. The texture photo was previously shown in [3]. Reproduced with permission from [3].

2.1.3. The viscous flow properties of nematics.

While nematics normally can flow easily, it makes a great difference if the flow is along or perpendicular to $\boldsymbol{n}$, and, for the latter case, it also makes a difference if the flow velocity gradient is along or perpendicular to $\boldsymbol{n}$. These three situations, depicted by three orthogonal director orientations $\boldsymbol{n}_1$, $\boldsymbol{n}_2$ and $\boldsymbol{n}_3$ in figure 3, were investigated by the Polish physicist Marian Męsowicz [4], after whom the three corresponding viscosities, η_1, η_2 and η_3 are now called the *Męsowicz viscosities*. The lowest viscosity is η_2, when the flow is along the director, whereas the highest is η_1, measured when the director is perpendicular to the flow but parallel to the flow velocity gradient. The third director orientation, $\boldsymbol{n}_3$, perpendicular to both flow and gradient, results in an intermediate viscosity η_3, generally much closer to η_2 than to η_1. The Męsowicz viscosities are typically on the order of a Poise, or 1kg m^{-1} s^{-1} in SI units.

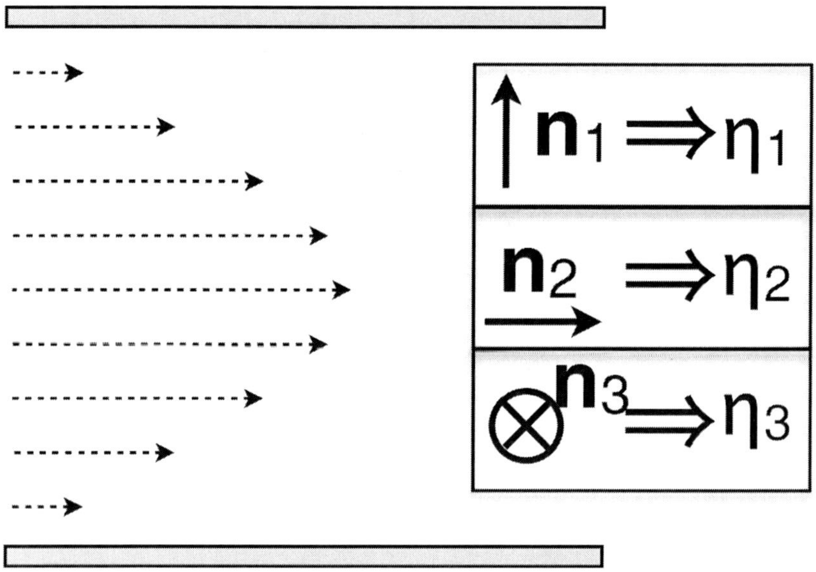

FIGURE 3. To define the three Mięsowicz viscosities, η_1, η_2 and η_3, one considers an experiment of flow of nematic liquid crystal through, e.g. a tube; the dashed arrows on the left indicate the flow velocity along the tube, as a function of location. The director $\boldsymbol{n}$ is blocked by an external magnetic field, giving three possible orthogonal director orientations, $\boldsymbol{n}_1$, $\boldsymbol{n}_2$ and $\boldsymbol{n}_3$, with respect to the flow direction, each giving rise to the respective Mięsowicz viscosity.

The Mięsowicz viscosities constitute the first approximation for understanding the viscous flow of LCs. If we make no effort to control the director orientation during flow, ***n*** will align nearly parallel to the flow to minimize viscous losses. The minimum viscosity is actually obtained for a slight offset, as a more rigorous treatment shows [5]. Flow alignment can be important in the study of drops, shells and fibers of LCs, as the sample preparation always involves flow. If the samples are prepared with the material in its nematic phase, the viscosity anisotropy will lead to a relatively uniform alignment during the process. Note, however, that for smectics (see section 2.2) the minimum dissipation is not along ***n*** but along the smectic layers. Moreover, in many cases the LC is brought into an isotropic state during sample preparation, by heating or by adding a solvent.

2.1.4. Liquid crystal elasticity and deformations in the director field.

Liquid crystals, with nematics forming the clearest example, exhibit a type of elasticity that is unique to this materials class. This 'nematic elasticity' is fundamentally different from solid or rubber elasticity, since in this case there is no restoring force arising in response to deformation of the sample shape, but instead a restoring *torque* that builds up in response to deformation of the *director field* $\mathbf{n}(\mathbf{r})$.

In the elastic ground state of an achiral nematic (ignoring thermally activated fluctuations [1]) the director field is uniform, i.e. $\mathbf{n}(\mathbf{r})$ is a constant, independent of location. A deviation from the uniform field costs energy, and this is what gives rise to nematic elasticity. This energy is quantified in the *Oseen–Frank theory,* based on Oseen's groundwork in the 1920s and the refinement, and introduction of modern terminology, by Frank in 1958 [2]. In the bulk of a nematic, any generic deformation from uniform $\mathbf{n}(\mathbf{r})$ can be described as a linear combination of three elementary elastic deformations: *splay*, *twist* and *bend*. These three fundamental distortions, defined graphically in figure 4, are independent of each other, such that no combination of twist and/or splay can compensate for a bend etc. The total free energy F_b related to bulk elastic deformations of the director field can thus be calculated by integrating a bulk energy density f_b, containing one term each for splay, twist and bend, over the volume of the sample (if S is not uniform within the volume considered, a fourth term, saddle-splay, to be discussed in a moment, must be included as well):

$$F_b = \int \mathrm{d}V f_b \, , \qquad \text{(2a)}$$

$$f_b = \frac{1}{2}K_1(\nabla \cdot \mathbf{n})^2 + \frac{1}{2}K_2[\mathbf{n} \cdot (\nabla \times \mathbf{n})]^2 + \frac{1}{2}K_3[\mathbf{n} \times (\nabla \times \mathbf{n})]^2 \, . \qquad \text{(2b)}$$

Reflecting its character as a source, the splay term in (2*b*) contains the divergence of the director field $(\nabla \cdot \mathbf{n})$, whereas the terms representing the two rotational distortions twist and bend contain the curl $(\nabla \times \mathbf{n})$. Because each term is squared, the bulk elastic constants, K_1 for splay, K_2 for twist and K_3 for bend, must all be positive, otherwise there would be no finite energy minimum.

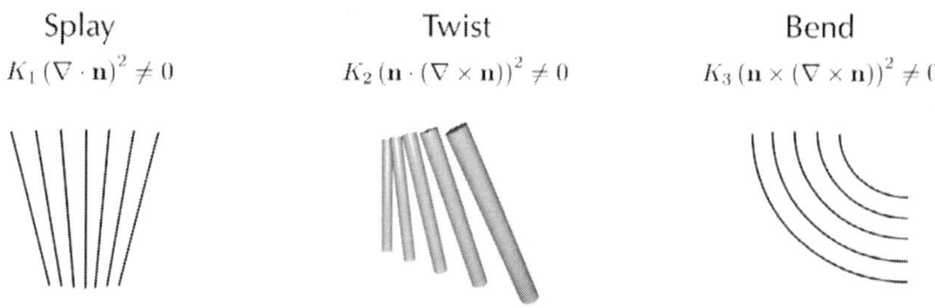

FIGURE 4. Graphical definitions of the three elementary director field deformations splay, twist and bend, together with their corresponding terms in the elastic free energy density and their respective elastic constants K_1, K_2 and K_3.

When considering strongly confined LC samples such as drops, shells and fibers, one must also consider the influence from the bounding interfaces. There are two contributions to the interfacial energy. First, as with any interface, there is an energy term given by the product of interfacial tension and the bounding area. However, since this is an LC the interfacial tension may depend on how $\boldsymbol{n}$ is aligned with respect to the boundary. We say that $\boldsymbol{n}$ is *anchored* in a certain preferred orientation at the interface. This terminology should not be interpreted too literally, however, as we may have anything from very strong to very weak anchoring. The balance between surface anchoring and bulk elasticity will be discussed for the case of droplets in section 3.3.1. The anchoring energy is quantified by a parameter W. A common expression for this contribution to the interfacial energy is the Rapini–Papoular energy density (an area density, thus J/m^2) [6, 7]:

$$f_a = \frac{W}{2}[1 - (\boldsymbol{n} \cdot \hat{\boldsymbol{\nu}})^2] \tag{3}$$

where the unit vector ν denotes the preferred orientation of the director at the interface, and we let the index 'a' stand for 'anchoring'.

The second interfacial energy contribution is the just mentioned saddle-splay term, which actually is an elastic energy related to director field distortion in the Oseen–Frank theory. As the name suggests, the term, quantified by the saddle-splay constant K_{24}, relates to a deformation of the director field such that $\boldsymbol{n}(\boldsymbol{r})$ is normal to a saddle (or potato chip) surface [8]. Such surfaces have negative Gaussian curvature, i.e. their two principal curvatures have opposites signs; a saddle curves upwards along the horse but downwards across the horse. The corresponding energy density is a pure divergence ($\nabla\cdot$) term:

$$f_{24} = -(K_{24}/2)\nabla\cdot(\boldsymbol{n}\nabla\cdot\boldsymbol{n} + \boldsymbol{n}\times\nabla\times\boldsymbol{n}) \quad (4)$$

This means that its volume integral, required to calculate the corresponding energy, can be converted into a surface integral via the divergence theorem (Gauss' theorem), provided that K_{24} is constant within the integration volume:

$$\begin{aligned} F_{24} &= \int \mathrm{d}V f_{24} \\ &= -\frac{K_{24}}{2}\int \mathrm{d}\boldsymbol{S}\cdot(\boldsymbol{n}\nabla\cdot\boldsymbol{n} + \boldsymbol{n}\times\nabla\times\boldsymbol{n}). \end{aligned} \quad (5)$$

Note that the integrand in the last term in (5) is not f_{24}, as defined in (4), which is a volume energy density, J/m^3. Compiling the above, we can obtain the total free energy (in the absence of fields and for constant K_{24} within the sample volume) as [7]:

$$F = F_{\mathrm{b}} + F_{24} + \int \mathrm{d}S f_{\mathrm{a}}. \quad (6)$$

A key difference between the F_{24} term and the other contributions to the free energy is that F_{24} enters with negative sign. In other words, while any deformation described by F_{b} raises the free energy, as does any increase in interfacial area, captured by $\int dsf_{a}$, a saddle-splay deformation of the bounding interface actually *decreases* the energy if $K_{24} > 0$. This seems to be the case for all materials that have been experimentally studied so far with respect to saddle-splay deformation.

Because the saddle-splay term in case of constant K_{24} does not influence the bulk energy far from boundaries it has often been ignored, as the sample size was reasonably large in many studies. Moreover, when the sample boundaries are flat, as in standard display-like devices, the F_{24}contribution vanishes entirely. In contrast, in our focus on small samples with strongly curved boundaries, where the interface influence can be considerable, the F_{24} contribution can play an important role. If K_{24} varies within the sample volume, which should be the case if the order parameter S is non-uniform, Gauss' theorem cannot be applied and the saddle-splay contribution must be treated as a bulk elastic energy. Such a situation is not as exotic as one might at first think; at temperatures near the clearing point a temperature gradient within the sample would lead to varying S and thus varying K_{24}, but even at uniform temperature, S decreases in the close vicinity of a topological defect or an impurity. As droplets, shells and fibers frequently comprise topological defects, the saddle-splay term may need to be treated with its full bulk character[2].

The four elastic constants are significant material parameters, with key relevance for understanding the behavior of strongly confined LCs. When the boundaries are strongly curved, as in the cases considered in this review, the boundary conditions often impose elastic distortions within the LC, thereby raising the free energy. The elastic constants have the dimension J/m = N, i.e. a force, with a magnitude typically in the pN range. Note that the saddle-splay energy density (4) contains no squared terms, hence the saddle-splay constant K_{24} can, in principle, take negative as well as positive signs, in contrast to the bulk elastic constants. For thermotropics, K_1 and K_2 often have about the same value whereas K_3 is about twice that value. A commonly applied convenient approximation is $\boldsymbol{k} = \boldsymbol{k_1} = \boldsymbol{k_2} = \boldsymbol{k_2}$ and $K_{24} = 0$, a simplification referred to as the 'one-constant approximation'.

Since a spatial derivative yields an inverse length, each term in (2*b*) is proportional to $1/R^2$, where $R$is the characteristic length of the respective deformation. This is the pitch in case of a twist, the radius of curvature of a bend, and the distance to the 'source' of a splay deformation. Note that R is independent of location for a twist of constant magnitude, while, in contrast, the characteristic length of bend and splay is only defined once we have chosen a certain location. A bend or splay gets less apparent the further away from the center of curvature or 'source' that we are.

The $1/R^2$ dependence and the small magnitude of the elastic constants means that elastic deformations with large characteristic length cost very little energy, so little in fact that large-scale deformations are thermally generated in a bulk nematic LC in its ground state. This means that large-scale non-uniformities arise spontaneously in bulk LC samples, giving them a characteristic turbid character. A uniform $\boldsymbol{n}(\boldsymbol{r})$ actually has higher free energy, which we may provide, e.g. by applying a magnetic field that aligns the LC [1].

2.2. Thermotropic smectics

The defining characteristic of smectic LCs is a layered structure, see figure 5. The layering should not be understood too literally, however. While we often speak of 'layer boundaries' for practical purposes, these are mainly help concepts: there are no sharp layer boundaries in smectics, but rather a quasi-long-range electron density modulation along one direction [9]. In the smectic-A phase, abbreviated SmA, the density modulation is along the director, i.e. ***n*** is identical to the smectic layer normal ***k***. The modulation can, however, also develop along an arbitrary angle with respect to ***n***, which means that the director tilts in the layers. The most imported tilted smectic is the smectic-C phase, SmC.

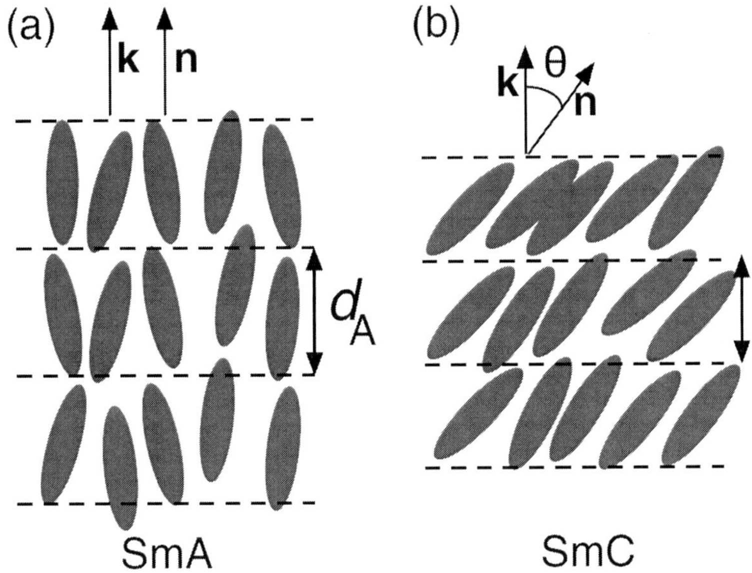

FIGURE 5. Schematic 2D drawings of the two most important smectic phases. Mesogens are drawn as ellipses, layer boundaries with dashed lines. Due to the tilt in the SmC phase of ***n*** with respect to the layer normal ***k***, by an angle θ, the layer thickness d_C is smaller than the layer thickness d_A in the SmA phase.

The SmC tilt angle, normally given the symbol θ, is generally temperature dependent, often approaching zero on heating towards a SmA phase, while it at lower temperatures may saturate in the range 20–30°. The exception is if the SmC phase follows on cooling directly from an N phase or from the isotropic liquid state; then θ is often large (around 45°) and close to temperature-independent. The same holds in case of a transition from a SmA phase with strong first-order character. There are several versions of the basic smectic-C-type structure, distinguished by different interlayer tilting direction correlation schemes (see e.g. [10]). In the fundamental SmC phase the tilt in adjacent layers is in the same direction.

The *magnitude* of tilt θ critically influences the free energy of a SmC phase, but the tilting *direction*has no impact on the energy. Any direction is thus equally good, yielding a range of degenerate SmC states that define a cone (the *SmC tilt cone*), with opening angle 2θ and symmetry axis along $\boldsymbol{k}$. Basically, the smectic layer thickness is the projection of the average mesogen length on $\boldsymbol{k}$ (taking conformational disorder into account [10]), see figure 5(b).

The Oseen–Frank elasticity theory was developed for nematics but it can be extended to phases with partial positional order. One must then take into account the special restrictions imposed on $\boldsymbol{n}(\boldsymbol{r})$ by the symmetry of each phase. Twist and bend are incompatible with the constant layer thickness in the SmA phase. In contrast, the phase is compatible with director splay, and rolling-up of SmA layers into a splayed director configuration frequently occurs.

Although SmC is less symmetric than SmA it allows more fluctuations in $\boldsymbol{n}(\boldsymbol{r})$. Tilting gives a new freedom (tilt direction) that enables deformations which do not interfere with the layered structure. In fact, the projection of $\boldsymbol{n}$ onto the layer plane (this projection is often called the C-director, $\boldsymbol{c}$) behaves similar to the ordinary director in the nematic. The characteristic texture of a quasi-homeotropic SmC sample therefore reminds of the schlieren texture of a planar-aligned nematic, the changing curvature of the schlieren reflecting variations in $\boldsymbol{c}$. There are some important differences in terms of which topological defects (see section 3.3) can develop, however, because the sign invariance of the director, $\boldsymbol{n} = \boldsymbol{n}$ does not apply to the C-director ($\boldsymbol{c} \neq -\boldsymbol{c}$). The director tilting directions for $\boldsymbol{c}$ and $-\boldsymbol{c}$ are opposite, hence they correspond to different physical states.

2.3. Cholesterics (chiral nematics)

In recognition of the first chiral nematic-forming compounds being cholesterol derivatives, we often use the term *cholesteric* as an alternative name for chiral nematic. The phase can consequently be abbreviated either Ch or N*, where the star signifies chirality. The key expression of chirality is a helical modulation of $\boldsymbol{n}$ in the N* ground state, see the schematic illustration in figure 6. The helix axis, often chosen as $\hat{\boldsymbol{z}}$, is perpendicular to $\boldsymbol{n}$. The period, generally referred to as pitch p, is defined as the distance over which $\boldsymbol{n}$ makes a 2π rotation. Due to the director sign invariance the physical properties repeat along $\hat{\boldsymbol{z}}$ with a period $p/2$.

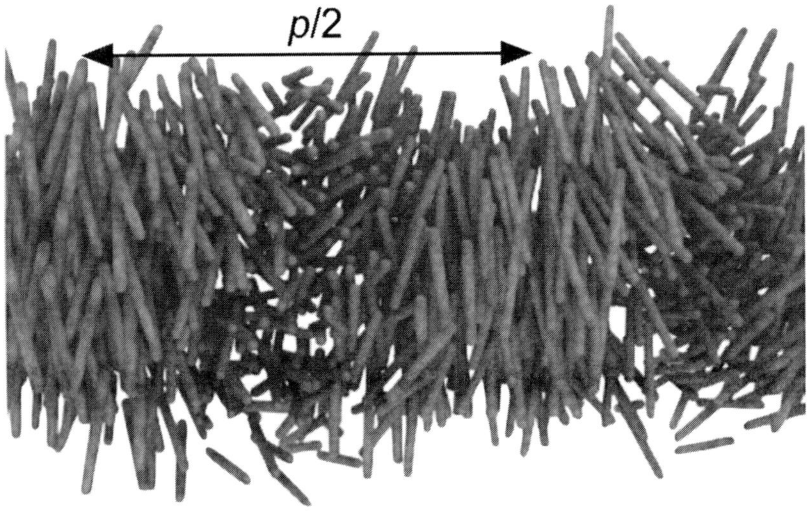

FIGURE 6. In chiral nematics the director is modulated helically, with a pitch *p*, along an axis $\hat{z}$ that is everywhere perpendicular to $\boldsymbol{n}$. Reproduced from [11]. CC BY 3.0.

In the ground state the pitch has a well-defined value p_0, sometimes called the natural pitch. This helical ground state is incompatible with (2*b*), which models a non-chiral nematic and thus indicates that any twist would raise the free energy. But for a cholesteric phase a *non-helical* state raises the free energy, as would any helix pitch different from p_0. The free energy density expression must therefore be modified, extending the achiral twist term slightly to express the spontaneous twist:

$$f_{\text{twist}}^{\text{N}^*} = \frac{1}{2}K_2[\boldsymbol{n} \cdot (\nabla \times \boldsymbol{n}) + q_0]^2 \tag{7}$$

The added term $q_0 = 2\pi/p_0$ is the wave vector of the spontaneous cholesteric helix. If the tendency to twist is weak, q_0 is small and the term in (7) does not significantly affect the elastic energy density. For strong spontaneous twist, however, typically with the pitch on the same order as the wavelength of visible light or smaller, the added term in (7) can have a considerable impact.

Chirality has several important consequences for the physical properties of LC phases, sometimes related to the helical modulation, sometimes in general to the loss of mirror symmetry [12, 13]. Here we will restrict ourselves to discuss the effect of the helix on the optical properties of cholesterics, which is probably the most striking chirality expression. For more detailed information on chiral LCs, see, e.g. the book by Kitzerow and Bahr [13]. A good derivation of the optical properties of cholesterics can be found in a text by Priestly [14].

If p_0 is short, on the order of 300 nm, $\boldsymbol{\Delta n}$ of the phase changes sign compared to the non-helical analog, and the helix axis $\hat{z}$ becomes the optic axis. Moreover, such a short-pitch cholesteric typically shows *selective reflection* of visible light incident along $\hat{z}$, turning the sample iridescently colored, the color depending on the viewing angle as well as on p_0. In modern terminology, the cholesteric LC intrinsically exhibits a photonic band gap. The reflected light is circularly polarized with the same handedness as the helix, whereas the opposite handedness of circularly polarized light is transmitted.

The helical modulation of $\boldsymbol{n}$ leads to a periodic modulation along the direction of light propagation of the refractive index, as long as the light does not enter perpendicular to $\hat{z}$. Although the helix periodicity is p_0 the optical periodicity along the helix axis is half of this, since already a 180° turn of $\boldsymbol{n}$ takes us back to an optically equivalent situation. We can apply Bragg's geometrical method (figure 7), well known from x-ray diffraction, to obtain an expression for the relation between the selectively reflected wavelength λ_{N^*} (as measured within the N* phase) and the viewing angle:

$$p_0 \sin\alpha = m\lambda_{N*} \tag{8}$$

where m is an integer. For normal incidence (along $\hat{z}$) we can set $m = 1$ because only the fundamental reflection is present. Higher-order reflections would be the result of higher harmonics of the periodic structure but a sinusoidal modulation has no harmonics, only the fundamental. For oblique incidence the modulation is no longer perfectly sinusoidal since we are not following the helix, hence we then get weak higher-order reflections.

Selective reflection occurs within a band of wavelengths that for normal incidence has the width $\boldsymbol{\Delta\lambda N^* = p_0\Delta n}$, where $\boldsymbol{\Delta n}$ is the birefringence the sample would have had if the helix were unwound. The light wavelength in the N* phase is shortened with respect to the wavelength in vacuum by a factor equal to the average refractive index. For typical LCs this is about 1.5. Thus, if an *N** phase appears green (~550 nm wavelength in air) at normal incidence, the helix pitch must be around 370 nm.

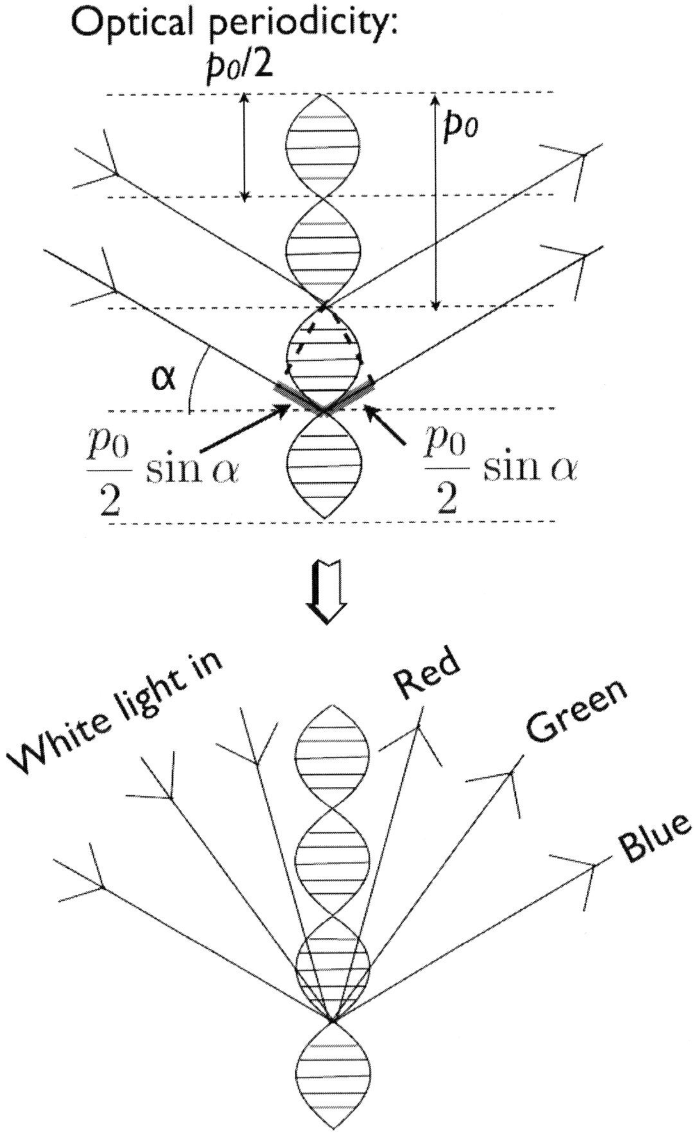

FIGURE 7. The viewing angle dependence of the color of a cholesteric liquid crystal, summarized in the bottom cartoon, can be understood following a simple graphical procedure analogous to Bragg's law for x-ray diffraction, sketched at the top. Reproduced with permission from [3].

A decrease in the angle of incidence α of the light leads, according to (8), to a reduced wavelength of the selectively reflected light. This explains the observation that the color changes towards shorter wavelengths (from red to blue) if we tilt a cholesteric sample away from us, as schematically summarized in the lower part of figure 7. If the pitch is slightly too long to give selective reflection at normal incidence, we might start seeing colors under oblique incidence.

2.4. Liquid crystal elastomers

Liquid crystal elastomers (LCEs) comprise a unique class of materials due to their remarkable feature of dramatic shape change induced by a phase transition (usually the clearing transition). Experimental studies have reported up to ~400% strain and stress of ~10–100 kPa [15–17]. They are sometimes portrayed as artificial muscles although soft actuator is perhaps more appropriate; in his plenary talk at the ILCC 2002 (Edinburgh, UK) Pierre-Gilles de Gennes pointed out that LCEs are unlikely to match the force and speed of real biological muscles. Nevertheless, there are many interesting application scenarios, e.g. in soft robotics or other areas where shape change is required but ordinary rigid mechanics are not an option.

When you stretch a piece of rubber (the layman's term for elastomer) you are working against entropy. The polymer chains in the rubber, which cannot flow due to a limited amount of crosslinks, yet still have considerable conformational freedom, will adopt an isotropic random coil conformation in the ground state. As you stretch the rubber you reduce the number of accessible conformations, imposing the much reduced set in which the polymer is extended along the stretching direction. It is the entropic cost of this reduced conformational freedom that provides the restoring force that brings the rubber back to its original shape when you release it.

In an LCE the polymer chains are stretched out into the extended subset of conformations even without any external force applied. The extension is along $\boldsymbol{n}$; the long-range orientational order of the LC state, with the resulting macroscopic anisotropic physical properties, restricts the conformational freedom of the polymer chains to fluctuate perpendicular to $\boldsymbol{n}$, and promotes alignment along $\boldsymbol{n}$. But just like when you release a stretched piece of ordinary rubber, the polymer chains will rearrange into their preferred set of isotropic conformations if the stretching force disappears. With an LCE this corresponds to heating above the clearing point of the LC phase, or otherwise inducing the transition into an isotropic state. The interesting thing is that, because the system is lightly crosslinked, this change of conformation brings with it a change in macroscopic sample shape, as schematically illustrated in figure 8. The process is fully reversible, so when the sample is brought back into the LC state, the shape reverts as well. There are a number of excellent reviews of LCEs and their actuation behavior [17–23]. As we here can only give a very brief introduction to the basic concept we refer the interested reader to these texts for more details.

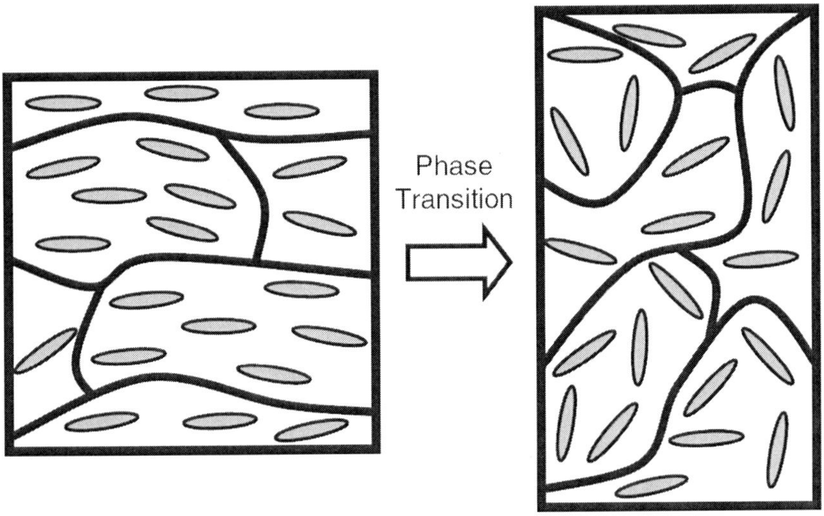

FIGURE 8. The working principle of actuation of a nematic liquid crystal elastomer (LCE). In the nematic ground state (left) the polymer chains are stretched out along ***n*** in an entropically unfavored reduced set of conformations. Upon inducing the phase transition into the isotropic phase (right), the anisotropic environment disappears and the polymer chains adapt the preferred isotropic random coil average conformation. The crosslinks couple this conformational change to a macroscopic sample shape change, with compression along the original ***n*** and expansion in both orthogonal dimensions. Reproduced from [21] John Wiley & Sons. Copyright © 2010 WILEY-VCH Verlag GmbH & Co. KGaA, Weinheim.

LCEs may be subdivided into three main classes [17, 21, 24], differing in the generic chemical structure, see figure 9. In *main chain* LCEs, the mesogens are part of the polymer chain, whereas they are attached as pendants to the polymer chain in *side chain* LCEs. The latter are further subdivided into *side-on* and *end-on* versions, referring to the attachment point within the mesogen to the polymer chain.

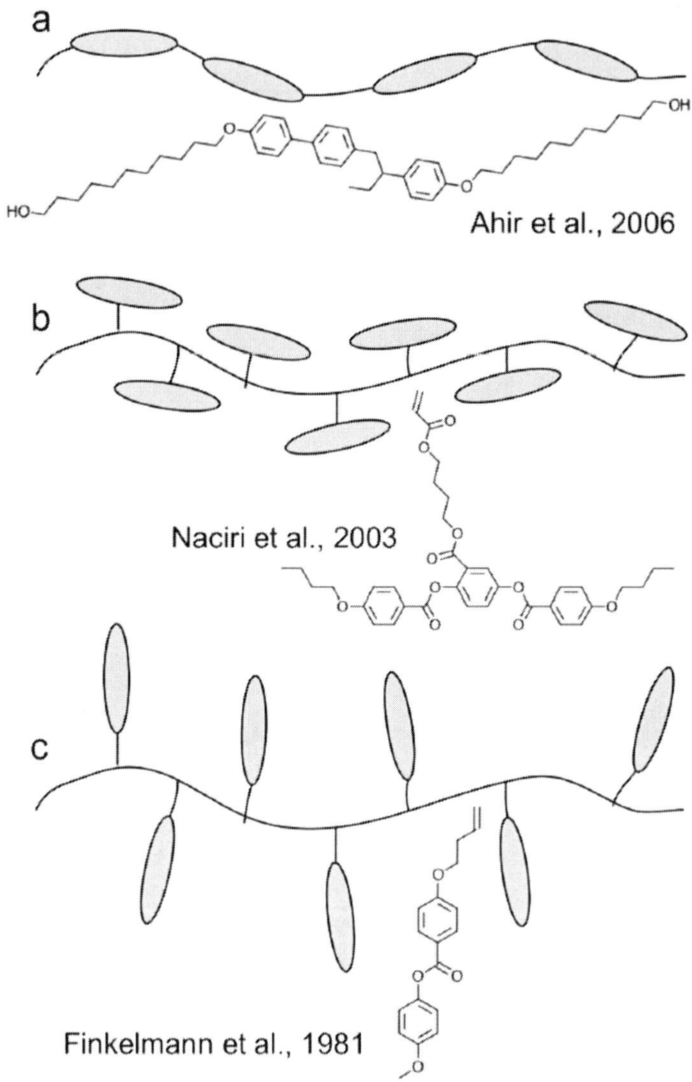

FIGURE 9. Schematic drawings of the three main types of LCEs, together with example mesogenic monomers for each case; main chain LCE (the example structure is studied in Ahir *et al* [25]), side-on side chain LCE (example structure studied in Naciri *et al* [26]) and end-on side chain LCE (example structure studied in Finkelmann *et al* [27]).

A prerequisite for the operation of LCEs is that $\boldsymbol{n}(\boldsymbol{r})$ is uniform in the ground state. Often this is achieved by starting with a liquid crystalline but non-polymeric mixture of low molar mass precursors. This mixture can be aligned by placing it in contact with substrates coated with an aligning layer [28–30], applying a magnetic[31–33] or electric field [34], or subjecting it to shear flow [35–37]. The precursor mixture typically consists of reactive mesogens, such as the acrylate mesogens exemplified in Figure 9, a small amount of diacrylate crosslinkers, and often a suitable photoinitiator. Then polymerization and crosslinking can be initiated by UV illumination after the desired uniform alignment has been ensured (in case of alignment by shear flow, photopolymerization must be initiated during shearing). Using photosensitive mesogenic components in the mixture, also the alignment can be achieved by UV illumination, in that case using linearly polarized UV light [23, 38, 39].

Another solution is the two-step polymerization introduced by the Finkelmann group, [40, 41] producing the very first LCEs after they had been predicted by de Gennes [42]. An intermediate main-chain polymer is produced in a first step of incomplete polymerization. Because the sample is now polymeric, uniaxial stretching can align $\boldsymbol{n}$ throughout the sample. Then, under the influence of the aligning mechanical field, the second polymerization and crosslinking step is triggered, which fixes the ground state orientation and produces the LCE film. This approach has the limitation that the resulting elastomer still remains somewhat mechanically constrained due to the presence of sharp domain boundaries, and it is difficult to remove excessive random crosslinks which persist [43].

Following the Finkelmann group principle, Yakacki and co-workers recently introduced a convenient approach to prepare LCE actuators that uses solely commercially available components [44]. The LCE formation here follows a two-stage thiol-acrylate Michael addition and photopolymerization reaction. The approach is attractive as it opens for low-cost and large-scale LCE synthesis without requiring in-house synthetic chemistry facilities.

When producing LCEs in shape of droplets, shells or fibers, the requirement on uniform alignment may be fulfilled in alternative ways, since the production method and/or the sample geometry impose a specific director field. External alignment fields or substrates may not be necessary (in fact, it may even be impossible to apply them) and some rather interesting curved LCE actuators can be realized with relative ease. This will be further discussed in section 3.4.

Heating is not the only means of inducing the phase transition that triggers actuation. A powerful method is to incorporate dyes that change shape between rod- and kink-shaped upon illumination with light of adequate wavelength. [19, 45] The most common choice is azo dyes, which switch from a rod-shaped *trans*-isomer in the dark state to a kinked *cis*-isomer upon illumination by UV-light. Visible light illumination speeds up the reverse isomerization, back to the *trans* state, allowing switching in both directions using light with different wavelengths [46]. Because the *cis*-isomer disturbs the liquid crystalline ordering, all LC-related phase transitions, including the clearing point, are brought down to lower temperatures by the presence of *cis*-isomer azo dye, compared to when the dye is in its rod-shaped *trans*-isomer. The latter isomer is fully compatible with the long-range order. With the right amount of azo dye one can thus modulate the clearing point between above room temperature without UV illumination and below room temperature during UV illumination. This means that light can trigger the LCE actuation, not by heating, but by dynamically changing the phase diagram.

3. LIQUID CRYSTAL SHELLS AND DROPLETS

We now have the required basic knowledge about LCs and LCEs to discuss the effects of varying types of confinement. This section is devoted to spherical confinement, in droplets and shells, whereas section 4 covers cylindrical confinement. In section 3.1 we introduce the microfluidic production technique, which has revolutionized our capability of producing droplets and shells with high precision and reproducibility. This is followed by a survey of options for making LC spheres, including but not limited to microfluidics, how to stabilize them and at the same time control the alignment (section 3.2.1), and how to tune the size and thickness of shells via osmosis (section 3.2.2).

Section 3.3 deals with topological defects in LC spheres. After defining the key concepts for describing and quantifying topological defects we continue with surveying the experimental and simulation data on nematic, cholesteric and SmA/SmC droplets and shells. We discuss actuators made by polymerizing LC droplets and shells into LCEs, followed by an overview of application opportunities of short-pitch N* spheres in security, ending the section with a quick look at the role of droplets and shells in the emerging field of active LCs.

3.1. Microfluidics in a nutshell

The emergence of microfluidics started around the 1980's from the miniaturisation of flow geometries through micro- and nano-fabrication technologies [47–49]. Microfluidics is a rapidly growing multidisciplinary field, where physical, biological, chemical and engineering sciences converge into a single platform. It deals with the process and manipulation of fluids in channels with dimensions of tens of microns [48, 50–52]. Consequently, the surface to volume ratio increases in microfluidic systems, thus inertial forces are often negligible in comparison to viscous, electrostatic/electrodynamic, or surface tension forces. Especially, the surface forces are crucial for regulating the fluid flow in microfluidic devices [53]. Typical values of the Reynolds number, *Re*, which is a dimensionless number that relates the inertial to the viscous forces, are less than 10 for microfluidic flows [54, 55].

The revolutionary growth of the microfluidics field was triggered by the advancement of a lithography technique originated from the Whitesides group, known as soft lithography. The term soft refers to the design of channels using a soft elastomer, typically polydimethylsiloxane, or PDMS. The PDMS based soft lithography system has several advantages. It is easy to design and fabricate various flow patterns, from simple to complex and multiphase. PDMS is inexpensive, optically transparent for microscopy and permeable to gases, the latter aspect crucial for microbiological culture [52]. Thus, soft lithography is rapidly spreading in various research fields like diagnostics and biological research [56], fluid physics [57–60], chemical synthesis [61], the study of infectious diseases and preventive medicine [62, 63] and many other interdisciplinary research fields [52].

When chemical and thermal stability is required, alternative materials to PDMS are needed, such as glass, steel or silica [52]. A completely different design for microfluidics, based on nested glass capillaries with circular and square cross sections, respectively, was developed in the Weitz group. The basic principle for this method, which allows easy production of multiple emulsions with excellent control, is shown in figure 10. Since the outer diameter of the tapered cylindrical capillaries matches the inner side length of the square one, all capillaries are easily aligned in a concentric fashion. An inner (often aqueous) phase, which is immiscible with the middle phase, is injected through the left cylindrical capillary. The middle phase is flown in the same direction, through the corner spaces between cylindrical and square capillaries. While the two phases are co-flowing, they meet a counter-flowing outer phase, also immiscible with the middle phase (also the outer phase is often aqueous), introduced from the opposite side, again via the corner spaces between cylindrical and square capillaries. The pressure of the outer fluid flow-focuses the co-flowing liquids as a narrow jet [64, 65]. The interfacial tension between the middle and outer fluids causes breaking of the stream into shells via the Rayleigh–Plateau instability. Several modifications of the basic device are possible, the one depicted in figure 10 yielding a double emulsion. If a simple emulsion of droplets is preferred, the same device can be used with the flow rate of the inner fluid reduced to zero.

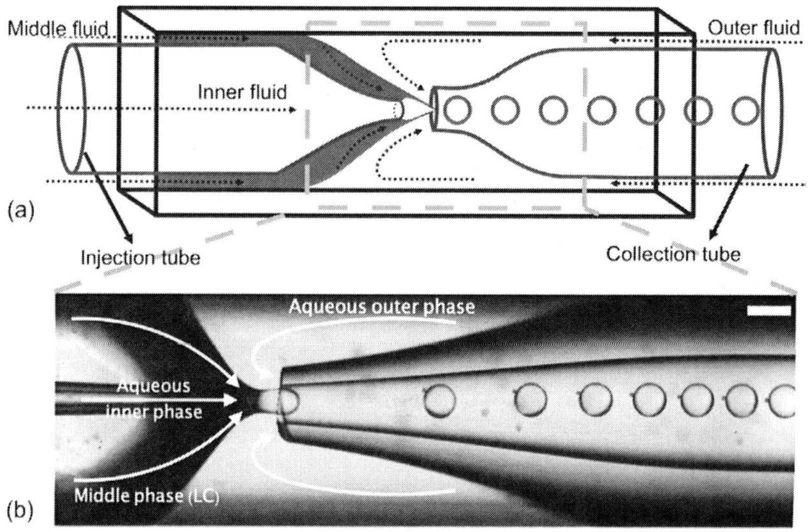

FIGURE 10. (a) Schematic illustration of the nested glass capillary microfluidic device for generating double emulsions from coaxial jets. Two tapered cylindrical glass capillaries are inserted into a square capillary, its inner side length identical to the outer diameter (~1 mm) of the cylindrical capillaries. Injection and collection tubes are tapered using a micropipette puller, in order to ensure the desired constriction size, and a micro forge is then used to cut off the tips and establish the orifices. The middle fluid must be immiscible with the inner as well as the outer fluid, and interface stabilisers are added to prevent coalescence of the dispersed phase. (b) Liquid crystal shell production. The shells in this example are about 130 μm in diameter and below 3 μm in thickness and there are small satellite droplets next to each individual shell. The scale bar is 200 μm.

Compared to the PDMS-based microfludic set-up, a glass micro capillary device has the advantage of reusability, thanks to its chemical resistance, and it does not suffer from swelling problems when working with organic solvents. A disadvantage is the difficulty in designing curved flow patterns. So far, the nested glass capillary technique dominates over PDMS-based microfluidics when the target is to prepare LC droplets and shells. Irrespective of the type of microfluidic device, the fluids are generally injected into the different channels by syringe pumps or pneumatic control units specially designed for microfluidics. The manipulation of multiphase flows in both systems enables generation of mono disperse bubbles [66], droplets [67], shells [68, 69], micro reactors for biology and medicine and various chemical reactions [56, 61–63, 70, 71].

To make LC droplets or shells, the LC is the disperse phase while the continuous phase is isotropic. It is worth mentioning that fascinating effects can also be seen when flowing LC as a continuous phase in microchannels. This brings novel flow profiles generated from the combination of surface anchoring and topological flow constraints within the channel, which are not observed with an isotropic continuous phase. As with conventional flat glass substrates, the alignment of an LC in a PDMS flow channel can be controlled by treating the channel walls chemically for either homeotropic or planar alignment, and even combinations of opposing alignment on different walls can be obtained by combining chemical alignment agents and plasma treatment and/or leave one side open to air [72, 73].

The different possibilities for surface anchoring combinations and the freedom to vary the channel aspect ratio (width/depth) open a playground for the creation of various topological constraints. In this way one can trigger the formation of topological point or line defects in nematics, which in turn are useful for guided transportation of micro-scale objects and tunable flow shaping [73–75]. Further interesting possibilities arise with SmA liquid crystals in rectangular channels, as reported by several research groups [76–78]. Here, the main focus is on inducing arrays of focal conic defects with controlled size and locations, again using different anchoring conditions and channel geometries to tune the final defect arrangements. Besides the channel-confined focal conic defect structures, Shojaei *et al* reported the formation of a two-dimensional network of the same defect structures in a thin film of the SmA-forming LC 8CB using cracked PDMS films [77]. Here, the crack pattern is controlled by stretching the plasma treated PDMS film.

3.2. How to make liquid crystal shells and droplets using microfluidics

A typical way of producing LC droplets is preparing emulsions by dispersing liquid crystals in an immiscible liquid. Among numerous methods for the emulsion preparation, stirring or vortexing constitute quick and low-cost approaches. The drawback of these methods, crudely breaking the bulk liquid crystal into small droplets by mechanical agitation, is the difficulty in controlling the size and obtaining uniformity. A common alternative is ultrasonication, in which a strong oscillating shear flow is induced by pressure waves at ultrasonic frequencies. This technique can provide a somewhat narrower size distribution with smaller droplets, even below the micron scale, in form of a miniemulsion [79]. Yet, also with ultrasonication one is far from achieving monodispersity, and the non-uniform size further leads to Ostwald ripening, which reduces emulsion stability. The high interface curvature of a small droplet raises the solubility of the disperse phase, triggering molecular diffusion from small to large droplets.

With the coaxial microfluidic technology developed in the Weitz group, a new alternative appeared that enables the production of highly controlled monodisperse simple as well as multiple emulsions [68], the liquid crystal taking the role of the middle phase. Now, not only LC droplets of selectable size could be produced, but even thin shells of LC, where a few microns thick spherical layer of the desired liquid crystal material is surrounded by isotropic immiscible phases (typically aqueous) on both sides. The shell/droplet dimensions, i.e. diameter and thickness (in case of shells), can be adjusted by tuning the production parameters: capillary/orifice size, flow rates, interfacial tension and viscosities of the fluids involved. The orifice sizes of the tapered cylindrical capillaries have a direct impact on the shell/droplet sizes. At low flow rate the resulting droplet/shell diameter can be estimated to be close to the orifice diameter [80].

However, the relative flow rates also play a critical role in controlling the shell/droplet sizes, even determining whether or not emulsion formation is at all possible. Utada *et al* distinguished between two different droplet-forming mechanisms in co-flowing Newtonian fluids, namely dripping and jetting modes [81]. Which of the two modes is active depends on the interplay between interfacial tension and viscous drag force, acting on the interface between middle and outer fluids. If the flow rates of the inner and middle phases are fixed, a high flow rate of the outer fluid increases viscous stress at the interface, eventually dominating over the interfacial tension such that a long stream of inner+middle phase is stabilized (jetting mode). In contrast, at lower flow rate of the outer fluid the inner complex stream is rapidly broken up into shells (dripping mode), because now the interfacial tension dominates. Even if liquid crystals are (weakly) non-Newtonian fluids, the basic drop-forming process is expected to be similar to that of Newtonian fluids [81]. On the other hand, the LC is most often brought to the isotropic state, for instance by heating, during production, turning it into a Newtonian fluid at this stage. Further tuning of shell/droplet size is possible via variations of interfacial tension (e.g. using surfactants) and viscosity [80, 82].

3.2.1. The connected problems of interface stabilization and liquid crystal alignment control.

The stabilization of liquid crystal shells/droplets can be ensured by adding interface stabilizers, also called emulsifiers, most often to the aqueous phases. Compared to water a thermotropic LC can be considered an oil phase, immiscible with water, thus the stabilizers are needed to reduce the interfacial tension between the phases. Without stabilizers, droplets would coalesce and shells would collapse into droplets. Importantly, the stabilizers may also play another key role, namely to determine the liquid crystal alignment.

The common choice for emulsifier/stabilizer is either an ionic surfactant, e.g. sodium dodecyl sulfate (SDS), or a water soluble polymer, e.g. polyvinylalcohol (PVA). At the water–LC interface, a surfactant adsorbs with its apolar aliphatic chain penetrating into the LC, promoting homeotropic alignment (see section 2.1.2) and thus a radial orientation of the director field in the shell/droplet. In contrast, a dissolved polymer like PVA in aqueous solution has a disordered random coil configuration which cannot be expected to promote a particular type of ordering in the LC. The polymer thus most likely acts solely as a stabilizer, leaving the LC-water interaction to determine $\boldsymbol{n}$ at the interface, which ends up being planar or possibly tilted. This is because the mesogens typically contain an aromatic core, and often also a cyano group, which can participate in hydrogen bonding with the surrounding water, whereas the aliphatic chains at the mesogen end are avoided by the water. In case of the shells, we can additionally achieve hybrid director alignment [83, 84] by imposing asymmetric boundary conditions at the interfaces to the LC. In this case we may use a surfactant in the inner phase and PVA in the outer, or vice versa.

As a first approximation, we can consider stabilizers to be localized to the LC–water interface and the immiscible phases to be truly immiscible, i.e. the LC is in its pure state within the actual shell. Because the phase sequence of LCs is so sensitive to contaminants, however, it is easy to find evidence that this simplified picture is not quite true in practice. We recently reported that the clearing point of nematic shells of the commonly used mesogen 5CB is distinctly lower than that of the bulk LC [85]. Interestingly, the study revealed that all constituents of the surrounding aqueous phases can enter the shell to some extent, water, surfactant and glycerol, which is often added to the aqueous phases for density and viscosity tuning. Moreover, an alignment change was detected in a long-term experiment with shells that were initially hybrid-aligned thanks to preparation with surfactant in one aqueous phase and PVA in the other. Over the course of a few days, the alignment was transformed into fully homeotropic, demonstrating that the surfactant can pass through the LC shell and adsorb on the opposite side.

3.2.2. Osmotic control of shell thickness and diameter.

By varying the orifice sizes of capillaries, interfacial tensions, viscosities, flow rates etc during the microfluidic production, we can tune the diameter and thickness of LC shells over a considerable range. However, it is challenging to produce the shells with a diameter smaller than about a 100 μm or a thickness below a few μm. If a thinner and larger or smaller and thicker shell is desired, an elegant trick can be applied for post processing, introduced by Lopez-Leon and co-workers [87], involving osmosis. Because the LC is not impenetrable to water, the LC shell can act like a semipermeable membrane separating the inner and outer phases, and if these have non-equal compositions, an osmotic pressure will develop that drives water into or out of the shell's inner isotropic droplet.

Although it was not properly elucidated until the mid 19th century, osmosis is a very common phenomenon on Earth, playing an important role in many vital and central processes in biological systems [88]. Originally osmosis is used to describe the movement of a solvent across a semipermeable membrane towards a higher concentration of solute, which is driven by the chemical potential difference between the two sides of the membrane [88, 89]. The membrane can be solid or liquid and it can be formed by layers of cells, particles and molecules [88]. The permeability of the membrane depends not only on the size of holes in the membrane, compared to the size of the solvent molecules, but also on the polarity of the solvent molecules and their miscibility with the material making up the membrane.

Lopez-Leon and co-workers introduced osmosis into LC shell studies as a method to reduce the shell thickness by increasing the size of the inner droplets [87]. Water based solutions were used as the inner and outer phase with different salt concentrations. Compared with the salt, water is more miscible with the liquid crystal, so the water can flow from low salt concentration side to the side with high concentration by passing through the LC shell. The shell thickness change can happen in both directions: by giving the outer phase a higher salt concentration than the inner phase, water would flow from inside the LC to the outside, making the shell thicker and the diameter smaller. By tuning the thickness of the nematic LC layer, the elastic free energy related to the defect arrangement is changed, allowing the authors to control which defects are formed and how they are distributed across the shell, see section 3.3.4.

Osmosis was also used in our recent studies to help remove the oily streak defects in cholesteric shells with planar anchoring on both sides of the LC layer [86]. Instead of using salt content difference to establish an osmotic pressure, we used higher concentration of PVA in the inner droplets (15wt.%) than in the outer continuous phase (10wt.%). As a result, the shells were expanded and became thinner over the course of several hours, with a consequent rapid disappearance of oily streak defects (figure 11). Already after 12 h the shells are free of visible defects, which should be compared to about a week without osmosis. In the process of osmotic expansion, the shell diameter increased from about 180 μm–240 μm, and the shell thickness decreased from 27 μm to 11 μm.

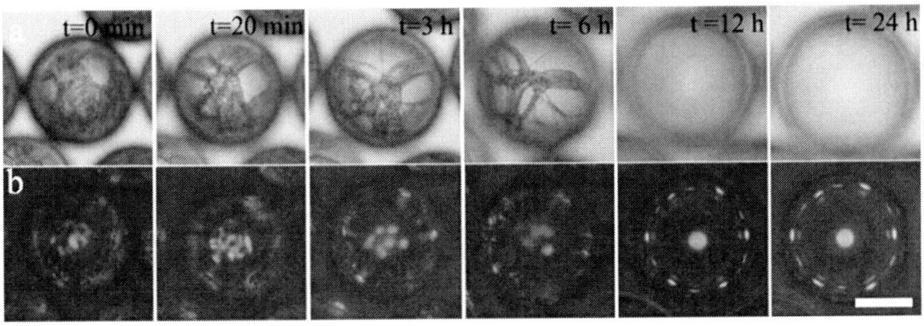

FIGURE 11. Removal of oily streak defects in short-pitch cholesteric shells by osmotic thinning. Optical microscopy textures in transmission without analyser (a, focus on top of shell) and in reflection between crossed polarisers (b) of a cholesteric shell subjected to expansion and thinning by osmosis. The patterns are explained in section 3.5. Scale bar 100 μm. Image adapted from [86]. CC BY 4.0.

3.3. Topological defects in liquid crystal shells and droplets

Liquid crystal shells and droplets exhibit a large variety of director field configurations with intriguing textures and defect structures. Numerous experimental, theoretical and simulation studies have focused on the origin of this structural diversity and established a detailed picture of what director configurations can occur in nematic, cholesteric or smectic droplets and shells. While the respective equilibrium director field results from a sensitive interplay between interfacial and bulk properties (e.g. spatial dimension of the spherical confinement, surface curvature, anchoring conditions at the confining interfaces, bulk and surface elasticity or density mismatch of the constituting fluids), the principle reason for the formation of topological defects is the sperical confinement itself. When a liquid crystal volume is confined by a spherical boundary, the anchoring conditions at this interface impose topological constraints on the director field which unavoidably force topological defects to be formed [90, 91].

A topological defect constitutes a singularity in the local orientational order of the LC, which is depicted by the director field $\boldsymbol{n}(\boldsymbol{r})$. In the close vicinity of this singularity the liquid crystalline order is reduced and the scalar order parameter S decreases. In the core of the defect the order breaks down completely and the LC is isotropic. A topological defect cannot decay and the director field in its surrounding cannot be continuously transformed into a uniform director field with no singularities.

An important distinction has to be made between topological defects confined to surfaces and topological bulk defects, see figure 12. Defects in

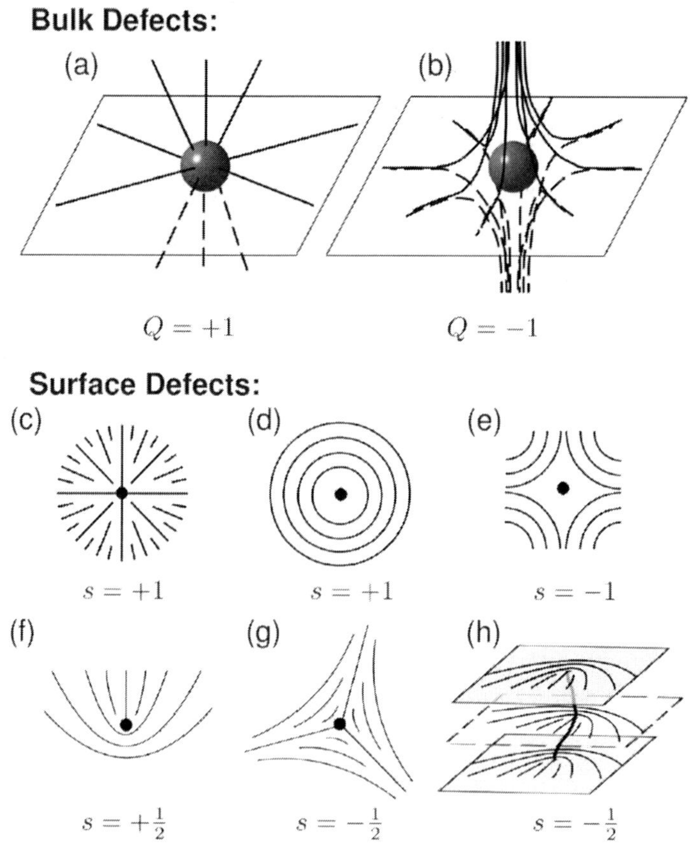

FIGURE 12. Schematic representations of topological bulk defects (a) + (b), singular points confined to interfaces (c)–(g) and a disclination (h) in a nematic director field. Bulk defects (*hedgehogs*) are characterized by the bulk charge Q. We distinguish between (a) radial hedgehog defects with $Q = +1$ and (b) hyperbolic hedgehogs with $Q = -1$. Defects confined to surfaces carry a topological charge s, which can be integer (c)–(e) or half-integer (f)–(g). Half-integer surface defects induce defect lines (*disclinations*) in the bulk which connect two surface defects of similar sign (h). A misunderstanding commonly arises since for half-integer defects (f) + (g) the director representation in the confining interfaces (shown as grey planes in (h)) appears identical to the director representation in an arbitrary plane cutting through the line defect (dashed plane in (h)). However, for integer charge defects (c)–(e) the point singularity only exists in the confining interface; these defects do not induce the formation of line defects in the bulk. One should therefore be aware that the sketches (c)–(g) represent the director orientation as it appears in the 2D confining interface of a bounded nematic.

the bulk are called hedgehogs and carry an integer topological charge of Q (see figures 12(a) and (b)) [91]. The magnitude of the bulk charge $|Q|$ is given by the fraction of all possible director orientations occuring within an imaginary sphere around the defect core. A detailed analytical description for the determination of the hedgehog charge can be found in references [91–93]. Since the nematic director is a sign-less pseudovector ($\boldsymbol{n} = -\boldsymbol{n}$), the sign of the hedgehog charge is degenerate in three-dimensional space, i.e. a positive defect can be continously transformed into a negative defect [94]. In this review we adopt the common practice that a radial hedgehog is given the topological charge $Q = +1$ whereas a hyperbolic hedgehog is designated with $Q = -1$ [95].

In contrast to hedgehogs, topological surface defects are confined to interfaces and cannot exist freely in the bulk (see figures 12(c)–(g)). This type of defect is characterized by a surface charge s, which is defined by the index of the director field projection onto the confining interface (see [91] or [93] for more details). The strict distinction between bulk charge Q and surface charge s is important: While hedgehogs with $Q = 1$ can exist both in the bulk and at the surface, their surface analogs with $s = 1$ cannot move inside the nematic bulk and are restricted to the interface [95]. Such surface defects with unity charge s are called *boojums*, inspired by the imaginary creature 'boojum' in Lewis Carroll's poem *The hunting of the Snark* [91]. Unlike bulk defects, surface defects can appear not only with integer but also with half-integer strength s. The strength of a surface defect is given by the ratio $s = \gamma/2\pi$ with γ being the angle by which the director in the surface projection rotates along a path encircling the singularity [93].

Following this definition, the strength s of a topological surface defect in LCs can easily be determined by analyzing the polarizing microscopy texture surrounding the defect, see e.g. figure 2. The relationship between $|s|$ and the number b of brushes pinned to the center of the singularity is given by $|s| = b/4$. Consequently, defects in the center of four brushes hold a strength $s = \pm 1$ and defects with only two brushes hold a strength of $s = \pm \frac{1}{2}$ [93]. The sign of the defect strength can be determined by rotating the crossed polarizers: if the brushes rotate in the same sense the defect is positive, if it rotates in the opposite, it is negative. It should be mentioned that the sign of a surface defect can only be derived by following the Volterra process and is not part of the topological stability theory [96].

Despite being pinned to an interface, surface defects also cause director distortions in the bulk. In the case of a radial defect with $s = 1$, which is shown in figure 12(c), these distortions resemble one half of the director field around a hedgehog defect [91]. Beyond the surface charge s, this defect can therefore also be characterized by a bulk charge Q. Upon changing the boundary conditions from planar to homeotropic, a surface defect continously disappears and, when carrying a bulk charge of $Q = +1$, is transformed into a hedgehog defect [97], as illustrated in figure 13. Other types of surface defects with integer strength are shown in figures 12(d) and (e). The bulk distortions caused by (d) and (e) induce virtual lines of strong director deformation along which the director field escapes into the third dimension [98]. Although often referred to as 'defect lines', these lines lack a defect core in their center, hence they are not 'defects' in a strict sense. They typically connect two integer surface defects through the bulk.

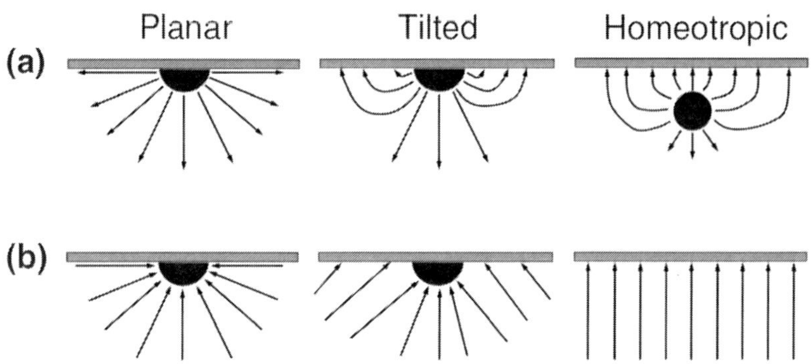

FIGURE 13. Schematic drawing of the two distinct boojum transformations under changing boundary conditions from planar to homeotropic (adapted with permission from [97]). If the boojum carries the topological charges $s = +1 \wedge Q = +1$, the defect dissolves from the interface and forms a hedgehog defect in the bulk. (a) In case the bulk charge Q of the boojum is zero $s = +1 \wedge Q = 0$, the defect gradually disappears upon reducing the tangential director field component at the interface (b).

Half-integer surface defects (f)–(h) play a special role among the configurations shown in figure 12: they induce stable line defects in the bulk of the liquid crystal, so called *disclinations*. Integer defect lines are not stable and the formation of a defect core is avoided by a director escape into the third dimension [98]. Disclinations can be subdivided into wedge and twist types [5], of which we will only consider the former in this review. The 'wedge' reflects the defect construction by insertion or removal of a wedge of matter following the Volterra process [96, 99]. Line defects occur in the bulk of the liquid crystal, but they typically end at a surface. While open-ended disclinations do not exist, defect lines can also close up into a loop, then existing separated from the surface. Such closed loops can be considered as hedgehogs with an extended core, an issue described in more detail in references [95, 100, 101] or with respect to defects in fibers in section 4.1. The determination of their topological charge then works accordingly to the determination of surface charges s in a virtual plane perpendicular to the disclinaton, while the bulk charge Q of the closed ring is unity [93].

The sum of all topological charges s and Q in a confined liquid crystal obey conservation laws, which are connected to the Euler characteristic of the confining surface [91]. For spherical confinement and planar or tilted alignment, the Poincaré-Hopf theorem requires a total topological charge of $s = +2$ on the surface, which corresponds to the Euler characteristic of a sphere [84, 102]. For homeotropic anchoring, having no director component in the interface, a total topological bulk charge of $Q = +1$ (half of the respective Euler characteristic) must be contained inside the spherical volume according to the Gauss-Bonnet theorem, while the total topological surface charge is zero [84].

Even without full knowledge of the bulk and interfacial properties of a liquid crystal shell or droplet, the thermodynamic ground state of the director configuration can be derived by energetic considerations. Any deformation in a uniform director field increases the elastic free energy as characterized by the Oseen-Frank free energy density (see (2*b*)). The director field is therefore driven towards a state that minimizes the total free energy given by (6) while still complying with the topological constraints imposed by the boundary of the spherical confinement [84, 103].

An important aspect of studying topological defects in LC shells and droplets by optical polarising microscopy has been emphasized by Liang *et al* [104]: while the optical setup can only be focused to a topological defect on one side of the shell or droplet, the birefringence of the other side, although being blurred, also contributes to the observed polarizing microscopy texture. This can render an analysis of the director field within the shell or droplet rather difficult. An ideal situation is given when the director field on the side currently out of focus is rather uniform, such that either the polarizer or the analyzer can be oriented in the same direction and the influence of birefringence from this side is minimized.

3.3.1. Nematic droplets.

Observable textures in nematic droplets differ for various droplet sizes and anchoring conditions and the respective equilibrium director field configuration strongly depends on bulk and surface energy contributions. The bulk elastic energy Fb scales linearly with the droplet radius R [91], while the surface energy Fs scales as $F_s \propto WR^2$ [105], where the coefficent W is the amplitude of the angular dependent anchoring energy from (3). Consequently, the energetic contribution of the interface dominates for large droplet radii. This relation might seem counterintuitive, since the ratio of surface area to bulk volume increases upon decreasing droplet radius and one might expect a dominating surface contribution for small droplets. The reason is the $1/R^2$ dependence of the bulk elastic energy density, and experimental observations confirm that influence of the topological constraints imposed by the interface vanishes for small droplet sizes. The director configuration is then determined solely by the bulk elasticity of the liquid crystal [105]. A critical radius $Rc = K/W$ (in the one-constant approximation) with a transition from small droplet to large droplet behavior is, depending on the anchoring energy W and bulk elasticity K, in the order of 1 μm [91].

In small droplets ($R < K/W$) with homeotropic anchoring the bulk elastic energy enforces a defect and deformation-free uniaxial director configuration, which disregards the homeotropic boundary condition and the topological constraints related to it [105, 106]. A schematic representation of such a director configuration is shown on the top left side of figure 14. Upon increasing the droplet size, the surface contribution also increases and for $R > Rc$ the homeotropic anchoring conditions impose a bulk charge Q of unity within the droplet [107]. The topological constraints imposed by the Gauss-Bonnet theorem are fullfilled by forming a radial director field with a hedgehog defect in the center of the droplet (see figure 14 *top-right*) [84, 107]. This centrosymmetric director field then typically relaxes away from the symmetric radial configuration to a lower energy twisted state, which appears as a spiraling cross between crossed polarizers [5, 107].

Nanometer-sized droplets with planar or tilted anchoring conditions exhibit a radial director configuration in equilibrium [108, 109], while a bipolar configuration is frequently observed in larger droplets (see figure 14, *bottom*). Tomar *et al* showed by numerical simulations that this transition from bipolar to radial upon decreasing droplet dimension is connected to the saddle-splay term (see (4) in section 2.1) in the elastic free energy [105]. For large nematic droplets ($R > K/W$) with planar or tilted anchoring the topological constraints expressed by the Poincaré-Hopf theorem require a total topological surface charge of $s = +2$ at the droplet interface. The liquid crystal typically meets these constraints by forming bipolar drops, which are characterized by two boojums with topological charge $s = +1$ separated by the maximum distance of a drop diameter [84]. The director field within the droplet is then aligned parallel between these two radial half hedgehogs, corresponding to pure splay deformation close to the defects and bend in the droplet bulk [103].

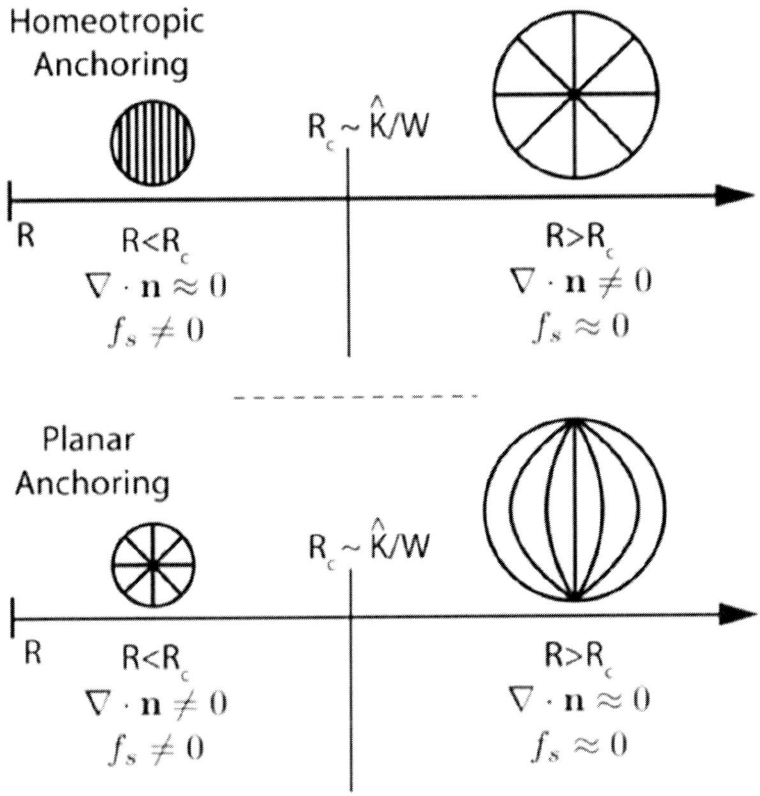

FIGURE 14. Schematic representation of the director configuration in nematic droplets. For homeotropic boundary conditions (top), decreasing the drop radius R below a critical value $R_c \approx K/W$ induces a transition from a radial to a uniform director field. For such small droplet size, the energy penalty for disregarding the anchoring conditions is smaller than the energy gain from decreasing the elastic energy density in the bulk. For planar boundary conditions (bottom), a decrease of droplet size induces a transition from a bipolar configuration to a radial configuration. Reproduced from [105] with permission of The Royal Society of Chemistry.

A second possible director configuration that fullfills the topological constraints, the concentric configuration with two bend disclinations of surface charge $s = +1$, is only observed for unusually low splay elastic constant, in comparison to the twist and bend constants [103]. The surface is here decorated by line rather than point defects, but—as stated in section 3.3—integer line defects are not stable. The line defect 'escapes in the third dimension' via a twisted configuration, yielding a virtual line defect connecting the two surface defects. There is no defect core, but the bulk is still much more distorted than in the usual bipolar configuration [110]. As shown by Fernández-Nieves and coworkers, this unstable concentric configuration is easily transformed into the stable bipolar configuration by applying external electric fields to align the director field [103]. The authors also show that a retransformation from bipolar to escaped concentric does not occur and contribute this observation to a significant energy barrier [111] between the two configurations and an overall lower energy for the bipolar configuration. According to their findings, only the large forces occuring during circular flow inside an LC droplet allow a transition from bipolar back to the escaped concentric configuration.

A continous transition between the bipolar and the radial director field configuration can be observed upon changing the anchoring conditions from planar to homeotropic or vice versa, as demonstrated by Volovik and Lavrentovich for large drops [91, 97]. The transformation may take place via a so-called axial configuration, in which the surface is already homeotropic but an equatorial defect ring allows a non-radial internal director field [91].

3.3.2. Cholesteric droplets. The first description of cholesteric droplets dates back to Lehmann and Friedel [**112**] at the beginning of the 20th century, but only from the 1980s onwards a systematic study of the complex configurations in higher ordered liquid crystal droplets began [**113**]. In contrast to the nematic phase, the equilibrium director configuration in cholesteric droplets depends not only on the boundary conditions at the interface and the bulk elasticity, but it is also influenced by a factor qR, the product of the cholesteric helix wave vector (see section **2.3**) and the droplet radius [**113, 114**]. The continuous twist allows for new types of disclinations in cholesteric liquid crystals, which can be categorized into the three classes $\lambda^{\pm m}$, $\tau^{\pm m}$ and $x^{\pm m}$ [**93, 115**]. Half-integer and integer values for m characterize the winding number of the respective defect type, schematic representations of which are shown in figure **15**.

Since a smooth unwinding of the cholesteric helix yields the uniaxial order of the nematic phase, a continuous transition from the director field configuration of a cholesteric droplet towards that of a nematic droplet can be expected. For cholesterics with a long pitch that are confined to small droplets ($qR < 1$) with planar or tilted boundary conditions, Bezić and Žumer predicted an unwinding of the cholesteric helix and the formation of a bipolar structure as frequently observed for nematics [114]. Their model yielded that the director lies solely in the meridional planes of the droplet, which could not be confirmed by later experimental studies [116]. Instead, Xu and Crooker established a twisted bipolar model for cholesteric droplets, which differs from the bipolar model for nematics by higher twist within the droplet (see figure 16(a)). While in nematic bipolar droplets the twist is not intrinsic and decays exponentially from the surface towards the bulk, the model introduced by Xu and Crooker assumes a linear decay of twist [116].

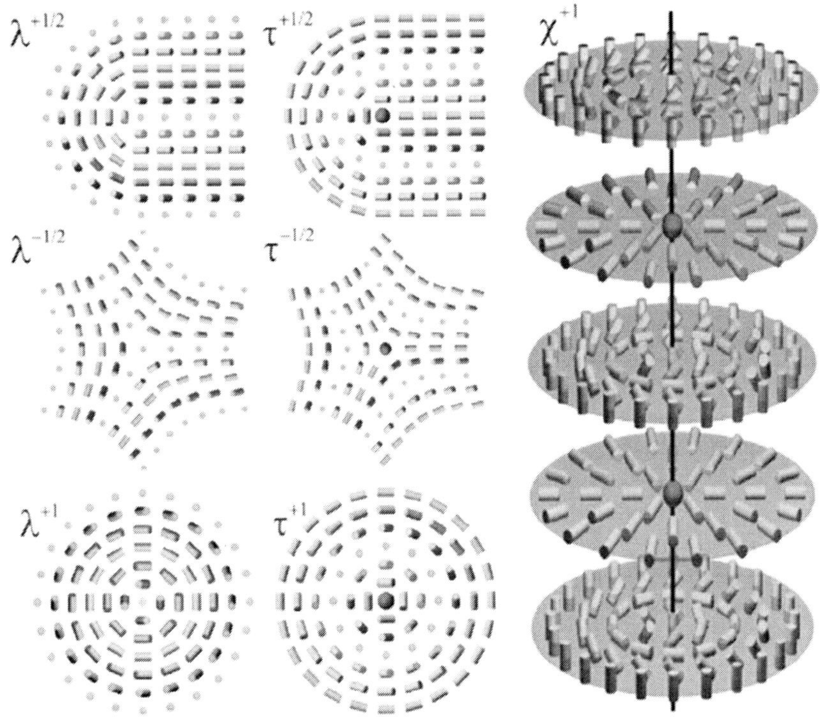

FIGURE 15. Schematic representation of $\lambda^{\pm m}$, $\tau^{\pm m}$ and $x^{\pm m}$ disclination lines in cholesteric liquid crystals. Half-integer and integer values for m characterize the winding number of the defect. The director orientation is represented by blue cylinders while the defect locations are indicated by red spheres. The core of the χ disclinations is depicted as a black line. Reproduced from [**115**] with permission of The Royal Society of Chemistry.

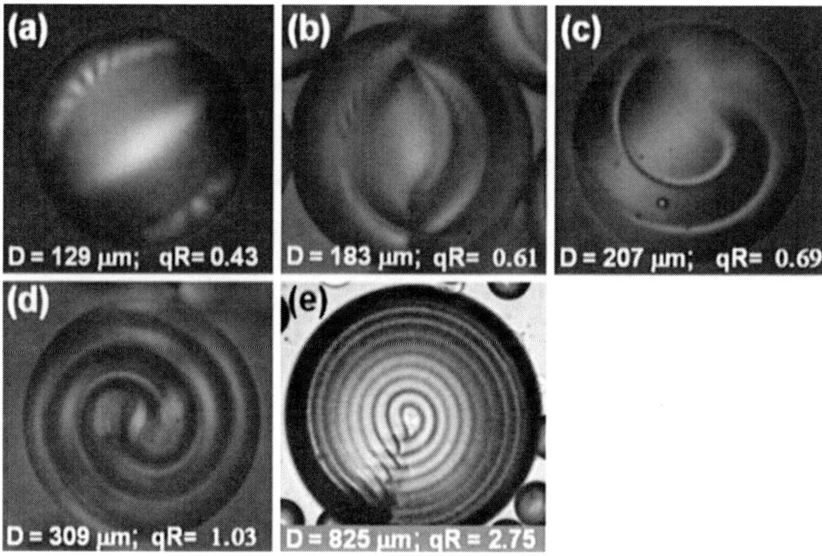

FIGURE 16. Polarising optical micrographs of cholesteric droplets with planar boundary conditions. The value of *qR*, the helix wave vector multiplied by the droplet radius, continously increases from (a)–(e). For low *qR* (a), a twisted bipolar texture is observed. As *qR* increases ((b)–(f)), the configuration evolves towards the Frank-Pryce structure (**E**). This structure is characterized by a central hedgehog with an attached $s = +2$ disclination line. Reproduced from [113] (© Springer-Verlag 2011). With permission of Springer.

For larger cholesteric droplets with planar/tilted anchoring and/or short helix pitch ($qR > 1$) the *Frank-Pryce structure* is the most commonly observed configuration. This structure is shown in figure 16(e) and consists of an $s = +2$ disclination line of length R that connects a hedgehog in the center of the droplet with its surface [114, 117]. The configuration is also referred to as *radial spherical structure* or *spherulitic texture* [115]. This structure also attracted attention beyond liquid crystal research, as it represents an intriguing analogy to an elementary magnetic charge known as Dirac monopole [91, 118]. Several studies report that the regular Frank-Pryce structure structure can relax into the so-called *escaped Frank-Pryce structure*, in which the director escapes along the direction of the disclination line [116, 119]. Bouligand and Livolant found that the Frank-Pryce structure with concentric layers can sometimes also occur with the defect line covering the whole diameter of the droplet [120]. Droplets with such structure show cylindrical symmetry as the disclination line connects two diametrical surface defects of topological charge $s = +1$ [114]. Frank-

Pryce structures with a curved defect line connecting two non-antipodal $s = +1$ surface defects have only rarely been reported [120]. When a cholesteric droplet with Frank-Pryce structure is observed along the disclination line, one sees either a double spiral texture (usually for a defect line of length R) or concentric circles (typically for defect lines of length $2R$ or curved singular defect lines) [120].

Besides these experimentally observed structures in cholesteric droplets with planar or tilted anchoring, a more recent numerical simulation study by Seć and coworkers suggested that the $s = +2$ disclination line of the Frank-Pryce structure is actually unstable and can decompose into two $s = +1$ defect lines [115]. Based on their simulation results, the authors introduced two new director field configurations in cholesteric droplets: the *Lyre-structure*, which resembles the shape of a lyre instrument, symmetric around the vertical axis, and the Yeti structure, with two diametrically positioned boojums and higher symmetry than the Lyre structure [115].

For cholesteric droplets with homeotropic boundary conditions and high values $qR > 10$, the conditions on the surface do not affect the packing of the helix. The textures in the droplets indicate that the Frank-Pryce structure prevails [113, 117]. For $qR < 5$ the appearance of an equatorial disclination ring is observed [117]. For smaller qR values, this ring relaxes to a surface point defect, which then for even smaller $qR < 1$ relocates to the center of the droplet, yielding the radial structure frequently observed in nematic droplets [117].

3.3.3. Smectic droplets. Director field configurations of smectic phases (SmA, SmC) in spherical confinement have been less intensively studied than for nematic or cholesteric phases. One reason for this is that smectic textures are typically highly complex with numerous macroscopic defects called *focal conic domains* [**121, 122**]. It is generally assumed that the smectic texture represents a packing of curved and equally spaced surfaces [**123**], as long as the dimension of confinement (typically in the μm range) is much larger than the spacing of smectic layers (typically < 10 nm). Although Kleman and Lavrentovich showed that the filling of space by smectic layers is far from complete [**124**], this simplifying assumption still allows a reasonable description of smectic textures in spherical confinement [**122**].

For spherical confinement and planar boundary conditions, a bipolar configuration represents the simplest SmA texture [122]. As in the bipolar configuration for nematic droplets, two boojums ($s = +1$) at the surface are separated by the diameter and $\boldsymbol{n}(\boldsymbol{r})$ connects the two defects along the shortest possible path throughout the sphere. The smectic layers are perpendicular to the director field and consequently run along the parallels of the sphere [125]. A topologically equivalent configuration is obtained by dividing the two boojums into four surface defects with topological charge s = $+\frac{1}{2}$, positioned along a great circle on the sphere [122]. Both

configurations exhibit the same free energy [126] and can by principle be transformed one into another by cutting the bipolar configuration in two halves, through a plane containing both boojums, and subsequently rotating one half with respect to the other by an angle of $\frac{\pi}{2}$ [125]. A schematic

representation of both configurations is shown in figure 17.

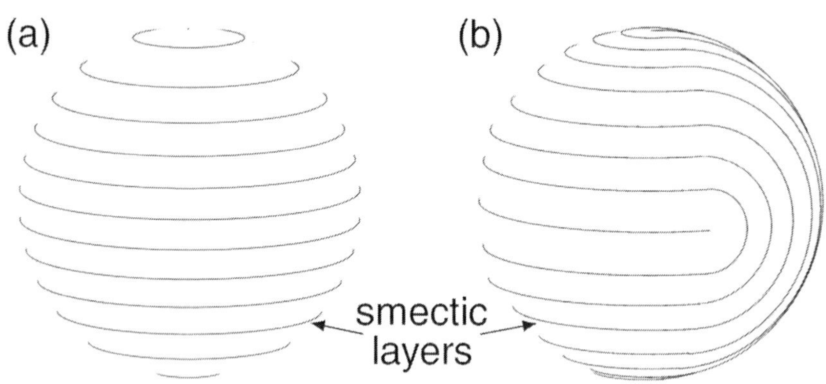

FIGURE 17. (a) Schematic representation of the smectic layer configuration in planar SmA droplets with two boojums $s = +1$ at the poles. While the surface director follows the meridians of the sphere, the smectic layers run perpendicular to ***n***, along the parallels of the sphere. (b) Schematic drawing of an equivalent configuration with four s = $+\frac{1}{2}$ defects. The two

configurations shown in (a) and (b) are equal in energy. Adapted with permission of Springer from [122] (2001) © EDP Sciences, Springer-Verlag, Società Italiana di Fisica 2001.

3.3.4. Nematic shells. Liquid crystal shells exhibit an even larger variety of equilibrium director field configurations than full droplets, as the presence of the inner interface imposes additional topological constraints to the director field. Beyond the classification of shells into homeotropic or planar/tilted shells, this second interface allows the formation of hybrid shells, i.e. shells with one interface imposing homeotropic alignment while the other interface imposes planar or tilted anchoring.

Early studies on the formation of topological defects in liquid crystal shells date back to the work by Lubensky *et al* [107] and Poulin *et al* [90] on multiple emulsions (i.e. water droplets dispersed in larger nematic droplets that in turn are dispersed in a continous aqueous phase). Due to a lack of a proper method for precise control over the shell dimensions during these early works, these studies derive information on $\boldsymbol{n}(\boldsymbol{r})$ from topological constraints only and do not study the formation of defects in detail. The introduction of a microfluidic-based method by Fernández-Nieves and coworkers [127], as described in detail in section 3.2, allowed the preparation of droplet-in-droplet double emulsions with well- controlled radii of outer (LC) and inner droplet (e.g. water) in an immiscible continuous phase. This setup also allowed studying the effects of the frequently observed thickness heterogeneity of shells in more detail: Depending on elastic forces in the nematic shell and density differences between the LC and the fluid of the inner droplet, the latter is not located in the center of the shell but shifted along one direction, in case of a density mismatch along the direction of gravity. More advanced topological studies on the formation of defects in nematic shells then focused on the effect of shell thickness [87, 104], heterogeneity in thickness [128] or variations of elastic constants [87, 128]. As discussed in more detail in section 3.2.2, dynamic tuning of shell thickness and diameter was achieved by utilizing osmosis [87, 104].

Homeotropic-aligned nematic shells. In shells with homeotropic boundary conditions at both interface, **n(r)** *adopts a radial configuration throughout the shell [**107**]. The inner droplet is thereby equivalent to a virtual radial hedgehog defect with unity charge and matches the topological requirements of the outer droplet [**90**]. The equilibrium director configuration in a simple shell with only one inner droplet therefore does not require additional topological defects. In multiple emulsions with homeotropic anchoring containing more than one inclusion, there must always be one hedgehog defect less than the number of inclusions, as was demonstrated by Poulin and Weitz [**90**]. When observed between crossed polarizers, the radial director configuration of a homeotropic shell appears as a rigid unrotated cross [**107**]. The regular orthoscopy texture resembles a conoscopy image of a flat slab of a uniaxial liquid crystal with homeotropic boundary conditions. The similarity arises from the analogy of image aquisition: While conoscopy probes all light directions through a flat sample, orthoscopy of a homeotropic shell probes a single light direction through all possible sample film orientations [**104**].*

*Planar-aligned nematic shells. Nematic shells with planar anchoring at outer and inner interfaces show a larger variety of defect structures. For thick shells with small inner droplet a bipolar defect configuration with two pairs of opposing surface defects at the outer and inner shell surfaces is expected [**129**]. Experimental observations show that for such shells the defect pairs* *are either located along a shell diameter [**87**, **127**], resembling the boojum structure of bipolar nematic droplets (see figure **18(i)**), or situated close to each other at one side of the shell [**84**, **128**], see figures **18(g)** + **(h)**. The latter defect configuration is observed for shells with high thickness heterogeneity, i.e. a large offset of the inner droplet from the center of the shell, and the defects are located at the thinnest part of the LC [**84**]. The equilibrium distance between the defects is given by the balance between elastic director field distortions in the bulk of the nematic and the repulsion between defects of same topological charge [**130**]. Within the shell, the bulk director deformation around each boojum core represents half hyperbolic hedgehogs on the inner surface, while the distortion close to the outer surface resembles half radial hedgehogs [**84**]. Each defect carries a topological surface charge of* $s = +1$ *and can be identified by four black brushes when observed between crossed polarizers [**84**].*

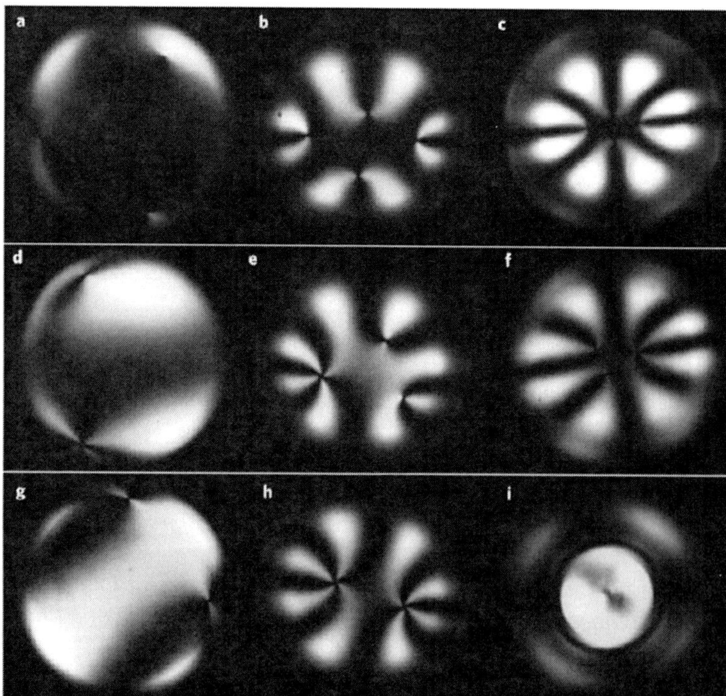

FIGURE 18. Polarizing optical microscopy images of planar shells with different thickness and thickness heterogeneity showing various defect configurations with four defects ((a)–(c)), three defects ((d)–(f)) or only two defects ((g)–(i)). The shell thickness and thickness heterogeneity increase along the rows from left to right, leading to different spatial arrangements of the defects. Reprinted by permission from Macmillan Publishers Ltd: Nature Physics [87], copyright 2011.

For thin planar shells a defect configuration with four surface defects constitutes the ground state [128, 131]. In the limit of infinitely thin shells the LC is confined to a two-dimensional surface and the disclinations reduce to points with no extension to the bulk. Each of the four defects carries a topological surface charge of $s = +1/2$, which reflects the π rotation of $\boldsymbol{n}(\boldsymbol{r})$ along a path encircling each defect. The sum of topological charges is then +2, as required by the Poincaré-Hopf theorem for the coinciding outer and inner interfaces [128]. Depending on the elastic anisotropy and the heterogeneity in thickness, the ground state with four defects can occur as tetrahedral arrangement of defects [131] (see figure 18(a)), as clustering of defects at the thinnest part of the shell [132] (figures 18(b) + (c)), or as arrangement of defects along a great circle [132, 133].

Fernández-Nieves *et al* [127] and Lopez-Leon *et al* [87] showed that the local defect clustering occurs in asymmetric shells only, as long as the average shell thickness is high enough that the reduction of the length of the defect cores can minimize the total free energy [132]. Later experimental studies showed that the four defects are then typically not symmetrically arranged around the thinnest part of the shell, but they adopt a skewed configuration of two pairs, where the defects in one pair are separated by a larger distance compared to the defects in the other pair [132, 133] (see figure 18(b)). By performing a basic polarization optics analysis, Liang *et al* noted that $\boldsymbol{n}(\boldsymbol{r})$ between the short pair of defects mainly exhibits bend deformation, while $\boldsymbol{n}(\boldsymbol{r})$ around the longer pair is mainly splay. Because bend defomations typically have greater energy densities than splay deformations ($K_3 \approx 2K_1$) the total elastic free energy is minimized by contracting the length of the bend disclinations at expense of the splay disclinations [132]. The difference in the free energy cost of director bend and splay thereby causes the frequently observed skewed distribution of surface defects.

Seč and coworkers later confirmed these observations by computer simulations [128]: The two closest defects are connected by a director field that runs along the shortest arc, whereas the other two defects are connected by the longest arc spanning a large fraction of a great circle (see figure 19). Increasing the anisotropy of elasticity further enhances the asymmetry between the two defect pairs. The simulations also showed that for larger ratios K_3/K_1 the defects connected by splay deformations move towards opposite sides of the sphere until they stabilize in the equatorial plane. Only after this first step, upon further increasing the elastic anisotropy, the defects of the bend pair progressively migrate towards the equatorial plane. This large ring configuration is expected to occur when the elastic splay deformation is strongly energetically favoured over bend deformation, as also confirmed by Monte Carlo simulations [126, 134]. Seč *et al* also showed that upon reducing the thickness heterogeneity of the shell, the defects increase their separation in a roughly symmetrical way and the ratio of separation between the two pairs of defects remains roughly constant [128].

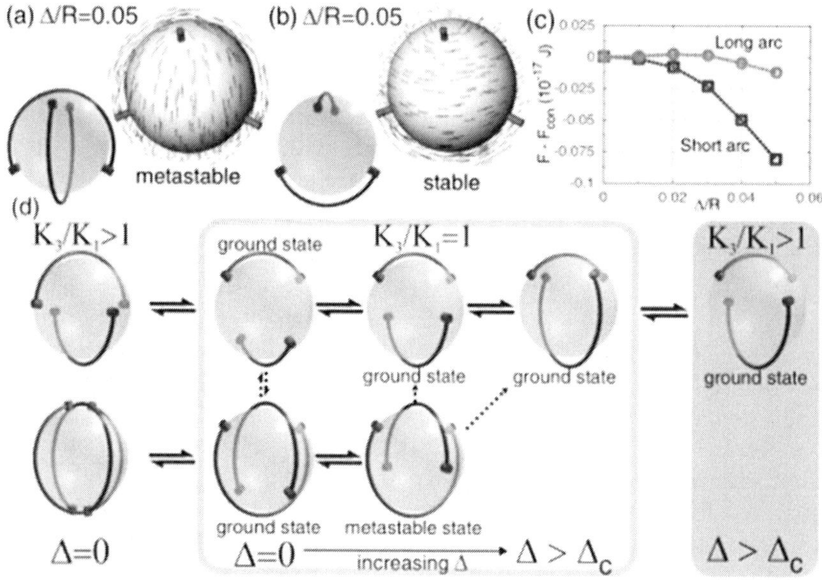

FIGURE 19. Schematic summary of the numerical simulation study by Seč *et al* [128]. (a) Metastable and (b) stable inhomogeneous shell states with equal defect positions and nonequal director: The director links defects over the long and over the short arc as indicated by the black lines in the schematics. (c) Free energy difference between the states shown in (a) and (b). (d) Scheme of transformation trajectories between director and defect shell states that can be controlled by varying the displacement Δ of the inner droplet from the center of the outer droplet and elastic constant anisotropy K_3/K_1. The metastable state becomes unstable at $\boldsymbol{\Delta}_c$ that depends substantially on the shell radius R and thickness h. Reprinted with permission from [128], Copyright 2012 by the American Physical Society.

Experimental observations near the N-SmA transition, see figure 20, confirm this relationship between the defect configuration and the relative elastic constant values [128, 132, 133], as the ratio between K_3 and K_1 increases rapidly on approaching a second-order transition to the SmA phase. Moreover, we recently demonstrated that this relationship can be used to tune the defect location by moving closer to or further away from the N-SmA transition and then making the configuration permanent by polymer-stabilization [135]. However, it turns out that polymerization of the reactive component very close to the transition actually triggers the phase transition, even though the temperature rises due to the heat released during polymerization, an effect of the smectic-promoting character of acrylate reactive mesogens. In addition to defect configurations with two or four surface defects per interface, a third possible configuration is found for intermediate shell thickness which contains three topological defects forming an isosceles triangle [87], see figures 18(d)–(f).

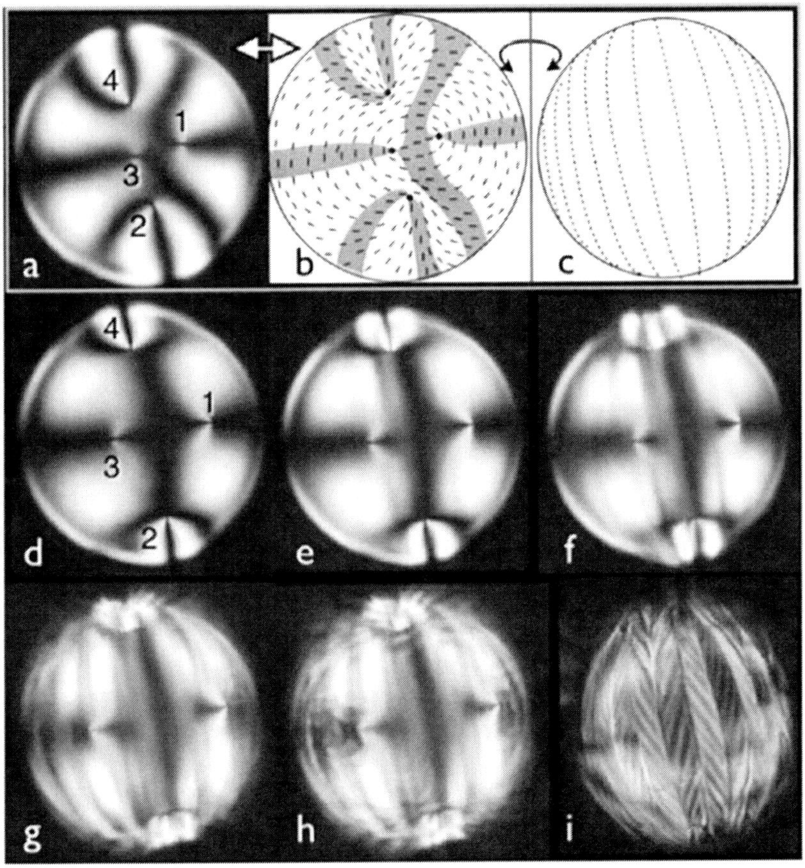

FIGURE 20. N-SmA transition in a planar-aligned shell $R \approx 11\mu$m, thickness $\approx$4 μm). In (a)–(h) the focus is on the thin shell side, where the defects are concentrated, thus the initial signs of the transition (c)–(g)—starting on the thick side—are blurry. In (i) the focus is on the thick side. Drawings (b) and (c) sketch $\boldsymbol{n}(\boldsymbol{r})$ corresponding to (a) on the thin and thick sides, respectively. Reprinted with permission from [132]. Copyright (2011) by the American Physical Society.

Buoyancy enhances the tendency for thickness heterogeneity in shells and favours vertical shifts along the field of gravity, so that only two scenarios (inner droplet shifted up or down) are experimentally observed. Seč and coworkers showed by numerical simulations that even in the absence of gravity heterogeneous shells are lower in energy compared to concentric shells [128]. These simulations also indicate that the total free energy would reach its minimum for a maximum displacement of the inner droplet. However, inner and outer surfaces of experimental shells have not been observed to coalesce [128]. The authors assume the existence of a disjoining pressure between outer and inner interface. This pressure is probably caused by the presence of PVA as a stabilizer. The PVA concentration increases as the nematic film gets thinner and this effect tends to homogenize the thickness [128]. In conclusion, buoyancy effects initiate and liquid crystal elasticity enhances a thickness heterogeneity of nematic shells by displacing the inner droplet from the center of the outer droplet. This symmetry breaking is irrespective of the elastic anisotropy [128].

Hybrid-aligned nematic shells. Hybrid shells feature one interface with homeotropic anchoring conditions and one interface with planar boundary conditions. Consequently, there are two types of hybrid shells, shells with planar anchoring at the outer interface and shells with planar anchoring at the inner interface. Since the Poincaré-Hopf theorem only requests a topological surface charge of s = *+2 at a planar or tilted interface, hybrid shells in equilibrium do not show topological defects at the homeotropic interface, but have two boojums of charge* s = *+1 at the planar interface [**84**]. The appearance of half-integer surface defects in hybrid shells is ruled out by the symmetry-breaking bend deformation through the bulk of the shell [**104**]. As discussed already for planar droplets, two possible director configurations at the planar interface of hybrid shells exist which fullfill these topological requirements: The bipolar configuration with two splay-type boojums and the concentric configuration with two bend-type defects [**83**, **113**]. The two configurations are topologically distinct by considering the director field within the shell, but give rise to identical textures between crossed polarizers. Liang* et al *showed that a clear distinction between the two configurations can be made by using a polarising microscope with compensator plate and confirmed that, as for droplets, the bipolar configuration is typically observed [**83**]. The two defects in the bipolar configuration cannot both be half-hyperbolic hedgehogs, since the hybrid alignment at the interfaces could then not be commensurate [**84**]. Instead, one of the boojums must be a half-radial hedgehog [**84**]. A continuous transition from a planar shell into a hybrid shell was reported by Lopez-Leon and Fernandez-Nieves by changing the additive in the outer aqueous continous phase to induce homeotropic alignment [**84**].*

Noh *et al* recently reported a two-step clearing transition for hybrid shells with surfactant-induced homeotropic anchoring at the outer shell interface and PVA-induced planar or tilted anchoring at the inner interface, or vice versa. The bipolar hybrid texture at first changes into the centrosymmetric homeotropic texture upon heating, before the nematic shell enters the isotropic phase. The authors argued that this behavior might be explained by a local stabilization of the nematic order caused by an adsorbed oriented monolayer of surfactant on one side [85].

3.3.5. Cholesteric shells. In a recent experimental study on cholesteric shells with planar anchoring, Darmon and coworkers discussed bipolar and radial director field configurations with a higher complexity than director fields in cholesteric droplets [**118**]. A bivalent configuration with two topological defects of surface charge $s = +1$ at the outer and inner shell interface is observed for cholesteric liquid crystals with long pitch p_0. The pairs of boojums on one interface appear close to each other and the defects on the outer and inner interface are linked to a stack of disclination rings through the shell.

Liquid crystals with shorter p_0 lead to monovalent shells with a radial defect. This structure is reminiscent of the Frank-Pryce structure described above for cholesteric droplets. Darmon *et al*showed by polarising optical microscopy that the disclination line in this structure actually consists of two λ+1 interwinded disclination lines forming a double helix [118]. The authors thereby provide an experimental confirmation for numerical simulations on cholesteric droplets by Seč *et al*, who were the first to suggest that the disclination line of the Frank-Pryce structure is not a x^{+2} disclination, but is built from two λ^{+1} disclinations, whose ends form a pair of close $s = +1$ defects at the outer surface while they form a monopole at the center of the droplet [115, 118]. A transition between the two possible configurations was shown to be of first-order. In a shell of thickness h the angular distance between the boojums is governed by the ratio h/R, while the transition itself depends on h/p_0 [118]. The optical behavior of cholesteric shells with sub-micron p_0 values will be discussed in section 3.5.

3.3.6. Smectic shells. Experimental textures of planar SmA shells can be quite complex [**104, 125, 132, 133**], see figure **20**. The equilibrium texture exhibits four topological defects of charge $s = +1/2$, located within the equatorial great circle, together with a set of additional lines which connect the defects by pairs [**125**]. The additional lines represent curvature walls that divide the sphere in crescent domains (spherical lunes) within which the smectic layers are undulated in a wavy fashion. Liang *et al* proposed that the undulation arises from a layer strain that is caused by the radially increasing spherical cross section area, coupled with the fact that the SmA phase is incompatible with director bend (see section **2.2**) [**132**]. Dislocations would be needed to obtain a larger number of smectic layers at the outer interface compared to the inner interface, but the high free energy cost of such a solution renders it unfavoured.

Instead, the layers buckle increasingly on approaching the outer interface, increasing the effective smectic layer distance a_1 along the original director orientation compared to the true layer spacing a_0(see figure 21). This reconciles the radially increasing surface area with the requirement for constant a_0. Lopez-Leon *et al* rather emphasized the possibility of a tilted instead of planar anchoring at the shell interfaces in order to release the layer strain [133]. In a subsequent theoretical treatment of SmA shells of varying thickness, Manyuhina and Bowick confirmed that both scenarios are possible [136]. In our recent study of polymer stabilization of LC shells [135] we demonstrated that the regular buckled geometry can be imprinted on the polymer network, rendering the instability-generated structure stable over a large temperature range, even after heating to the isotropic range of the host LC.

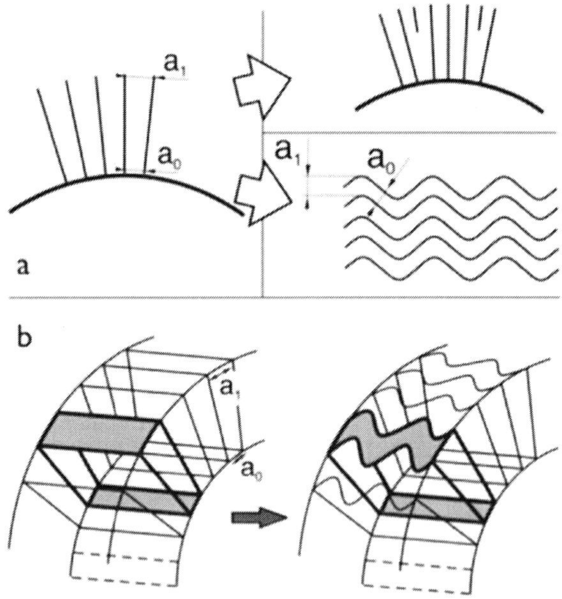

FIGURE 21. (a) Smectic layering in a curved shell poses problems for retaining the smectic layer thickness, which can be solved by dislocations (upper path) or layer buckling (lower path). (b) Sketch of how the effective layer distance along a cut can increase from a_0 at the inside to $a_1 > a_0$at the outside through a gradually increasing layer undulation. Reprinted figure with permission from [132]. Copyright (2011) by the American Physical Society.

In SmA shells with homeotropic anchoring at the inner and outer interfaces the characteristic cross-like texture as for nematic shells is observed, the only difference between nematic and SmA shells being a reduction of thermal fluctuations in SmA [104]. Liang *et al* also proposed that, compared to nematics, smectic LCs reduce the shell thickness heterogeneity and moves the inner droplet towards the shell centre [137]. In a homeotropic shell with high thickness heterogeneity, many smectic layers in the thicker part of the shell cannot be closed since fewer layers fit in the thinner part of the shell.

Smectic-A shells with hybrid alignment show focal conic domains as equilibrium director field configuration, mediating between planar/tilted alignment at the outer interface and homeotropic alignment at the inner interface, or vice versa. After the phase transition from the nematic phase, an unstable pattern of stripes, connecting the two poles of the shell, appears [104]. These stripes represent the boundaries of a folding membrane of remaining nematic order, which connects the smectic regimes with opposite alignment that grow from the two interface regions of the shell [83, 104]. Torza and Cladis observed the same type of texture much earlier for hybrid alignment in flat smectic samples, explaining it as a result of director bend being expelled from SmA [138]. The stripes are unstable since they come with an interfacial energy between two fluid phases. Hence they undergo a Rayleigh instability and break up into toric focal conic domains within which the layers bend, the SmA phase now spanning the full sample. In this configuration the SmA phase can accommodate the hybrid boundary conditions by only showing splay deformations of the director field [83].

A pecularity occurs for homeotropic SmC shells, as despite homeotropic anchoring the tilt in the smectic layers provides a tangential director field component at the interfaces and therefore requires the formation of topological defects [104]. Half-integer defects are ruled out because the C-director field is not sign invariant (see section 2.2), so a bipolar defect configuration with two opposing boojums at the outer and inner interface constitutes the ground state. Upon further cooling the SmC biaxiality becomes apparent and the texture distorts such that it does not allow to identify the number and charge of topological defects [104]. Liang *et al* propose that at low temperatures the SmC phase avoids the formation of defects by locally reducing the tilt to zero and shifting the system to a SmA state [104].

3.4. Droplet- and shell-shaped liquid crystal elastomer actuators

The above drops and shells were all made from low molar mass LCs, with limited lifetime and stability. By polymerizing the liquid crystal or a reactive component mixed into it, robust and durable particles can be produced, templated from the LC. Apart from providing robustness and enable long-term storage, the polymerization can lead to new properties, in particular if the polymerization is done such as to obtain an LCE (see section 2.4). In this section, we review the existing experimental methods for creating micron-scale droplet-/shell-shaped LCEs and their actuation mechanism.

In order to allow the production of droplets/shells with good control of diameter and thickness (in case of shells) one must start with low molar mass precursors and carry out the polymerization and cross-linking after the microfluidic production process. Thus, the first step in designing an experiment is to select the appropriate initial mesogenic monomers along with cross-linkers, keeping in mind that they must be miscible with each other, but not with any of the isotropic phases that form the external continuous phase and (in case of shells) the internal phase. The two-stage Finkelmann approach [41] may be feasible, provided that also the first step of polymerization is done after droplet/shell production and that an appropriate means of aligning the intermediate polymer state can be found.

The microfluidic platform is well established for the synthesis of monodisperse non-LC polymer beads [139, 140]. Zentel and later on others transferred this approach to the case of side-on side chain LCE micro droplet synthesis [21, 37, 141, 142]. A complication is that all starting materials for the LCE synthesis are in solid phase at room temperature, hence they used fused silica capillaries assembled in a form of T-junction immersed in a hot bath of silicone oil as shown in figure 22. Syringe pumps were used to flow both the LC monomer mixture (dispersed phase) and a silicone oil stream, which was used as continuous phase, through the hot bath. The LC monomer mixture stream breaks up into droplets after the T-junction due to the Rayleigh instability, driven by the interfacial tension between the two fluids. The shear flow within the droplet as it flows downstream ensures director alignment, although—as discussed in section 3.3.1—one should expect a continuous transformation between bipolar and concentric configuration [103].

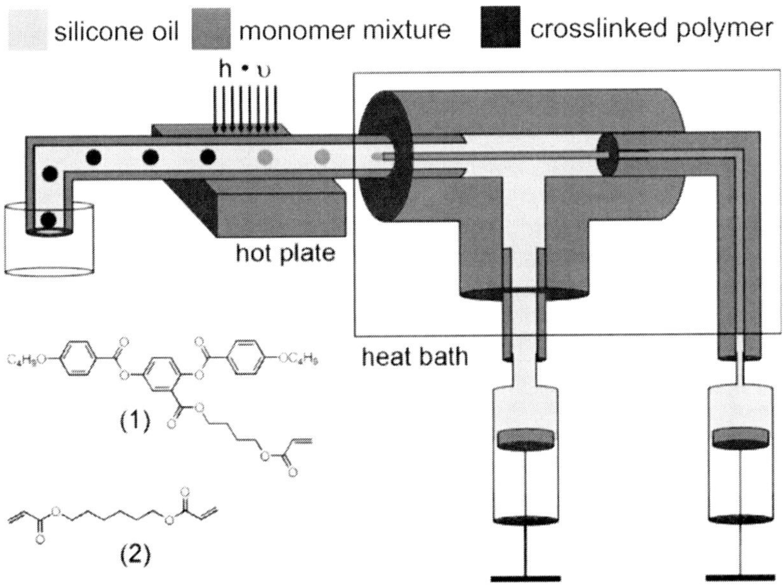

FIGURE 22. Schematic drawing of microfluidic setup for LCE particle synthesis:(1) shows the structure of the LC-monomer used and (2) the structure of the crosslinker. Image adapted, with permission, from [37] John Wiley & Sons. Copyright © 2009 WILEY-VCH Verlag GmbH & Co. KGaA, Weinheim.

The size of the droplets could be controlled by varying the flow rates of the disperse and the continuous phases. With optimized flow ratio between the phases, they achieved monodisperse droplets, with a coefficient of variation in size that was less than 0.7%. It is important to keep the monomer mixture in the nematic phase during the polymerization and crosslinking stage. This was achieved by maintaining the nematic phase temperature downstream from the point of droplet break-up using a separate hot plate as shown in figure 22.

The actuation of the LCE particles produced in this way involved a sphere–ellipsoid shape transformation upon heating to the isotropic phase and back. A similar approach was used to produce planar-aligned LCE particles by Kim *et al* [143], and in the production of LCE particles using thiol-ene 'click' chemistry by Fleischmann *et al* [144]. Recently, Marshall *et al* demonstrated an alternative approach for the production of this type of LCE particles, separating in time the droplet production from the UV crosslinking step [145]. Here, a PDMS based microfluidic device was used rather than the capillary device. Contrary to the Zentel group approach,

Marshall *et al* used mechanical stretching as a tool for aligning the droplet of LCE precursor mixture into a mono-domain. This was possible by collecting the spherical droplets of precursor mixture in a high-concentrated aqueous PVA solution, from which the water was then evaporated. The thermoplastic matrix that thus developed was then stretched, resulting in elongation of the encapsulated precursor droplets with consequent uniform alignment of the director, and at that stage the sample was polymerized and cross-linked by UV illumination. Afterwards, the produced LCE ellipsoids were separated from the PVA film by dissolving PVA in water. This approach too is rather straightforward and useful for bulk production of LCE particles in micron scale. The authors reported a reversible shape change with 30% strain and 30 kPa stress upon heating from nematic to isotropic and back, see figure 23 [145].

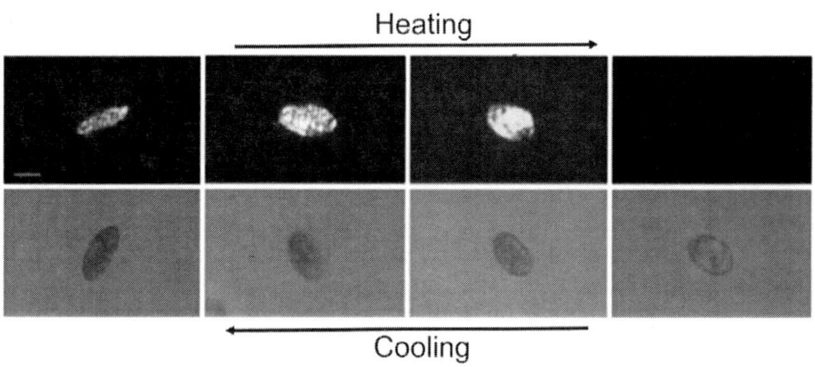

FIGURE 23. Anisotropic shape change of LCE particle in glycerol upon heating and cooling. The top row shows the texture between crossed polarizers; the LCE gradually looses birefringence and eventually turns dark at the far right, which indicates that the particle is now optically isotropic, although its shape has not fully relaxed to spherical. The latter aspect is clear from the second-row images, taken without crossed polarizers. The scale bar represents 15 μm. Reprinted with permission from [145]. Copyright (2014) American Chemical Society.

A very interesting new avenue for applying LCE microspheres, opening an entirely new direction of research, was recently introduced by the group of Elda Hegmann [146]. Here a network of polydisperse LCE microspheres produced by emulsion polymerization was investigated as a scaffold for muscle cell growth. Rather than using microfluidic production, the authors dispersed the LC monomer mixture in toluene along with a small quantity of aqueous surfactant solution, leading to formation of droplets under vigorous agitation. The same approach was used by the Zentel group to produce nanosized shape-changing LCE particles [147]. Once the desired emulsion is produced, polymerization and crosslinking of the precursor mixture into LCE droplets is triggered by UV illumination. In the Hegmann study a 3D network of LCE particles formed after removing the solvent and surfactant.

In a joint study by the Lagerwall and Zentel groups, a modified nested capillary microfluidic device was used to produce LCE shells, which upon actuation functioned as one-piece micropumps [35]. Again the inconveniently high LC temperature range of the precursor mixture called for some modifications of the shell production set-up. Rather than aqueous surrounding phases, the authors used silicone oil for the outer phase and glycerol for the inner phase, and no stabilizers of any kind were included. In order to allow stable production of the triple emulsion, with glycerol inside the LCE precursor mixture inside the silicone oil, a three-fold coaxial capillary geometry was chosen, allowing unidirectional flow of all liquids. The microfluidic device comprised three capillaries; a tapered square glass capillary carried the LC monomer mixture, and into this was nested a tapered cylindrical capillary, carrying the inner phase. Finally, an outer teflon tubing was used as collection tube as well as for carrying the outer phase as shown in figure 24(a).

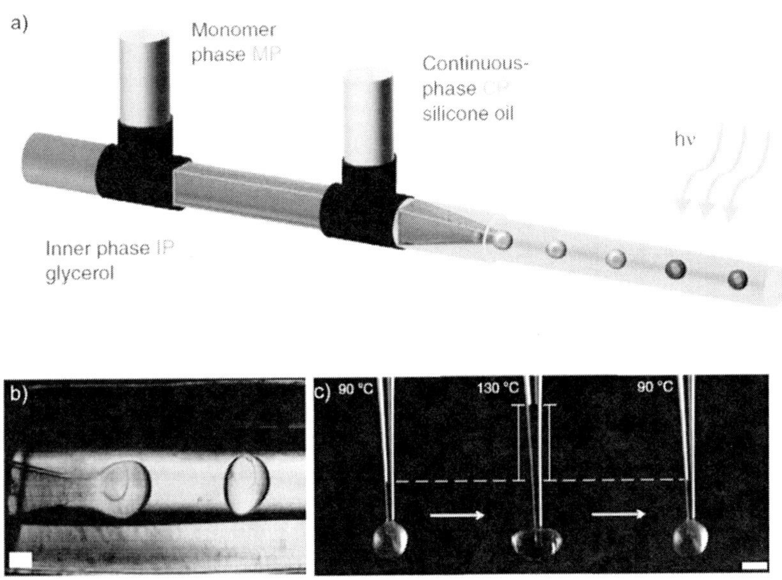

FIGURE 24. Nested capillary setup for LCE shell production. (a) A triple microfluidic co-flow geometry allows fabrication of LCE shells with glycerol core. (b) Shell formation at the capillary tip. (c) Demonstration of micropump: when heated to 130 °C (isotropic), the force exerted by the actuating shell pushes the glycerol core upwards into an inserted capillary.

The glycerol sinks to its original level when the shell resumes its former shape on cooling to 90 °C (nematic). Scale bars, 100 μm. Reprinted by permission from Macmillan Publishers Ltd: Nature Communications [35], Copyright (2012).

The precursor mixture was heated to its isotropic phase at 90 °C and the temperature of the silicone oil flowing through the collection tube was adjusted to the nematic range at 65 °C, using different hot plates. Note that this set-up for shell production involves two consecutive Rayleigh instabilities, first the glycerol stream breaking up into droplets within the LCE precursor stream, then this compound stream breaking up into shells within the silicone oil stream. The second of these instabilities is shown in figure 24(b). UV illumination was then used to induce polymerization and cross-linking, effectively freezing the flow induced planar director orientation within the nematic shell. This was confirmed by wide angle x-ray scattering [35].

Probably due to the dynamic production procedure, the shells were biaxial ellipsoidal in shape. Upon actuation by heating them above the clearing point of the LCE, the ellipsoid changed the relation between its three axes, the shortest axis becoming the longest, extending by a factor of about 2. The most interesting effect was obtained by punching a hole in the shell after heating it above the LCE glass transition temperature and inserting a pointed capillary, until it reached the glycerol core. Actuating the punctured shell by heating above T_{NI} now yielded a micropump action, as the shell pushed out the glycerol core into the capillary (against gravity) on heating, sucking it back in when cooled back into the nematic phase, see figure 24(c). Approximately 65% of the inner fluid can be actively pumped out and sucked back in, in a total of $\approx$0.4 s time.

In summary, micro/nano LCE droplets and shells have been generated using various designs of microfluidic systems as well as by regular emulsion polymerization. The LCE network alignment was generally achieved by the shear induced flow fields during the production, and the alignment was fixed by UV illumination before the director field could relax. Considering the excellent control of $\boldsymbol{n}(\boldsymbol{r})$ achieved with low molar mass LC shells where no conversion into an LCE was attempted, it is a promising direction for future research to move away from flow-induced alignment. Instead one may rely on the control of topological defects to control $\boldsymbol{n}(\boldsymbol{r})$ throughout the shell. The actuation behavior of LCE droplets/shells was so far demonstrated only with external heating as a stimuli. Also here there are many openings for future improvements, e.g. by tuning the shell composition such that light can be used as a trigger.

3.5. Tunable photonic crystal microspheres made from cholesteric shells and droplets

As introduced in section 2.3, cholesteric LCs are recognized as extraordinary photonic materials, providing UV, visible or IR selective (Bragg) reflection due to the spontaneously formed helical structure with a pitch p_0 that is equal to the reflected light wavelength within the LC. This turns cholesterics into dynamically tunable, self-assembled photonic crystals [2]. When confining a short-pitch cholesteric LC within a small spherical shape, like a droplet or a shell, the combination of periodic internal structure and spherical geometry can lead to many intriguing and useful phenomena. Humar and Musevic [148] succeeded in turning cholesteric droplets into omnidirectional lasers, and later Uchida *et al* [149] and Chen *et al* [150] achieved the same with cholesteric shells. Cipparrone and co-workers demonstrated chiral optomechanics involving the interaction of cholesteric droplets and optical tweezers [151].

We recently showed that the selective reflection in combination with the spherical geometry of cholesteric droplets [152, 153] or shells [86] gives rise to a very interesting new behaviour when many such cholesteric spheres are brought close to each other: a photonic cross communication takes place between adjacent spheres, along specific channels defined by Bragg's law and the directions of communication. This generates intricate multicolored patterns that can be dynamically tuned by varying the illuminated area (see figure 25). Panel (b) in the figure schematically illustrates the cross communication for the case of a normally red-reflecting cholesteric. A red spot seen at the center of each sphere corresponds to normal incidence selective reflection. A first set of peripheral blue cross communication spots appears when the illumination beam meets the cholesteric neighbor sphere at an angle of 45° with the helical axis. The light is reflected horizontally until it reaches the central sphere, which reflects the light back up to the observer. This constitutes the most basic type of cross communication, which we refer to as *direct* communication.

A second category of cross communication arises in uncovered samples, where the continuous phase is followed by air, with lower refractive index. This boundary condition allows for total internal reflection (TIR) at the continuous phase–air interface, enabling an indirect cross communication that is recognized through green-yellow spots. They have the same hexagonal symmetry as the main green spots, a bit further towards the centre of each sphere. Several additional series of spots can arise from cross communication with cholesteric spheres situated further away, in particular in uncovered samples and with rather large spheres [86, 152].

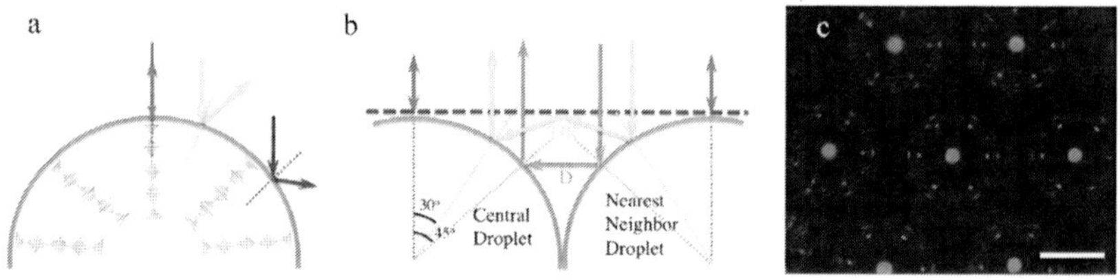

FIGURE 25. (a) The selective reflection from N* spheres with planar anchoring. Because of the radial arrangement of the helix the usual dependence of the color on the viewing angle does not appear: in any direction that you view a planar-aligned N* droplet you see reflected light only from the center, and here you are looking along the helix. In an isolated sphere, you thus see only a central spot with the normal selective reflection color. (b) Schematic of direct (D) photonic cross communication between nearest neighbor (nn) droplets (same pitch) and of the communication path mediated via total internal reflection (TIR) against the surface of the continuous phase. (c) Hexagonally close-packed array of droplets of a normally red-reflecting cholesteric. The sample is uncovered, thus with air following the continuous phase, and it is observed in reflection between crossed polarizers. The scale bar is 100 μm.

In our first paper on the topic [152], we discussed a number of potential applications of cholesteric spheres based on their peculiar reflection properties, including photonic couplers, all-optical switches/distributors, or autonomous sensors. Currently, we believe that the most interesting application opportunity is in creating secure authentication tokens, relying on the uniqueness and unclonability of the patterns [86]. The helical N* self-assembly provides an intrinsic protection from cloning, as the colored patterns arise from reflection, changing dynamically depending on how large area is illuminated and where this area is located. As the light additionally is circularly polarized, it would be exceptionally difficult to reproduce the pattern in a copy. While an adversary with the same microfluidic N* sphere production technique would be able to produce a new token that is qualitatively similar, it would be extremely difficult to reproduce a particular random arrangement, and thus the particular pattern. This makes each token unique to the point that even the authorized manufacturer could not make an identical copy.

Further opportunities for introducing a unique character into the cholesteric spheres and their resulting patterns were demonstrated by the Li and Katsonis groups, by incorporating light-sensitive chiral dopants [150, 154–156]. By illuminating their spheres with UV light they could dynamically and locally tune the pitch of the cholesteric helix and consequently the reflection wavelength and the resulting colored patterns.

With the molecular motors used as chiral dopants in the Katsonis group work, they could even dynamically change the handedness of the helix [156]. The Li group added additional detail to the pattern by producing dual layers of cholesteric droplets [155].

Before one can realistically exploit any novel applications of cholesteric spheres, however, the optical quality needs to be improved and long-term stability and durability needs to be ensured. The durability issue is best addressed by polymerization and several groups have explored this approach. The Kim group produced cholesteric droplets inside a shell of polymerizable immiscible liquid, thereby protecting the LC either by a hydrogel shell [157] or an elastomer shell [158]. The latter technique is particularly attractive, since the same elastomer precursor as was used to produce the sheath protecting the droplets after production can be used as a new continuous phase, which can then in turn be polymerized. In this way the authors could tune the communication color by changing the refractive index of the continuous phase, and a flexible photonic film functionalized by N^* droplets could be realized [158]. Moreover, the droplets sustained substantial mechanical deformation without loss of selective reflection. The Katsonis group attempted polymerization of a reactive component mixed into the LC [156]. While this helped to stabilize the droplets, the polymerization reaction negatively affected the optical quality. In transmission the optical appearance was severely degraded while the reflection pattern was still visible, but with a substantial quality decrease even with only 5% of reactive component.

In our recent work we moved from droplets to shells, thereby solving the optical quality as well as the stability problems [86]. With droplets of $R \approx 100\ \mu m$ the influence of the single aligning interface is too weak to prevent thermal fluctuation of the helix axis. This renders the cholesteric rich in so-called oily streak defects [159], giving the droplets poor optical quality. In several reports on N^*droplets the reflection is non-uniform [152, 153, 157]. By using shells instead of droplets we could greatly improve the optical quality since the inner interface provides a second aligning influence, that together with the alignment imposed by the outer interface reduces the helix thermal fluctuations. Moreover, because the shell contains a much smaller volume of LC, with the two boundaries supporting the planar alignment, we were also able to polymerize as much as 20% of reactive mesogen within the shells without loss of optical quality. The stabilizing action was enough to ensure that the shells sustained substantial mechanical deformation without rupture, perfectly retaining their selective reflection properties throughout the process [86].

While the shells can be considered thin in comparison to the diameter, they are still thick enough to ensure effective selective reflection. In fact, as was first demonstrated by Fan *et al* [155], a set of well-aligned cholesteric droplets exhibit photonic cross communication at different depths, leading to defocused lines rather than sharp spots in the generated pattern. By moving step by step from a full droplet to increasingly thin shells, we gradually removed the unfocused communication spots, enhancing the optical quality and sharpening the reflection pattern [86]. An additional benefit of working with shells is that we can use the osmotic process discussed in section 3.2.2 to thin the shells [87]. As demonstrated in figure 11 the thinning of the shell greatly reduced the time required to remove oily streak defects, as compared to the week-long annealing required without osmosis.

Based on the simplest level of analysis presented in our original work, shells with different pitch should not communicate, and one would thus expect missing spots in the communication patterns. In reality, cross communication does occur, as shown in figure 26. The communication between shells with different p_0 is possible because the asymmetry in N^* pitch can be compensated by an asymmetry in ray directions and reflection planes [86]. The microscope used for illuminating and imaging the sample integrates light over a certain range of propagation directions, depending on the aperture diaphragm and numerical aperture of the objective used.

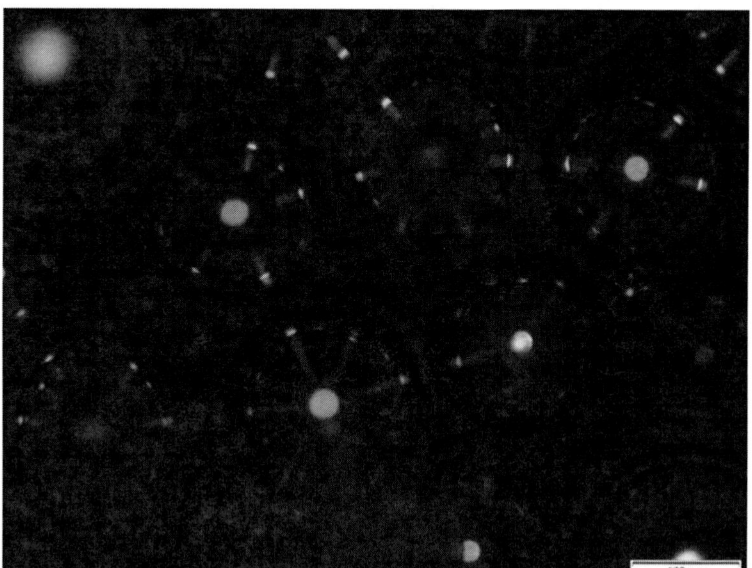

FIGURE 26. Several cholesteric microspheres with different pitch length p_0 and varying sizes arranged randomly, showing photonic cross communication with different wavelengths in different directions. The scale bar is 100 μm. Reproduced from [86]. CC BY 4.0.

As figure 26 demonstrates, the photonic cross communication between shells with different p_0 values means that one and the same shell can exhibit many different colors of communication spots along its perimeter. The possibility to randomly combine shells with different photonic band gaps in a single token, with bi- as well as uni-directional communication at intermediate wavelengths, thus vastly extends the possibilities of creating unique and highly distinct patterns using these shells, with great potential in the field of secure authentication.

While most of the attention concerning N*spheres has been focused on their appearance in reflection, we recently analyzed their optical properties also in transmission [160]. The combination of optical activity, birefringence and selective reflection (in this case having a subtractive effect) gives rise to a quite peculiar and complex appearance of planar-aligned short-pitch cholesteric shells also when they are viewed in transmission between crossed polarizers. Pitch, shell thickness and diameter all influence the texture, but in contrast to the reflection behavior, there is no apparent cross communication seen in transmission. An isolated shell looks essentially similar to shells in an assembly.

3.6. Active liquid crystal droplets and shells

A recent strong trend in LC research is to study active liquid crystals, i.e. non-equilibrium fluid systems with long-range orientational order that continuously dissipate energy. Droplets and shells appear also in this field, hence we end this section by briefly discussing this development. In nature, we come across several 'active LC' systems like flocks, fish swarms and bacterial colonies, in which the 'particles' actively interact with each other during their collective motion, leading to the emergeance of complex patterns [161]. In order to understand such real-life systems one has to deal with the complex collective behavior, mutual interactions, and pattern formation [162–165]. Bacterial colonies are convenient as they constitute a live system that can be studied experimentally in the lab, rendering an elucidation of the origin of their collective behavior viable. On the other hand, there are several inanimate but energy dissipating systems that also develop collective behavior, making them suitable for lab experiments.

In this view, autonomously swimming LC droplets, recently introduced as model units from the Bahr and Herminghaus groups, have opened a new avenue for investigating life-like non-equilibrium systems [166]. Considering the size of the droplets in the order of micrometers and average speed about one diameter per second, this leads to typical Reynolds numbers smaller than unity. This means that viscous forces are dominant over inertial forces. Herminghaus *et al* used LC droplets in comparatively high-concentrated aqueous surfactant solutions [167] and demonstrated that the swimming originated from the generation of non-uniform interfacial tension around the droplet interface. In detail, nematic 5CB droplets were placed in a continuous phase of water-dissolved tetradecyl-trimethyl-ammonium bromide (TTAB). The surfactant concentration is maintained well above the critical micellar concentration (CMC). The dissolution of mesogens into the surfactant solution, and conversely the inflow of water into the LC droplets, creates a gradient in interfacial tension. This leads to Marangoni stresses which propel the 5CB droplets. Interestingly, the phenomenon lasts for hours to days depending on the reserve concentration of LC and surfactant micelles in the continuous water phase.

The mechanism of self-propulsion is explained by two possible dissolution scenarios for the 5CB phase into the aqueous surfactant phase: either by micellar pathway or by molecular pathway. The solubilization of mesogens at the interface is captured by micelles in the micellar pathway, whereas in the molecular pathway mesogens freely diffuse away from the interface and are then captured by micelles. In both cases, the 5CB solubilization creates a gradient in surfactant concentration at the interface, with consequent spatial variation of interfacial tension. The Marangoni stress at the interface in turn creates a convective flow inside the LC droplet. The collective motion of several convective LC droplets arranged into a two dimensional colloidal-crystalline raft in a shallow well is shown in figure 27(a) and a colloidal crystal deformed into the third dimension against buoyancy during swimming in a deep well is shown in figure 27(b).

The surfactant layer at the surface of each droplet results in a homeotropic alignment at the interface, thus a radial director field, hence a topological point defect at the center of each drop appears. Since LCs are birefringent, it is easy to visualize the convection profile inside the droplet using a polarizing microscope [166, 167]. The coupling between the swimming behavior, internal convection and nematic director profile leads to continuous rotation of the droplet during swimming [168]. The detailed theoretical analysis on LC droplet swimmers is beyond the scope of this paper. The interested reader is referred to the review of Maass *et al* [166].

The chiral version of LC droplet swimmers was also reported recently [169]. Here, the authors extended the above mentioned Marangoni flow induced self-propulsion concept to cholesteric LC droplets. The authors observed the time evolution of director profile in 2D and 3D swimming droplets with different handedness of the chiral dopant. The swimming trajectories of the droplets exhibit a helical path which is opposite to the handedness of the chiral dopant. The theoretical description was formulated based on translational velocity from Marangoni flows, an angular velocity from the rotation of the droplet and the internal helical director field [169].

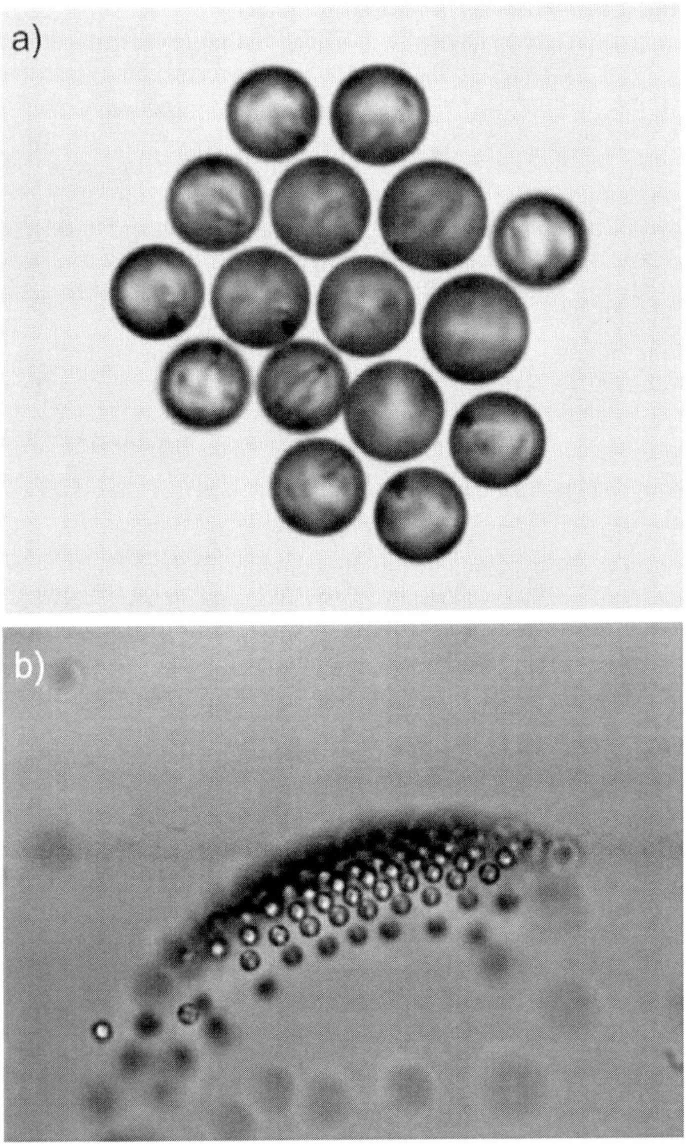

FIGURE 27. Snapshots of collective motion of swimming LC droplets. (a) Crystalline raft formation of nematic droplets during collective motion. (b) A part of a crystalline raft of swimming droplets is out of focus, which shows that it propagates in three dimensions as an integral structure. Reproduced from [167]. CC BY 3.0.

While the above examples constitute active systems of LC droplets developing collective behavior, it is also possible to incorporate an active liquid crystal system, with very different composition from that of thermotropic nematics, within a shell. In a large-scale collaboration between the Giomi, Bowick, Marchetti, Dogic and Bausch groups, Keber *et al* succeeded in preparing a vesicle (a shell consisting of a bilayer of phospholipids) containing an active nematic phase of motile microtubules driven by kinesin motors [170]. As for the case of shells of passive nematic, the spherical topology implies that defects are formed on the surface, but the fact that the phase is active leads to entirely different dynamics. The defects move spontaneously and the authors observed, on the one hand, oscillations of defect configurations and, on the other hand, changes in shape of the entire vesicle driven by the motility of the microtubules.

4. CORE-SHEATH FIBERS WITH ENCAPSULATED LC

The focus now shifts to LCs inside fibers, where they are subject to strong cylindrical confinement. We first review the classic works—as well as some recent additions—on LCs in containers like glass capillaries or porous inorganic membranes. Thus, sections 4.1.1–4.1.4 will be of more fundamental physics character, describing how nematics, smectics and cholesterics respond to cylindrical containment. In section 4.2 we survey the relevant literature regarding electrospinning, beginning with how the process fundamentally works, what experimentalists should generally expect to observe during the process, and the key components of the setup. This discussion is primarily for physicists and chemists who are unfamiliar with the experimental technique. This section will also further elucidate important solution specific parameters such as polymer concentration, molecular weight, and viscosity (section 4.2.3–4.2.4). Section 4.2.5 focuses on some aspects of coaxial electrospinning, useful for introducing low molar mass LCs into the fiber cores.

In section 4.3.1 we survey the reports so far of successful incorporation of LCs into fibers. We cover fibers from liquid crystalline polymers (LCPs) as well as low molar mass LCs incorporated in the core of non-liquid crystalline polymer sheaths. The latter topic is, in fact, the main focus throughout section 4. Finally, in section 4.4 we also relate to the application areas which we feel that LC core fibers and textile could help to advance in today's world. In this context we cover fields ranging from visibly responsive volatile gas sensors (section 4.4.1) to smart flexible actuators (section 4.4.2).

4.1. Nematics, cholesterics and smectics in cylinders

While containment inside fibers is a relatively recent addition to LC research, being easily accessible since the introduction of LC core electrospinning (section 4.2.1), the confinement of LCs inside cylindrical cavities has been studied since long, theoretically, in simulation as well as experimentally. In reviewing the key outcomes of these studies, we limit ourselves to the N, N* and SmA phases, in cylinders with planar and homeotropic surface anchoring, respectively. We do not consider spontaneously forming LC filaments [171–174], typically occurring in certain smectics and discotics subjected to tensile strain, although this is indeed a fascinating phenomenon.

Unfortunately the study of cylindrical confinement presents a case of confusing terminology mix-up. In section 2 we introduced the terms *homeotropic* and *planar* as referring to the orientation of ***n*** with respect to the sample interface, the former referring to orthogonal alignment whereas the latter means that ***n*** is in the plane of the interface. This is the modern definition of the terms and they are widely used as such in current literature. However, in the early studies of LCs in cylinders, the term 'planar' was often used to indicate that ***n*** lies in the cross section plane perpendicular to the fiber axis. In order to avoid a self-contradicting terminology, we thus adjust the notation for cylindrical ***n***(***r***) slightly, replacing the problematic old-style 'planar' with *perpendicular*, referring to the fact that the director lies fully within the perpendicular cross section plane. This choice has the advantage that we are able to stick to the original abbreviation P in our text. When comparing with the problematic original articles, the reader needs to keep in mind that 'planar' in those texts may correspond to 'perpendicular' in ours.

4.1.1. Nematics in cylinders with homeotropic surface anchoring.

The attention early on focused on the interesting frustration caused when a nematic is confined in a narrow cylindrical space with homeotropic anchoring to the surface. The intuitive idea that an $s = +1$ wedge disclination line would form along the core, with a uniformly splayed, or Radial (R), director field within the Perpendicular (P) cross section (PR, see figure 28a) was early on recognized to be incorrect; the LC can in most cases reduce the elastic distortion energy by the escape into the third dimension (see section 3.3). In the cylinder this means that ***n***(***r***) bends into the cylinder axis as one moves from the periphery to the center. Indeed, this Escaped Radial (ER) configuration was confirmed by optical microscopy in 1972 by Williams, Pieranski and Cladis [110] and by Meyer [98]. Both studies used the nematic compound MBBA in glass capillaries where the inner wall was coated with a surfactant for homeotropic alignment. The capillary radius in these studies varied from 20 μm to 200 μm. Since the direction of escape ('up' or 'down' the cylinder axis) is arbitrary, Williams *et al*also found a regular set of $Q = \pm 1$ point defects where an upward escape encountered a downward escape and vice versa. Considering a certain length of the cylinder we may call this configuration Escaped Radial with Point Defects (ERPD, figure 28(b)).

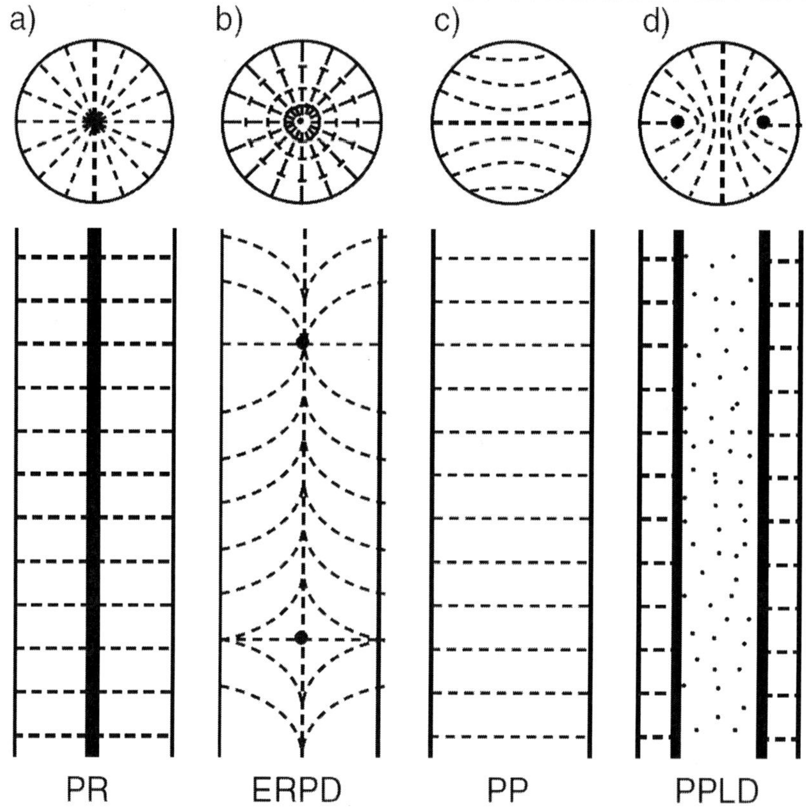

FIGURE 28. Schematic illustrations of the four different director configurations considered for cylindrical confinement with homeotropic surface anchoring. The top drawings show a circular cross section perpendicular to the cylinder and the bottom drawings show a cross section through the middle of the cylinder, along the cylinder axis. Nail heads point down below the paper plane. PR = perpendicular cross section radial; ERPD = escaped radial with point defects; PP = perpendicular cross section polar; PPLD = perpendicular cross section polar with line defects.

Some 20 years later the focus shifted to cylindrical cavities of radically smaller diameter. LCs were now filled into the pores of Nucleopore (polycarbonate) or Anopore (alumina) membranes, where the radius is frequently below 1 μm. Classic optical microscopy does not provide a resolution high enough to interrogate $\boldsymbol{n}(\boldsymbol{r})$ in such thin channels, thus most studies were done using nuclear magnetic resonance (NMR). Crawford, Allender and Doane [175] filled selectively deuterated 5CB into a series of Nucleopore membranes, varying the pore radius from 0.05 μm to 0.5 μm. In some cases they left the pristine polycarbonate in contact with the LC, in others they coated the pore walls with lecithin, a classic agent for promoting homeotropic alignment. The key question was whether the ERPD configuration would still be energetically favorable in these narrow cylindrical cavities, as the required bend deformation would now take place over such a small range that the elastic distortion cost would be substantial.

The result depends on the anchoring strength: for strong homeotropic anchoring—surprisingly given by the naked polycarbonate—the ERPD configuration developed down to 0.05 μm pore radius. Lecithin actually gave weaker homeotropic anchoring, such that ERPD was found only at 0.5 μm radius. For thinner cavities another configuration developed, which we here refer to as Perpendicular cross section Polar (PP, figure 28(c))3. The homeotropic anchoring breaks down at two antipodes on the cylinder periphery. This allows a defect-free $\boldsymbol{n}(\boldsymbol{r})$, entirely within the cross section perpendicular to the cylinder axis, with only moderate splay-bend deformation. Interestingly, the authors also succeeded in establishing the saddle-splay constant K_{24} of 5CB, finding that it is comparable to the bulk elastic constants. Kralj and Zumer later proposed [176], based on numerical simulation, that a modification of the PP configuration may arise for strong homeotropic alignment. In this case two $s = +\frac{1}{2}$ disclination lines,

developing along the cylinder axis, allows the homeotropic anchoring to be retained everywhere on the surface, while the deformations in the perpendicular circular cross section are moderate. We refer to this configuration as perpendicular cross section polar with line defects, PPLD (figure 28(d)).

In a complementary experimental study, Crawford *et al* studied Anopore membranes with 0.2 μm diameter pores, in which the alumina surface was coated with fatty acids of varying chain length, acting as surfactant [177]. The naked alumina promotes planar alignment whereas the fatty acid increasingly promotes homeotropic alignment the longer the chain. This allowed them to detect an alignment transition from the PP configuration for long fatty acids, to fully planar alignment for chain lengths of 6–7 carbon atoms. The LC thus adopts a uniform axial (UA) configuration, where ***n*** points along the cylinder axis everywhere (figure 30(a)).

Crawford, Stannarius and Doane conducted an interesting study where the high surface-to-volume ratio of Anopore membranes was used to induce nematic order at temperatures high above the bulk clearing point T_{NI}^{B} [178]. Depending on the surface treatment slightly different results were obtained, but in all cases nematic order was detected close to the solid surfaces at least 15 K above T_{NI}^{B}. The same phenomenon was later detected by Lagerwall *et al* using Raman spectroscopy on a nematic mixture contained within electrospun fibers with a PVP-titania composite sheath [179]. Here anisotropic order was detected even 20 K above T_{NI}^{B}.

While the majority of experiments have been conducted on uniform confining cylinders, i.e. with constant radius (and the corresponding theory was developed for the same situation), Matthias and Kitzerow conducted a combined experimental-simulation study of the behavior of a nematic inside channels of sinusoidally varying diameter [180]. The channels were electrochemically etched out of a silicon substrate, and the walls were then coated with surfactant to ensure strong homeotropic alignment, hence the ERPD configuration was expected. The pore diameter was about an order of magnitude larger than in Anopore/Nucleopore membranes, varying between 2.2 μm at the necks and 3.3 μm at the bellies.

A smart choice was to work with a glass forming LC polymer. This allowed the researchers to fill the pores at high temperature, anneal the sample in the N phase until an equilibrium configuration is stable, and then quench-cool the sample into the glassy state. The silicon substrate was removed chemically, allowing a confocal microscopy characterization of the nematic glass. While the basic ER configuration was indeed confirmed, the hedgehog point defects split into disclination rings, the cross section of which is illustrated on the right in figure 29. Because of the sinusoidal diameter modulation the ring defects were regularly spaced, the positive-signed disclination rings appearing at the bellies and the negative-signed ones at the necks. These observations of LC behavior in cylinders with modulated diameter are highly relevant for the understanding of LCs in fibers, formed e.g. by electrospinning, since a certain modulation of fiber diameter is not uncommon, see section 4.2.

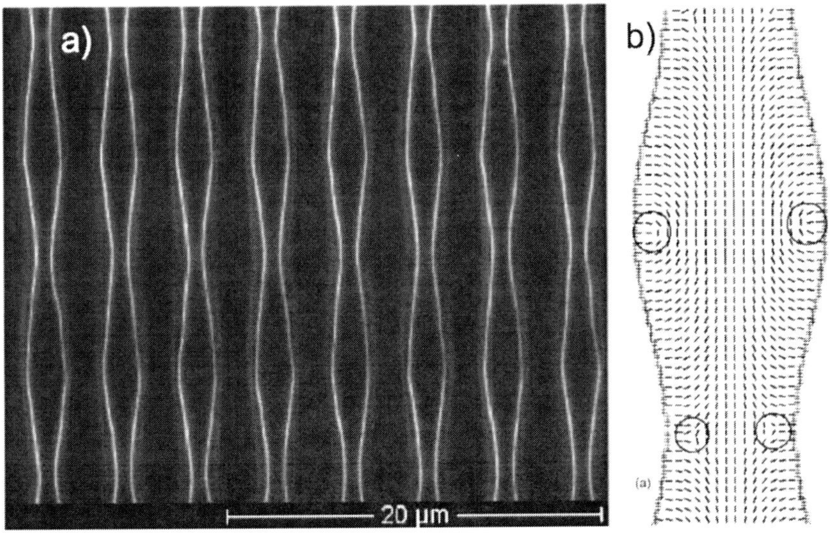

FIGURE 29. Escaped Radial with Defect Rings (ERDR) configuration (right) detected when a glass-forming nematic LCP was filled into homeotropic-aligning pores with sinusoidally varying diameter (left). Compilation of two images, reprinted from [180], with the permission of AIP Publishing.

This appears to be the first experimental report of hedgehog point defects splitting into disclination rings in cylinders, but it turns out that the modulated pore diameter is not necessary for this to occur. In fact, a number of theoretical papers predicted that point defects are unstable and should always be replaced by nanometer-sized disclination rings [95, 100, 105, 181–183]. In an elegant recent experimental study [184], the Abbott group verified that this is in fact the case for $Q = +1$ radial hedgehogs, replaced by an $s = +\frac{1}{2}$ line disclination bent into a nano-sized loop, but not for $Q = -1$ hyperbolic hedgehogs.

They filled a 400 μm diameter capillary, prepared with homeotropic-inducing aligning agent, with 5CB to which was added fluorescently labeled and photopolymerizable phospholipids, respectively. These lipids spontaneously aggregate within defect lines above a critical concentration [185], allowing the researchers to localize defects and even form solid objects templated by them. When studying the capillary using fluorescence microscopy the authors found lipid aggregation in the radial $Q = +1$ hedgehog but not in the hyperbolic $Q = -1$. Moreover, imaging the solids formed by polymerizing the aggregated lipids in transmission electron microscopy (TEM), they found the objects to be not spheres, but tori (rings). A calculation of the free energies of $Q = \pm 1$ point defects and $s = \pm \frac{1}{2}$ disclination rings confirmed that the positive-signed hedgehog was unstable with respect to the disclination ring, but not the negative-signed hedgehog. A question that may be interesting to investigate in the future is why these defect rings were orders of magnitude smaller than those found by Matthias *et al* [180].

4.1.2. Nematics in cylinders with planar surface anchoring.

With planar surface anchoring, the intuitive expectation is that the director aligns without deformation along the cylinder axis, in a so-called uniform axial (UA) configuration (figure 30(a)), at least for thin cylinders. Interestingly, Ondris-Crawford *et al* succeeded in preparing a sample where planar anchoring was promoted, yet with the preferred direction in the interface being azimuthal [186]. They coated the polycarbonate surface of Nucleopore membranes with polyimide, benefiting from a production-induced inner corrugated topography of the pores to create a preference for circumferential rather than axial alignment. Having achieved this unusual boundary condition for nematics in cylindrical confinement, they contemplated three possible configurations (figures 30(b)–(d)): perpendicular cross section concentric (PC), perpendicular cross section bipolar (PB) and escaped twisted (ET). Unsurprisingly, the PC configuration, with its extremely strong bend deformation at the core, was never observed. For small cavity sizes, with radii of 0.2–0.3 μm, the PB configuration was detected, whereas for radius greater than 0.3 μm a transition to the ET configuration was found.

The ET configuration is intriguing, as it means that an achiral LC spontaneously develops a twist. As in any such case of spontaneous chiral symmetry breaking, the twisting sense can be left- or right-handed, each occurring with equal probability, hence interesting phenomena happen where left- and right-handed regimes meet [187]. The twist can either be unwound at the meeting point, resulting in a wall that locally violates the boundary condition, or a hedgehog point defect can form at the center. This costs much energy locally but it respects the boundary condition, hence it may be preferable in case of strong anchoring at the cylinder interface. Considering the results discussed above, suggesting that a $Q = +1$ point defect is unstable with respect to a nanoscopic ring-shaped $+\frac{1}{2}$ defect line, it may be that—in practice—the point defect (PD) connection sketched towards the bottom of figure 30(d)should rather be a defect ring.

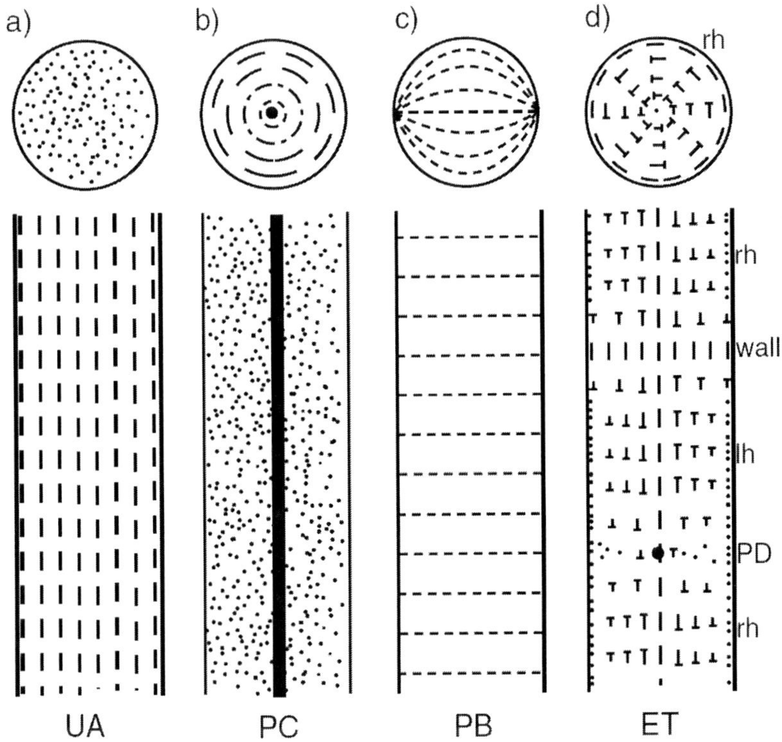

FIGURE 30. Schematics of the four director configurations considered for cylindrical confinement with planar surface anchoring. The top drawings show a circular cross section perpendicular to the cylinder and the bottom drawings shows a cross section through the middle, along the cylinder axis. Nail heads point down below the paper plane. UA = uniform axial; PC = perpendicular cross section concentric; PB = perpendicular cross section bipolar; ET = escaped twisted; rh = right-handed; lh = left-handed; PD = point defect.

The saddle-splay term F_{24} of the elastic energy, (5), was important for explaining the transition to the escaped twisted state in the study by Ondris-Crawford *et al.* Recently, in three separate studies [187–189], F_{24} was found to generate twist also in planar-aligning cylinders of much greater thickness, and without any tangential symmetry breaking promoting concentric to axial alignment of $\boldsymbol{n}$. In the first case [188] the cylinder was bent into a torus, following an ingenious preparation procedure where 5CB nematic is injected into a yield-stress material subjected to continuous rotation. Because a torus has a topological Euler characteristic $\chi = 0$ (see section 3.3) the total topological defect charge on the surface of a planar-aligned nematic torus must be zero. Hence no surface defect formation is expected, in contrast to the case of a planar-aligned shell or droplet. Indeed, no defects were formed on a simple torus, but the surprising observation was that $\boldsymbol{n}(\boldsymbol{r})$ was always twisted, although no chiral dopant had been added.

It is perhaps less surprising that F_{24} generates twist in a torus than in a cylinder, since the inside of a torus is indeed saddle-shaped. Importantly, $\boldsymbol{n}(\boldsymbol{r})$ only experiences this saddle shape if the surface director is neither parallel nor perpendicular to the local symmetry axis of the torus. In these two special cases it would experience pure bend, but for any other orientation, i.e. in a twisted configuration, it will experience the saddle-splay on the inside. If this is difficult to visualize, it becomes clear by simply drawing lines in different directions on a donut. Thus, if $K_{24} > 0$, making F_{24} negative and thereby promoting saddle-splay, a twisted ground state seems plausible for a torus. Pairam *et al* complemented their experimental study with numerical simulations, finding that, indeed, the tighter the torus, the more a twist will reduce the free energy. However, even as the torus gets larger and larger, eventually to an infinite major radius, corresponding to a straight cylinder, they found that a twist can be promoted if $K_{24} > K_2$ [188].

The twist in straight cylinders with degenerate planar boundary conditions was found last year in two independent studies [187, 189]. In these cases, the thermotropic 5CB was replaced by lyotropic chromonic nematics, formed by aqueous solutions of the dyes sunset yellow (SSY) [187] and disodium chromoglycate (DSCG) [189], respectively. Chromonics are convenient because they have exceptionally low twist constant K_2. Thus, $K_{24} >> K_2$, ensuring a twisted ground state in the cylinder although both chromonic nematics are non-chiral. The theoretical analysis of the effect of F_{24} on $\boldsymbol{n}(\boldsymbol{r})$is quite involved and the two papers follow slightly different approaches, coming to qualitatively identical conclusions but a quantitative offset: while Nayani *et al* [189] reproduce the $K_{24} > K_2$ condition found in the study of tori by Pairam *et al* [188], Davidson *et al* [187] require $K_{24} > K_2$ to stabilize a twisted $\boldsymbol{n}(\boldsymbol{r})$.

Both author teams conclude that most thermotropics such as 5CB will generally not have a sufficiently large K_{24} compared to K_2 to generate a twisted configuration in degenerate planar-aligning cylinders. However, very recently such twisted states appear to have been detected in electrospun fibers containing 5CB, without any chiral dopant present [190]. These fibers are substantially smaller (diameter on the order of a μm) than the glass capillaries used in the chromonics studies (100 and 400 μm, respectively) but both theoretical analyses predict that the effect should be scale invariant. While the results on the electrospun fibers are preliminary, and further investigations both of $\boldsymbol{n}(\boldsymbol{r})$ and of possible deviations from a degenerate planar-aligning cylindrical cavity are required, the observations may indicate that K_{24} is still slightly larger than K_2 in 5CB, and that this difference is sufficient to stabilize a twisted ground state.

4.1.3. Chiral nematics in cylinders.

While twisted structures arising in non-chiral nematics are surprising at first sight, this is what we expect when a chiral LC is used, whether by using chiral mesogens or by adding a chiral dopant. The non-trivial question is in what way $\boldsymbol{n}(\boldsymbol{r})$ will twist within the cylinder, and again it depends sensitively on the boundary conditions. One of the first studies, by Cladis *et al*, focused on an N* phase formed on heating from SmA*, contained in a cylindrical capillary imposing radial boundary conditions [191]. The authors also studied the effect of an applied magnetic field. Because neither twist nor bend is compatible with the SmA/SmA* geometry, the authors considered PR (figure 28(a)) as starting configuration (later theoretical studies, discussed below, indicate that this may not have been quite appropriate). As mentioned above, the PR configuration is unstable in N/N* and consequently the authors detected a splitting directly after the phase transition into two $s = \frac{1}{2}$ line defects, essentially developing the PPLD configuration of figure 28(d), with the modification that the disclinations spiral due to the chirality. Moreover, they found an additional helically modulated line, which they postulated to be an $s = +2$ non-singular line [191].

Close to the N*-SmA* phase transition, heating/cooling yields strong simultaneous variations in elastic constants and helix pitch p_0 (on cooling towards a second order N*-SmA* transition p_0 must diverge and become infinite), rendering this study rather complex. Moreover, kinetic aspects are important, hence the speed of heating influenced the behavior strongly. An elegant way to systematically study $\boldsymbol{n}(\boldsymbol{r})$ as a function of p_0 in the long-pitch regime without the complication of simultaneous changes in the elastic constants was presented by Kitzerow and co-workers [192]. They used a chiral dopant known to exhibit a helical twisting power inversion at a specific temperature, such that it induces helices with opposite handedness above and below this temperature. At the inversion point p_0 is infinite, thus there is no difference in $\boldsymbol{n}(\boldsymbol{r})$ from a non-chiral nematic at exactly this temperature.

The authors prepared capillaries with diameters ranging from 25 to 200 μm with agents for hometropic and planar boundary conditions, respectively, in the latter case also imposing (weak) director anchoring along the cylinder by means of photoalignment. In homeotropic-aligning cylinders at the helix inversion temperature they reproduced the ERPD configuration (figure 28(b)) well known from studies of achiral nematics under the same type of confinement. A peculiar observation was the behavior of the hedgehog point defects as the temperature was slightly raised or lowered from the helix inversion point: while p_0 was still too long for any sign of helix formation to be detected in the capillary, the radial hedgehog defects moved towards one capillary wall whereas the hyperbolic defects stayed in the middle of the capillary, see figure 31. At the time of publication, the authors explained the difference by the inability of the hyperbolic defect to satisfy the homeotropic boundary conditions. In light of the recent experimental results of Wang *et al* [184] mentioned above, demonstrating that the radial hedgehog defects are in fact no point defects but nanoscopic defect rings, it would be worthwhile to revisit this difference between radial and hyperbolic hedgehog in weakly chiral systems.

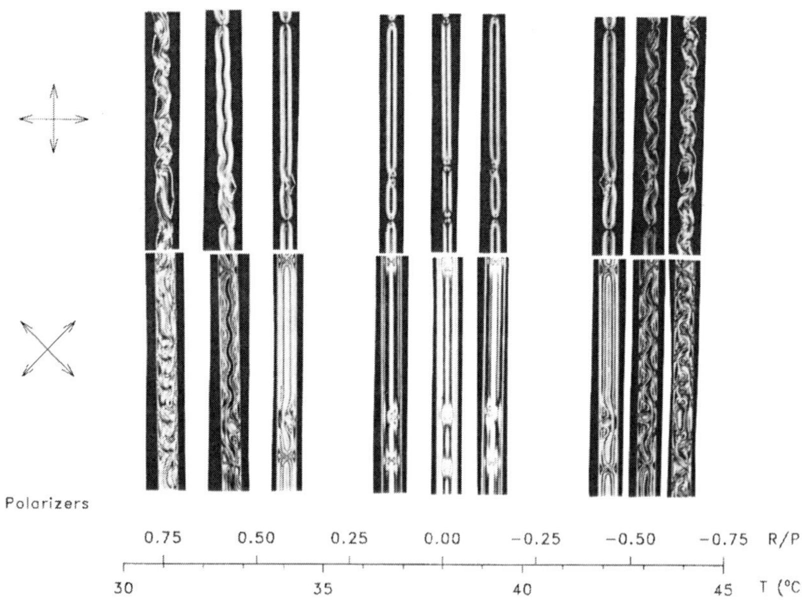

FIGURE 31. Polarizing microscopy textures in a homeotropic-aligning cylindrical capillary filled with an N* mixture exhibiting a helix inversion at 38°C, as a function of temperature T and ratio between the capillary radius R and the pitch (consistently denoted p_0 in the main text but labelled P in this figure). Reprinted figure with permission from [192], Copyright (1996) by the American Physical Society.

As p_0 decreased further on heating or cooling from the inversion point, the defect pattern gradually increased in complexity, first with a sinusoidal periodicity remaining visible, finally succeeded by a chaotic pattern. Representative snapshots for different p_0 are shown in figure 31. The pitch in this study was never very short, its minimum value being several tens of micrometers. Within the cylinder diameter range investigated, it appears that only the ratio of pitch and capillary diameter was important, hence one may anticipate a similar behavior for short-pitch cholesterics in very thin cylinders, such as those formed by electrospinning (see below), if strong homeotropic anchoring can be ensured.

In the planar-aligning cylinders a perfect UA configuration was found at the inversion point, without any sign of twist, see figure 32. While this may be interpreted as a sign of F_{24} playing a less important role here than in the cylinder-enclosed chromonics discussed above [187, 189], we note that in this case the boundary conditions were not degenerate planar, but with anchoring for axial director alignment. It would be interesting to revisit also this experiment with degenerate planar anchoring conditions.

Far away from the inversion point, where p_0 is much smaller than the capillary diameter, a periodically striped 'fingerprint' texture was observed (figure 32). The lines ran along the capillary, indicating that $\boldsymbol{n}(\boldsymbol{r})$ adopted a radially twisted arrangement, as known from the so-called double-twist (DT) cylinders (figure 33(c)) characteristic of blue phases. The director must be aligned along the capillary axis at the boundaries and at the center, to comply with the boundary conditions, hence an integer number of helix pitches must fit within the diameter. As the temperature was shifted towards the inversion point the fingerprint lines therefore did not move away from each other continuously, but the pitch quantization imposed by the boundary conditions forced them to do so in discrete steps, each time with a pair of lines disappearing. The confinement locks the pitch into non-equilibrium values $p \neq p_0$. The same effect was later described by Enz *et al* [193] for selectively reflecting short-pitch cholesterics, confined inside the much thinner cylindrical cavity provided by electrospun fibers, as discussed in section 4.3.1.

Surprisingly, when p_0 tightened again after having passed the inversion point, the fingerprint texture reflecting a double-twist cylinder configuration did not reappear. Instead, the texture transitioned into a complex but regular spiral pattern, see the right side of figure 32. Supported by computer- simulated polarizing microscopy textures the authors conjectured that this pattern was due to an Eccentric Double-twisted (ED) cylinder configuration (figure 33(b)). As p_0 decreased further, instabilities in the helical arrangement were detected once p_0 was below the radius of the cylinder. These helical instabilities, as well as the double-twist configurations at slightly longer p_0, were also detected in an earlier study by Lequeux and Kleman [195].

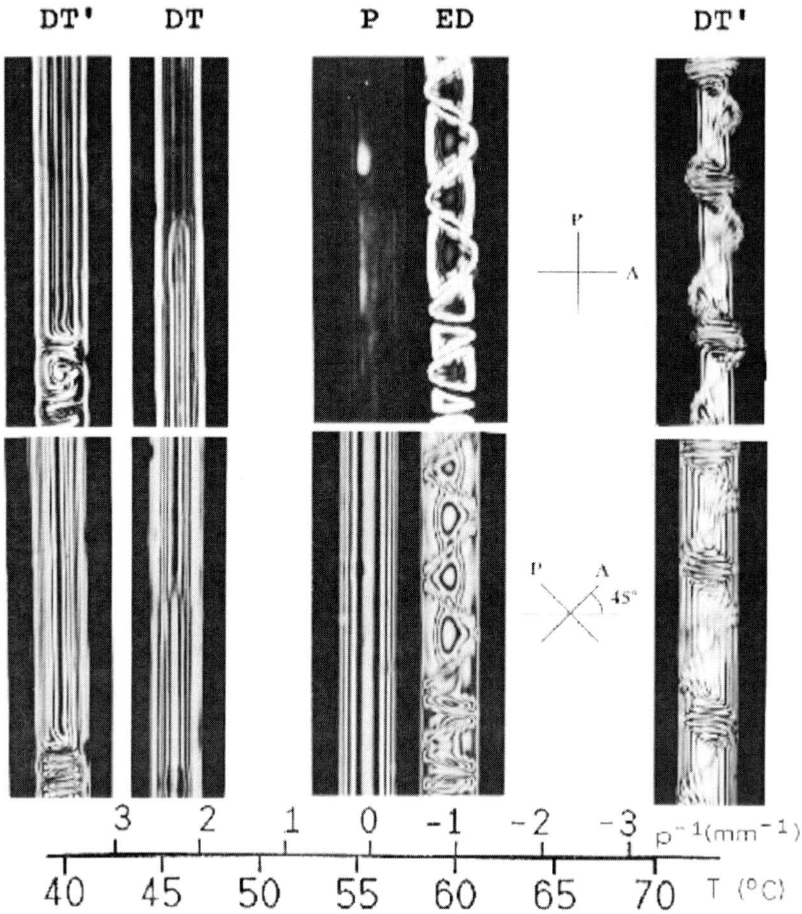

FIGURE 32. Polarizing microscopy textures in a planar-aligning cylindrical capillary filled with a cholesteric mixture exhibiting a helix inversion at 55.6°C, as a function of temperature T and inverse pitch p^{-1}. Reprinted figure with permission from [192], Copyright (1996) by the American Physical Society.

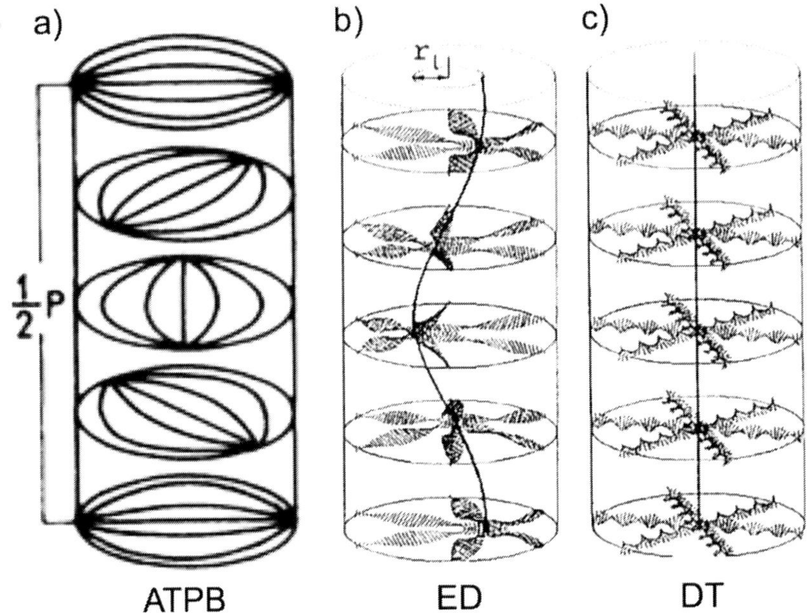

FIGURE 33. Drawings of possible director fields of cholesterics in cylindrical capillaries with planar-aligning boundary conditions. ATPB = axially twisted Perpendicular cross section Bipolar; ED = eccentric double-twist; DT = double-twist. Adapted with permission from (a) [194] and (b–c) [192], Copyright (1994) and (1996), respectively, by the American Physical Society.

The first study of cholesterics in smaller cylindrical confinement than what glass capillaries offer was reported by Ondris-Crawford *et al* in 1994 [194]. NMR was again used to characterize the configurations in the cylindrical pores of alumina anopore membranes, with diameter of 0.1 μm. As mentioned above, the uncoated Anopore membrane induces planar alignment, and the anchoring is not degenerate but concentric, perpendicular to the cylinder axis. In these very thin cylinders the authors found an alignment transition for $p_0 < 1\ \mu$m from the DT configuration to a configuration in which the helix develops along the cylinder. The Perpendicular cross section is of the Bipolar type, with the bipolar axis direction being helically modulated (figure 33(a)); we may call this configuration the axially twisted perpendicular cross section bipolar (ATPB) configuration.

This alignment transition was not seen by Kitzerow *et al* in their study of much thicker planar-aligning cylinders, for corresponding pitch-to-diameter ratios [192]. It has also not been detected in the first optical studies of cholesterics in very thin cylinders, conducted on coaxially electrospun fibers [193, 196, 197]. In these cases cholesterics with pitch short enough to yield selective reflection were confined inside channels as narrow as 0.5 μm [196], and selective reflection was always detected, suggesting a (possibly distorted) DT configuration rather than an ATPB configuration, which should not be reflective when studied in a direction orthogonal to the fiber axis.

Matthias *et al* studied also cholesterics in the cylinders with modulated diameter [198]. As with the study of non-chiral nematics they chose a glass-forming LC, which they could then investigate by fluorescence polarizing microscopy in the glassy state after the template had been removed. The experimental study was complemented with a thorough theoretical treatment. For very long pitch they again found disclination rings rather than point defects in an escaped radial structure, positioned regularly along the channel thanks to the sinusoidally modulated diameter. For pitch comparable to the pore size, a twisted escaped radial structure was found.

4.1.4. The SmA phase in cylinders.

A variety of director fields might be expected when different smectics are confined in a cylindrical space. Here we will discuss only the response of SmA/SmA*, with planar as well as homeotropic boundary conditions. While the chiral SmA* phase has reduced symmetry and thus a different response to electric fields than non-chiral SmA [12], the structures of the chiral and non-chiral versions are identical. Some studies of other smectics exist and it may be a stimulating topic for future work to investigate different smectic phases within cylinders with varying boundary conditions.

The fact that SmA/SmA* cannot sustain twist or bend means that no other deformations than splay should appear in an equilibrium director field configuration for pure SmA/SmA*. Specifically, this would mean that the director escape in the third dimension that favors the ERPD over the PR configuration (figure 28) is no longer expected. As a consequence, the PR configuration was early on thought to be seen in homeotropic-aligned SmA/SmA* in cylinders [191], as mentioned above.

Ondris-Crawford *et al* did a corresponding study with achiral 8CB in the thin pores of a Nucleopore polycarbonate membrane to follow the effect of diverging K_3 on cooling the N phase towards the second-order transition to SmA, potentially extracting information also about surface anchoring W and the saddle-splay constant K_{24} [199]. Their theoretical analysis of the elastic deformations in these very narrow cylinders suggested that a direct transition from ERPD to PR should be found in case of strong anchoring, and that the temperature of the transition can be used to establish K_{24}.

However, the homeotropic anchoring in their nucleopore membranes was too weak, bringing the system into a regime where also the PP configuration is stable. They were not able to give an absolute numerical value of the saddle-splay constant but they could conclude that $K_{24} > K_1$ in 8CB.

A more elaborate theoretical study of possible smectic arrangements inside narrow cylinders with homeotropic surface anchoring was carried out a few years later by Kralj and Zumer [200]. They expanded the viewpoint first by considering that the confinement-induced frustration may lead to an extremely thin prevailing nematic core that releases the elastic stress by letting $\boldsymbol{n}(\boldsymbol{r})$ escape in the third dimension through a centrally localized ERPD configuration, although the temperature is below $T_{\text{smA-N}}$ of a bulk sample. This is similar to the model of Cladis and Torza discussed in section 3.3.4, for explaining the striped patterns forming during a N-SmA transition of a sample subject to hybrid boundary conditions, with homeotropic and planar anchoring on the two surfaces [138].

Second, Kralj and Zumer contemplated that, at least for weak homeotropic anchoring, the core may adopt layers perpendicular to the fiber or tilted in a chevron geometry. Their analysis revealed that the PR configuration is in fact extremely unlikely, and that rather a supercooled nematic ERPD core exists in case of strong homeotropic anchoring, whereas weak anchoring yields chevron or perpendicular smectic layers. This would suggest that the core of *any* smectic arrangement with concentric layers, for instance in the Dupin cyclides that often form spontaneously in bulk samples with little control of alignment [5], should be nematic in an ERPD configuration. Considering today's possibilities to confirm such structures on the nanoscale, as demonstrated by Wang *et al* [184], this motivates new studies of smectics in frustrated geometries.

Using calorimetry, Iannacchione and Finotello studied the N-SmA transition of 8CB in the 0.2 μm pores of Anopore membrane (alumina), prepared for planar and homeotropic surface anchoring, respectively [201]. They found a profound influence on the N-SmA as well as N-isotropic transitions, both broadening considerably and reducing in transition enthalpy as an effect of the confinement, for either alignment geometry. Moreover, the clearing point was always reduced in confinement, whereas the N-SmA transition was shifted to lower temperatures with planar alignment but to higher temperatures with radial alignment. Unfortunately no information was given on sample history during the experiments, in particular concerning whether the measurements were done on cooling or heating, and the time for which the sample had stayed in the preceding phase prior to the experiment.

These aspects can be very important for the behavior of LCs in strong cylindrical confinement, as shown in a later calorimetry study by Enz *et al* of 8CB inside electrospun fibers with PVP sheath [202]. While this study was restricted to planar alignment (as confirmed by x-ray scattering on the fibers), increase as well as decrease of the N-SmA transition could be found, depending on the sample history. On heating, the SmA-N transition was considerably increased compared to the bulk value of $T_{smA\text{-}N}$, up to some 5K for the thinnest fibers studied (outer fiber diameter less than 2 μm; inner core diameter was not assessed). However, when cooling back from the isotropic phase, through the nematic, the N-SmA transition was absent throughout the experiment. The N-SmA transition thus exhibits extreme supercooling inside the planar-aligning fibers. The encapsulated LC stayed in the N state well below room temperature, as confirmed by a subsequent heating experiment where only the clearing transition was detected. However, if the fibers were annealed at room temperature for several weeks, the smectic phase reformed and the SmA-N transition peak was recovered in the calorimetry curves, shifted to higher temperatures compared to the bulk value.

The authors explained these surprising observations as a result of the mesogens at the interface to the polymer sheath being strongly anchored at that interface, in an arrangement reflecting the history. After long annealing at room temperature, allowing the SmA phase to develop, the sheath-anchored mesogens are in a smectic arrangement, promoting the layering and thereby raising $T_{smA\text{-}N}$. However, if the positional order is lost by heating into the nematic or isotropic state, then the mesogens close to the polymer sheath are anchored without a layered arrangement, promoting the N state in a narrow cylinder where the influence of surface molecules can penetrate throughout the volume of the sample. The broadening and weakening of transition peaks seen by Iannacchione and Finotello was reproduced in this study, but the clearing point on heating was always raised (the shift was much smaller than for $T_{smA\text{-}N}$) in the fibers.

4.2. Electrospinning basics

4.2.1. A brief note on the development & popularity of electrospinning.

Electrospinning is the process by which solidified filaments are drawn from a viscous, continuously flowing polymer solution by the action of an electric field. While the modern term 'electrospinning' was popularized by Doshi, Reneker, and many other researchers in the latter end of the 20th century [203–207], the process had in fact already appeared in many patents filed during the period 1900–1940 by John Francis Cooley, and Anton Formhals. It was then referred to as 'electrostatic' or 'electrical spinning' [208–210].

During the mid-1900s, much of the world moved towards a fast paced industrially and commercially developing economy, and many were attempting to exploit the electrospinning process for commercial gains. This explains the appearance mainly of patents rather than fundamental research articles on the topic during this time [211–214]. It wouldn't be until the 1960s, and then again, in the 1990s, when the rigorous theoretical and experimental frameworks for truly understanding the process would be developed. Sir Geoffrey Ingram Taylor contributed to the former by first accurately proposing a model for the characteristic shape of the prolate polymer solution droplet (now referred to as the *Taylor cone*) known to emerge from the charged spinneret proceeding the fiber forming process [215, 216].

Since then, a large number of theoretical and experimental works for comprehensively explaining the intricate level of instabilities observed during electrospinning, such as the onset of jet formation and the final bending trajectories of the fiber formed from the jet, have been published [217–222]. At the forefront, Reneker and Yarin were the leaders in revitalizing modern 21st century interest in the process by not only further characterizing the bending instabilities of the fluid jet, but also systematically studying the parameters affecting various types of fiber formation [217, 223, 224]. Ultimately, while these parameters have succumbed to countless debates and scrutiny by both experimentalists and theoreticians over the years, it has all culminated into the identification of a select handful of both processing parameters (set-up and apparatus related), and solution specific parameters (such as concentration, polymer type, polymer molar mass M, and solvent type) which influence fiber production the most. These parameters will be discussed in section 4.2.2.

The 2000s saw an exponential increase in the research devoted to electrospinning [208]. The technique has proved to be extremely useful for the development of advanced materials, as well as the advanced functioning of other processes in many fields, from air filtration [225, 226] and biotechnology [227] to sensors [228] and waste remediation [229]. The introduction of LCs to the field of electrospinning is rather recent. Interestingly, the two discoveries—the electrospinning method and the liquid crystalline state of matter—were formally first reported by C.V. Boys, and Friedrich Reinitzer, respectively, within only one year of each other in 1887, and 1888. And yet, the first study on combining LCs with electrospinning was published only just nine years ago. Since then, though, it has been steadily growing in popularity.

The first electrospun LCE fibers were presented by Wendorff and Finkelmann *et al* [230]. About one year later Godinho and co-workers demonstrated electrospun fibers from lyotropic LCP solutions [231], and within the same year Lagerwall, Xia and co-workers demonstrated the incorporation of low molar mass LCs inside the core of coaxially electrospun fibers [179]. Our group continuously developed the coaxial approach [190, 193, 202, 232, 233], while two years later West and co-workers [234] demonstrated that LC core fibers could be spun from mixing LC and polymer in a common solvent, without separating them into two channels. This single-phase electrospinning relies on *in situ* phase separation during solvent evaporation for acquiring the final coaxial LC-polymer fiber morphology.

Ultimately, the realization that LC functionalized fibers may prove to be useful in the future for the development of highly functional, environmentally responsive textile fabrics is steadily moving toward the forefront. In the sections following the discussion of how the electrospinning process works, and the parameter considerations that should be considered during the fiber development process, we discuss how electrospun LC-fibers, and LC-elastomeric fibers are already showing high potential for becoming the next non-electronic functional gas sensors.

4.2.2. How does the electrospinning process work?

Despite the many complex and interdependent variables that must be accounted for in order to properly analyze an electrospinning experiment, it is extensively used by researchers in many independent scientific, and even design-related [235], fields. The simple reason is that the process provides immense flexibility for experimenting with fiber production more quickly than other fiber spinning processes and that the set-up (see figure 34) is small-scale, not very complex and comparatively inexpensive. An experiment starts with dissolving a chosen polymer with appropriate molar mass in a compatible solvent until it is homogenously dispersed. This solution is mechanically pumped via an external device (often a syringe pump) through a metal spinneret which is connected to a high voltage power supply. In case of coaxial electrospinning a thin second capillary runs at the center of the main spinneret, allowing ejection of a core fluid inside the polymer solution, see figure 35.

A strong electric field (~kV m^{-1}) is then established between spinneret and a grounded counter electrode some 10–20 cm away. As the solution in the spinneret becomes charged, the droplet protruding from the tip undergoes distortion into a conical shape (the Taylor cone) as a result of the conflict between surface tension—promoting a spherical droplet shape—and the electrostatic attraction to the grounded counter electrode acting on the high density of charges in the droplet. Once the voltage passes a threshold, usually 10–30 kV (depending on solution viscosity, type of polymer and solvent, and the ionic content), the combined action of attraction to the ground electrode and self-repulsion between the concentrated like-charged ions overpowers the surface tension and a thin jet is expelled from the tip of the Taylor Cone.

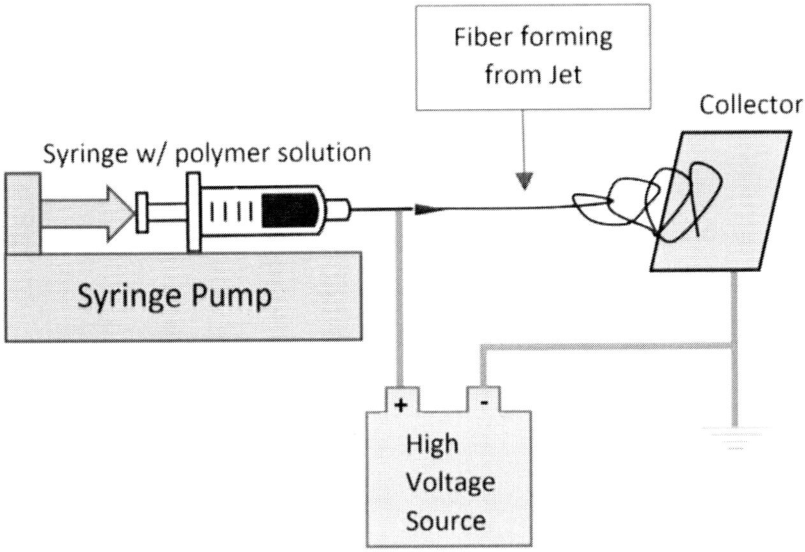

FIGURE 34. A schematic of the generalized electrospinning process showing the most basic components needed, and the connections necessary, to create nano- and micro- fibers.

The polymer solution carried by the jet undergoes a series of helical whipping and bending instabilities as solvent evaporates. The initial jet formation and the subsequent instabilities (figure 36) are triggered, in part, by the fluid jet attempting to lower the charge density by extending [236], in part by the electrostatic self-repulsion between like-charged ions along the jet [217]. The equations describing the initial stage of the process have been derived by Bugarski *et al* [237, 238], Loeb *et al*[239] and Yarin *et al* [240], respectively. The final result is a solidified, continuous fiber collected as a non-woven mat on the grounded electrode or on another collector inserted between spinneret and counter electrode [217, 236].

All components required to devise a basic electrospinning setup are shown in figure 34. Typically, the dissolved polymer solution is placed into a syringe, with an appropriately sized needle, hereafter referred to as the *spinneret* (often blunt ended needles between 22 gauge and 18 gauge are employed). The solution is pumped (via a syringe pump or similar device) to the needle orifice at constant rate on the order of 0.1– 1 ml h^{-1}. The viscosity of the polymer solution is often near that of corn syrup or molasses, about 30–100 Ps (or, 3–10 Ns m^{-2}).

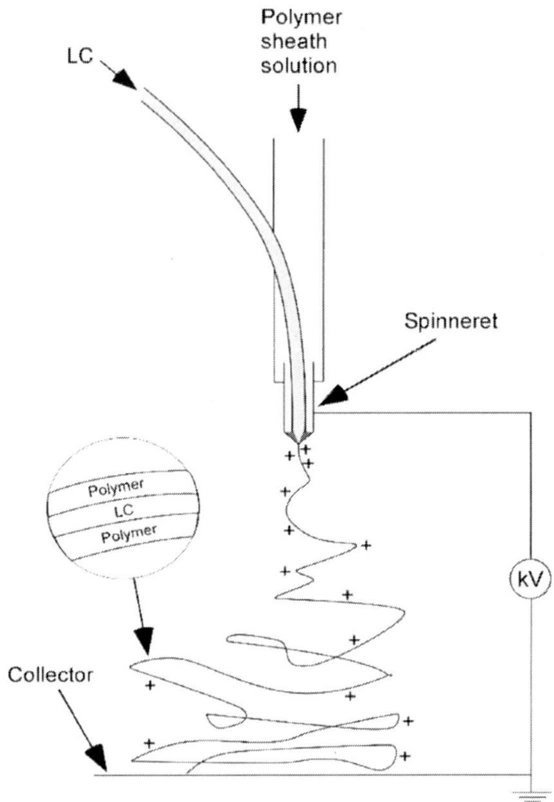

FIGURE 35. Schematic illustration of coaxial electrospinning with LC core. Reproduced from [202]. CC BY 2.0.

If the dissolved polymer has insufficient molar mass and/or concentration, the stretching within the jet is so large that polymer coils are disconnected (there are too few entanglements) and then nothing prevents the Rayleigh–Plateau instability (already mentioned in section 3.1) to break the jet up into droplets. Such an experiment results in the *electrospraying* phenomenon. Electrospray involves further interesting phenomena as the solvent evaporates, leading to break-up of the initial droplets into much smaller ones, but a deeper discussion of electrospray is outside the scope of this review.

However, if the polymer solution employed contains enough molecular chain entanglements, then electrospraying will not occur. The Rayleigh–Plateau instability is now counteracted by a continuous chain of stretched-out but entangled polymer molecules along the jet. To ensure continuous spinning, polymers of $M \approx 10–100$ kg mol^{-1} are used, with concentrations varying from 5% (for high M polymers) up to 30% (for low M) [242, 243].

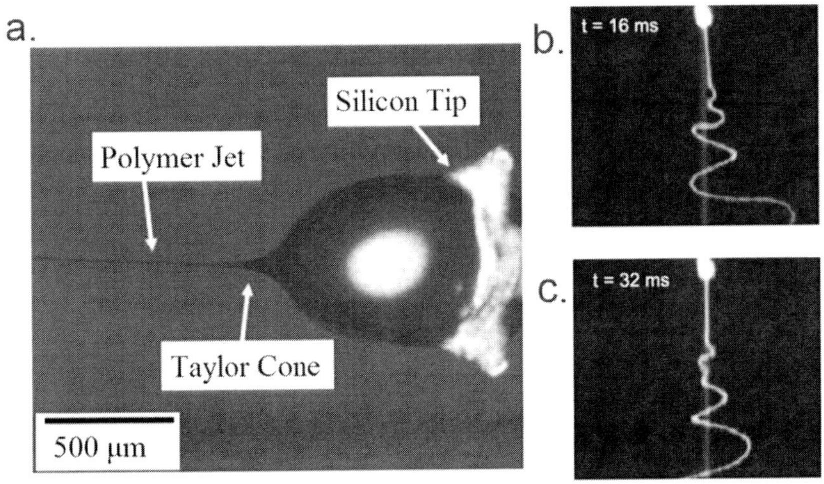

FIGURE 36. Key features of the Taylor cone and polymer solution jet development. An electrified polymer solution droplet is shown in (a) (reproduced from [241]. © IOP Publishing Ltd. All rights reserved). Two high-speed camera images of the polymer jet undergoing helical whipping at the early stages of spinning are shown in (b) and (c), at 16 and 32 milliseconds after jet ejection, respectively (reprinted from [221] with the permission of AIP Publishing).

As the jet is stretched considerably due to the whipping instability, it will get considerably thinner before enough solvent has evaporated to solidify the polymer, and the diameter of electrospun fibers can thus reach values as low as tens of nanometers. In the case of electrospinning PAN (poly(acrylonitrile)) fibers from a DMF (N,N-dimethylformamide) solution, according to Wu *et al*[244], a jet emanating from a Taylor cone that is 50 μm in diameter, can be reduced during solvent evaporation to about 100 nm diameter in only a few seconds. For jets that start at sub-micron diameters, their reduction to the nanometer scale can occur within milliseconds.

While fibers collected onto the grounded collector have a landing speed of about 10 m s^{-1}, the initial acceleration of the straight part of the jet from the Taylor cone can be up to 600 m s^{-2} [242]. The fact that this value is about two orders of magnitude larger than the acceleration due to the gravitational force plainly indicates that the effects gravity acting on the electrospinning jet are nearly negligible—it is also the reason why we have the flexibility of creating setups which can function horizontally or vertically without the jet being perturbed in the electrospinning process. However, in case of coaxial electrospinning, where the Taylor cone is comprised of two different fluids, generally with different density, gravity may have an influence on the symmetry of the Taylor cone. One may speculate that this could influence the structure of the produced fibers, but this is hardly investigated.

Viscoelastic shearing in the jet is largely absent as well since the speed at which fluid flows inside of the launching jet can be up to 5 m s^{-1} at a 2 cm distance from the Taylor cone, much lower than the speed with which the whole jet is moved due to the stretching. Once the whipping jet impacts the collector it should by this point be a solidified fiber. Depending on the type of polymer and the solvent used, the fiber can have varying morphologies, but typically, continuous and bead free fibers with circular cross-section are desired [236, 242, 245, 246].

4.2.3. The right polymer, with the right solvent, gives the right electrospun fiber.

There are many polymers which, when dissolved in the right solvent and electrospun under certain conditions, can form a very large variety of fiber morphologies, from the standard cylindrical to fibers with internal pores and cavities [247, 248] to barbed fibers [249]. Although the processing related parameters, such as the spinneret—ground electrode distance, the polymer solution flow rate and the applied voltage all can have significant effects on the final fiber quality, they come secondary to the choices of polymer and solvent. These two components dictate the solution properties, determining whether or not fibers can be spun.

When we wish to incorporate a low molar mass LC in the fiber core, whether by using a coaxial spinneret or by relying on *in situ* phase separation, choosing the polymer for the sheath can be non-trivial. If we use a crystallizing polymer, for instance, the elongational forces acting on the solidifying jet during spinning will cause the polymer molecules to be much better aligned than if we used an amorphous polymer. While beneficial for fiber strength, this gives the sheath strong birefringence, competing with the birefringence of the incorporated LC. In case of coaxial electrospinning, also the polymer solvent can be a tricky decision to make: if the solvent is miscible with the LC it may significantly diffuse into the core during spinning, a loss of control that may have considerable impact on the end result.

In applied research, the choice of polymer is largely limited by the intended use of the fibers. Biodegradable polymers such as poly(ethylene glycol) (PEG), and poly(l-lactide) (PLLA) may be better suited for producing tissue scaffolds due to their lack of toxicity effects [227, 250], than polystyrene (PS) or polyurethane (PU). If high tensile strength is critical, then crystallizing polymers like Nylon-6 may be useful, although we then get problems with sheath birefringence. Moreover, electrospun nylon fibers tend to be very thin, most likely a consequence of the rather low maximum values of *M* that are commercially available. A large number of literature reviews, books and book chapters exist, discussing how choice of polymer and solvent influence the final fiber performance [217, 242, 251].

As far as the solvents go, polar aprotic or nonpolar solvents are often recommended. While it is possible to electrospin polymers dissolved in polar protic solvents, which tend to have lower vapor pressure (e.g. water: 17.5 hPa; aqueous acetic acid: 15.3 hPa), special care needs to be taken when working with such solvents [252]. Their strong tendency for hydrogen bonding increases ionic dissociation, thus raising the conductivity of the polymer solution, with the possible consequence of rapidly fluctuating charge densities during electrospinning. This can destabilize the Taylor cone, which may dissociate into multiple, small side cones and result in the formation of branched fibers [253]. Moreover, the limited rate of solvent evaporation can lead to swollen or wet fibers landing on the collector.

Even if no water is in the original polymer solution, the hydrophilic character of non-aqueous polar protic solvents can lead to accumulation of water from the vapor in the atmosphere [254], hence high humidity should generally be avoided. Usually, electrospinning at 20–40% relative humidity works well for most polymer solutions. Higher humidity can influence electrospinning dramatically, both for water-soluble and water-insoluble polymers. For instance, at 50% relative humidity PS, being hydrophobic, will tend to form porous fibers [248, 255, 256], whereas PEG and PLLA, when spun together, will create hydrophilic fibers that can easily bead or mesh together [227, 254].

Solvents with higher vapor pressures such as 1,1-dichloroethane (240 hPA), acetone (240 hPA), tetrahydrofuran (THF) (200 hPA), and ethyl acetate (97 hPA), are commonly used, often yielding smooth cylindrical fibers [236]. Many researchers use guides such as the Hildebrand solubility parameter for solvents [257], and the more comprehensive Teas Plots, first developed by Jean Teas [258] and made popular by John Burke [259]. A Teas plot is particularly useful if one is attempting to find multiple solvents for a particular polymer, as the graph is able to more specifically show the fractional contributions of dispersive, hydrogen, and polar bonding that make up the total Hildebrand solubility parameter [236, 259].

4.2.4. The critical roles of polymer molar mass and polymer concentration.

As already mentioned, the molar mass of the polymer is critical for successful electrospinning. The molecules must be large enough, and present at sufficient concentration, to ensure that a continuous chain of entangled molecules persist throughout spinning, otherwise the Rayleigh–Plateau instability will disrupt the jet and no fibers are produced. We thus have two critical values that must be considered for the polymer: the molar mass M and the concentration, which must both be greater than a respective minimum value [236, 260]. Raising the polymer concentration beyond the minimum will not necessarily have an increased positive effect in the spinning process. In fact, very high concentration could lead to premature drying at the spinneret tip, and a solidified Taylor cone before fibers can even be ejected [241, 261].

Typical M values range from 10 to 10^4 kg mol^{-1}. To give a few reference values, the isoprene compound (commonly called Latex) that naturally pours out from rubber trees as a thick white syrup generally has $M > 100$ kg mol^{-1} [262], while Nylon ranges between 3 to 20 kg mol^{-1} [263]. While molar mass and concentration are clearly related, no increase in concentration can compensate for too low molar mass. This precludes spinning of certain polymers, which may be impossible to obtain at the high molar mass required for electrospinning.

The viscosity of a polymer solution can provide a first indication of whether or not sufficient entanglements are present to allow stable fiber spinning [243, 264–266]. However, it can be misleading, since also the solvent can strongly influence the viscosity. According to Agarwal *et al*, for a certain polymer type and molar mass, low vapor pressure solvents tend to raise the viscosity of the solution, whereas high vapor pressure solvents yield lower viscosity [261].

Close to the critical values of molar mass and concentration, fibers may be produced but the Rayleigh–Plateau instability is not entirely hindered. Some polymer molecules disentangle from the overall chain, coiling up more or less spherically as they are no longer subject to the stretching force. This results in periodic bulging-out of the fibers, a phenomenon often referred to as 'beading', since the final fiber morphology may resemble beads on a string. Koski and co-workers found that electrospinning a solution of PVA with $M \approx 10$ kg mol^{-1} yielded beaded fibers, while raising M to 30–50 kg mol^{-1} produced non-beaded smooth long fibers, with all other parameters kept constant, and an equal concentration of 25% PVA in solvent for both situations [267].

4.2.5. A note regarding multi-core & coaxial electrospinning.

Even though many aspects of the coaxial electrospinning process are insufficiently understood, this process for making core-shell nano- and micro-fibers has gained popularity as it can allow low viscosity fluids to be incorporated within polymer fibers. Numerical simulations of coaxial electrospinning suggest that charges rapidly escape to the outer surface of the compound flow, meaning that it is fundamentally the sheath that is acted upon by the electric field, and the extension of the core flow into the jet thus relies on traction between sheath and core flows during spinning [268–270].

As the sheath solution is forming the Taylor cone and the jet is ejected from it, the core fluid, flowing at a similar rate as that of the sheath solution, is being 'carried' inside. Processing parameters, such as applied voltage and spinneret design, influence coaxial fiber formation strongly. Too high applied voltage will result in jet splitting whereby both the core fluid and sheath fluid separate and spin as two Taylor cones side by side [271]. Experiments have shown that core incorporation into the jet is favored by having the inner capillary of the coaxial spinneret protruding slightly from the outer sheath fluid nozzle [268].

4.3. LCs in polymer fibers

4.3.1. Low molar mass LCs in fibers.

Coaxial electrospinning is a very convenient and adaptable technique for introducing LCs into micro- and nanosized fibers. The spinning components can be easily assembled, most of the materials are readily available, and small amounts are sufficient to produce macroscale non-woven fiber mats which can be easily handled. Moreover, by replacing syringe pumps with a pneumatic microfluidic control system [137, 193, 202], very small vials can be used, the dead volume minimized, and temperature control is greatly simplified. These aspects makes the coaxial electrospinning approach for incorporating LCs into fibers very attractive, as LCs can be very expensive. The set-ups used in our group offers a simple, low-cost and versatile solution for coaxially introducing nematic [179, 190, 202, 272], smectic [202] and cholesteric [193, 197, 232, 233, 272] LCs into polymer fiber cores for obtaining fibers with varying functionalities.

Since a non-crystallizing polymer is preferred when spinning together with LCs, to avoid competition in birefringence when they are characterized via polarized light microscopy, many experiments were done with polyvinylpyrrolidone (PVP) as sheath. This amorphous polymer is available at very high molar mass. The polymer sheath serves the double function of encapsulation and alignment layer, because it appears that the stretching during spinning gives a slight preferential axial alignment of the polymer chains (not so strong as to induce significant birefringence) that orients $\boldsymbol{n}$ along the fiber [196]. As the polymer sheath often becomes quite thin during the spinning, and as the core is filled with the LC component, responsive core-sheath fibers can be readily obtained, the macroscopic properties being governed by the type of LC used.

In the very first report by Lagerwall *et al* [179] core-sheath fibers were obtained by coaxially spinning a room temperature nematic LC mixture inside a solution of PVP in ethanol containing dissolved Ti(O *i*Pr)$_4$, a sol-gel precursor for TiO_2. The diameters of the non-woven and aligned fibers were on the order 0.1-1 μm, see figure 37. Because the large amount of TiO_2 in the sheath precluded polarizing microscopy characterization, Raman spectroscopy was used to establish that LC was present in the fiber core, with $\boldsymbol{n}$ aligned along the fiber axis. As discussed in section 4.1.1 a considerable increase in LC clearing point as a result of encapsulation was detected.

In a subsequent study this coaxial LC-polymer electrospinning approach was extended for spinning coaxial fibers with the smectic-forming 8CB in the core [202]. The fibers were spun at room temperature, with 8CB in the SmA phase. Finding a correct LC flow rate was found to be key to ensure continuous LC filling, yet maintaining good overall fiber morphology, see figure 38. At very low LC flow rates, the fibers were discontinuously filled, whereas at too high LC flow rate, the fibers became deformed and appeared 'smeared out'. X-ray diffraction on oriented fibers showed that the smectic layers were perpendicularly oriented to the fiber axis, a rather unexpected finding considering that SmA flows the easiest along the layers (section 2.1.3). As discussed in section 4.1.4, an interesting hysteresis effect was observed upon heating and cooling between the N and SmA phases.

Scalia and co-workers [197] studied the effects of varying spinning distance, polymer concentration and the presence/absence of sol-gel precursors on the morphology of coaxially electrospun PVP fibers with LC core. The morphology changed considerably, with high polymer concentration and/or large spinning distance yielding premature solvent evaporation with a consequent flat ribbon fiber morphology, whereas the opposite conditions resulted in soft fibers landing on the substrate, where they collapsed into a hemicylinder shape. In some fibers a distinctly porous sheath was detected, a phenomeon that the authors explained as a result of the spinning atmosphere being saturated with solvent. The solvent then re-condenses as liquid droplets on the fiber, leaving a porous surface after final evaporation.

The optical properties and responsiveness of LC core polymer sheath fibers can be further improved by introducing short-pitch cholesterics in the core as shown by Enz *et al* [193]. The PVP fibers, with diameters in the range of a few microns, showed selective reflection in the visible spectrum. Similar to the case when smectic LCs were used as the core, the LC filling greatly depends on flow rates which in turn affect the overall fiber thickness. When

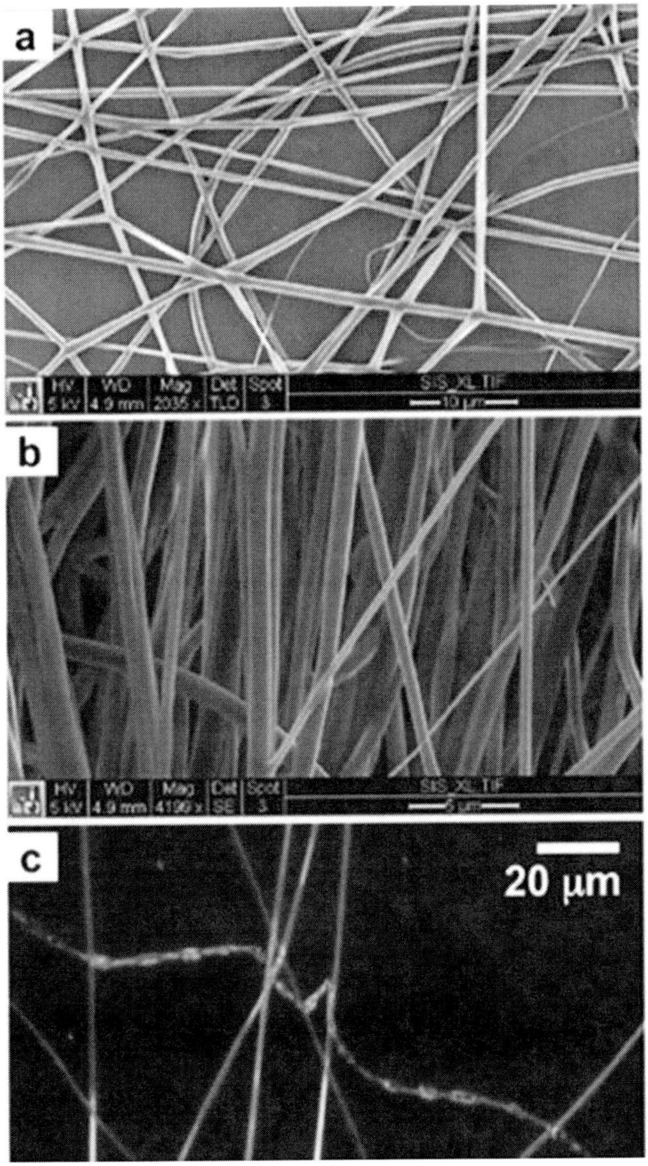

FIGURE 37. Fibres consisting of LC core and TiO_2-PVP sheath observed by SEM, (a) non-woven sample, (b) aligned sample, and by differential interference optical microscopy (c). The sheath solution flow rate was 0.4 ml h^{-1} and the LC flow rate 0.2 ml h^{-1} ((a), (b)) and 0.3 ml h^{-1} (c). Reproduced from [179] by permission of The Royal Society of Chemistry.

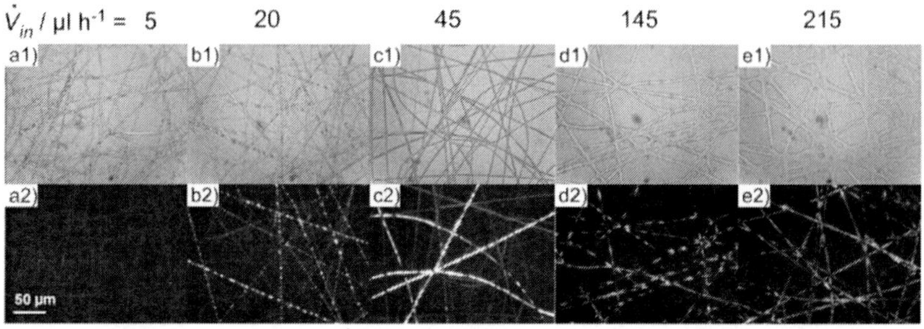

FIGURE 38. Microscopy photographs of characteristic samples when varying the flow rate of the LC while retaining the flow rate of the polymer solution constant. Upper row: sample seen without polarisers. Lower row: same sample position seen between crossed polarisers. Reproduced from [202]. CC BY 2.0.

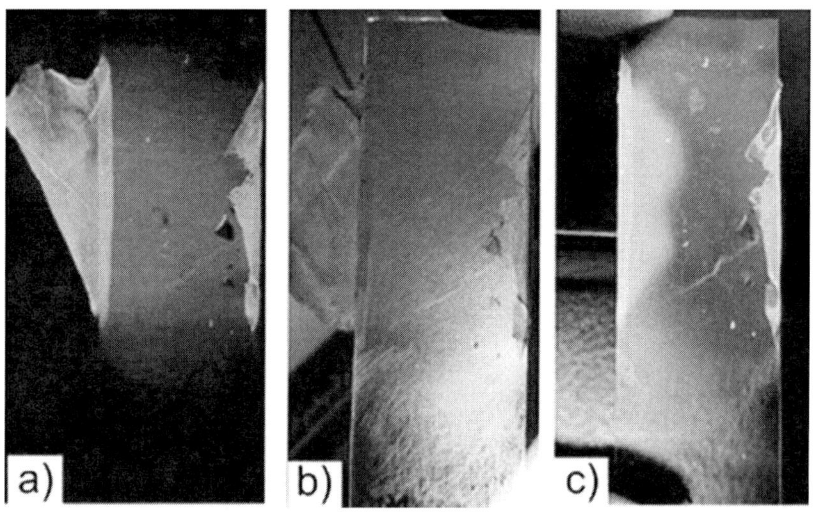

FIGURE 39. Fiber mat with the selective reflection of an encapsulated short-pitch N^* LC visible even macroscopically by the naked eye, in reflection (a), transmission (b), and reflection after heating the LC to the isotropic phase (c). Reproduced from [193] by permission of The Royal Society of Chemistry.

the LC flow rate was high enough, close to that of the sheath solution, the selective reflection from the macroscopic fiber mats could be observed by the naked eye, see figure 39. For thinner fibers, less than ~7 μm in diameter, light scattering from the sheath dominates over the selective reflection, such that the reflected colors are only detectable with a microscope.

Interestingly, the selective reflection color of the fibers with diameter in the micron range varied across the full visible spectrum, although the reflection color of the bulk N* phase was blue. This was explained as a result of quantization of the helix pitch p (not necessarily at its equilibrium value p_0) within the fiber core, due to the cylindrical boundary conditions imposed by the polymer sheath [193]. The effect is identical to that described by Kitzerow and co-workers [192], discussed in section 4.1.3. Because there is a certain random variation in the diameter of the electrospun fibers, we see the full spectrum of colors. Even variations on the scale of tens of nanometers yield visible color changes, as p must always adjust to the actual fiber diameter. The relation between the selective reflection color and the core size and shape was investigated in much greater detail in a follow-up study by Scalia and co-workers [196], in which Focused Ion Beam (FIB) was used to dissect the core-sheath fibers at selected locations, allowing nanoscale imaging of the interior of the fibers.

Kye *et al* [272] further demonstrated that spinning with dual core channels containing two different types of LCs within the same polymer sheath held further potential to increase the multifunctionality of LC-fibers fabricated through the coaxial electrospinning method, see figure 40. The authors produced fibers with adjacent N* and N cores, the former showing visible selective reflection, as well as fibers with adjacent cores filled with the nematic LCs 5CB and E7, respectively, clearing at distinctly different temperatures. The greatest challenge with this multi-core spinning concept was to counteract mixing of the two core fluids and to maintain a wall between the cores that was intact throughout the fiber, even after heating. To this end the fibers were collected on hydrophobic substrates, because Kim and Lagerwall showed in an earlier study that the wetting of water condensed from the atmosphere during spinning promotes fiber spreading and loss of core content [273].

This allowed Kye *et al* to maintain core integrity, see figure 40. Moreover, the authors noted that, if a surfactant had been added to the PVP sheath solution (to reduce surface tension), then even in fibers with intact wall between the cores the two LCs appeared to be contaminated. The authors proposed that this was due to diffusion of surfactant and condensed water from the sheath to the core solutions during spinning. This assumption was corroborated when single-core fibers without surfactant in the sheath solution were spun with 5CB and E7, and on heating clearing points were close to the bulk values of the pure LCs.

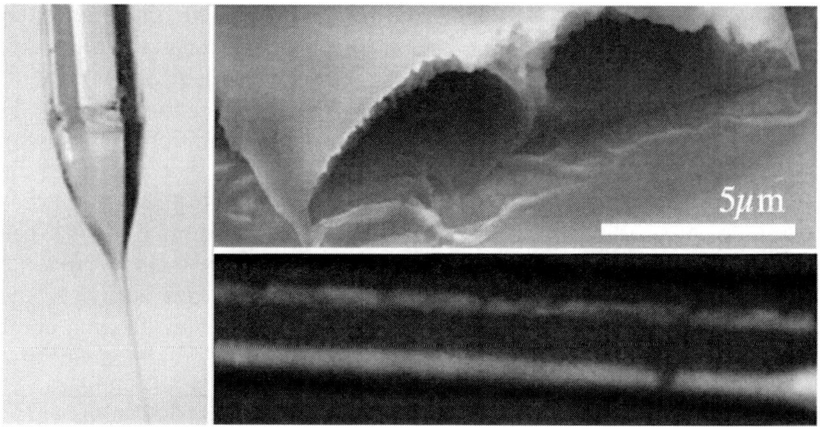

FIGURE 40. (Left:) photo of Taylor cone during spinning of fibers with N* LC in one core and a black dye-doped nematic in the other. (Top right:) Scanning Electron Microscopy image of a fractured dual-core fiber (coated with platinum for imaging purposes) deposited on a hydrophobic substrate, after the LC has been evaporated. (Bottom right:) reflection polarizing microscopy image of a single dual core-fiber, with N* core along the top and N core along the bottom. Reproduced from [272] by permission of The Royal Society of Chemistry.

Reyes *et al* [190] showed that coaxially electrospun fibers with PVP and 5CB could be used as effective optically responsive platforms for gas sensing, as will be discussed in detail in section 4.4.1. The authors investigated aligned and randomly-aligned mats of fibers with morphology varying from uniform cylindrical with continuous LC-filled core to strongly beaded character, where the LC is mainly in the beads. Beading was promoted either by spinning at high relative humidity or by increasing the LC flow rate in relation to that of the polymer sheath. The LC inside partially beaded fibers with continuous LC core show interesting textures, see figure 41. Defect lines running perpendicular to the fiber axis are seen repeatedly along the fibers, in particular at the widest diameter points. The investigation of these defects is still on-going. Considering that we here have essentially cylindrical fibers with modulated diameter in the μm range, but inducing planar rather than homeotropic LC alignment, it is interesting to compare with the related situation for homeotropic boundary conditions studied by Matthias *et al* [180] (see section 4.1.1).

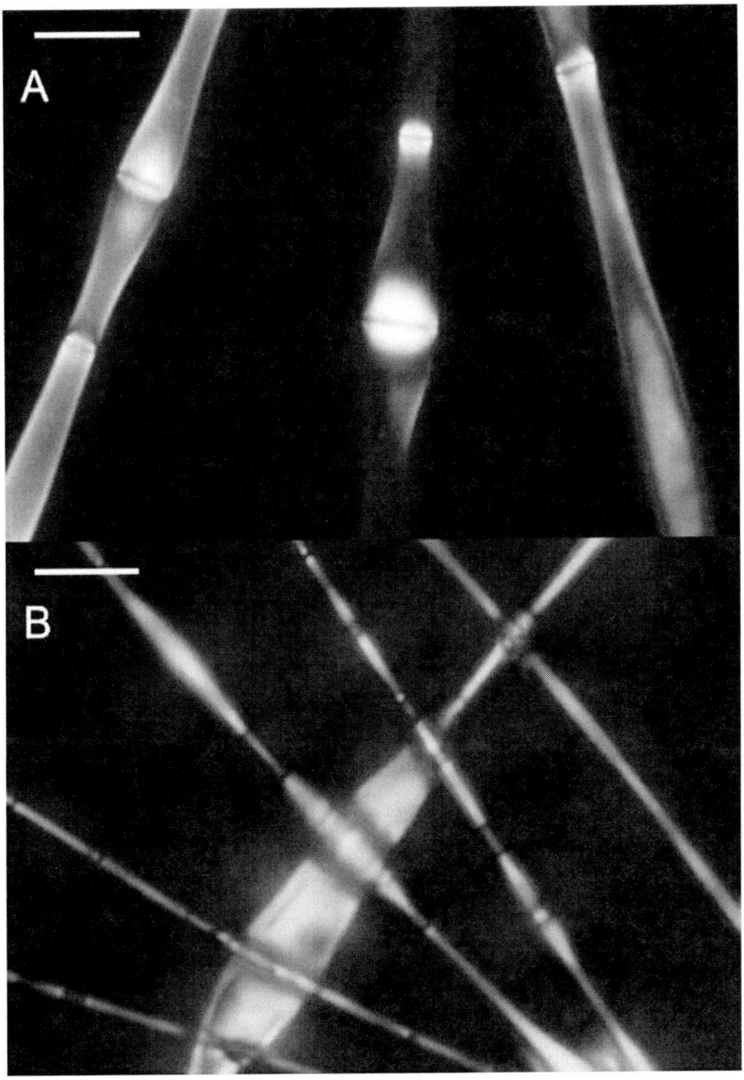

FIGURE 41. Thick (A) and thin (B) coaxially electrospun fibers with 5CB core and PVP sheath can show a continuous LC filling but partially beaded character if spun under the appropriate conditions. In this case defects, running roughly perpendicular to the fiber axis, appear, in particular in every bead. The origin of the defect structure is under investigation. These samples were electrospun in 25% humidity, 23°C, with 12.5% PVP in ethanol sheath solution flowing at 0.9 ml h^{-1}, and the 5CB at 0.5 ml h^{-1}. Scale bar 20 μm.

When co-electrospinning LCs that are crystalline at room temperature, other factors such as miscibility with the sheath's solvent must be considered. Although finding a solvent for a non-room temperature LC may be trivial, its compatibility with the solvent of the sheath fluid to be coaxially electrospun alongside it, may prevent fiber formation. If the core and sheath solvents are miscible, but the sheath solvent is a non-solvent for the LC, a rapid coagulation of the LC results, analogous to the coagulation of polymer fibers during wet spinning. Examples are shown in figure 42.

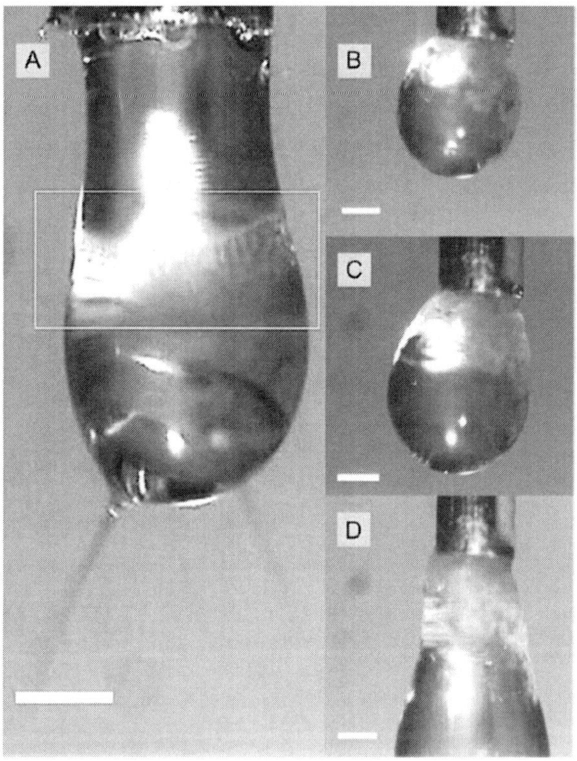

FIGURE 42. Coaxial electrospinning of LC materials that are crystalline at room temperature are challenging to spin, as they must be either heated or dissolved prior to spinning. Here are examples of chloroform-dissolved LC falling out of solution in the Taylor cone when the core fluid comes in contact with ethanolic PVP sheath solution. The LC is the nematic single compound EBBA in (A) and the N* single compound cholesteryl nananoate in (B)–(D). A wrinkled 'skin' may form (yellow highlight) in the Taylor cone after voltage is applied. In these situations (the humidity was above 25%) it was either not possible to obtain fibers containing LC, or no fibers were formed. Scale bars: 0.5 mm.

An alternative way of introducing LCs into electrospun fibers was devised by Buyuktanir *et al* [234]. A chloroform-acetone solution of 5CB as LC and polylactide (PLA) was electrospun to obtain birefringent fibers (diameter ~0.5–2 μm). The PLA concentration was kept constant while 5CB to PLA ratio (7–86%) was varied for optimum spinnability. During spinning, the LC phase separates from the polymer, giving a core–sheath morphology to the fiber. However, phase separation was not always complete and the sheath was in some cases a mixture of PLA and 5CB. The director was generally aligned along the fiber but there was strong beading. Recently, the same group made a more systematic investigation of the balance between beaded and cylindrical morphology in electrospun LC core fibers produced via the phase separation route [274]. By controlling humidity and optimizing flow rate and solution composition, the authors succeeded in reproducibly spinning LC-filled fibers with morphologies varying from uniform cylindrical to beaded.

The simplicity of the phase separation route is attractive, as a standard spinneret can be used. The recent success is therefore important for the further development of LC-functionalized fibers. At the same time, coaxial electrospinning offers more detailed and dynamic control, and it is more versatile in the LCs and polymer solutions that can be combined. The two methods can be considered largely complimentary, each with its benefits and limitations.

The West and Jakli team also demonstrated air brushing as a facile (non-spinning) way of producing LC core–polymer sheath fibers with average diameter of 5–10 μm [275]. Two types of polymers, PLA and PVP, were dissolved together with E7 nematic LC, in a chloroform/acetone solvent mixture for PLA or in isopropanol for PVP. The obtained fibers were investigated for optical, electro-optical, and morphological characteristics by POM and SEM, respectively. The core-sheath morphology was confirmed as a pipe structure in SEM, demonstrating that LC was phase separated and forms a core.

The fibers obtained were uniformly birefringent with planar anchoring conditions of E7 at the polymer sheath and average LC alignment along the fiber axis. The authors also applied an electric field perpendicular to the fiber axis with the aim to control the optic axis direction; since E7 has positive dielectric anisotropy, ***n*** should reorient away from the fiber axis, reducing the effective birefringence, ultimately yielding a dark state between crossed polarizers. However, the latter condition could not be reached with the available fields, a result the authors explained with the voltage drop over the polymer sheath. Fibers were also produced with short-pitch cholesteric LC core, and they showed visible selective reflection.

4.3.2. High molar mass LCs and LCPs in fibers.

The first approach to incorporate liquid crystallinity into fibers that might come to mind is to electrospin liquid crystal polymers (LCPs). Although there are only a few examples of lyotropic LCPs that have been known to form fibers using electrospinning, several types of thermotropic nematic, smectic, and cholesteric LCPs have been electrospun. For instance, Canejo *et al* [231] showed that electrospinning a lyotropic LC solution of 60 wt.% acetoxypropylcellulose (APC) with dimethylacetamide (DMAc) as solvent successfully formed chiral birefringent fibers with an average diameter of 3 μm and a helically twisted morphology. Fibers with different diameters had different helical pitch and the authors observed spontaneous helical twisting with both left- and right-handed twist. About 75% of the fibers were right-handed. Interestingly, according to the theory developed to support the experiments, the presence of more right-handed fibers did not come from the natural twisting sense of APC.

Rather, it was attributed to the spontaneous helical winding of the polymer chain in the presence of the electrical field, with the twisting sense determined by a chiral coupling constant depending on the enantiomer of the polymer chain, when the solution concentration is sufficiently high. This finding further establishes the importance of tuning electrospinning parameters; the properties of the solution do not only influence the choice of processing parameters, the processing parameters can also influence the development of the fibers at sub-molecular levels (refer to section 4.2.2).

Tsuboi *et al* [276] showed another example of fibers electrospun from a lyotropic LCP; poly(γ-benzyl- L-glutamate) (PBLG), a synthetic polypeptide that forms a rod-like α-helical conformation in organic solvents. Nanofibers were electrospun from an isotropic and biphasic PBLG /dichloromethane-pyridine solution under various spinning conditions. For fibers averaging sizes from 0.65 to 1.67 μm, it was counter- intuitively observed that Δn increased with decreasing fiber diameter, which was attributed to improved uniaxial molecular orientation of PBLG from undulated molecular orientation. A further interesting finding was that the fiber orientation and internal structure could be controlled by changing the spinning solution from isotropic to biphasic solution, which was confirmed by wide-angle x-ray diffraction (WAXD) analysis, and polarized micro-Raman spectroscopy. The nanofibers spun from an isotropic solution generally showed lower orientational order than nanofibers spun from the biphasic solutions, in which the the hexagonal lattice of αhelices was uniaxially aligned along the fibers.

Commercially available LCPs, such as Vectra and Kevlar, have also been used to obtain birefringent nanofibers via electrospinning [277–279]. In preliminary work shown by Reneker's group, an isotropic solution of poly(*p*-phenylene terephthalamide) (PPTA), commercially known as Kevlar 49®, in concentrated sulfuric acid, was electrospun into a water-bath as a collection target [277]. Using transmission and scanning electron microscopy, TEM and SEM, respectively, they found the average fiber sizes ranged from 40 nm to a few hundreds of nanometers, much smaller than the fibers obtained from conventionally spun Kevlar®. Although this method produced thermally stable ordered fibers with smaller diameter, the surface roughness of the fibers was of the same order as of commercially obtained fibers.

For obtaining Vectra fibers, a melt electrospinning setup equipped with a CO_2 laser was used by Ogata *et al* [278] to obtain micron size fibers of polyalirate commercially known as Vectra®. The microfibers analyzed by x-ray diffraction were found to have ***n*** oriented along the fibers. Interestingly, the diameter of the fibers decreased upon increasing the laser intensity. Due to limitation of the set-up fibers with diameters smaller than 1 μm could not be formed.

Araujo and co-workers [279] continued the work on electrospinning Vectra® and employed a host-guest approach using blends of Vectra with polyethylene oxide (PEO) to electrospin Vectra-PEO nanofibers. Rather than employing melt spinning and the required strong heating, a two-solvent system of chloroform and pentafluorophenol (70:30 ratio) was used to dissolve the commercial Vectra sample. This solution was mixed with a chloroform solution containing PEO, and then the authors electrospun fibers with a Vectra:PEO ratio of 8.19:1, by weight. After spinning, a heat treatment was used to remove the PEO portion of the fibers, and to impose mechanical stability and ordering onto the amorphously spun Vectra fibers.

The heat treatment also improved the fibers' thermal stability and increased their tensile strength by more than 300% and elongation at break points by around 150%. The tensile modulus of the fibers, however, was found to be similar to that of conventional Vectra microfibers. It was hypothesized that the increase in tensile strength is due to PEO acting as plasticizer during Vectra fiber spinning, and contributed to the ductile behavior of the untreated mats. Annealing after the complete removal of the PEO enhanced the structural order and resulted in stiffer fibers, hence, improved mechanical properties.

Summarizing the results from electrospinning of Vectra and Kevlar fibers, the overall conclusion is that the functionality of the produced fibers is limited, disappointing in the sense that the fibers do not demonstrate superior properties compared to current commercial Kevlar or Vectra fibers, produced following melt spinning or extrusion. These limitations have ultimately dissuaded many researchers from continuing studies to produce nanoscale fibers made from Kevlar and Vectra.

Rather than using these famous high-performance LCPs, which are difficult to dissolve in compatible solvents, Wu and co-workers chose to electrospin blends of polysiloxane based side-chain cholesteric LCPs with triethylamine and PEO, respectively, dissolved in chloroform [280]. The PEO was added to increase the spinnability of LCPs. In order to obtain uniform fibers another LC was added, triethylamine (Et_3N), and different parameters were optimized, such as the PEO concentration, along with the processing parameters like applied voltage and collector distance. Although the addition of a small amount of LC successfully increased the spinnability of the LCP solution, initially only beaded fibers were obtained.

The authors also studied the effect of conductivity of the spinning solution, which affects the net charge density of the jet and hence the quality of fibers produced. As observed by POM the addition of Et_3N in the presence, and absence, of trace amounts of water had a great effect on morphology and optical properties of the obtained fibers. It was ultimately found that at a low added volume of Et_3N (0.05 ml), uniform fibers formed, while beaded fibers were seen after the addition of more Et_3N (0.1 ml). In the presence of trace water Et_3N (0.05 ml) is protonated and converts to Et_3NH^+, which acts an electrolyte. This increases the birefringence and the morphology of the fibers improved as high enough electrolyte concentration increases the conductivity and hence the spinnability. However, the fibers, with an average size of 1 μm, were brittle and hence not suitable for mechanical characterization. When anhydrous Et_3N was used, without trace water present, the LCP solution could not be electrospun.

WXRD analysis showed that while the polymer chains within the fibers were aligned through the electrospinning process, the degree of orientational order was lower compared to fibers prepared by conventional drawing processes. A significant amount of annealing was necessary to increase the level of crystallinity and mechanical stability in the thin electrospun LCP fibers; the Young's modulus of a single fiber increased from 2.00 to 3.93 GPa. In addition to finding optimal processing parameters needed to electrospin LCPs, varying degrees of after treatment processes (annealing, drying...etc) may thus be required to reach satisfactory performance.

Finally, attempts were also made to electrospin main-chain smectic [281] and nematic [230] LCPs. Nakashima *et al* [281] studied the effects that changing fiber diameter had on the internal structure and molecular orientation of smectic LCP, BB-5(3-me) spun from HFIP solvent, or 1,1,1,1,3,3,3- hexafluro-2-propanol. With concentrations varying from 3–25wt.%, fibers with diameters ranging from 0.13 to 4.71 μm were observed.

Under optimal spinning conditions, fibers with diameters ranging from 0.11 to 0.69 μm were found. Interestingly, this polymer was known to form an anticlinic tilted smectic phase, SmC_a, in melt spun fibers, but the pristine electrospun fibers showed only SmA order. However, this SmA phase was metastable, and after annealing it was found to transform into the stable SmC_a phase also in the electrospun fibers.

Alternatively, while electrospinning main-chain nematic LCPs with photocrosslinkable moieties, Krause *et al* [230] were able to obtain aligned LCE microfibers after exposing them with UV light for *in situ* crosslinking. A 5 wt.% solution of LCP in chloroform yielded beaded fibers with average diameter sizes of 450 nm, but upon increasing the concentration to 7 wt.%, uniform fibers were observed with average diameter 1.5 μm. Although the fibers were found to be insoluble in chloroform, proving that the fiber mat was successfully crosslinked, and had uniform director alignment along the fiber axis as confirmed by polarizing microscopy, no tests of actuation were reported. The authors tested actuation only on films prepared in parallel from the same LCE material, without commenting on the lack of actuation data for the fibers. One may speculate that this may have been due to the fibers being fixed to the substrate on which they were collected, thus precluding actuation. It remains an important challenge to electrospin free-hanging LCE fibers and study their actuation behavior.

4.4. Applications of liquid crystal-functionalized fibers

If liquid crystals can be integrated within textile-grade fibers, their extraordinary responsiveness could be put to use in realizing a new generation of *wearable technology* devices. 'Wearable tech' has been used as a buzz-word throughout many fields and industries looking to capitalize on society's growing interests in seamlessly integrating sensors, tracking devices and other electronics into apparel and accessories. While wearable tech has been successful in creating multifunctional Bluetooth integrated smart-accessories, like smartwatches and fitness bands [282–284], and showing that such accessories and clothing can be fashionable [285], the field has in many respects disappointed. Its products do not appear to be as functionally necessary or as irreplaceable as many hoped they could be [286]. A smart watch can be replaced by a less expensive smart phone containing the same level of functionality, and haute couture fiber-optic fashion, while stylish, is not justifiable in neither cost nor function to many consumers.

In fact, consumer conscious wearable tech creates more wearable electronic *accessories* right now, than it does true *clothing* based technology, involving responsive fibrous textiles which actually seek to minimize external electrical sources and dependence on other mobile devices in order to function. Clearly, this is where liquid crystals embedded within individual fibers, introduced already during the actual fiber fabrication, can potentially fill an important niche in creating the next generation of wearable technology. Wrist watches have been with us for two centuries, so in many respects smart watches and the likes can be considered evolution rather than revolution. While important challenges remain to be addressed for bringing LC-functionalized fibers to applications—in particular in terms of transitioning from the research polymers used in today's lab experiments to true textile-grade polymers—they offer some very interesting opportunities, with potentially revolutionary potential. They introduce an entirely different form factor, perfectly adapted for textile garments, and they offer autonomous, non-electronic, functionality. This opens for clothing-integrated devices that are fundamentally different from today's dominating technology, fully independent of batteries and power supplies. We here cover two illustrative examples of the possibilities: gas sensors made from LC-core fibers and actuating fibers made from LCEs.

4.4.1. Gas sensing with liquid crystal-filled fibers.

Based on the observation that light scattering by an electrospun mat of PVP sheath–5CB core fibers is visibly reduced (change easily seen by naked eye) upon exposure to toluene or other organic solvent vapor, Kim *et al* proposed that LC-core fibers could be used as wearable sensors/detectors of hazardous gases [287]. A recent detailed follow-up study by Reyes *et al* [190] showed that the morphology of the fibers (beaded or uniformly cylindrical) is very important for the behavior. Interestingly, it was found that beaded fiber mats had visible scattering responses without the need for polarizers, while the response of non-beaded fiber mats could only be visualized with polarizers. Although the latter response thus requires additional components (polarizers) it is very interesting, because it is extremely fast (response and relaxation time below seconds) and highly sensitive [190]. While the alignment effects of the fiber sheath encapsulating the 5CB are still under investigation, as well as the limits to the fiber mats' functionality, the concept of integrating LCs into fibers for furthering ideas in wearable technology is steadily advancing. In terms of developing LC-fiber based gas sensors as a viable concept, the two key challenges are to ensure specificity (response only to the analyte gas) and enhance sensitivity. To reach these critical goals, a good starting point is to survey the work done on gas sensing with liquid crystals in flat non-fibrous samples.

Liquid crystals, on their own, have been used for gas sensing in a research context since the 1990s [288–291]. If issued patents on the usage of LCs for gas sensing are taken into account, then the idea was present decades before, since the 1970s [292–294]. Many of the early works regarding detection of aromatic and halogenated organic vapors at room temperature exploited the use of short-pitch cholesterics because of their ability to change p_0, and therefore their color, depending on how much organic vapor came in contact with the samples. The early pioneering works from Dickert and his group [288, 289, 295] demonstrated quantitative correlation between analyte concentration, determined *in situ* using quartz crystal microbalance (QCM), and p_0, calculated from the optical reflection band as determined by UV-VIS spectroscopy. More recent work from the TU Eindhoven group demonstrated that the principle can be used even with fluctuations in atmospheric humidity and temperature [296, 297]. These and other reports [291, 298, 299] have ensured that the idea that LCs could be used for optically detecting harmful volatile organic compounds (VOCs) has grown in popularity throughout the years.

The critical problem of specificity was, however, largely left unaddressed in the early work. Likewise, it remained an open question how LC-based gas sensors can be tailored to respond to different gaseous analytes of interest. During the last decade, the interest in LC based gas sensing applications has been revitalized, largely thanks to a new approach that solves the specificity problem, presented by the Abbott group at the University of Wisconsin-Madison. In their work, they characterized organophosphate and organic gas vapor responses in thin films of 5CB with one side exposed to the air, the other supported on a substrate decorated by perchlorate salt. Due to complexation of the salt by 5CB as well as by the analyte gas, the latter forming a much stronger complex than the former, a sensitive and specific response can be ensured. The substrates vary from polystyrene [300] to photolithography masked PDMS and polyurethane micro-wells [301–303]. The team successfully quantified the chemo-optical transitions induced upon gas exposure from as low as ppb level concentrations under various humidity settings [300, 302–305]. Thus, throughout the years Abbott's group has influenced other groups to extend, and they themselves have extended, their initial micro-level LC-VOC sensing concept. Apart from gas sensing they study sensing of biomolecular phenomena such as protein binding energies [305, 306] and enzymatic activities using nematic LCs [307].

While the Abbott group VOC sensing approach is based on an analyte-triggered change in boundary conditions, resulting in a change of LC alignment through the bulk, other groups have instead focused on sensitizing the bulk LC to specific analytes. This may be disadvantageous for the sensitivity and speed of the sensor, but it obviously gives more freedom to how the boundaries are designed. Moreover, it is an interesting alternative that serves to expand the range of VOCs that can be detected by LCs. Dirk Broer's group demonstrated high selectivity using functionalized chiral dopants added to the standard nematic mixture E7. Depending on the choice of chiral dopant they could detect carbon dioxide or oxygen [308]. Later on the same concept was used to sense increases in temperature [296] and humidity [296, 297]. The Broer group has also experimented with different form factors of LC containment in creating thin LCP responsive films for sensing [296].

If the above strategies, or alternative ones, to ensure specificity can be transferred to LCs in fibers, the new form factor should be most attractive, in particular in implementing LC sensors into non-laboratory environments. While Abbott *et al* demonstrated VOC sensitivity of perchlorate-sensitized 5CB films down to the ppb level, microscope and video equipment are still necessary to detect the anchoring rearrangement upon gas exposure. Since LC-filled fibers can yield a response that is visible to the naked eye, they may offer a means to take the concept out of the lab. Moreover, with a textile polymer sheath, a mat of appropriately LC-functionalized fibers can easily be incorporated in clothing, for instance uniforms of persons who risk exposure to hazardous gases in their working environment. Because the sensing requires no electrical power, the garment does not need to carry any power supply, battery or electrical wiring to drive the sensor. And with the large surface-to-volume ratio that thin fibers provide, we may expect both a sensitive and rapid response if all parameters are optimized.

Table 1 gives an overview of available and currently developing gas sensing devices, with their respective advantages and disadvantages. This can give a more succinct perspective of the progress and highlight the possibilities LCs have for integration into wearable sensing technology. The primary aspects of gas selectivity, robust form factor containment, and gas sensitivity are the main three issues which need to be addressed for creating commercially viable LC-based gas sensors. Fundamentally, as applied LC research has primarily targeted displays and related technologies, requiring electrodes and assuming hard impermeable (usually glass) substrates to encase the LCs, this novel type of LC application poses a challenge quite different from the ordinary.

Table 1. Current state of the art regarding portable/wearable gas sensing technologies, with and without LCs.

Sensing device	Pros	Cons
Traditional gas sensors	+ Large variety commercially available and thorough understanding of	– Each type has flaws:

	functionality [314]	
	+ Metal oxide (MOX) sensors: low fabrication cost, high sensitivity, wide analyte range [315, 316]	– MOX sensors: high energy consumption, unstable in uncontrolled environments [316, 317]
	+ Polymer sensors: low energy consumption, low fabrication cost, very sensitive	– Polymer sensors: issues with selectivity and long-term stability [318]
	+ CNTs: high surface-to-volume-ratio, low cost, portable	– CNT sensors: high costs, reproducibility problems [319]
LC Based Gas Sensors	+ High sensitivity and selectivity demonstrated for relevant VOCs (DMMP [304, 320], nerve agents [299, 302], toluene [190, 298, 300], THF [295], chloroform [289])	– Generic LC sensors have issues with specificity [287, 302]
	+ Various types of LCs may be employed; dopants can ensure specificity	– Form factor challenges: flexible yet robust LC containment allowing exposure to environment
	+ Require no electrical components or power source to function; room temperature operation	– Many current solutions require polarizing microscopy analysis
	+ Small devices possible	

	+ N*-sensors enable immediate detection by human sight (color change)	
LC Gas Sensors in Fiber Form	+ Potential solution to form factor challenge: demonstrated function in non-woven mats [190, 287]	– Reproducibility problems [190]
	+ Response visible at room temperature, with/without additional optics [190]	– Response function needs further investigation [190, 287]
	+ Response and relaxation within seconds [190]	– Sensitivity needs development [190]
	+ Long shelf life possible (minimum two months demonstrated [190])	– Transition to robust textile-grade polymers needed; washability issues
Current non-LC-based Wearable Gas Sensors	+ Conductive liquid/interdigitated electrodes for capacitive/resistive sensing: can be woven into garments; some level of flexibility [321, 322]	– Reproducibility problems; time-consuming production for non-printed sensors [323, 324]
	+ Inkjet printed CNT-based electrodes: fast, low-cost production; integration into garments; good lifetime with coin	– Capacitive/resistive response speed can be slow [321]

	cell batteries due to low power requirements	
		– Read-out of sensing response requires further electronics
		– Sensing limited (so far) to water vapor, sweat, and ammonia [325, 326]
		– Form factor issues: not true textiles but rather PDMS rubber; breathability and reusability issues [326, 327]
		– High cost

4.4.2. Liquid crystal elastomer fiber actuators.

There are a number of LCE materials and fabrication methods that have been explored for making actuators that are capable of either thermal, photo-mechanical, or photo-thermo-mechanical actuation. Undoubtedly, these types of LCE actuators show promise in fields like soft robotics [21, 309], and performance comparable to that of skeletal muscles has been demonstrated [16]. As discussed in section 2.4, a commonly used method for obtaining monodomain LCE networks is the two-step crosslinking approach pioneered by Küpfer and Finkelmann [40]. When the target is to make LCE fibers, this approach is convenient, the fibers being stretched during the second crosslinking stage. However, connected to the fiber production another unique way to form LCE fibers arises, based on mechanically orienting a non-crosslinked but reactive LC polymer by flow while drawing fibers, and then quickly crosslinking the fibers to prevent the orientational loss. While some have tried simply drawing fibers using tweezers [26], and toothpicks [310], others have used more sophisticated methods like melt extrusion [25], and even wet-spinning driven by microfluidic setups [311–313].

Naciri *et al* [26] showed the first examples of LCE fibers obtained through the simple approach of drawing from a melted viscous mixture of a nematic LC terpolymer and crosslinker. The samples were left in the stretched state for 120 hours for crosslinking to complete. The obtained fibers, with average diameter of 300 μm, were found to be birefringent with $\boldsymbol{n}$ along the fiber axis. Thermoelastic measurements to determine the mechanical properties were conducted on the elastomeric fibers in the nematic phase at 50 °C, yielding a Young's modulus of 4.37 MPa. Furthermore, under constant applied stress the length was monitored, and the resulting strain calculated, as the fibers were heated from the nematic into the istoropic phase. The length did not change in the nematic phase, it started to reduce on approaching the clearing transition at T_{NI}, where it sharply decreased, revealing a transition-induced strain up to 35%. The actuation behavior of the LCE fiber was also demonstrated by mounting it inside a heating coil and attaching a 200 mg weight to one end. When the fiber was heated above the N-isotropic transition, the fiber contracted and the weight was lifted.

Ohm *et al* [311] introduced a technique similar to the classic wet-spinning fiber formation process by instead using a microfluidics based mechanism (shown in figure 22) to obtain oriented main-chain LCE fibers from a main-chain photocrosslinkable smectic LCP. A solution of LCP (30 wt.%) and photo-initiator (3wt.%) in dichloromethane (DCM), was injected into a stream of silicone oil within a glass capillary. The co-flowing 'bath' of silicone oil, similar in one of its two functions to the precipitation bath that wet-spun fibers are extruded into prior to being collected, coagulated the fibers. Due to the miscibility of DCM and silicone oil, the solvent is removed from the LCP, which thus enters its LC state (the co-flowing system was kept at 40 °C, within the SmA phase). The result is a continuous LCP filament that is pulled and aligned by the silicone oil stream (its second function), prior to UV exposure for initiating the crosslinking.

The resulting LCE filaments were collected on a winding spool and checked for their optical and actuation behaviors. The fibers obtained had thickness between 20 and 50 μm, which was varied by changing the flow rate of the oil phase. With POM experiments it was also confirmed that the director orientation was along the fiber axis. After heating above 80 °C, to the isotropic phase, the fiber lost its birefringence and a reduction in fiber length was observed. Surprisingly, although the fiber regained its birefringence upon cooling back to 50 °C, the deformation in length was irreversible. The authors proposed that the irreversibility was due to the high activation energy required to form new smectic layers.

Theoretically, this activation energy could be overcome by an external force. To check this, the authors did another actuation experiment in which a single LCE fiber was glued at both ends, with a 'u'-shaped 25 mg weight hanging from the center, the entire setup kept inside a thermo-regulating box, see figure 43. Upon heating the air inside the box above T_{NI}, the weight was lifted as the fiber length reduced from 19.4 mm to 16.3 mm. On cooling, the fiber now regained its original length at a temperature of 50 °C, indicating a reversible actuation. This work showed that the continuous microfluidics based spinning method has the capability of making LCE fibers with reproducible and regular thickness which can act as actuators and lift objects more than one thousand times its own weight. Ultimately, it demonstrates a method for producing versatile and useful soft actuators.

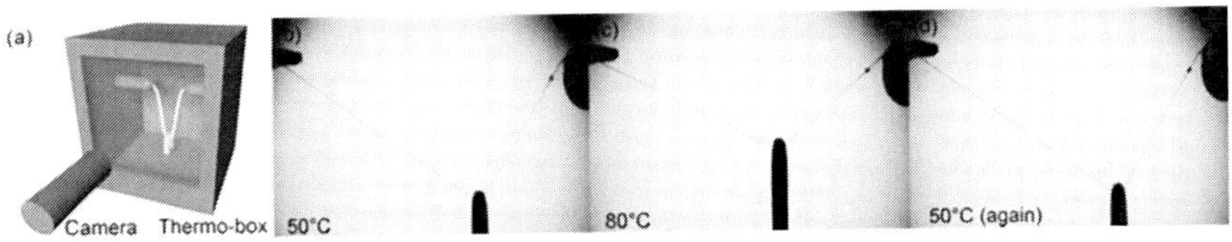

FIGURE 43. Schematic setup (a) and results (b) of a heating experiment performed on a single SmA LCE filament, glued between two static substrates inside a heatable metal box. A small, U-shaped weight was clipped to the sample. The temperature in the box was cycled while observing the sample with a camera. Between 50 (b) and 80 °C (c) the weight was lifted up while the substrates remained stationary. Upon cooling back to 50 °C (d) the sample went back to it's original height, proving the reversibility of the actuation. Reproduced from [311] with permission of The Royal Society of Chemistry

A more detailed mechanical and optical study on LCE fibers with thickness of 30 ± 1 μm, based on the same material, was done by Stannarius *et al* [312]. One of the findings was strain induced birefringence; Δn could either be increased or decreased upon stretching and compressing of the samples in an axial direction. However, the effect was reversible only once the fiber had been heated into the SmA state at 40 °C. The slow dynamics of the room temperature glassy state does not allow the dynamic coupling between strain and orientational order parameter that the varying birefringence signifies.

Mechanical studies indicated that the relaxation of the fibers after stretching at room temperature shows glassy behavior and plastic deformation, as the initial length is not recovered. However, if the fiber is then heated above its glass transition temperature, it reaches its original relaxed length within one minute. When the fiber is further heated to T_{NI}, it shortens and a suspended weight is lifted accordingly. A very small stress of 80 μN is required to stretch the sample during the transition to the smectic phase. This shape memory effect is attributed to the formation of a polydomain structure in the SmA phase and a structural reorganization of smectic layers during cooling, similar to chiral SmC* LCEs [328, 329] and nano-sized shape-changing LCE colloids [147]. However, the authors did not provide any x-ray data to confirm this particular hypothesis.

To the best of our knowledge, there has been only one report, so far, on side-chain nematic LCE fibers [313]. This was fabricated in a microfluidic set-up using a side-chain UV-crosslinkable LCP exhibiting a nematic phase between 40 and 56 °C. The LCP was mixed with 0.4 wt.% chlorobenzene and 5 wt.% photoinitiator and injected through a fused silica tube into co-flowing silicone. The set-up was immersed in a water bath for temperature control. Crosslinking was done at different temperatures within the nematic phase. POM analysis showed that fibers crosslinked deep in the N phase were uniaxially oriented whereas fibers crosslinked at a temperature closer to T_{NI} were not. Depending on the exact combination of precursor LCPs and on the crosslinking time, some fibers did not actuate upon heating above the clearing point, whereas others actuated several 100%. The relevant parameter is the crosslinking density, which must not be too high, as this prevents actuation. Thus, fibers obtained from LCPs with lower numbers of crosslinkable units showed the best actuation upon heating, see figure 44.

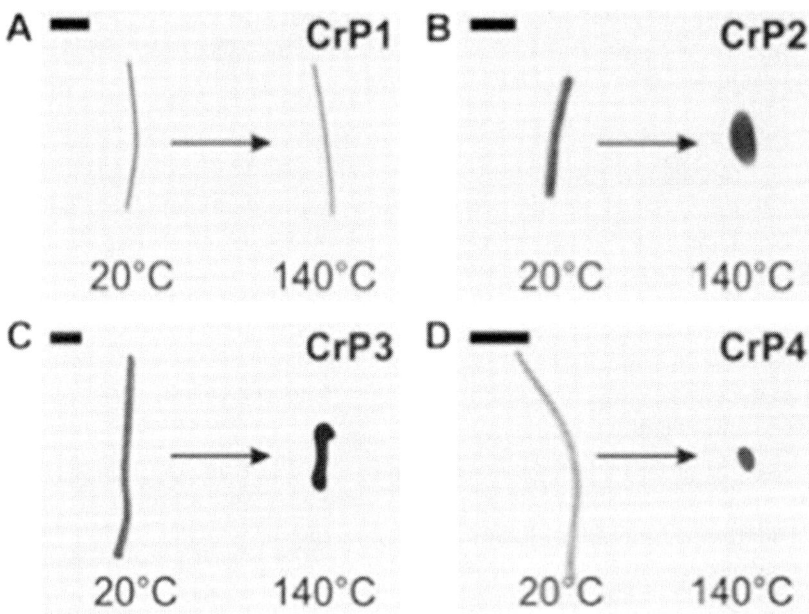

FIGURE 44. Overview of the contraction behavior of nematic side-chain LCE fibers crosslinked for 13 s. (A) Fibers made from polymer CrP1 (45 mol%) do not contract because this prepolymer yields too many crosslinks. ((B)–(D)) Fibers made from polymers CrP2-CrP4 with reduced amounts of crosslinkable units (31 to 9 mol%) contract when heated to $T_{NI} = 140$ °C. Scale bars: 100 μm. Reproduced from [313] John Wiley & Sons. © 2014 WILEY-VCH Verlag GmbH & Co. KGaA, Weinheim.

Conclusively, for actuation, it seems like fiber orientation and a softer network that allows deformation are the key factors. The orientation of the LC phase in the fibers can also be optimized by crosslinking at low temperatures in the nematic phase, for example. Actuation can be tuned by either varying the degree of crosslinking, or varying the UV exposure time when the crosslinker amount is lowered.

5. OUTLOOK

Based on the steadily increasing interest in droplets, shells and fibers of liquid crystals, together with the continuous innovation in production and characterization technologies adapted to these fascinating geometries for LC research and applications, we can expect a stimulating development over the next few years. As described above, we have already seen several examples of unexpected director field configurations arising by strong confinement with curved interfaces, and the new opportunities for characterizing topological defects even with nanoscale resolution, for instance via polymerization of reactive mesogens that aggregate in defects, may yield further intriguing surprises. While much valuable information on spheres and cylinders of LCs can be found in the pioneering studies from the 70s–90s, today's ability to produce monodisperse LC droplets, and even shells, with controlled diameter has opened a colorful physics toolbox that is far from fully explored. Likewise, the introduction of electrospinning allows us to scale down the cylindrical confinement to the μm range and below, still using optical methods to probe the response of the LC. This may well reveal confinement phenomena that do not take place in the capillaries of the early studies, one or two orders of magnitude larger in diameter. We may also expect interesting new phenomena as lyotropic LCs start to be explored in droplets, shells and fibers.

So far the emphasis has been primarily on fundamental physics while the applied research has just started to take off. Some very promising results have been put forward in terms of soft actuators based on microfluidics-produced curved LCEs, unclonable pattern formation from regular arrangements of cholesteric shells, or gas sensing from LC-functionalized fiber mats. The challenges until commercial exploitation can be realistically considered are, however, still considerable. For wearable technology applications of LC core fibers a transition to textile-grade polymers is essential, and for many applications it may be wise to move from electrospinning, convenient on the lab scale, to conventional high-throughput spinning methods like wet spinning. As the interest in soft robotics is surging world-wide, the opportunities for LCE-based actuators are enormous, but so far the soft robotics community knows very little about the promise of LCEs. This is an area where we hope LCs will fill an important new role. Apart from advancements in LCE chemistry, which is still very limited in terms of what is commercially available, such a scenario will require a fruitful cross fertilization between the LC and robotics communities.

It is our hope that this review article can contribute in fostering such boundary crossing, also in the other relevant application areas, like wearable technology, materials science or security research. If this happens, the currently unconventional LC geometries that we have discussed above may well become mainstream technology, with potential to revolutionize many aspects of our lives.

ACKNOWLEDGMENTS

We acknowledge financial support from the European Research Council ERC (project INTERACT, grant code 648763, for CGR, AS, VSRJ and JPFL), from the Fonds National de la Recherche (project DISGONA, grant code 9170337, for MU; project ULISCO, grant code 6992111, for JN) and the University of Luxembourg (project UNIQUE, for YG and JPFL).

FOOTNOTES

[2] We thank Ralf Stannarius for pointing out this rarely recognized impact of local variations in K_{24}.

[3] The *polar* configuration in LC cylinder research is comparable to the *axial* configuration for LC droplets (section 3.3.1).

REFERENCES

1. Lagerwall S T 2013 On some important chapters in the history of liquid crystals Liq. Cryst. **40** 1698–729
2. Sluckin T J, Dunmur D A and Stegemeyer H 2004 Crystals that Flow: Classic papers from the History of Liquid Crystals (London: Taylor and Francis)
3. Lagerwall J P F 2010 Three facets of modern liquid crystal science Habilitation Thesis Martin-Luther-Universität Halle-Wittenberg
4. Męsowicz M 1946 The three coefficients of viscosity of anisotropic liquids Nature **158** 27
5. de Gennes P-G and Prost J 1993 The Physics of Liquid Crystals (Oxford: Clarendon)
6. Rapini A and Papoular M 1969 Surface anchoring of nematic liquid crystals J. Phys. Colloq. **30** 54–66

7. Stark H 1999 Director field configurations around a spherical particle in a nematic liquid crystal Eur. Phys. J. B **10** 311–21
8. Frank F C 1958 On the theory of liquid crystals Discuss. Faraday Soc. **25** 19–28
9. Chistyakov I 1975 Ordering and structure of liquid crystals Advances in Liquid Crystals(1) ed G Brown (London: Academic) pp 143–68
10. Lagerwall J P F and Giesselmann F 2006 Current topics in smectic liquid crystal research ChemPhysChem **7** 20–45
11. Lagerwall J P F, Schütz C, Salajkova M, Noh J-H, Park J H, Scalia G and Bergström L 2014 Cellulose nanocrystal-based materials: from liquid crystal self-assembly and glass formation to multifunctional thin films NPG Asia Mater. **6** e80
12. Lagerwall S T 1999 Ferroelectric and antiferroelectric liquid crystals (Weinheim: Wiley)
13. Kitzerow H and Bahr C 2000 Chirality in Liquid Crystals (Partially Ordered Systems) (Berlin: Springer)
14. Priestly E B 1974 Introduction to the optical properties of cholesteric and chiral nematic liquid crystals An Introduction to the Science and Technology of Liquid Crystals—III(35) (Princeton, NJ: RCA) pp 584–99
15. De Gennes P G, Hébert M and Kant R 1997 Artificial muscles based on nematic gels Macromolecular Symp.
16. Thomsen D, Keller P, Naciri J, Pink R, Jeon H, Shenoy D and Ratna B 2001 Liquid crystal elastomers with mechanical properties of a muscle Macromolecules **34** 5868–75
17. Warner M and Terentjev E M 2007 Liquid Crystal Elastomers (Oxford: Oxford University Press)
18. Li M H and Keller P 2006 Artificial muscles based on liquid crystal elastomers Phil. Trans. A: Math. Phys. Eng. Sci. **364** 2763–77
19. Corbett D and Warner M 2009 Changing liquid crystal elastomer ordering with light—a route to opto-mechanically responsive materials Liq. Cryst. **36** 1263–80
20. Yang H, Ye G, Wang X and Keller P 2011 Micron-sized liquid crystalline elastomer actuators Soft Matter **7** 815–23
21. Ohm C, Brehmer M and Zentel R 2010 Liquid crystalline elastomers as actuators and sensors Adv. Mater. **22** 3366–87
22. Fleischmann E K and Zentel R 2013 Liquid-crystalline ordering as a concept in materials science: from semiconductors to stimuli-responsive devices Angew. Chem. Int. Ed. **52** 8810–27
23. White T J and Broer D J 2015 Programmable and adaptive mechanics with liquid crystal polymer networks and elastomers Nat. Mater. **14** 1087–98
24. Brömmel F, Kramer D and Finkelmann H 2012 Preparation of liquid crystalline elastomers Liquid Crystal Elastomers: Materials and Applications ed W de Jeu (Berlin: Springer)
25. Ahir S V, Tajbakhsh A R and Terentjev E M 2006 Self-assembled shape-memory fibers of triblock liquid-crystal polymers Adv. Funct. Mater. **16** 556–60
26. Naciri J, Srinivasan A, Jeon H, Nikolov N, Keller P and Ratna B R 2003 Nematic elastomer fiber actuator Macromolecules **36** 8499–505
27. Finkelmann H, Kock H-J and Rehage G 1981 Investigations on liquid crystalline polysiloxanes 3. liquid crystalline elastomers—a new type of liquid crystalline material Die Makromolekulare Chem. Rapid Commun. **2** 317–22
28. Sánchez C, Verbakel F, Escuti M J, Bastiaansen C W M and Broer D J 2008 Printing of monolithic polymeric microstructures using reactive mesogens Adv. Mater. **20** 74–8
29. van Oosten C L, Bastiaansen C W and Broer D J 2009 Printed artificial cilia from liquid-crystal network actuators modularly driven by light Nat. Mater. **8** 677–82
30. Ware T H, McConney M E, Wie J J, Tondiglia V P and White T J 2015 Actuating materials. Voxelated liquid crystal elastomers Science **347** 982–4
31. Li M H, Keller P, Yang J and Albouy P A 2004 An artificial muscle with lamellar structure based on a nematic triblock copolymer Adv. Mater.

32. Yang H, Buguin A, Taulemesse J-M, Kaneko K, Mery S, Bergeret A and Keller P 2009 Micron-sized main-chain liquid crystalline elastomer actuators with ultralarge amplitude contractions J. Am. Chem. Soc. **131** 15000–4
33. Schuhladen S, Preller F, Rix R, Petsch S, Zentel R and Zappe H 2014 Iris-like tunable aperture employing liquid-crystal elastomers Adv. Mater. **26** 7247–51
34. Ringsdorf H and Zentel R 1982 Liquid crystalline side chain polymers and their behaviour in the electric field Die Makromolekulare Chem. **183** 1245–56
35. Fleischmann E-K, Liang H-L, Kapernaum N, Giesselmann F, Lagerwall J P F and Zentel R 2012 One-piece micropumps from liquid crystalline core-shell particles Nat. Commun. **3** 1178
36. Ohm C, Kapernaum N, Nonnenmacher D, Giesselmann F, Serra C and Zentel R 2011 Microfluidic synthesis of highly shape-anisotropic particles from liquid crystalline elastomers with defined director field configurations J. Am. Chem. Soc. **133** 5305–11
37. Ohm C, Serra C and Zentel R 2009 A continuous flow synthesis of micrometer-sized actuators from liquid crystalline elastomers Adv. Mater. **21** 4859
38. Yu Y, Nakano M and Ikeda T 2003 Directed bending of a polymer film by light—miniaturizing a simple photomechanical system could expand its range of applications Nature **425** 145–5
39. Ikeda T, Nakano M, Yu Y, Tsutsumi O and Kanazawa A 2003 Anisotropic bending and unbending behavior of azobenzene liquid-crystalline gels by light exposure Adv. Mater. **15** 201–5
40. Küpfer J and Finkelmann H 1994 Liquid crystal elastomers: influence of the orientational distribution of the crosslinks on the phase behaviour and reorientation processes Macromol. Chem. Phys. **195**1353–67
41. Küpfer J and Finkelmann H 1991 Nematic liquid single crystal elastomers Die Makromolekulare Chem. **12** 717–26
42. De Gennes P G 1969 Possibilites offertes par la reticulation de polymeres en presence d'un cristal liquide Phys. Lett. A **28** 725–6
43. Fridrikh S V and Terentjev E M 1999 Polydomain-monodomain transition in nematic elastomers Phys. Rev. E **60** 1847–57
44. Yakacki C M, Saed M, Nair D P, Gong T, Reed S M and Bowman C N 2015 Tailorable and programmable liquid-crystalline elastomers using a two-stage thiol-acrylate reaction RSC Adv. **5**18997–9001
45. Iamsaard S, Aßhoff S J, Matt B, Kudernac T, Cornelissen J J, Fletcher S P and Katsonis N 2014 Conversion of light into macroscopic helical motion Nat. Chem. **6** 229–35
46. Yamada M, Kondo M, Mamiya J I, Yu Y L, Kinoshita M, Barrett C and Ikeda T 2008 Photomobile polymer materials: towards light-driven plastic motors Angew. Chem. Int. Ed. **47** 4986–8
47. Gravesen P, Branebjerg J and Søndergård Jensen O 1993 Microfluidics-a review J. Micromech. Microeng. **3** 168
48. Manz A, Jed Harrison D, Verpoorte E M J, Fettinger J C, Paulus A, Lüdi H and Michael Widmer H 1992 Planar chips technology for miniaturization and integration of separation techniques into monitoring systems: capillary electrophoresis on a chip J. Chromatogr. A **593** 253–8
49. Shoji S and Esashi M 1994 Microflow devices and systems J. Micromech. Microeng. **4** 157
50. Xia Y N and Whitesides G M 1998 Soft lithography Ann. Rev. Mater. Sci. **28** 153–84
51. Anderson J R, Chiu D T, Wu H, Schueller O J and Whitesides G M 2000 Fabrication of microfluidic systems in poly (dimethylsiloxane) Electrophoresis **21** 27–40
52. Whitesides G M 2006 The origins and the future of microfluidics Nature **442** 368–73
53. Darhuber A A and Troian S M 2005 Principles of microfluidic actuation by modulation of surface stresses Annu. Rev. Fluid Mech. **37** 425–55
54. Tabeling P 2010 Introduction to microfluidics (Oxford: Oxford University Press)
55. Chakraborty S 2010 Microfluidics and microfabrication (Berlin: Springer)
56. Sackmann E K, Fulton A L and Beebe D J 2014 The present and future role of microfluidics in biomedical research Nature **507** 181–9

57. Squires T M and Quake S R 2005 Microfluidics: Fluid physics at the nanoliter scale Rev. Mod. Phys.**77** 977–1026
58. Stroock A D, Dertinger S K W, Ajdari A, Mezić I, Stone H A and Whitesides G M 2002 Chaotic mixer for microchannels Science **295** 647–51
59. Stone H A, Stroock A D and Ajdari A 2004 Engineering flows in small devices: microfluidics toward a lab-on-a-chip Annu. Rev. Fluid Mech. **36** 381–411
60. Beebe D J, Mensing G A and Walker G M 2002 Physics and applications of microfluidics in biology Ann. Rev. Biomed. Eng. **4** 261–86
61. Elvira K S et al 2013 The past, present and potential for microfluidic reactor technology in chemical synthesis Nat. Chem. **5** 905–15
62. Hood L, Heath J R, Phelps M E and Lin B 2004 Systems biology and new technologies enable predictive and preventative medicine Science **306** 640–3
63. Chin C D et al 2011 Microfluidics-based diagnostics of infectious diseases in the developing world Nat. Med. **17** 1015–9
64. Ganan-Calvo A M and Gordillo J M 2001 Perfectly monodisperse microbubbling by capillary flow focusing Phys. Rev. Lett. **87** 4
65. Anna S L, Bontoux N and Stone H A 2003 Formation of dispersions using 'flow focusing' in microchannels Appl. Phys. Lett. **82** 364–6
66. Garstecki P, Gitlin I, Willow DiLuzio, Whitesides G M, Kumacheva E and Stone H A 2004 Formation of monodisperse bubbles in a microfluidic flow-focusing device Appl. Phys. Lett. **85** 2649–51
67. Baroud C N, de Saint Vincent M R and Delville J-P 2007 An optical toolbox for total control of droplet microfluidics Lab Chip **7** 1029–33
68. Utada A, Lorenceau E, Link D R, Kaplan P D, Stone H A and Weitz D A 2005 Monodisperse double emulsions generated from a microcapillary device Science **308** 537–41
69. Romanowsky M B, Abate A R, Rotem A, Holtze C and Weitz D A 2012 High throughput production of single core double emulsions in a parallelized microfluidic device Lab Chip **12** 802–7
70. Song H, Chen D L and Ismagilov R F 2006 Reactions in droplets in microfluidic channels Angew. Chem. Int. Ed. **45** 7336–56
71. Franke T A and Wixforth A 2008 Microfluidics for miniaturized laboratories on a chip ChemPhysChem**9** 2140–56
72. Yaroshchuk O and Reznikov Y 2012 Photoalignment of liquid crystals: basics and current trends J. Mater. Chem. **22** 286–300
73. Sengupta A 2014 Topological constraints in a microfluidic platform Liquid Cryst. **41** 290–301
74. Na Y J, Yoon T Y, Park S, Lee B and Lee S D 2010 Electrically programmable nematofluidics with a high level of selectivity in a hierarchically branched architecture ChemPhysChem **11** 101–4
75. Sengupta A, Tkalec U, Ravnik M, Yeomans J M, Bahr C and Herminghaus S 2013 Liquid crystal microfluidics for tunable flow shaping Phys. Rev. Lett. **110** 048303
76. Choi M C, Pfohl T, Wen Z, Li Y, Kim M W, Israelachvili J N and Safinya C R 2004 Ordered patterns of liquid crystal toroidal defects by microchannel confinement Proc. Natl Acad. Sci. USA **101** 17340–4
77. Shojaei-Zadeh S and Anna S L 2006 Role of surface anchoring and geometric confinement on focal conic textures in smectic-a liquid crystals Langmuir **22** 9986–93
78. Kim Y H, Yoon D K, Choi M C, Jeong H S, Kim M W, Lavrentovich O D and Jung H-T 2009 Confined self-assembly of toric focal conic domains (the effects of confined geometry on the feature size of toric focal conic domains) Langmuir **25** 1685–91
79. Tongcher O, Sigel R and Landfester K 2006 Liquid crystal nanoparticles prepared as miniemulsions Langmuir **22** 4504–11
80. Umbanhowar P B, Prasad V and Weitz D A 2000 Monodisperse emulsion generation via drop break off in a coflowing stream Langmuir **16** 347–51

81. Utada A S, Fernandez-Nieves A, Stone H A and Weitz D A 2007 Dripping to jetting transitions in coflowing liquid streams Phys. Rev. Lett. **99** 094502
82. Abate A R, Thiele J and Weitz D A 2011 One-step formation of multiple emulsions in microfluidics Lab Chip **11** 253–8
83. Liang H-L, Zentel R, Rudquist P and Lagerwall J 2012 Towards tunable defect arrangements in smectic liquid crystal shells utilizing the nematic-smectic transition in hybrid-aligned geometries Soft Matter **8** 5443–50
84. Lopez-Leon T and Fernandez-Nieves A 2009 Topological transformations in bipolar shells of nematic liquid crystals Phys. Rev. E **79** 021707
85. Noh J-H, De Sousa K R and Lagerwall J P F 2016 Influence of interface stabilisers and surrounding aqueous phases on nematic liquid crystal shells Soft Matter **12** 367–72
86. Geng Y, Noh J-H, Drevensek-Olenik I, Rupp R, Lenzini G and Lagerwall J P F 2016 High-fidelity spherical cholesteric liquid crystal bragg reflectors generating unclonable patterns for secure authentication Sci. Rep. **6** 26840
87. Lopez-Leon T, Koning V, Devaiah K B S, Vitelli V and Fernandez-Nieves A 2011 Frustrated nematic order in spherical geometries Nat. Phys. **7** 391–4
88. Borg F G 2003 What is osmosis? explanation and understanding of a physical phenomenon (arXiv preprint physics/0305011)
89. Kramer E M and Myers D R 2012 Five popular misconceptions about osmosis Am. J. Phys. **80** 694–9
90. Poulin P and Weitz D A 1998 Inverted and multiple nematic emulsions Phys. Rev. E **57** 626–37
91. Lavrentovich O D 1998 Topological defects in dispersed liquid crystals, or words and worlds around liquid crystal drops Liq. Cryst. **24** 117–25
92. Kléman M 1973 Defect densities in directional media, mainly liquid crystals Phil. Mag. **27** 1057–72
93. Kleman M and Lavrentovich O D 2002 Soft Matter Physics: an Introduction (Berlin: Springer)
94. Stark H 1999 Physics of inhomogeneous nematic liquid crystals: Colloidal dispersions and multiple scattering of light Habilitation Thesis University of Stuttgart Stuttgart, Germany
95. Kleman M and Lavrentovich O D 2006 Topological point defects in nematic liquid crystals Phil. Mag.**86** 4117–37
96. Kleman M and Friedel J 2008 Disclinations, dislocations and continuous defects: a reappraisal Rev. Mod. Phys. **80** 61–115
97. Volovik G E and Lavrentovich O D 1983 Topological dynamics of defects: boojums in nematic drops Zh. Eksp. Teor. Fiz. **85** 1997–2010
98. Meyer R B 1973 On the existence of even indexed disclinations in nematic liquid crystals Phil. Mag.**27** 405–24
99. Oswald P, Pieranski P and Constantin D 2005 Nematic and Cholesteric Liquid Crystals: Concepts and Physical Properties Illustrated by Experiments (Boca Raton, FL: Taylor & Francis)
100. Terentjev E M 1995 Disclination loops, standing alone and around solid particles, in nematic liquid crystals Phys. Rev. E **51** 1330
101. Senyuk B, Liu Q, He S, Kamien R D, Kusner R B, Lubensky T C and Smalyukh I I 2013 Topological colloids Nature **493** 200–5
102. Kamien R D 2002 The geometry of soft materials: a primer Rev. Mod. Phys. **74** 953
103. Fernandez-Nieves A, Link D R, Marquez M and Weitz D A 2007 Topological changes in bipolar nematic droplets under flow Phys. Rev. Lett. **98** 087801
104. Liang H L, Noh J, Zentel R, Rudquist P and Lagerwall J P F 2013 Tuning the defect configurations in nematic and smectic liquid crystalline shells Phil. Trans. A: Math. Phys. Eng. Sci. **371** 20120258
105. Tomar V, Hernandez S I, Abbott N L, Hernandez-Ortiz J P and de Pablo J J 2012 Morphological transitions in liquid crystal nanodroplets Soft Matter **8** 8679–89
106. Erdmann J H, Zumer S and Doane W J 1990 Configuration transition in a nematic liquid crystal confined to a small spherical cavity Phys. Rev. Lett. **64** 1907

107. Lubensky T C, Pettey D, Currier N and Stark H 1998 Topological defects and interactions in nematic emulsions Phys. Rev. E **57** 610
108. Gupta J K, Zimmerman J S, de Pablo J J, Caruso F and Abbott N L 2009 Characterization of adsorbate-induced ordering transitions of liquid crystals within monodisperse droplets Langmuir **25**9016–24
109. Gupta J K, Sivakumar S, Caruso F and Abbott N L 2009 Size-dependent ordering of liquid crystals observed in polymeric capsules with micrometer and smaller diameters Angew. Chem. Int. Ed. **48**1652–5
110. Williams C, Pierański P and Cladis P E 1972 Nonsingular s = +1 screw disclination lines in nematics Phys. Rev. Lett. **29** 90–2
111. Xu F, Kitzerow H-S and Crooker P P 1994 Director configurations of nematic-liquid-crystal droplets: Negative dielectric anisotropy and parallel surface anchoring Phys. Rev. E **49** 3061
112. Friedel M G 1922 The mesomorphic states of matter Ann. Phys. **18** 273–474
113. Lopez-Leon T and Fernandez-Nieves A 2011 Drops and shells of liquid crystal Colloid Polym. Sci. **289**345–59
114. Bezic J and Zumer S 1992 Structures of the cholesteric liquid crystal droplets with parallel surface anchoring Liq. Cryst. **11** 593–619
115. Sec D, Porenta T, Ravnik M and Zumer S 2012 Geometrical frustration of chiral ordering in cholesteric droplets Soft Matter **8** 11982–8
116. Xu F and Crooker P P 1997 Chiral nematic droplets with parallel surface anchoring Phys. Rev. E **56**6853–60
117. Kurik M V and Lavrentovich O D 1982 Negative-positive monopole transitions in cholesteric liquid-crystals JETP Lett. **35** 444–7
118. Darmon A, Benzaquen M, Dauchot O and Lopez-Leon T 2015 Waltzing route towards double-helix formation in cholesteric shells (arXiv:1512.06039)
119. Candau S, Roy P L and Debeauvais F 1973 Magnetic field effects in nematic and cholesteric droplets suspended in a isotropic liquid Mol. Cryst. Liq. Cryst. **23** 283–97
120. Bouligand Y and Livolant F 1984 The organization of cholesteric spherulites J. Phys. **45** 1899–923
121. Kleman M and Lavrentovich O D 2000 Curvature energy of a focal conic domain with arbitrary eccentricity Phys. Rev. E **61** 1574–8
122. Blanc C and Kleman M 2001 The confinement of smectics with a strong anchoring Eur. Phys. J. E **4**241–51
123. Fournier J B 1993 Curvature elasticity of smectic-a textures with virtual surface singularities Phys. Rev. Lett. **70** 1445–9
124. Kleman M and Lavrentovich O D 2000 Grain boundaries and the law of corresponding cones in smectics Eur. Phys. J. E **2** 47
125. Lopez-Leon T, Fernandez-Nieves A, Nobili M and Blanc C 2012 Smectic shells J. Phys. Condens. Matter **24** 284122
126. Shin H, Bowick M J and Xing X J 2008 Topological defects in spherical nematics Phys. Rev. Lett. **101**037802
127. Fernandez-Nieves A, Vitelli V, Utada A S, Link D R, Marquez M, Nelson D R and Weitz D A 2007 Novel defect structures in nematic liquid crystal shells Phys. Rev. Lett. **99** 157801
128. Seč D, Lopez-Leon T, Nobili M, Blanc C, Fernandez-Nieves A, Ravnik M and Zumer S 2012 Defect trajectories in nematic shells: role of elastic anisotropy and thickness heterogeneity Phys. Rev. E **86**020705
129. Vitelli V and Nelson D R 2006 Nematic textures in spherical shells Phys. Rev. E **74** 021711
130. Skacej G and Zannoni C 2008 Controlling surface defect valence in colloids Phys. Rev. Lett. **100**197802
131. Nelson D R 2002 Toward a tetravalent chemistry of colloids Nano. Lett. **2** 1125–9
132. Liang H-L, Schymura S, Rudquist P and Lagerwall J 2011 Nematic-smectic transition under confinement in liquid crystalline colloidal shells Phys. Rev. Lett. **106** 247801

133. Lopez-Leon T, Fernandez-Nieves A, Nobili M and Blanc C 2011 Nematic-smectic transition in spherical shells Phys. Rev. Lett. **106** 247802
134. Bates M A 2008 Nematic ordering and defects on the surface of a sphere: A monte carlo simulation study J. Chem. Phys. **128** 104707
135. Noh J, Henx B and Lagerwall J P 2016 Taming liquid crystal self-assembly: the multifaceted response of nematic and smectic shells to polymerization Adv. Mater. **28** 10170–4
136. Manyuhina O V and Bowick M J 2015 Thick smectic shells Int. J. Nonlinear Mech. **75** 87–91
137. Liang H-L, Enz E, Scalia G and Lagerwall J 2011 Liquid crystals in novel geometries prepared by microfluidics and electrospinning Mol. Cryst. Liq. Cryst. **549** 69–77
138. Cladis P E and Torza S 1975 Growth of a smectic-a from a bent nematic phase and smectic light valve J. Appl. Phys. **46** 584–99
139. Bouquey M, Serra C, Berton N, Prat L and Hadziioannou G 2008 Microfluidic synthesis and assembly of reactive polymer beads to form new structured polymer materials Chem. Eng. J. **135** S93–8
140. Serra C A and Chang Z 2008 Microfluidic-assisted synthesis of polymer particles Chem. Eng. Techno.**31** 1099–115
141. Yu Y and Ikeda T 2006 Soft actuators based on liquid-crystalline elastomers Angew. Chem. Int. Ed.**45** 5416–8
142. Buguin A, Li M-H, Silberzan P, Ladoux B and Keller P 2006 Micro-actuators: when artificial muscles made of nematic liquid crystal elastomers meet soft lithography J. Am. Chem. Soc. **128** 1088–9
143. Kim C, Mukherjee S, Luchette P and Palffy-Muhoray P 2014 Director orientation in deformed liquid crystal elastomer microparticles Soft Mater. **12** 159–65
144. Fleischmann E-K, Romina Forst F, Köder K, Kapernaum N and Zentel R 2013 Microactuators from a main-chain liquid crystalline elastomer via thiolene 'click' chemistry J. Mater. Chem. C **1** 5885–91
145. Marshall J E, Gallagher S, Terentjev E M and Smoukov S K 2014 Anisotropic colloidal micromuscles from liquid crystal elastomers J. Am. Chem. Soc. **136** 474–9
146. Bera T, Freeman E J, McDonough J A, Clements R J, Aladlaan A, Miller D W, Malcuit C, Hegmann T and Hegmann E 2015 Liquid crystal elastomer microspheres as three-dimensional cell scaffolds supporting the attachment and proliferation of myoblasts ACS Appl. Mater. Interfaces **7** 14528–35
147. Haseloh S, Ohm C, Smallwood F and Zentel R 2011 Nanosized shape-changing colloids from liquid crystalline elastomers Macromol. Rapid Commun. **32** 88–93
148. Humar M and Musevic I 2010 3D microlasers from self-assembled cholesteric liquid-crystal microdroplets Opt. Express **18** 26995–7003
149. Uchida Y, Takanishi Y and Yamamoto J 2013 Controlled fabrication and photonic structure of cholesteric liquid crystalline shells Adv. Mater. **25** 3234–7
150. Chen L, Li Y, Fan J, Bisoyi H K, Weitz D A and Li Q 2014 Photoresponsive monodisperse cholesteric liquid crystalline microshells for tunable omnidirectional lasing enabled by a visible light-driven chiral molecular switch Adv. Opt. Mater. **2** 845–8
151. Hernández R J, Mazzulla A, Pane A, Volke-Sepúlveda K and Cipparrone G 2013 Attractive-repulsive dynamics on light-responsive chiral microparticles induced by polarized tweezers Lab Chip **13** 459–67
152. Noh J-H, Liang H-L, Drevensek-Olenik I and Lagerwall J P F 2014 Tuneable multicoloured patterns from photonic cross communication between cholesteric liquid crystal droplets J. Mater. Chem. C **2**806–10
153. Noh J-H, Drevensek-Olenik I, Yamamoto J and Lagerwall J P F 2015 Dynamic and complex optical patterns from colloids of cholesteric liquid crystal droplets Proc. SPIE **9384** 93840T
154. Chen J, Lacaze E, Brasselet E, Harutyunyan S R, Katsonis N and Feringa B L 2014 Textures of cholesteric droplets controlled by photo-switching chirality at the molecular level J. Mater. Chem. C **2**8137–41

155. Fan J, Li Y, Bisoyi H K, Zola R S, Yang D K, Bunning T J, Weitz D A and Li Q 2015 Light-directing omnidirectional circularly polarized reflection from liquid-crystal droplets Angew. Chem. Int. Ed. **54**2160–4
156. Asshoff S J, Sukas S, Yamaguchi T, Hommersom C A, Le Gac S and Katsonis N 2015 Superstructures of chiral nematic microspheres as all-optical switchable distributors of light Sci. Rep. **5** 14183
157. Lee S S, Kim B, Kim S K, Won J C, Kim Y H and Kim S H 2015 Robust microfluidic encapsulation of cholesteric liquid crystals toward photonic ink capsules Adv. Mater. **27** 627–33
158. Lee S S, Kim S K, Won J C, Kim Y H and Kim S H 2015 Reconfigurable photonic capsules containing cholesteric liquid crystals with planar alignment Angew. Chem. Int. Ed. **54** 15266–70
159. Dierking I 2003 Textures of Liquid Crystals (Weinheim: Wiley)
160. Geng Y, Noh J-H and Lagerwall J P F 2016 Transmission polarized optical microscopy of short-pitch cholesteric liquid crystal shells Proc. SPIE **9769** 97690U
161. Vicsek T and Zafeiris A 2012 Collective motion Phys. Rep. **517** 71–140
162. Hatwalne Y, Ramaswamy S, Rao M and Aditi Simha R 2004 Rheology of active-particle suspensions Phys. Rev. Lett. **92** 118101
163. Peruani F, Deutsch A and Bär M 2006 Nonequilibrium clustering of self-propelled rods Phys. Rev. E **74** 030904
164. Ramaswamy S 2010 The mechanics and statistics of active matter Ann. Rev. Condens. Matter Phys.**1** 323–45
165. Marchetti M C, Joanny J F, Ramaswamy S, Liverpool T B, Prost J, Rao M and Aditi Simha R 2013 Hydrodynamics of soft active matter Rev. Mod. Phys. **85** 1143
166. Maass C C, Krüger C, Herminghaus S and Bahr C 2016 Swimming droplets Ann. Rev. Condens. Matter Phys. **7** 171–93
167. Herminghaus S, Maass C C, Krüger C, Thutupalli S, Goehring L and Bahr C 2014 Interfacial mechanisms in active emulsions Soft Matter **10** 7008–22
168. Krüger C, Klös G, Bahr C and Maass C C 2016 Curling liquid crystal microswimmers: a cascade of spontaneous symmetry breaking (arXiv:1605.03396)
169. Yamamoto T and Sano M 2016 Chirality-induced helical self-propulsion of cholesteric liquid crystal droplets (arXiv:1604.00298)
170. Keber F C, Loiseau E, Sanchez T, DeCamp S J, Giomi L, Bowick M J, Marchetti M C, Dogic Z and Bausch A R 2014 Topology and dynamics of active nematic vesicles Science
171. Eremin A, Nemes A, Stannarius R, Schulz M, Nadasi H and Weissflog W 2005 Structure and mechanical properties of liquid crystalline filaments Phys. Rev. E **71** 031705
172. Bailey C, Murphy M, Eremin A, Weissflog W and Jàkli A 2010 Bundles of fluid fibers formed by bent-core molecules Phys. Rev. E **81** 031708
173. Morys M, Trittel T, Eremin A, Murphy P and Stannarius R 2012 Tension of freely suspended fluid filaments Phys. Rev. E **86** 040501 (R)
174. Peddireddy K, Jampani V S, Thutupalli S, Herminghaus S, Bahr C and Musevic I 2013 Lasing and waveguiding in smectic A liquid crystal optical fibers Opt. Express **21** 30233–42
175. Crawford G, Allender D and Doane J 1992 Surface elastic and molecular-anchoring properties of nematic liquid-crystals confined to cylindrical cavities Phys. Rev. A **45** 8693–708
176. Kralj S and Zumer S 1993 The stability diagram of a nematic liquid-crystal confined to a cylindrical cavity Liq. Cryst. **15** 521–7
177. Crawford G, Ondris-Crawford R, Zumer S and Doane J 1993 Anchoring and orientational wetting transitions of confined liquid-crystals Phys. Rev. Lett. **70** 1838–41
178. Crawford G P, Stannarius R and Doane J W 1991 Surface-induced orientational order in the isotropic phase of a liquid-crystal material Phys. Rev. A **44** 2558–69

179. Lagerwall J P F, McCann J T, Formo E, Scalia G and Xia Y 2008 Coaxial electrospinning of microfibres with liquid crystal in the core Chem. Commun. **42** 5420–2
180. Matthias H, Röder T, Wehrspohn R B, Kitzerow H-S, Matthias S and Picken S J 2005 Spatially periodic liquid crystal director field appearing in a photonic crystal template Appl. Phys. Lett. **87**241105
181. Mori H and Nakanishi H 1988 On the stability of topologically non-trivial point defects J. Phys. Soc. Japan **57** 1281–6
182. Bradac Z, Kralj S, Svetec M and Zumer S 2003 Annihilation of nematic point defects: postcollision scenarios Phys. Rev. E **67** 050702 (R)
183. De Luca G and Rey A D 2007 Point and ring defects in nematics under capillary confinement J. Chem. Phys. **127** 104902
184. Wang X, Kim Y K, Bukusoglu E, Zhang B, Miller D S and Abbott N L 2016 Experimental insights into the nanostructure of the cores of topological defects in liquid crystals Phys. Rev. Lett. **116** 147801
185. Wang X, Miller D S, Bukusoglu E, de Pablo J J and Abbott N L 2015 Topological defects in liquid crystals as templates for molecular self-assembly Nat. Mater. **15** 106–12
186. Ondris-Crawford R J, Crawford G P, Zumer S and Doane J W 1993 Curvature-induced configuration transition in confined nematic liquid crystals Phys. Rev. Lett. **70** 194–7
187. Davidson Z S, Kang L, Jeong J, Still T, Collings P J, Lubensky T C and Yodh A G 2015 Chiral structures and defects of lyotropic chromonic liquid crystals induced by saddle-splay elasticity Phys. Rev. E **91**050501 (R)
188. Pairam E, Vallamkondu J, Koning V, van Zuiden B C, Ellis P W, Bates M A, Vitelli V and Fernandez-Nieves A 2013 Stable nematic droplets with handles. Proc. Natl Acad. Sci. USA **110** 9295–300
189. Nayani K, Chang R, Fu J, Ellis P W, Fernandez-Nieves A, Park J O and Srinivasarao M 2015 Spontaneous emergence of chirality in achiral lyotropic chromonic liquid crystals confined to cylinders Nat. Commun. **6** 8067
190. Reyes C, Sharma A and Lagerwall J 2016 Non-electronic gas sensors from electrospun mats of liquid crystal core fibers for detecting volatile organic compounds at room temperature Liq. Cryst. **43**1986–2001
191. Cladis P E, White A E and Brinkman W F 1979 The cholesteric defect structure near the smectic A transition J. Phys. **40** 325–35 (Paris)
192. Kitzerow H S, Liu B, Xu F and Crooker P P 1996 Effect of chirality on liquid crystals in capillary tubes with parallel and perpendicular anchoring Phys. Rev. E **54** 568–75
193. Enz E and Lagerwall J 2010 Electrospun microfibres with temperature sensitive iridescence from encapsulated cholesteric liquid crystal J. Mater. Chem. **20** 6866–72
194. Ondris-Crawford R J, Ambrozic M, Doane J W and Zumer S 1994 Pitch-induced transition of chiral nematic liquid crystals in submicrometer cylindrical cavities Phys. Rev. E **50** 4773–9
195. Lequeux F and Kleman M 1988 Helicoidal instability in cholesteric capillary tubes J. Phys. **49** 845–55 (Paris)
196. Enz E, La Ferrara V and Scalia G 2013 Confinement-sensitive optical response of cholesteric liquid crystals in electrospun fibers ACS Nano **7** 6627–35
197. Scalia G, Enz E, Calò O, Kim D K, Hwang M, Lee J H and Lagerwall J P F 2013 Morphology and core continuity of liquid-crystal-functionalized, coaxially electrospun fiber mats tuned via the polymer sheath solution Macromol. Mater. Eng. **298** 583–9
198. Matthias H, Schweizer S L, Wehrspohn R B and Kitzerow H S 2007 Liquid crystal director fields in micropores of photonic crystals J. Opt.: A Pure Appl. Opt. **9** S389–95
199. Ondris-Crawford R J, Crawford G P, Doane J W, Zumer S, Vilfan M and Vilfan I 1993 Surface molecular anchoring in microconfined liquid crystals near the nematic–smectic-A transition Phys. Rev. E **48** 1998–2005

200. Kralj S and Zumer S 1996 Smectic-A structures in submicrometer cylindrical cavities Phys. Rev. E **54**1610–7
201. Iannacchione G and Finotello D 1993 Confinement and orientational study at liquid crystal phase-transitions Liq. Cryst. **14** 1135–42
202. Enz E, Baumeister U and Lagerwall J 2009 Coaxial electrospinning of liquid crystal-containing poly(vinyl pyrrolidone) microfibers Beilstein J. Org. Chem. **5** 58
203. Reneker D H and Chun I 1996 Nanometre diameter fibres of polymer, produced by electrospinning Nanotechnology **7** 216–23
204. Doshi J and Reneker D H 1995 Electrospinning process and applications of electrospun fibers J. Electrostat. **35** 151–60
205. Bergshoef M M and Vancso G J 1999 Transparent nanocomposites with ultrathin, electrospun nylon-4,6 fiber reinforcement Adv. Mater. **11** 1362–5
206. Jaeger R, Bergshoef M M, Batlle C M I, Schönherr H and Julius Vancso G 1998 Electrospinning of ultra-thin polymer fibers Macromol. Symp. **127** 141–50
207. Deitzel J M, Kleinmeyer J, Harris D and Beck Tan N C 2001 The effect of processing variables on the morphology of electrospun nanofibers and textiles Polymer **42** 261–72
208. Nascimento M L F, Araujo E S, Cordeiro E R, de Oliveira A H P and de Oliveira H P 2015 A literature investigation about electrospinning and nanofibers: historical trends, current status and future challenges Recent Patents Nanotechnol. **9** 76–85
209. Cooley J F 1900 Improved methods of and apparatus for electrically separating the relatively volatile liquid component from the component of relatively fixed substances of composite fluids (patent) Great Britain PO
210. Formhals A 1940 Artificial thread and method of producing same US Patent SpecificationUS2187306 A
211. Hagiwara K 1929 Process for manufacturing artificial silk and other filaments by applying electric current (patent) US Patent Specification 1699615
212. Formhals A 1934 Process and apparatus for preparing artificial threads (patent) USPTO US1975504
213. Formhals A 1944 Method and apparatus for spinning (patent) US Patent Specification 2349950
214. Subbiah T, Bhat G S, Tock R W, Parameswaran S and Ramkumar S S 2005 Electrospinning of nanofibers J. Appl. Polym. Sci. **96** 557–69
215. Taylor G 1964 Disintegration of water drops in an electric field Proc. R. Soc. A: Math. Phys. Eng. Sci.**280** 383–97
216. Taylor G 1969 Electrically driven jets Proc. R. Soc. A: Math. Phys. Eng. Sci. **313** 453–75
217. Reneker D H and Yarin A L 2008 Electrospinning jets and polymer nanofibers Polymer **49** 2387–425
218. Yarin A L, Koombhongse S and Reneker D H 2001 Bending instability in electrospinning of nanofibers J. Appl. Phys. **89** 3018–26
219. Hohman M M, Shin M, Rutledge G and Brenner M P 2001 Electrospinning and electrically forced jets. I. Stability theory Phys Fluids **13** 2201–20
220. Hohman M M, Shin M, Rutledge G and Brenner M P 2001 Electrospinning and electrically forced jets. II. Applications Phys Fluids **13** 2221–36
221. Yarin A L, Kataphinan W and Reneker D H 2005 Branching in electrospinning of nanofibers J. Appl. Phys. **98** 064501
222. Shin Y M, Hohman M M, Brenner M P and Rutledge G C 2001 Experimental characterization of electrospinning: the electrically forced jet and instabilities Polymer **42** 9955–67
223. Theron S A, Zussman E and Yarin A L 2004 Experimental investigation of the governing parameters in the electrospinning of polymer solutions Polymer **45** 2017–30
224. Thompson C J, Chase G G, Yarin A L and Reneker D H 2007 Effects of parameters on nanofiber diameter determined from electrospinning model Polymer **48** 6913–22

225. Graham K, Ouyang M, Raether T, Grafe T, McDonald B and Knauf P 2002 Polymeric nanofibers in air filtration applications 15th Annual Technical Conf. & Expo of the American Filtration & Separations Society (Galveston,) pp 9–12
226. Podgórski A, Bałazy A and Gradoń L 2006 Application of nanofibers to improve the filtration efficiency of the most penetrating aerosol particles in fibrous filters Chem. Eng. Sci. **61** 6804–15
227. Zong X, Bien H, Chung C Y, Yin L, Fang D, Hsiao B S, Chu B and Entcheva E 2005 Electrospun fine-textured scaffolds for heart tissue constructs Biomaterials **26** 5330–8
228. Park J, Wang S, Li M, Ahn C, Hyun J K, Kim D S, Kim do K, Rogers J A, Huang Y and Jeon S 2012 Three-dimensional nanonetworks for giant stretchability in dielectrics and conductors Nat. Commun.**3** 916
229. Kumar P S, Sundaramurthy J, Sundarrajan S, Babu V J, Singh G, Allakhverdiev S I and Ramakrishna S 2014 Hierarchical electrospun nanofibers for energy harvesting, production and environmental remediation Energy Environ. Sci. **7** 3192–222
230. Krause S, Dersch R, Wendorff J H and Finkelmann H 2007 Photocrosslinkable liquid crystal main-chain polymers: thin films and electrospinning Macromol. Rapid Commun. **28** 2062–8
231. Canejo J P, Borges J P, Godinho M H, Brogueira P, Teixeira P and Terentjev E M 2008 Helical twisting of electrospun liquid crystalline cellulose micro-and nanofibers Adv. Mater. **20** 4821–5
232. Lagerwall J P F 2012 Switchable and responsive liquid crystal-functionalized microfibers produced via coaxial electrospinning Proc. SPIE **8279** 82790N
233. Enz E 2013 Electrospun polymer-liquid crystal composite fibers PhD Thesis Martin-Luther-Universität Halle-Wittenberg Halle, Germany
234. Buyuktanir E A, Frey M W and West J L 2010 Self-assembled, optically responsive nematic liquid crystal/polymer core-shell fibers: formation and characterization Polymer **51** 4823–30
235. Trejo N K, Reyes C G, Sanchez V, Zhang D and Frey M W 2016 Developing composite nanofibre fabrics using electrospinning, ultrasonic sewing and laser cutting technologies Int. J. Fashion Des. Technol. Educ. **3266** 1–9
236. Andrady A L 2008 Science and Technology of Polymer Nanofibers Description of Electrostatic Spinning (Hoboken, NJ: Wiley)
237. Bugarski B, Li Q, Goosen M F A, Poncelet D, Neufeld R J and Vunjak G 1994 Electrostatic droplet generation: mechanism of polymer droplet formation AIChE J. **40** 1026–31
238. Bugarski B, Amsden B, Goosen M F A, Neufeld R J and Poncelet D 1994 Effect of electrode geometry and charge on the production of polymer microbeads by electrostatics Can. J. Chem. Eng. **72** 517–21
239. Loeb L B, Kip A F, Hudson G G and Bennett W H 1941 Pulses in negative point-to-plane corona Phys. Rev. **60** 714–22
240. Yarin A L, Koombhongse S and Reneker D H 2001 Taylor cone and jetting from liquid droplets in electrospinning of nanofibers J. Appl. Phys. **90** 4836–46
241. Kameoka J, Orth R, Yang Y, Czaplewski D, Mathers R, Coates G W and Craighead H G 2003 A scanning tip electrospinning source for deposition of oriented nanofibres Nanotechnology **14** 1124–9
242. Wendorff J H, Agarwal S and Greiner A 2012 Electrospinning—Materials, Processing and Applications Nature of the Electrospinning Process—Experimental Observations and Theoretical Analysis(Weinheim: Wiley)
243. Shenoy S L, Bates W D, Frisch H L and Wnek G E 2005 Role of chain entanglements on fiber formation during electrospinning of polymer solutions: good solvent, non-specific polymer–polymer interaction limit Polymer **46** 3372–84
244. Wu X F, Salkovskiy Y and Dzenis Y A 2011 Modeling of solvent evaporation from polymer jets in electrospinning Appl. Phys. Lett. **98** 2014–7
245. Bhardwaj N and Kundu S C 2010 Electrospinning: a fascinating fiber fabrication technique Biotechnol. Adv. **28** 325–47

246. Greiner A and Wendorff J H 2007 Electrospinning: a fascinating method for the preparation of ultrathin fibres Angew. Chem. Int. Ed. **46** 5670–703
247. McCann J T, Li D and Xia Y N 2005 Electrospinning of nanofibers with core-sheath, hollow, or porous structures J. Mater. Chem. **15** 735–8
248. Casper C L, Stephens J S, Tassi N G, Chase D B and Rabolt J F 2004 Controlling surface morphology of electrospun polystyrene fibers: effect of humidity and molecular weight in the electrospinning process Macromolecules **37** 573–8
249. Holzmeister A, Yarin A L and Wendorff J H 2010 Barb formation in electrospinning: Experimental and theoretical investigations Polymer **51** 2769–78
250. Bhattarai N, Cha D I, Bhattarai S R, Khil M S and Kim H Y 2003 Biodegradable electrospun mat: Novel block copolymer of poly (p-dioxanone-co-l-lactide)-block-poly(ethylene glycol) J. Polym. Sci. B: Polym. Phys. **41** 1955–64
251. Ramakrishna S, Fujihara K, Teo W-E, Lim T-C and Ma Z 2005 An Introduction to Electrospinning and Nanofibers Electrospinning Process vol 3 (Singapore: World Scientific)
252. Reichardt C 2003 Solvents and Solvent Effects in Organic Chemistry (Weinheim: Wiley)
253. Koombhongse S, Liu W X and Reneker D H 2001 Flat polymer ribbons and other shapes by electrospinning J. Polym. Sci. B **39** 2598–606
254. Nezarati R M, Eifert M B and Elizabeth Cosgriff-Hernandez 2013 Effects of humidity and solution viscosity on electrospun fiber morphology Tissue Eng. Part C: Methods **19** 810–9
255. Liu W, Huang C and Jin X 2014 Tailoring the grooved texture of electrospun polystyrene nanofibers by controlling the solvent system and relative humidity Nanoscale Res. Lett. **9** 350
256. Lu P and Xia Y 2013 Maneuvering the internal porosity and surface morphology of electrospun polystyrene yarns by controlling the solvent and relative humidity Langmuir **29** 7070–8
257. Hansen C M 1969 The universality of the solubility parameter Ind. Eng. Chem. Prod. Res. Dev. **8** 2–11
258. Teas J P 1968 Graphic analysis of resin solubilities J. Paint Technol. **40** 19–25
259. Burke J 1984 Solubility parameters: theory and application Book Paper Gourp Ann. **3** 33–6
260. Graessley W W 1967 The shear-rate dependence of viscosity in concentrated solutions of narrow-distribution polystyrene J. Rheol. **11** 267
261. Agarwal S, Burgard M, Greiner A and Wendorff J 2016 Electrospinning: a Practical Guide to Nanofibers (Berlin: Walter de Gruyter)
262. Greve H-H 2000 Rubber, 2. Natural Ullmann's Encyclopedia of Industrial Chemistry **31** 583–94
263. Waltz J E and Taylor G B 1947 Determination of molecular weight of nylon Anal. Chem. **19** 448–50
264. McKee M G, Layman J M, Cashion M P and Long T E 2006 Phospholipid nonwoven electrospun membranes Science **311** 353–5
265. McKee M G, Wilkes G L, Colby R H and Long T E 2004 Correlations of solution rheology with electrospun fiber formation of linear and branched polyesters Macromolecules **37** 1760–7
266. Gupta P, Elkins C, Long T E and Wilkes G L 2005 Electrospinning of linear homopolymers of poly(methyl methacrylate): exploring relationships between fiber formation, viscosity, molecular weight and concentration in a good solvent Polymer **46** 4799–810
267. Koski A, Yim K and Shivkumar S 2004 Effect of molecular weight on fibrous PVA produced by electrospinning Mater. Lett. **58** 493–7
268. Yarin A L 2011 Coaxial electrospinning and emulsion electrospinning of core-shell fibers Polym. Adv. Technol. **22** 310–7
269. Bazilevsky A V, Yarin A L and Megaridis C M 2007 Co-electrospinning of core-shell fibers using a single-nozzle technique Langmuir **23** 2311–4

270. Reznik S N, Yarin A L, Zussman E and Bercovici L 2006 Evolution of a compound droplet attached to a core-shell nozzle under the action of a strong electric field Phys. Fluids **18** 062101

271. Moghe A K and Gupta B S 2008 Co-axial electrospinning for nanofiber structures: Preparation and applications Polym. Rev. **48** 353–77

272. YooKye M, Kim C and Lagerwall J 2015 Multifunctional responsive fibers produced by dual liquid crystal core electrospinning J. Mater. Chem. C **3** 8979–85

273. Kim D K and Lagerwall J P F 2014 Influence of wetting on morphology and core content in electrospun core-sheath fibers ACS Appl. Mater. Interf. **6** 16441–7

274. Wang J, Jákli A and West J L 2016 Morphology tuning of electrospun liquid crystal/polymer fibers ChemPhysChem **17** 3080–5

275. Wang J, Jákli A and West J L 2015 Airbrush formation of liquid crystal/polymer fibers ChemPhysChem**16** 1839–41

276. Tsuboi K, Marcelletti E, Matsumoto H, Ashizawa M, Minagawa M, Furuya H, Tanioka A and Abe A 2012 Preparation of poly(gamma-benzyl-l-glutamate) nanofibers by electrospinning from isotropic and biphasic liquid crystal solutions Polym. J. **44** 360–5

277. Srinivasan G and Reneker D H 1995 Structure and morphology of small diameter electrospun aramid fibers Polym. Int. **36** 195–201

278. Ogata N, Shimada N, Yamaguchi S, Nakane K and Ogihara T 2007 Melt-electrospinning of poly(ethylene terephthalate) and polyalirate J. Appl. Polym. Sci. **105** 1127–32

279. Araujo T M, Sinha-Ray S, Pegoretti A and Yarin A L 2013 Electrospinning of a blend of a liquid crystalline polymer with poly(ethylene oxide): vectran nanofiber mats and their mechanical properties J. Mater. Chem. C **1** 351–8

280. Wu Y, An Q, Yin J, Hua T, Xie H, Li G and Tang H 2008 Liquid crystal fibers produced by using electrospinning technique Colloid Polym. Sci. **286** 897–905

281. Nakashima K, Tsuboi K, Matsumoto H, Ishige R, Tokita M, Watanabe J and Tanioka A 2010 Control over internal structure of liquid crystal polymer nanofibers by electrospinning Macromol. Rapid Commun. **31** 1641–5

282. Stein S 2016 Apple watch review CNET (www.cnet.com/products/apple-watch/) (Accessed: 3 July 2016)

283. Rawassizadeh R, Tomitsch M, Nourizadeh M, Momeni E, Peery A, Ulanova L and Pazzani M 2015 Energy-efficient integration of continuous context sensing and prediction into smartwatches Sensors**15** 22616–45

284. Poselad S 2009 Ubiquitous Computing: Smart Devices, Environments and Interactions Smart Mobiles, Cards and Device Networks vol 4 (New York: Wiley)

285. CuteCircuit 2012 The Aqua Dress by CuteCircuit (http://cutecircuit.com/the-aqua-dress/) (Accessed: 3 July 2016)

286. Pierce D 2016 Everyone Wants to Make Smart Rings—But No One Knows What For www.wired.com/2016/02/everyone-wants-to-make-smart-rings-but-no-one-knows-what-for/(Accessed 3 July 2016)

287. Kim D K, Hwang M and Lagerwall J P F 2013 Liquid crystal-functionalization of electrospun polymer fibers J. Polym. Sci. B: Polym. Phys. **51** 855–67

288. Dickert F L, Haunschild A, Hofmann P and Mages G 1992 Molecular recognition of organic solvents and ammonia: shapes and donor properties as sensor effects Sensors Actuators B: Chem. **6** 25–8

289. Dickert F L, Haunschild A and Hofmann P 1994 Cholesteric liquid-crystals for solvent vapor detection—elimination of cross-sensitivity by band shape-analysis and pattern-recognition Fresenius J. Anal. Chem. **350** 577–81

290. Clements J, Boden N, Gibson T D, Chandler R C, Hulbert J N and Ruck-Keene E A 1998 Novel, self-organising materials for use in gas sensor arrays: beating the humidity problem Sensors Actuators B: Chem. **47** 37–42

291. Winterbottom D A, Narayanaswamy R and Raimundo I M 2003 Cholesteric liquid crystals for detection of organic vapours Sensors Actuators B **90** 52–7

292. David D J and Hardy E E 1977 Organic vapor dectection with liquid crystals US Patent Specification4040749
293. Jacobs J E 1975 Liquid crystal gas analyzer US Patent Specification 3927977
294. Nicholas P and Mccollum T 1999 Liquid crystal sensors WIPO Patent Specification WO1999032878 A1
295. Mujahid A, Stathopulos H, Lieberzeit P A and Dickert F L 2010 Solvent vapour detection with cholesteric liquid crystals-optical and mass-sensitive evaluation of the sensor mechanism Sensors**10** 4887–97
296. Herzer N et al 2012 Printable optical sensors based on h-bonded supramolecular cholesteric liquid crystal networks J. Am. Chem. Soc. **134** 7608–11
297. Saha A, Tanaka Y, Han Y, Bastiaansen C W M, Broer D J and Sijbesma R P 2012 Irreversible visual sensing of humidity using a cholesteric liquid crystal Chem. Commun. **48** 4579–81
298. Chang C-K, Kuo H-L, Tang K-T and Chiu S-W 2011 Optical detection of organic vapors using cholesteric liquid crystals Appl. Phys. Lett. **99** 073504
299. Zhao Y-B, Yu J-H, Zhao H-F, Tong C-Y and Wang P-H 2011 A novel method for label-free detection of ricin using liquid crystals supported on chemically functionalized surfaces Sensors Actuators B: Chem. **155** 351–6
300. Bedolla-Pantoja M and Abbott N L 2016 Surface-controlled orientational transitions in elastically strained films of liquid crystal that are triggered by vapors of toluene ACS Appl. Mater. Interfaces **8**13114–22
301. Cheng D, Sridharamurthy S S, Hunter J T, Park J-S, Abbott N L and Jiang H 2009 A sensing device using liquid crystal in a micropillar array supporting structure J. Microelectromech. Syst. **18** 973–82
302. Cadwell K D, Lockwood N A, Nellis B A, Alf M E, Willis C R and Abbott N L 2007 Detection of organophosphorous nerve agents using liquid crystals supported on chemically functionalized surfaces Sensors Actuators B: Chem. **128** 91–8
303. Kim S-R, Teixeira A I, Nealey P F, Wendt A E and Abbott N L 2002 Fabrication of polymeric substrates with well-defined nanometer-scale topography and tailored surface chemistry Adv. Mater. **14** 1468–72
304. Hunter J T, Abbott L and Nicholas P 2013 Dynamics of the chemo-optical response of supported films of nematic liquid crystals Sensors Actuators B: Chem. **183** 71–80
305. Carlton R J, Hunter J T, Miller D S, Abbasi R, Mushenheim P C, Tan L N and Abbott N L 2013 Chemical and biological sensing using liquid crystals Liq. Cryst. Rev. **1** 29–51
306. Govindaraju T, Bertics P J, Raines R T and Abbott N L 2007 Using measurements of anchoring energies of liquid crystals on surfaces to quantify proteins captured by immobilized ligands J. Am. Chem. Soc. **129** 11223–31
307. Zhang W, Ang W T, Xue C-Y and Yang K-L 2011 Minimizing nonspecific protein adsorption in liquid crystal immunoassays by using surfactants ACS Appl. Mater. Interf. **3** 3496–500
308. Han Y, Pacheco K, Bastiaansen C W M, Broer D J and Sijbesma R P 2010 Optical monitoring of gases with cholesteric liquid crystals J. Am. Chem. Soc. **132** 2961–7
309. Jiang H, Li C and Huang X 2013 Actuators based on liquid crystalline elastomer materials Nanoscale**5** 5225–40
310. Yoshino T, Kondo M, Mamiya J-I, Kinoshita M, Yu Y and Ikeda T 2010 Three-dimensional photomobility of crosslinked azobenzene liquid-crystalline polymer fibers Adv. Mater. **22** 1361
311. Ohm C, Morys M, Romina Forst F, Braun L, Eremin A, Serra C, Stannarius R and Zentel R 2011 Preparation of actuating fibres of oriented main-chain liquid crystalline elastomers by a wetspinning process Soft Matter **7** 3730–4
312. Stannarius R, Eremin A, Harth K, DeMiglio M M A, Ohm C and Zentel R 2012 Mechanical and optical properties of continuously spun fibres of a main-chain smectic a elastomer Soft Matter **8** 1858–64
313. Fleischmann E K, Romina Forst F and Zentel R 2014 Liquid crystalline elastomer fibers prepared in a microfluidic device Macromol. Chem. Phys. **215** 1004–11

314. Liu X, Cheng S, Liu H, Hu S, Zhang D and Ning H 2012 A survey on gas sensing technology Sensors **12** 9635–65
315. Fine G F, Cavanagh L M, Afonja A and Binions R 2010 Metal oxide semi-conductor gas sensors in environmental monitoring Sensors **10** 5469–502
316. Bochenkov V E and Sergeev G B 2010 Sensitivity, selectivity and stability of gas-sensitive metal-oxide nanostructures Metal Oxide Nano-structures and Their Applications ed A Umar and Y-B Hahn (New York: Springer) pp 31–52
317. Yamazoe N 1991 New approaches for improving semiconductor gas sensors Sensors Actuators B: Chem. **5** 7–19
318. Yoon B, Lee S and Kim J-M 2009 Recent conceptual and technological advances in polydiacetylene-based supramolecular chemosensors Chem. Soc. Rev. **38** 1958–68
319. Jacobs C B, Jennifer Peairs M and Jill Venton B 2010 Review: carbon nanotube based electrochemical sensors for biomolecules Anal. Chim. Acta **662** 105–27
320. Bungabong M L, Ong P B and Yang K-L 2010 Using copper perchlorate doped liquid crystals for the detection of organophosphonate vapor Sensors Actuators B: Chem. **148** 420–6
321. Ataman C, Kinkeldei T, Mattana G, Vásquez Quintero A, Molina-Lopez F, Courbat J, Cherenack K, Briand D, Tröster G and De Rooij N F 2013 A robust platform for textile integrated gas sensors Sensors Actuators, B: Chem. **177** 1053–61
322. Frutiger A, Muth J T, Vogt D M, Menguc Y, Campo A, Valentine A D, Walsh C J and Lewis J A 2015 Capacitive soft strain sensors via multicore-shell fiber printing Adv. Mater. **27** 2440–6
323. Courbat J, Briand D and de Rooij N F 2010 Foil level packaging of a chemical gas sensor J. Micromech. Microeng. **20** 055026
324. Bandodkar A J, Molinnus D, Mirza O, Guinovart T, Windmiller J R, Valdes-Ramirez G, Andrade F J, Schoning M J and Wang J 2014 Epidermal tattoo potentiometric sodium sensors with wireless signal transduction for continuous non-invasive sweat monitoring Biosens. Bioelectron. **54** 603–9
325. Guinovart T, Bandodkar A J, Windmiller J R, Andrade F J and Wang J 2013 A potentiometric tattoo sensor for monitoring ammonium in sweat Analyst **138** 7031–8
326. Bandodkar A J, Jeerapan I and Wang J 2016 Wearable chemical sensors: present challenges and future prospects ACS Sens. **1** 464–82
327. Vilela D, Romeo A and Sanchez S 2016 Flexible sensors for biomedical technology Lab Chip **16** 402–8
328. Rousseau I A and Mather P T 2003 Shape memory effect exhibited by smectic-c liquid crystalline elastomers J. Am. Chem. Soc. **125** 15300–1
329. Sanchez-Ferrer A and Finkelmann H 2011 Polydomain-monodomain orientational process in smectic-c main-chain liquid-crystalline elastomers Macromol. Rapid Commun. **32** 309–15

CHAPTER 4

Ionic Liquid Crystals Based on Pyridinium Salts

Viorel Cîrcu*

Inorganic Chemistry Department, University of Bucharest, Bucharest, Romania

ABSTRACT

This chapter describes the liquid crystalline properties of the ionic liquid crystals (ILC) based on pyridinium salts as well as their metal-containing compounds with an emphasis on the recent systems described in literature. The main factors that influence the liquid crystalline properties of pyridinium ILC are discussed. Selected thermal data are given according to mesogenic group employed and its position (either N-substitution or pyridinium ring substitution) and the number of structural cationic units (mono-, di-, or polycationic pyridinium ILC).

Keywords: ionic liquids, liquid crystals, pyridinium salts

1. INTRODUCTION

Ionic liquid crystals (ILCs) are extensively studied nowadays due to their unique properties resulting from the combination of liquid crystal (LC) and ionic liquid (IL) properties. Several reviews covering this topic were published in the recent years [1–3]. The field of ILCs is continuously growing as many recent applications were found: solar cells, membranes, battery materials, electrochemical sensors, or electroluminescent switches. Different factors are responsible for governing the nature of ILC phases, such as the molecular shape, location, and size of ionic groups, intermolecular interactions, and microphase segregation. Thus, the hydrophobic interactions between the long alkyl chain groups, ionic, dipole-

dipole, anion–cation hydrogen bonding, and cation-π interactions together with the π-π stacking of the aromatic rings, all have a contribution to the stabilization of the liquid crystalline phase. For instance, there is a strong tendency to stabilize lamellar (smectic) phases, with SmA the most common phase for ILC, due to electrostatic interactions and ion-ion stacking in ILC. The combination of all these factors leads to LC behavior ranges from typical calamitic materials to discotic. The imidazolium or pyridinium derivatives, one of the most studied classes of ILC, with weakly coordinating anions, such as tetrafluoroborate (BF_4^-) and hexafluorophosphate (PF_6^-), are well known for their high thermal and electrochemical stabilities. It is worth to mention here that the pyridinium-based ILC has been known for a long time, displaying very similar properties with the related imidazolium-based ILC.

The most common types of LC phases displayed by ILCs are represented in **Figure 1**. Their identification relies upon three characterization techniques: polarizing optical microscopy (POM), differential scanning calorimetry (DSC), and powder X-ray diffraction (XRD). The usual textures seen by POM for different mesophases are presented in **Figure 2**. It is very common to see the SmA phase of ionic mesogens as a so-called oily streak texture by POM, in particular during heating runs (**Figure 2**). When cooling

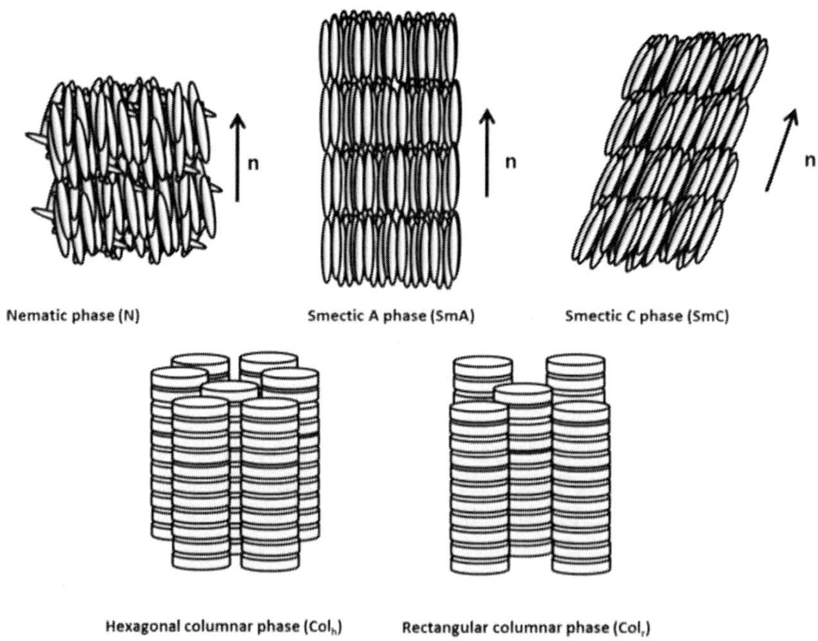

FIGURE 1. Schematic representation of the most common liquid crystalline phases encountered for ionic liquid crystals.

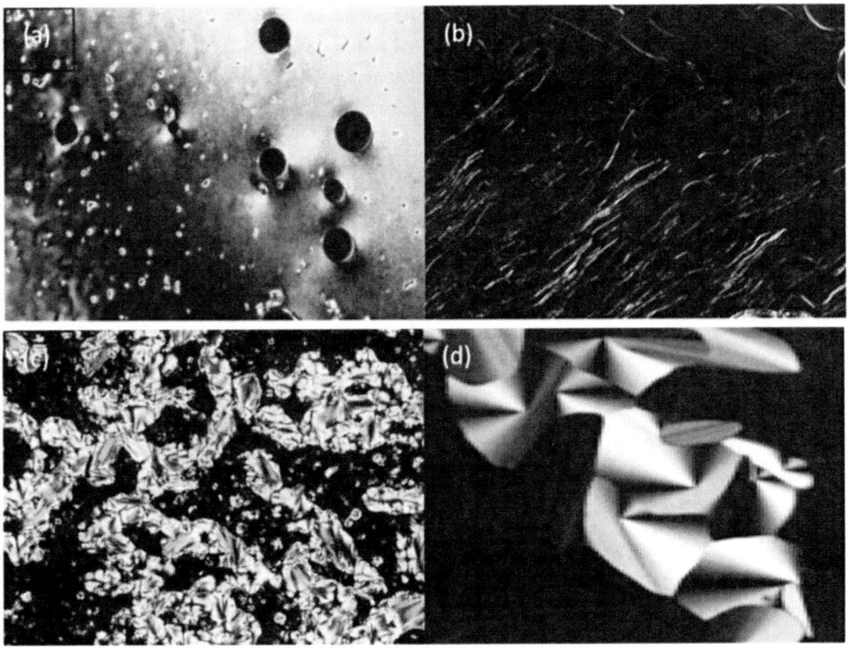

FIGURE 2. Typical textures for LC phases: marbled texture of a nematic phase (a), oily-streaks texture for a SmA phase on heating run (b), focal conical fan-shaped texture for SmA phase (c), and fan-shaped texture of a hexagonal columnar phase Col_h (d).

from the isotropic state, very often the SmA phase can develop spontaneous homeotropic alignment due to interactions developed between the cationic groups and the glass substrate surface (the samples are sandwiched between two microscope slides), and further orientation of the mesogenic groups. As a result, the microscopy image contains large dark regions corresponding to these homeotropic alignments.

The POM technique is an important tool for mesophase identification, but an ultimate technique that confirms unequivocally the phase type is the XRD method. The later one can give also important information regarding the internal organization within the mesophase resulting from the different molecular packing related to the chemical structure of the ionic mesogens and the nature of interactions between them.

The thermal behavior as well as the mesophase type of pyridinium-based ILCs depends on several factors: the position and the nature of the mesogenic group attached to the pyridinium ring and the counterion employed. These structural factors will be discussed further and selected examples will be presented to illustrate their influence on the mesomorphic behavior. The pyridinium-based ILC classification was made according to mesogenic group used and its position (either N-substitution or pyridinium ring substitution) and the number of cationic units contained in their structure (mono-, di-, or polycationic pyridinium ILC).

2. MONOCATIONIC PYRIDINIUM IONIC LIQUID CRYSTALS

2.1. Simple Protonated Pyridinium Salts

The simplest 4-alkyl substitution protonated pyridinium salts with halide as counterions (**1**) show no liquid crystalline properties as they melt straight to the isotropic phase. Such products have been synthesized by reaction between the corresponding hydrogen halide with their corresponding 4-alkyl pyridine compounds [4].

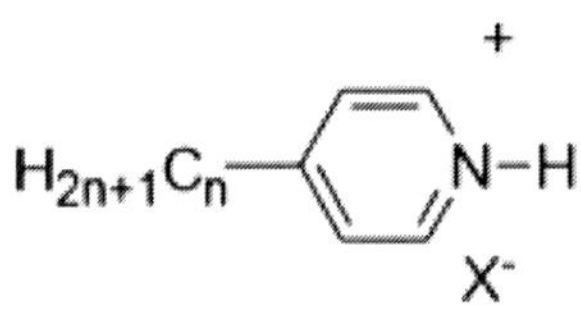

n = 9, 12

X^- = Cl^-, Br^-, I^-

1

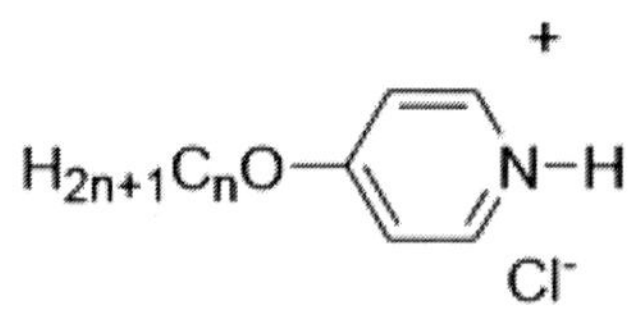

n = 12, 14, 16, 18

2

2a, n = 12 Cr 92 Cr' 122 SmA 133 dec
2b, n = 14 Cr 95 Cr' 122 SmA 139 dec
2c, n = 16 Cr 95 Cr' 124 SmA 142 dec
2d, n = 18 Cr 93 Cr' 124 SmA 145 dec

By changing the alkyl group with an alkoxy group, the resulting protonated chloride pyridinium salts show a SmA phase, with decomposition near the isotropization process. The melting and isotropization temperatures of these products were very insensitive to the chain lengths, suggesting that the hydrophobic interactions are the least significant factor in the thermal behavior of such protonated pyridinium salts [5].

A series of more elaborated protonated pyridinium salts have been prepared by the reaction between pyridine derivatives and phosphoric acid with the aim of studying their anhydrous proton conduction. X-ray diffraction measurements suggested that the pyridinium salts formed a bilayer structure with head-to-head configuration in the SmA phase [6].

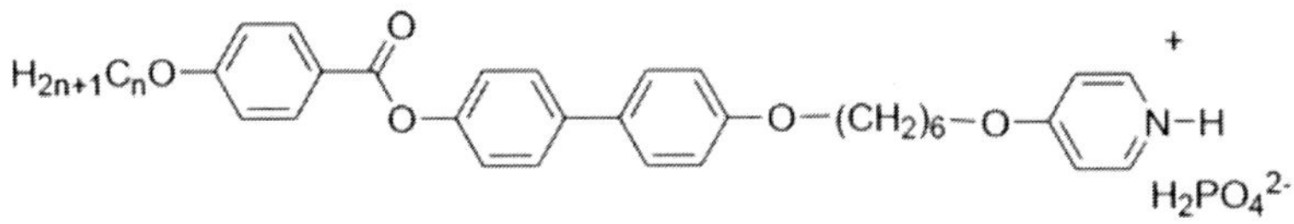

n = 6, 8, 10

3

3a, n = 6; Cr 170 SmA 235 dec
3b, n = 8; Cr 169 SmA 232 dec
3c, n = 10; Cr 159 SmA 213 dec

The protonated pyridinium cation was employed to generate liquid crystalline materials having as counterions a series of wedge-shaped benzenesulfonate mesogens **4–6** [7, 8]. Compound **4**, having a rather unusual substitution pattern, 2,3,4-tris(dodecyloxy)benzenesulfonate, exhibits a reversible transition from a columnar disordered phase into an ordered columnar during the slow heating and cooling cycles. The thermal behavior of pyridinium salts **5** and **6** was compared to thermal properties of their corresponding benzenesulfonic acid and its sodium salts. It was found that **5** and **6** exhibit much lower transition temperatures than the sulfonic acids (the corresponding sulfonic acid of **6** was not stable upon drying) and sodium salts, respectively.

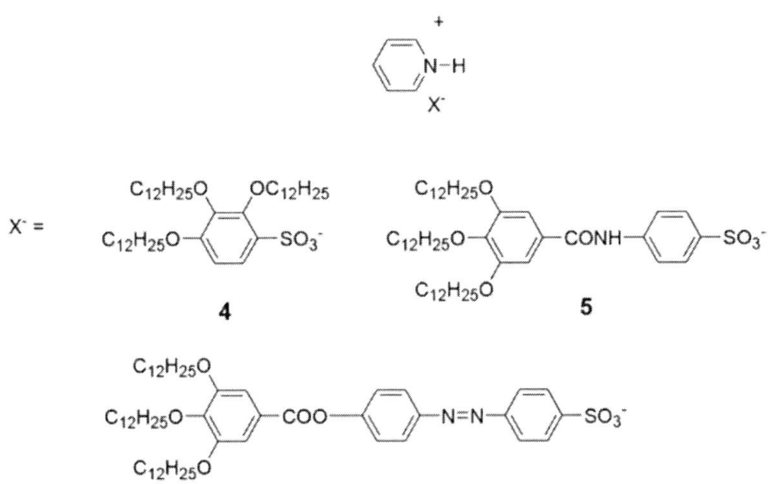

6

4, Cr 30 Cr' 53 Cr'' 80 Col_{hd} 122 Iso
5, Cr 39 M 65 Iso
6, Cr 94 Col 200 Iso (dec)

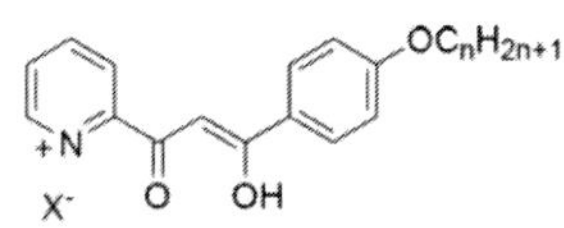

n = 10, 12, 14, 16, 18
X- = Cl⁻, $[ZnCl_4]^{2-}$

7

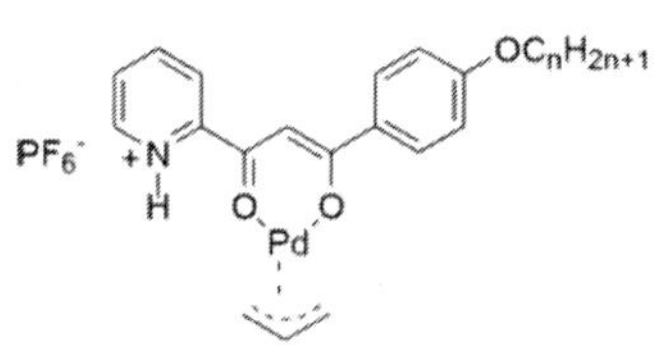

8

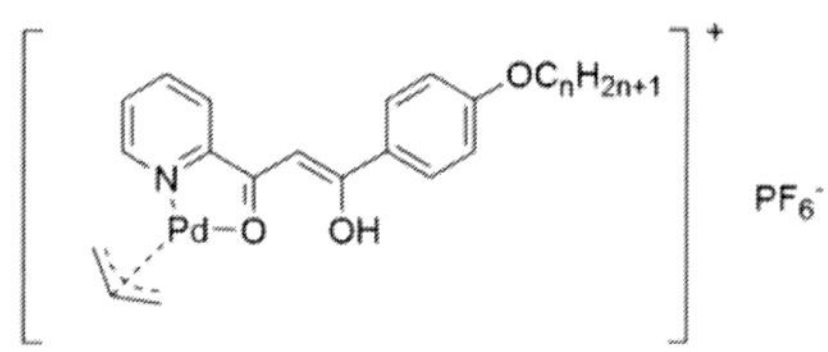

9

n = 12, 14, 16, 18

8, n = 14, Cr 84 Cr' 158 SmC 170 Iso **9**, n = 14, Cr 76 SmC 173 Iso

When the protonated pyridinium moiety is part of a α-diketone compound, the resulting β-diketone pyridinium chloride salts are not mesomorphic. Replacement of chloride anion with the tetrachlorozincate ion produced mesomorphic ionic salts displaying a SmA phases over a broad temperature range [9]. The same authors reported the thermal behavior and photophysical properties of the metal-free β-diketone pyridinium ligands, and their allyl-palladium(II) complexes (**8, 9**) [10, 11]. Interestingly, an equilibrium between **8** and **9** exists in solution. Another peculiar aspect about these complexes is that normally, the β-diketone ligands are first deprotonated before coordination to the metal center. Pd(II) metallomesogens with tetradecyloxy groups display a SmC phase, with reduced mesomorphic range for the protonated compound. The emission properties of Pd(II) complexes were preserved in the LC state.

2.2. N-Alkylated Pyridinium Salts

2.2.1. N-Methyl Pyridinium Salts

A clear trend for izotropization temperatures was seen for N-methyl pyridinium salts, meaning that, based on strong dependence of these temperatures on the alkyl chain length, longer alkyl chain length in position 4 of the pyridinium ring lead to higher transition temperatures. Moreover, it was observed that the 1-methyl-4-alkoxycarbonylpyridinium iodides salts have thermochromic properties. Color changes were observed on heating at both crystal to crystal and crystal to mesophase transitions, while such an observation could not be made for the 1-methyl-4-alkylpyridinium iodides salts [4, 12, 13].

10 **11**

10a, n = 8, X^- = I^-: Cr 103 Iso
10b, n = 10, X^- = I^-: Cr 113 Iso
10c, n = 11, X^- = I^-: Cr 116 SmA Iso
10d, n = 12, X^- = I^-: Cr 113 SmA 153 Iso
10e, n = 12, X^- = ClO_4^-: Cr 93 SmA 95 Iso
10f, n = 13, X^- = I^-: Cr 112 SmA 188 Iso
10g, n = 14, X^- = I^-:Cr 116 SmA 209 Iso

11a, n = 10, X^- = I^-: Cr 86 SmA154 Iso
11b, n = 12, X^- = I^-: Cr 94 SmA 178 Iso
11c, n = 14, X^- = I^-: Cr 99 SmA 182 Iso
11d, n = 16, X^- = I^-: Cr 103 SmA 182 Iso

2.2.2. Other N-Alkylated Pyridinium Salts

The simple N-alkylated chloride pyridinium salts **12** were obtained by reacting the corresponding alkyl chloride and pyridine, while the tetrachlorometalates, where M = Co(II) or Ni(II), were prepared by the reaction between these chloride pyridinium salts and anhydrous metallic chlorides [14]. The N-alkylated pyridinium chloride salts with a chain length below 12 units do not display liquid-crystal behavior.

Importantly, these salts (**12**) exhibit a SmA phase with relatively large mesophase ranges that become relatively independent of the chain length for alkyl chain length greater than 14 carbon atoms. On the contrary, for the tetrachlorometalate(II) salts, the mesophase stability increases monotonically with an increase in chain length. While initially, the n-dodecyl derivative has been found to show only a SmA phase, later it was reported that this compound showed cubic and smectic A phases on the heating run, and columnar (16–46°C), cubic (46–66°C), and smectic A (46–143°C) phases on the cooling run. The authors found that the phase transition temperatures were strongly influenced by a glass surface and on the thermal history [15].

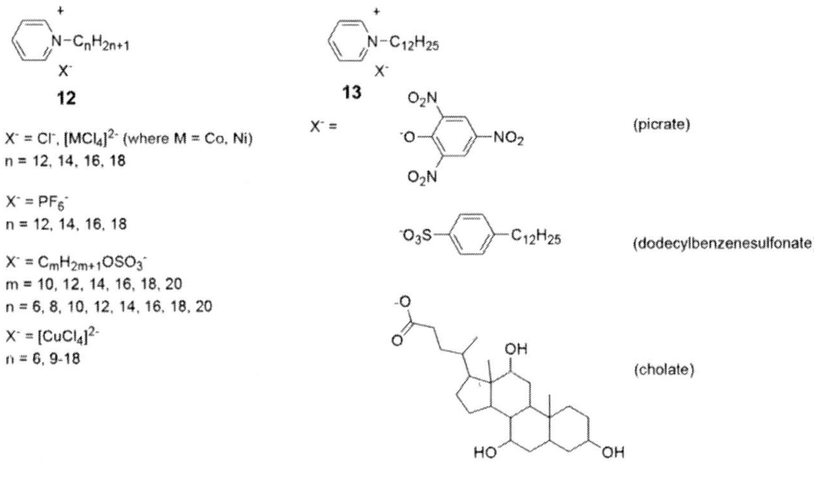

12a, n = 16, $X^- = PF_6^-$: Cr 104 Cr' 126 SmA 138 Iso
12b, n = 18, $X^- = PF_6^-$: Cr 107 Cr' 126 SmA 176 Iso
12c, n = 12, $X^- = [CuCl_4]^{2-}$: Cr 54 Col_h 76 Iso

Replacement of chloride anion with bulkier hexafluorophosphate anion led to mesophase destabilization for pyridinium salts with alkyl chain shorter than 16 carbon atoms. Extensive studies were performed for N-alkyl pyridinium salts with alkylsulfates as counterion, all of them displaying a typical SmA phase, the mesophase temperature range depending on the combination of chain lengths of both N-alkyl and the aliphatic chain connected to the sulfate group [16, 17].

The tetrachlorocuprate pyridinium salts show a very interesting and rich polymorphism for alkyl chain longer than or equal to 12 carbon atoms, with hexagonal columnar, cubic, and SmA phases appearing in the order of increasing the chain length and/or temperature [18].

By exchanging the chloride anion with biologically active picrate, dodecylbenzenesulfonate, or cholate anions, different thermal behavior was found [19]. Thus, only the compound with dodecylbenzenesulfonate anion displayed a SmA phase stable up to 152°C, while the other two showed no liquid crystalline properties. The absence of mesomorphic properties in the case of picrate anion was attributed to the formation of interlayer 3-D hydrogen bond network between pyridinium and picrate ions.

2.2.3. Chiral N-Alkylated Pyridinium Salts

Recently, a series of N-alkylated pyridinium salts containing a chiral center in the four positions with respect to nitrogen atoms have been prepared and investigated by Laschat et al. [20, 21]. Their thermal properties were compared to their imidazolium counterparts. Generally, these salts show a SmA phase, with melting and isotropization temperatures lower than the temperatures of their corresponding imidazolium counterparts. The mesophase stability range depends on the alkyl group length as well as on the counterion and it was found to decrease in the following order: Br^- > OAc^- > BF_4^- > I^- > SCN^-. The pyridinium salts with hexafluorophosphate anion show no liquid crystalline properties.

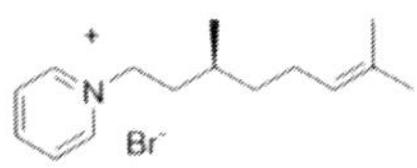

14

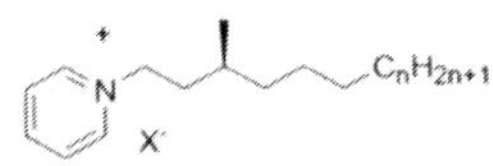

X^- = Br^-, I^-, SCN^-, BF_4^-, PF_6^-, OAc^-
n = 6-9

15

15a, n = 9, X^- = Br^-: Cr -8 Cr' -1 SmA 173 Iso
15b, n = 9, X^- = I^-: Cr -6 Cr' 41 SmA 171 Iso
15c, n = 9, X^- = SCN^-: Cr 2 Cr' 21 SmA 135 Iso
15d, n = 9, X^- = BF_4^-: Cr -19 Cr' 1 SmA 140 Iso
15e, n = 9, X^- = PF_6^-: Cr 63 Cr' 80 Cr" 126 Iso
15f, n = 9, X^- = OAc^-: Cr 0 SmA 162 Iso

For chiral 1-citronellylpyridinium bromide salt **14**, only a glass transition was detected and no mesomorphism was evidenced.

2.2.4. N-Alkylated Pyridinium Salts Derived From Picoline

The pyridinium salts derived from either 4- or 3-picoline, **16** and **17**, have been extensively studied [12, 22–24]. Their mesophase stability (SmA phase) depends strongly on the alkyl chain length as well as on the position of methyl group on the pyridinium ring. The mesophase temperature range increases significantly with increasing the alkyl chain length, while the alkyl substitution at the 3- and 4-positions on the pyridinium ring leads to a decrease in the melting point compared with the corresponding unsubstitution pyridinium salts. The effect of pyridinium ring substitution was also seen on the mesophase broadness, the pyridinium salts derived from 4-picoline having a much broader mesomorphic range compared to those derived from 3-picoline. Regarding the internal mesophase structure, the molecules are considered to pack in an interdigitated fashion, with the anions sandwiched between the pyridinium rings.

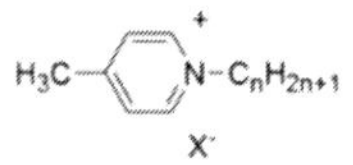

16

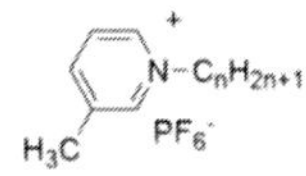

17

$X^- = Br^-, I^-$
n = 12, 14, 16, 18, 20, 22

$X^- = PF_6^-$
n = 12, 14, 16, 18

n = 12, 14, 16, 18

16a, n = 18, $X^- = Br^-$: Cr 92 SmA 158 Iso
16b, n = 20, $X^- = Br^-$, Cr 84 SmA 197 Iso
16c, n = 22, $X^- = Br^-$, Cr 91 SmA 207 Iso
16d, n = 22, $X^- = I^-$: Cr 99 SmA 198 Iso
17a, n = 12, $X^- = PF_6^-$: Cr 56 Iso
17b, n = 14, $X^- = PF_6^-$: Cr 71 Iso
17c, n = 16, $X^- = PF_6^-$: Cr 75 Iso 60 SmA 55 Cr
17d, n = 18, $X^- = PF_6^-$: Cr 88 Iso 84 SmA 77 Cr
16e, n = 12, $X^- = PF_6^-$: Cr 55 Iso
16f, n = 14, $X^- = PF_6^-$: Cr 68 Iso
16g, n = 16, $X^- = PF_6^-$: Cr 74 Iso 61 SmA 58 Cr
16h, n = 18, $X^- = PF_6^-$: Cr 87 SmA 94 Iso

A series of N-alkylated pyridinium salts, having an alkoxy group in 4-position of pyridinium ring, were synthesized by the reaction of either N-alkyl-4-pyridones or 4-alkoxypyridines with corresponding alkyl bromides. Further, hexafluorophosphate and tetrafluoroborate pyridinium salts could be prepared by a metathesis reaction with the ammonium salts.

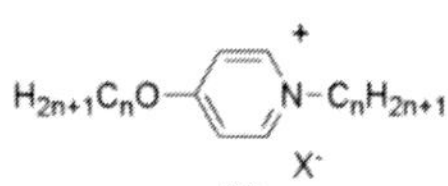

18

n = 10, 12, 14, 16, 18
$X^- = Br^-, PF_6^-, BF_4^-$

18a, n = 12, $X^- = Br^-$: Cr 70 SmA 120 Iso
18b, n = 12, $X^- = PF_6^-$: Cr 72 Iso 53 SmA 17 Cr
18c, n = 12, $X^- = BF_4^-$: Cr 39 SmA 77 Iso
18d, n = 14, $X^- = Br^-$: Cr 77 SmA 140 Iso
18e, n = 14, $X^- = PF_6^-$: Cr 67 SmA 94 Iso
18f, n = 14, $X^- = BF_4^-$: Cr 51 SmA 98 Iso

These salts exhibit smectic A phase. The transition temperatures, both melting and clearing, as well as the mesophase stability were influenced by the alkyl chain length and the counterion type. The hexafluorophosphate anion, PF_6^-, produced smaller mesophase ranges, followed by tetrafluoroborate anion, BF_4^-, when compared to bromide pyridinium salts as a consequence of a weaker cation-anion interaction for the BF_4^- or PF_6^- than for the Br^- [5].

Pyridinium salts with a 1,3-dioxane ring attached in 4-position were prepared and their liquid crystalline properties were investigated [25, 26]. A very wide range for the SmA mesophase (between −24 and 150°C) was found for ionic thermotropic liquid crystal system having two rings in its central core (**19**) reported by Haramoto et al. [25]. Based on the same design, chiral pyridinium liquid crystals **20** show an interesting thermal behavior [26].

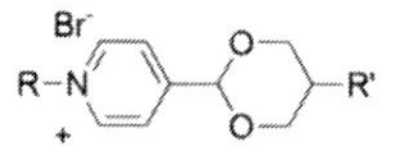

19

$R = C_{10}H_{21}$, $CH_2{=}CH\text{-}(CH_2)_8\text{-}$
$R' = C_2H_5$, $CH_2{=}CH\text{-}CH_2\text{-}$

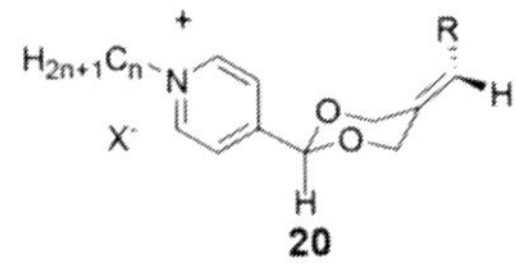

20

n = 3, 10
$X^- = Br^-, I^-, BF_4^-, PF_6^-, N(Tf)_2^-$
R = Br, Ph, p-MeO-C_6H_4, p-BuO-C_6H_4, p-OctO-C_6H_4, C_8H_{17}, $C_{18}H_{37}$

19a, Cr -24 SmA 150 Iso ($R = C_{10}H_{21}$, $R' = C_2H_5$)
20a, n = 10, $R = C_{18}H_{37}$, $X^- = I^-$: SmX 41 N 90 Iso (dec.)
20b, n = 10, R = p-BuO-C_6H_4, $X^- = I^-$: N 47 Iso
20c, n = 10, $R = C_8H_{17}$, $X^- = Br^-$: SmC 53 Iso

First, LC phases were observed only for long alkyl chains on both sides of the molecule. Furthermore, the authors found that the enantiomeric purity influences significantly the nature of the mesophase (chiral nematic phase N* for pure single enantiomeric compounds and nematic and smectic phases for racemic mixture) and only slightly the transition temperatures between solid state and LC phase and LC phase and isotropic state. Moreover, the LC phase stability depends on the counterion size (higher stability for halide and tetrafluoroborate anions) due to the contribution of bigger anions to decreasing the packing of the mesomorphic cationic units. Yousif et al. found an enantiotropic cholesteric phase (N*) for a series of quaternized cholesteryl isonicotinates with tosylate counterions [27]. The assignment of such a mesophase was made based on polarizing optical microscopy when a planar texture with bright oily streaks was identified for such compounds. No additional XRD studies were undertaken to characterize this phase.

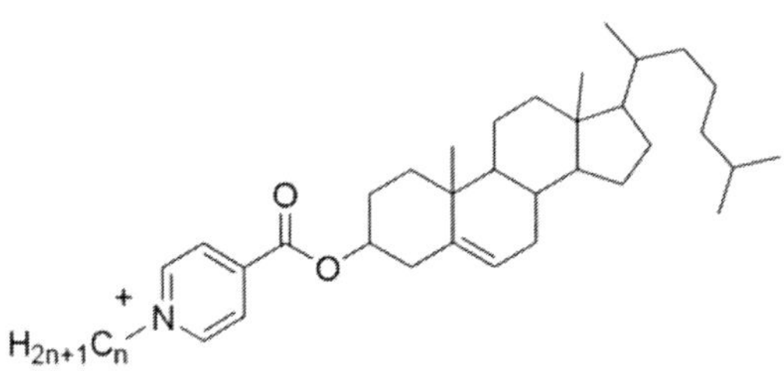

n = 2-8

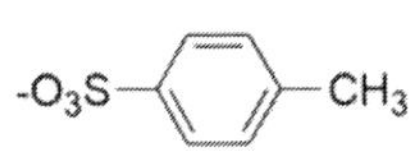

21

21a, n = 2, Cr 225 N* 241 Iso
21b, n = 3, Cr 176 N* 228 Iso
21c, n = 4, Cr 194 N* 225 Iso
21d, n = 5, Cr 181 N* 219 Iso
21e, n = 6, Cr 183 N* 236 Iso
21f, n = 7, Cr 177 N* 237 Iso
21g, n = 8, Cr 149 N* 218 Iso

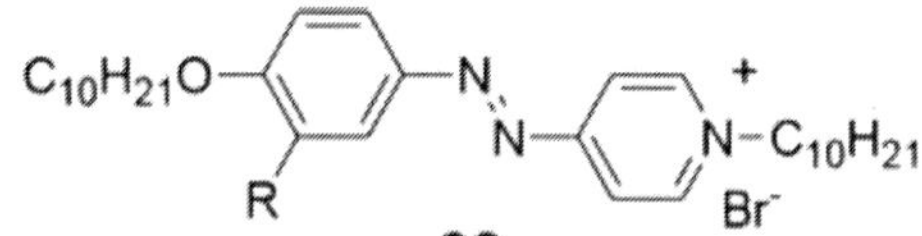

R = H, $OC_{10}H_{21}$

22

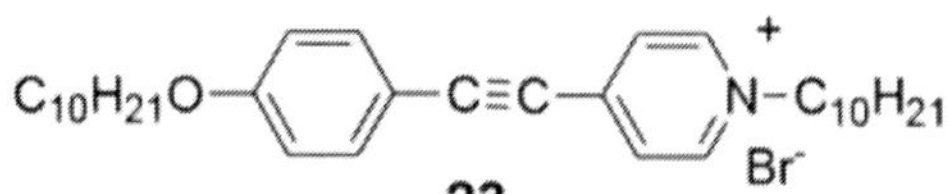

23

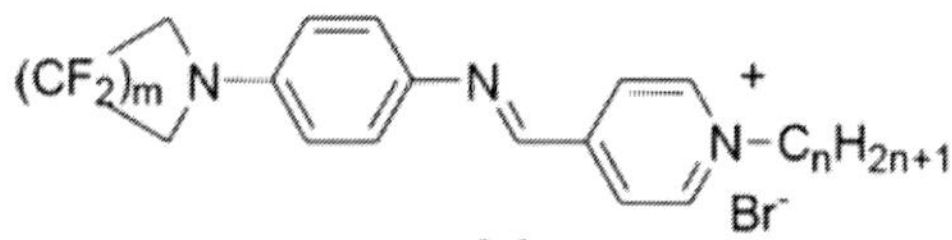

m = 2, 4
n = 14, 16, 18

24

22a, R = H, Cr 119 SmA 138 Iso
22b, R = $C_{10}H_{21}$, Cr 86 SmA 119 Iso
23, Cr 174 SmA 183 dec
24a, m =2, n = 14, Cr 87 M 156 Iso (dec.)
24b, m = 2, n = 16, Cr 90 M 159 SmA 175 Iso (dec)
24c, m = 4, n = 14, Cr 80 Iso
24d, m = 4, n = 16, Cr 80 Iso

Several other pyridinium ILCs having a second aromatic ring connected to the 4-position of the pyridinium ring via a linking group (azo, **22**, acetylene, **23**, or imine, **24**) were designed and investigated [28, 29]. Their thermal properties are influenced by the alkyl chain length, the nature of linking groups, while the dominant factor for **24** was the size of perfluorinated ring [29].

2.2.5. N-Alkyl-4'-Substitution-Stilbazolium Halides

N-alkyl-stilbazolium ILC, having different groups at the 4'-position of the stilbazolium head unit, were investigated by several authors [23, 30, 31]. The 4'-substitution at the stilbazolium unit has a great influence on the LC properties. For instance, the compounds with NO_2 or CN groups and bromide as counterion show a relatively narrow temperature range of the LC phase [30]. By introduction of chloride anions, the LC phase stability increases and the compounds decompose before reaching the isotropic phase. The use of dialkylamino group at the 4'-position in the stilbazolium core led to a significant decrease of the temperature range of the mesophase. For compounds **25a-f**, the substituents at the 4'-position of stilbazolium core increase the stability of the SmA phase in the following order: OCH_3 > OH > H, meaning that the mesomorphic behavior is also influenced by dipolar interactions resulting from OH and OCH_3 groups [23]. All these salts show an increase of the SmA phase range with increasing the alkyl chain length. On the other hand, the stability of the SmA phase is influenced by the size of the counterion; lower melting and clearing temperatures were seen for iodide compared to bromide stilbazolium salts.

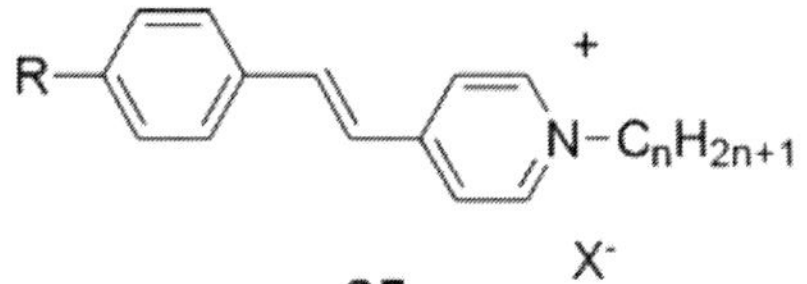

25

n = 5 - 10	n = 14, 16, 18	n = 14, 16, 18, 20, 22
R = NO_2, CN	R = $N(CH_3)_2$	R = OH, OCH_3, H
X- = Cl^-, Br^-, I^-	X^- = Br^-	X^- = Br^-, I^-

25a, R = H, n = 18, X^- = Br^-: Cr 153 SmA 217 Iso (dec)
25b, R = H, n = 18, X^- = I^-: Cr 189 Iso
25c, R = OH, n = 18, X^- = Br^-: Cr 172 SmA 230 Iso (dec)
25d, R = OH, n = 18, X^- = I^-: Cr 130 SmA 184 Iso
25e, R = OCH_3, n = 18, X^- = Br^-: Cr 63 SmA 255 Iso (dec)
25f, R = OCH_3, n = 18, X^- = I^-: Cr 62 Cr' 245 Iso (dec)

2.2.6. Monocationic Oxadiazole Pyridinium Salts

The pyridinium bromide salts **26a-d**, all of them show a SmA phase with a slight increase of clearing temperatures on increasing the alkyl chain length. By replacing the bromide ion with alkyl sulfate ions, a significant depression of clearing temperature was achieved (about 40°C) [32]. On the other hand, a total suppression of liquid crystalline phase was observed when the bis(2-ethylhexyl)-sulfosuccinate ion (DOSS) was employed. Interestingly, the pyridinium bromide salts display thermochromic properties [33]. Thus, the crystal to SmA transition of such compounds was accompanied with a color change from pale yellow to bright red that is almost fully reversible upon cooling. Such thermochromic effect was not observed on changing the bromide anion for alkylsulfate or DOSS ions. The origin of the thermochromic properties was attributed by authors to a charge-transfer couple formation between the bromide and the pyridinium ring.

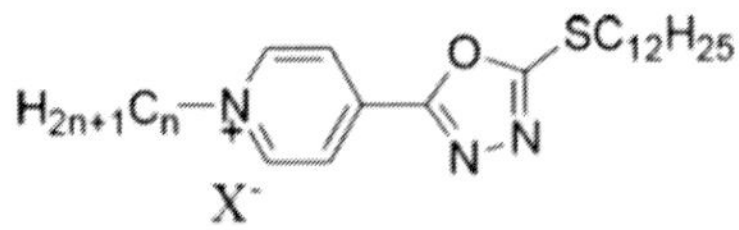

26

n=10,12,14,16; X^-=Br^-

n=12 $X^- = C_mH_{2m+1}$-SO_3^- m=10, 14

X^- =

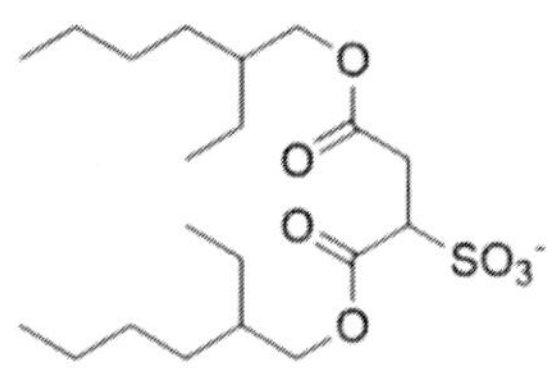

DOSS
bis(2-ethylhexyl)sulfosuccinate

26a, n = 10, X^- = Br^-: Cr 64 Cr' 117 SmA 172 Iso

26b, n = 12, X^- = Br^-: Cr 74 Cr' 119 SmA 185 Iso

26c, n = 14, X^- = Br^-: Cr 70 Cr' 118 SmA 186 Iso

26d, n = 16, X^- = Br^-: Cr 85 Cr' 115 SmA 189 Iso

More recently, a systematic study of liquid crystalline and photophysical properties of a series of pyridinium salts was reported [34, 35], where the 1,3,4-oxadiazole unit connects one pyridinium ring and one mono- or di-substitution phenyl ring with alkoxy groups.

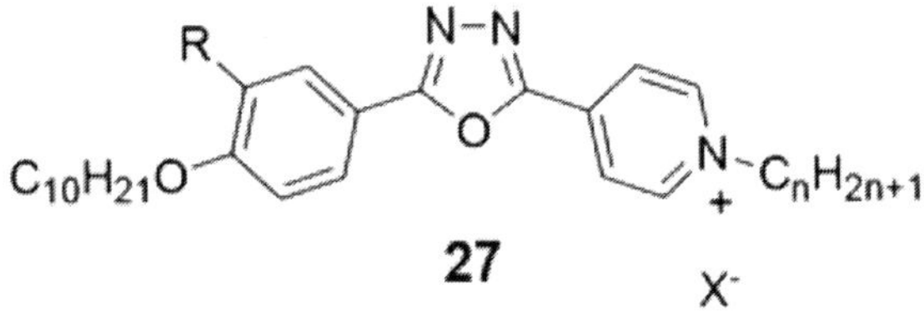

27

R=H n=1,10

R=$C_{10}H_{21}$

X-=Br^- (I^-), NO_3^-, BF_4^-, ClO_4^-, dodecylsulfate (DS)

27a, n = 1, R = H, X^- = BF_4^-: Cr 193 SmA 256 dec
27b, n = 1, R = H, X^- = ClO_4^-: Cr 202 SmA 257 dec
27c, n = 1, R = H, X^- = DS: Cr 122 SmA 208 dec
27d, n = 10, R = H, X^- = Br^-: Cr 182 SmA 193 dec
27e, n = 10, R = H, X^- = NO_3^-: Cr 126 SmA 171 dec
27f, n = 10, R = H, X^- = BF_4^-: Cr 153 SmA 213 Iso
27g, n = 10, R = H, X^- = ClO_4^-: Cr 167 SmA 217 Iso
27h, n = 10, R = H, X^- = DS: Cr 135 SmA 172 Iso
27i, n = 10, R = $C_{10}H_{21}$, X^- = Br^-: Cr 148 SmA 194 dec

The N-decyl pyridinium salts exhibit SmA mesomorphism, regardless of the counterion, while their N-methyl counterparts show decomposition and no liquid crystalline properties for iodine and nitrate anions, and, for the remaining anions (BF_4^-, ClO_4^-, and DS), the salts decompose before reaching the isotropic phase from the previous SmA phase, suggesting that the thermal properties (melting point, mesophase range, and clearing or decomposition temperature) are sensitive to counterion exchange. Moreover, the two pyridinium salts with DS as counterion are similar in terms of thermal behavior, and this was explained by XRD studies. These compounds have a monolayered SmA phase, instead of a bilayered SmA phase found for the others compounds of the series, and the same molecular lengths (due to the presence of the same counterion), and such a different mesophase structure could account for their similar behavior.

An interesting example showing how the 3'- or 4'-substitution pattern of the pyridinium ring affects the thermal behavior is represented by the iodides and trifluoromethanesulfonates salts derived from perfluoroalkylated 1,2,4-oxadiazolylpyridines 28–31 [36]. Thus, the 3'-

substitution derivatives exhibited liquid crystalline properties (SmX phase) on a narrow temperature range, while the corresponding 4'-substitution derivatives were found to pass from the crystalline state straight to the isotropic phase. Such a behavior was attributed to a charge delocalization over the entire structure, including the oxadiazole ring, which makes the cation/anion electrostatic interactions weaker and, thus, leading to the mesophase detabilization for 4'-substitution compounds.

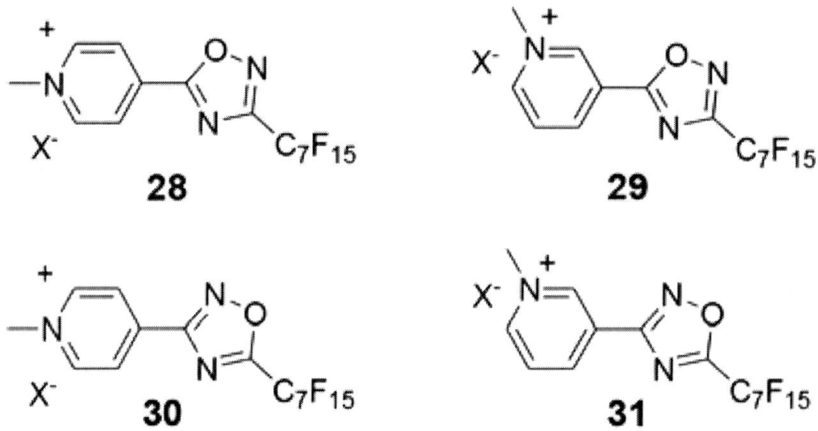

X- = I^-, OTf^-

29a, X^- = I^-: Cr 97 SmX 150 Iso

31a, X^- = I^-: Cr 113 SmX 135 Iso

The fact that only the iodide salts showed liquid crystalline properties was explained by the greater coordinating ability of the iodide anions with respect to the trifluoromethanesulfonates anions. Moreover, all iodide salts showed thermochromism phenomena suggesting prospective applications in optoelectronics.

2.3. Monocationic Pyridinium Salts With Mesogenic Groups Attached To Nitrogen Atom

A series of pyridinium salts with the 3,4,5-tridodecyloxybenzyl moiety attached to nitrogen atom and different counterions (bromide, nitrate, tetrafluoroborate, and hexafluorophosphate) **32** were prepared starting from 4-hydroxypyridine [37].

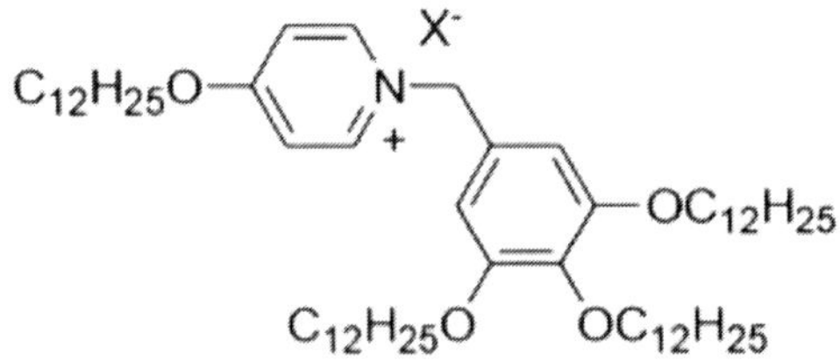

32

$X^- = Br^-, NO_3^-, BF_4^-, PF_6^-$

32a, $X^- = Br^-$: Cr 63 Col_1 108 Iso 105 Col_1 74 Col_2

The pyridinium bromide salt shows one enantiotropic columnar mesophase and one additional monotropic columnar phase at lower temperatures. The size of counterion has a significant contribution to the LC phase stability. Surprisingly, when the bromide ion (Br−) was replaced with bulkier counterions (NO_3^-, BF_4^-, and PF_6^-), the resulting products showed no liquid crystalline behavior. The photoluminescent properties of these pyridinium salts were investigated both in solution and solid state and it has been shown that their emission is only slightly influenced by the nature of counterion employed.

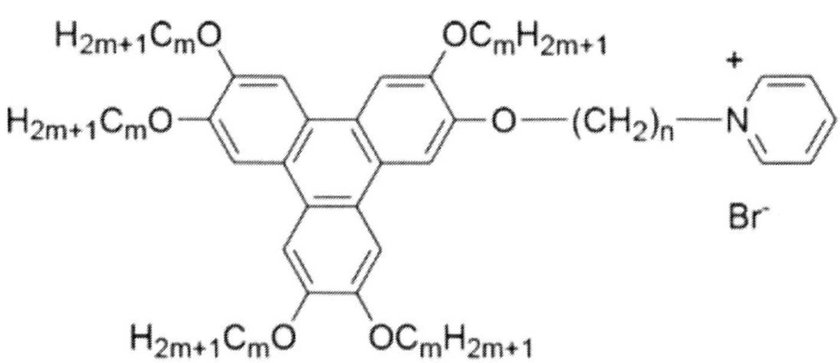

33

m = 4, n = 5, 6, 7
m = 5, 6, n = 5

33a, m = 4, n = 6, Cr 84 Col 95 Iso

The classical triphenylene unit was connected to pyridinium ring via a flexible methylene flexible chain to give discotic liquid crystals [38]. The mesophase was identified to be a columnar phase based on microscopy observations of optical textures. The stability of this columnar phase was found to depend both on the peripheral alkyl chain length as well as the methylene spacer length. Thus, the stability of columnar phase increases by increasing the number of carbon atoms on the peripheral chains of the triphenylene core, while longer spacer connecting the triphenylene unit with the pyridine ring destabilized the mesophase.

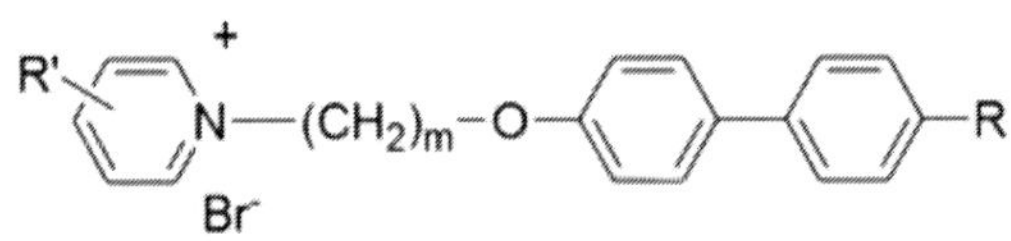

34

m = 6, 10

R = CN, $-OC_nH_{2n+1}$, n = 3, 10

R' = H, 4-ethyl, 2-ethyl, 3,5-dimethyl

34a, m = 6, n = 3, R' = H, Cr 151 SmA 159 Iso

34b, m = 10, n = 10, R' = H, Cr 136 SmC 165 Iso

34c, m = 10, n = 10, R' = 2-ethyl, Cr 75 SmC 116 SmA 180 Iso

Interesting results were obtained for pyridinium bromides salts containing a biphenyl core and alkyl chains of different lengths **34** [39]. The substitution of the pyridine ring greatly influences the thermal behavior of these salts. While the unsubstitution pyridinium groups promote mesomorphism (SmA and SmC phases), the 2- and 4-ethyl-substitution pyridinium groups give rise to liquid crystalline phases only with sufficiently long alkyl chains (decyl chains on both sides of the biphenyl core). The salts having a 3,5-dimethyl substitution pyridine ring do not show any liquid crystalline properties. Moreover, the substitution pattern at the pyridinium groups resulted in different types of smectic phases (SmA, SmC, and SmE).

When an additional ring was added, the new pyridinium bromide salts **35** with a triphenylene moiety displayed tilted smectic mesophases (SmC) with molecules stacked in a double layer morphology, as indicated by XRD studies [40]. The tilt angle of these compounds shows a great temperature dependence, and, in fact, at lower temperature being closed to 0°.

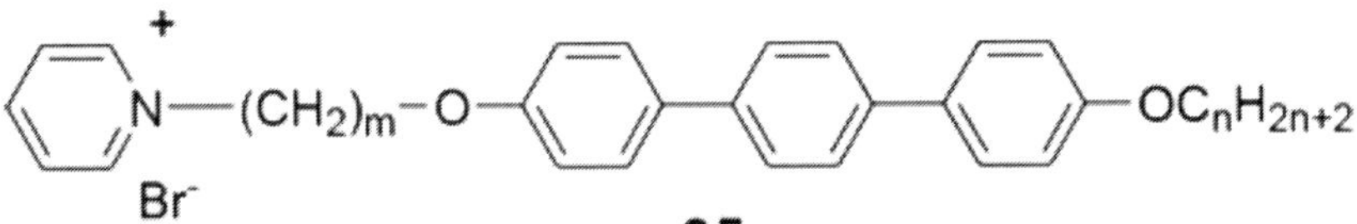

35

m = 8, 12
n = 14, 16

The XRD studies together with molecular modeling indicated that π-π interactions counterbalance the strong ionic forces leading to a full segregation of molecular parts in the smectic structures.

2.3.1. Dendritic Pyridinium Ionic Liquid Crystals

Pyridinium chloride salts having dendritic building blocks connected to the nitrogen atom have been investigated by Percec et al. [41]. Depending on the number of peripheral alkoxy groups, these molecules were found to form columns or spheres, leading to 2D hexagonal columnar phases or a 3D cubic phase, respectively.

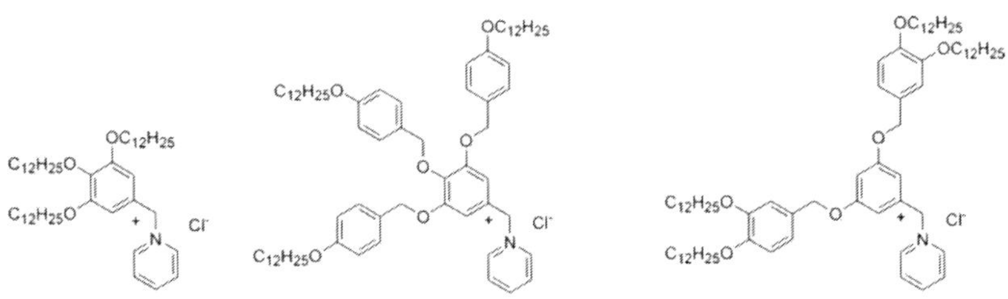

36, Cr 65 Col_h 166 Iso
37, Cr 15 Cr' 35 Cr" 57 Col_h 159 Iso
38, Cr 17 Cr' 92 Col_h 129 Cub 164 Iso

2.3.2. Nematic Pyridinium Ionic Liquid Crystals

Starting from 4-hydroxypyridine, it is possible to attach two cyanobiphenyl mesogenic units, via flexible alkyl spacer, on both sides of the pyridinium ring giving rise to a series of nematic pyridinium liquid crystals [42]. The nematic phase has numerous technological applications due to its highest fluidity of all LC phases and hence the possibility to align it by applying an external electric/magnetic field. Moreover, the nematic phase is commonly used in electro-optical devices. The nematic phase is quite rare in the case of ILCs, several examples have been reported so far for ammonium [43, 44], imidazolium [45–48], pyridinium-based ILC [49], or miscellaneous type of ILC [50–54]. Generally, as the smectic phases are the most common phases for ILCs, especially due to electrostatic interactions, the nematic phase can be seen rather as an exception. It has to be reminded here the cholesteryl-containing compounds 21a-g that display a cholesteric phase (chiral nematic phase N*) at relatively high temperatures and on a narrow temperature range. Compounds **39**, all of them with the exception of **39d** and **39e**, displayed a monotropic nematic phase on cooling from the isotropic state. This is the first example of a series of ILC that exhibit a long range nematic phase, no matter the counterion or spacer length between the ionic core and the mesogenic group.

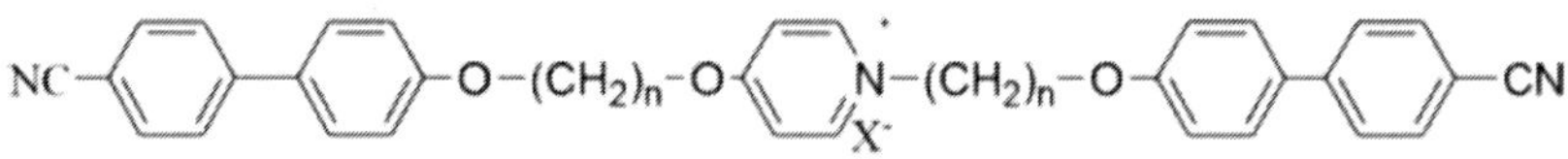

39

n = 6, 9, 10

X^- = Br^-, NO_3^-, BF_4^-, PF_6^-, OTf^-, SCN^-

39a, n = 6, X^- = Br^-: Cr 120 Cr'144 Iso 70 N 23 g
39b, n = 6, X^- = NO_3^-: Cr 81 Iso 65 N 22 g
39c, n = 6, X^- = BF_4^-: Cr 57 Cr'85 Cr"102 Iso 60b N 24 g
39d, n = 6, X^- = PF_6^-: Cr 119 Iso 21 g
39e, n = 6, X^- = OTf^-: Cr 62 Cr'89 Iso 23 g
39f, n = 6, X^- = SCN^-: Cr 121 Iso 70 N 24 g
39g, n = 9, X^- = Br^-: Cr 128 Iso 89 N 28 g
39h, n = 9, X^- = NO_3^-: Cr 92 Cr' 103 Iso 92 N 26 g
39i, n = 9, X^- = BF_4^-: Cr 84 Cr'97 Iso 76 N 22 g
39j, n = 9, X^- = PF_6^-: Cr 114 Iso 41 N 21 g
39k, n = 9, X^- = OTf^-: Cr 104 Iso 45 N 20 g
39l, n = 9, X^- = SCN^-: Cr 92 Cr'110 Iso 89 N 21 g
39m, n = 10, X^- = Br^-: Cr 114 Iso 83 N 47 Cr
39n, n = 10, X^- = NO_3^-: Cr 99 Iso 86 N 25 g
39o, n = 10, X^- = BF_4^-: Cr 91 Iso 68 N 20 g
39p, n = 10, X^- = PF_6^-: Cr 120 Iso 40 N 21 g
39r, n = 10, X^- = OTf^-: Cr 72 Iso 44 N 10 g
39s, n = 10, X^- = SCN^-: Cr 95 Iso 82 N 40 Cr

The nematic phase was identified by using the combination of the three techniques: POM, DSC, and XRD, as well as, miscibility studies with 5 CB and doping with a chiral compound. A marbled texture or thread-like texture could be seen by POM, which flashed brightly under pressure, while several samples also exhibited regions with not well developed Schlieren texture. Previous examples of pyridinium ILC showing a Schlieren texture were assigned to a SmC phase [39]. For this reason, additional XRD studies were undertaken to rule out the possibility of misinterpretation of experimental data. Indeed, the diffractograms showed no sharp peaks in the low-angle region, but just a broad signal centered at 4.5 Å assigned to the average intermolecular separation, close to the typical value for liquid crystalline phases, confirming the nematic phase nature. Additional confirmation came from miscibility studies with common nematic LC, a mixture of **39i** 40% wt. and 5 CB displayed the N-Iso transition at 54°C. Furthermore, doping **39i** with a chiral dopant, a typical fingerprint texture could be seen by POM, a good indication of nematic phase. It has been shown that the temperature range of the liquid crystalline phase is greatly influenced by the spacer length and the nature of counterion employed. In fact, the very bulky PF_6^- and OTf^- anions have a great contribution to the destabilization of nematic phase.

The NMR data are strongly dependent on the anion-cation interaction specific to ILC, and they can be related to the presence of hydrogen-bonding interactions that cause downfield chemical shift of the related H atoms for imidazolium- and pyridinium-based ILC. Stronger interactions with anions lead to further downfield chemical shifts, in particular for C-H adjacent to nitrogen atom. The chemical shifts of these protons, from the ^1H NMR spectra of pyridinium salts **39a-s**, were found to follow the order: $Br^- > SCN^- \sim NO_3^- > OTf^- > BF_4^- > PF_6^-$.

3. DICATIONIC BIS(PYRIDINIUM) AND POLYCATIONIC PYRIDINIUM SALTS

3.1. 4,4'-Bipyridinium-Based Ionic Liquid Crystals

The 4,4'-bipyridinium salts or viologens are a special class of materials that show interesting properties, such as electrochromism and electrical conductivity, which were successfully employed in producing liquid crystalline materials. The reports dealing with liquid crystals based on 4,4'-bipyridinium salts significantly grew after 2000. It is important to mention that such materials can be part of a two steps reduction process, and this process is reversible, as depicted below.

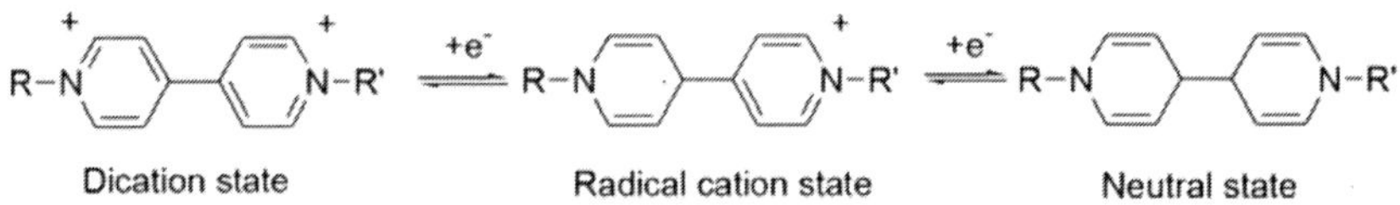

Additionally, it is of interest to note here that many dicationic and tetracationic ILC based on the 4,4'-bipyridinium rigid core (viologen-based ILC) have been investigated so far, displaying mesomorphic properties typically of calamitic and discotic materials [55–59].

The mesogenic properties of compounds based on asymmetric viologen salts of bis(trifluoromethanesulfonyl)amide ($[NTf_2]^-$) **40** were influenced by their symmetry. It was found that the strongly asymmetric system does not display mesomorphic behavior, having low melting points (below 40°C). The less asymmetric products exhibit a smectic phase (SmB) with a wide range of stability, from about 0°C up to above 140°C. More structural changes on these 4,4'-bipyridinium systems, in general, give rise to more ordered smectic phases. 4,4'-bipyridine group can also be employed to prepare polymeric liquid crystals **41**, and one example bearing a pentaphenylene connecting group via flexible chain length is presented below [60]. This polymeric ionic liquid crystal shows single layer smectic-type mesophases, both tilted (SmC) and orthogonal (SmA) phases.

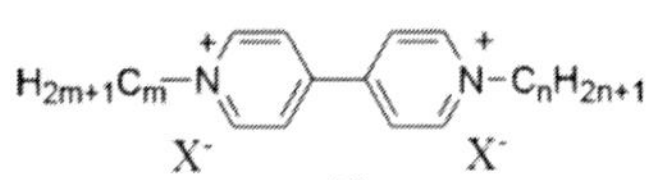

40

$X^- = Br^-, [NTf_2]^-$

m = 3-8, n = 8-11

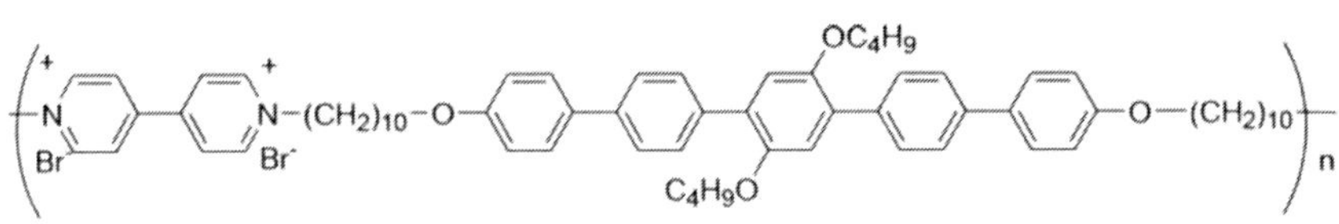

41

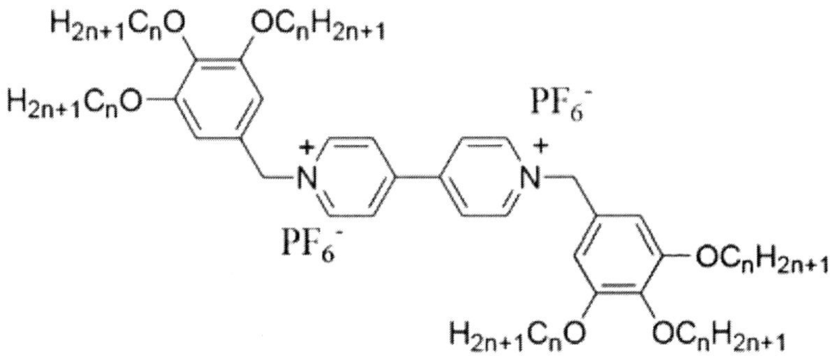

42

n = 8,12, 16

42a, n = 8, Cr -15 Col_h 160 dec
42b, n = 12, Cr 47 Col_r 160 dec
42c, n = 16, Cr 64 Col_r 160 dec

Columnar phases could be obtained for liquid crystals bearing two 3,4,5-tris(alkoxy) benzyl units attached to the 4,4'-bipyridinium moiety [61]. The viologen product bearing six octyloxy chains shows a hexagonal columnar (Col_h) phase, while the analogues with longer alkoxy chains (12 or 16) show rectangular columnar (Col_r) phases. For these products, the clearing temperatures could not be measured due to thermal degradation before reaching the isotropic state. Different molecular packing (Col_h or Col_r) was explained by formation of more elliptical molecular structures for compounds with elongated terminal alkoxy chains that may prefer to form the Col_r phases.

3.2. Dicationic Bis(Pyridinium) Salts With Flexible Linker

Bis(pyridinium) salts with flexible spacers **43** derived from 4-hydroxypyridine containing mesogenic 3,4,5-tris(alkyloxy)benzyl moieties (alkyl = dodecyl or tetradecyl) on each side and various counterions, such as bromide (Br^-), hexafluorophosphate (PF_6^-), tetrafluoroborate (BF_4^-), and triflate (OTf^-), were reported recently [62]. While there are numerous examples of dicationic pyridinium ILC, this one represents the first example of this type. These dicationic pyridinium salts display an enantiotropic liquid crystalline behavior with a hexagonal columnar (Col_h) phase assigned on the basis of their characteristic texture, pseudo-focal conic and spherulitic textures, when observed by POM and XRD studies. The temperature range of the hexagonal columnar phase is greatly influenced by the terminal chain length (12 or 14) and the nature of counterion employed. For example, the triflate salt with 12 carbon atoms in the terminal chains, **43d**, has the lowest melting point, 15°C, below ambient temperature and does not show any mesogenic behavior. It was found that the thermal behavior resembles the general trend found for pyridinium ILC, where transition temperatures show a decreasing tendency with an increase in the size of anions. Furthermore, the compounds with higher number of carbon atoms have higher melting points and isotropization temperatures with broader mesophase ranges (31°C for **43a** and 40°C for **43e**).

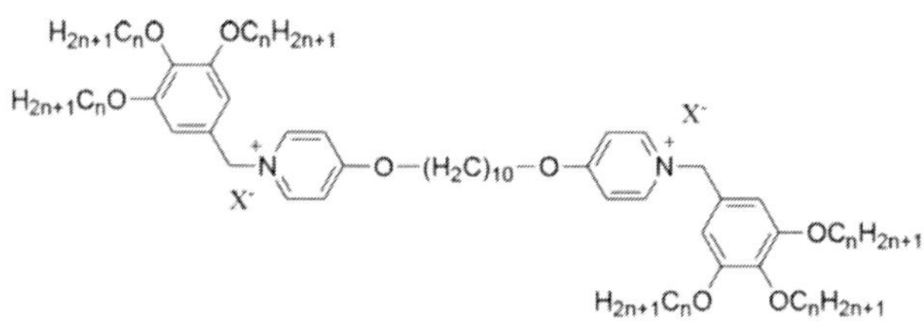

43

n = 12, 14

X^- = Br^-, PF_6^-, BF_4^-, OTf^-

43a, n = 12, X^- = Br^-: Cr 21 Cr' 39 Col_h 70 Iso
43b, n = 12, X^- = PF_6^-: g 8 Col_h 87 Iso
43c, n = 12, X^- = BF_4^-: Cr 14 Col_h 84 Iso
43d, n = 12, X^- = OTf^-: Cr 15 Iso
43e, n = 14, X^- = Br^-: Cr 39 Col_h 79 Iso
43f, n = 14, X^- = PF_6^-: Cr 34 Col_h 100 Iso
43g, n = 14, X^- = BF_4^-: Cr 37 Col_h 77 Iso
43h, n = 14, X^- = OTf^-: Cr 36 Col_h 47 Iso

The study of emission properties of these luminescent bis(pyridinium) salts revealed a weak emission in dichloromethane solutions at room temperature, with quantum yields up to 4.4%. Their solid-state emission is significantly red shifted by comparison to solution emission spectra recorded in dichloromethane, suggesting a more complex emission mechanism, probably an aggregate-type emission in solid state [62].

3.3. Dicationic Bis(Pyridinium) Salts With A Rigid Core

Dicationic pyridinium salts with an anthracene moiety connecting two pyridine rings substitution with tris(alkoxy)benzyl groups **44** were designed and investigated for their liquid crystalline properties and stimulus responsive luminescence [63].

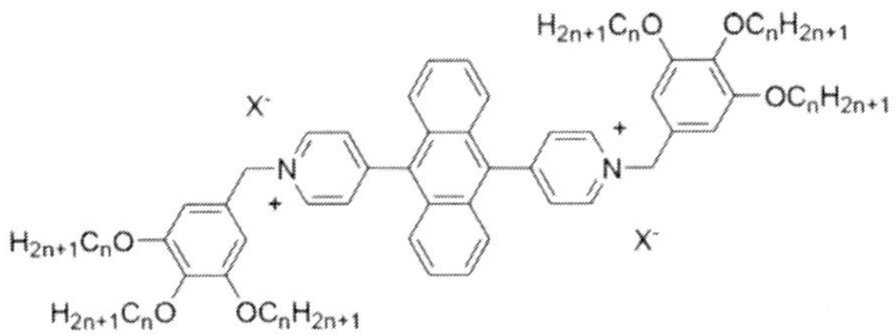

44

n = 8, 12
X- = Br⁻, BF_4^-, PF_6^-

44a, n = 8, X^- = Br^-: Cr 197 Col_r 228 dec
44b, n = 8, X^- = PF_6^-: Cr 216 Iso 151 Col_r
44c, n = 8, X^- = BF_4^-: Cr 181 Col_r 218 dec
44d, n = 12, X^- = Br^-: Cr 203 Col_r 235 dec
44e, n = 12, X^- = PF_6^-: Cr 220 Iso 148 Col_r
44f, n = 12, X^- = BF_4^-: Cr 185 Col_r 205 dec

The anthracene moiety act as a fluorophore in the center of molecules, and the solvate of pyridiniums salts **44**, *n* = 1, show piezoluminescence by grinding. The anion has an influence on the LC phase stability and thus, the long alkyl chain derivatives **44a-f** show either enantiotropic (for bromide and tetrafluoroborate salts) or monotropic (in the case of hexafluorophosphate salts) rectangular columnar phases.

3.4. Tripodal Pyridinium ILC

Photoluminescent tripodal pyridinium-based ionic liquid crystals were reported by Kato et al. [64, 65]. These tricationic pyridinium salts show thermotropic hexagonal or rectangular columnar phases or cubic phases with a large temperature ranges. The columnar or cubic phase stability is given by the difference in the anion size. For small-size anions (bromide), the cationic pyridinium cores should be packed more closely through electrostatic interactions and hence, the molecules prefer to self-assemble in cubic structures at higher temperatures (**45a**). Upon photoirradiation with UV light, these pyridinium salts emit blue-green or green light with photoluminescence quantum yields rather low, 3–6% in solution [65].

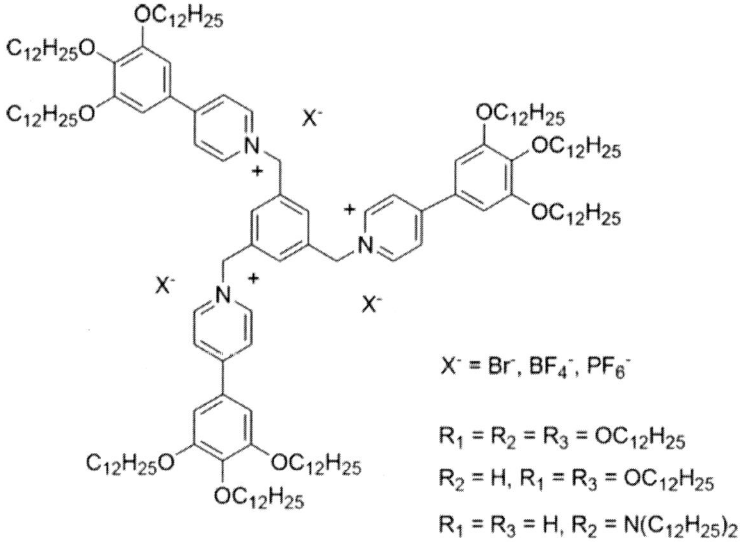

45

45a, X^- = Br^-: Cr 77 Col_h 179 Cub 245 Iso
45b, X^- = BF_4^-: Cr -3 Col_h 160 Iso
45c, X^- = PF_6^-: Cr 13 Cr' 24 Col_r 245 Iso

4. ACKNOWLEDGEMENTS

This work was supported by a grant of the Romanian Authority for Scientific Research, CNCS-UEFISCDI, and project number PN-II-ID-PCE-2011-3-0384.

REFERENCES

1. K. V. Axenov, S. Laschat, Thermotropic ionic liquid crystals, Materials, 2011, 4, 206–259. DOI: 10.3390/ma4010206
2. K. Binnemans, Ionic liquid crystals, Chem. Rev., 2005, 105, 4148–4204. DOI: 10.1021/cr0400919
3. K. Goossens, K. Lava, C. W. Bielawski, K. Binnemans, Chem. Rev., 2016, 116, 4643–4807. DOI: 10.1021/cr400334b
4. E. J. R. Sudholter, J. B. F. N. Engberts, W. H. de Jeu, Thermotropic liquid-crystalline behavior of some single- and double-chained pyrdinium amphyphiles, J. Phys. Chem., 1982, 86, 1908–1913. DOI: 10.1021/j100207a035

5. J.-T. Lu, C.-K. Lee, I. J. B. Lin, Ionic liquid crystals derived from 4-hydroxypyridine, Soft Matter, 2011, 7, 3491–3501. DOI: 10.1039/c0sm01376e
6. X. Yang, S. Tan, T. Liang, B. Wei, Y. Wu, Synthesis, characterization, and electrochemical properties of smectic pyridinium salts with inorganic dihydrogen phosphate ions, Ionics, 2016, 22, 85–92. DOI: 10.1007/s11581-015-1524-x
7. X. Zhu, B. Tartsch, U. Beginn, M. Müller, Wedge-shaped molecules with a sulfonate group at the tip–a new class of self-assembling amphiphiles, Chem. Eur. J., 2004, 10, 3871–3878. DOI: 10.1002/chem.200400050
8. U. Beginn, L. Yan, S. N. Chvalun, M. A. Shcherbina, A. Bakirov, M. Möller, Thermotropic columnar mesophases of wedge-shaped benzenesulfonic acid mesogens, Liq. Cryst., 2008, 35, 1073–1093. DOI: 10.1080/02678290802376107
9. M. J. Mayoral, P. Ovejero, J. A. Campo, J. V. Heras, E. Pinilla, M. R. Torres, M. Cano, Ionic liquid crystals from β-diketonyl containing pyridinium cations and tetrachlorozincate anions, Inorg. Chem. Commun., 2009, 12, 214–218. DOI: 10.1016/j.inoche. 200 8.12.016
10. M. J. Mayoral, P. Ovejero, J. A. Campo, J. V. Heras, E. Oliveira, B. Pedras, C. Lodeiro, M. Cano, Exploring photophysical properties of new boron and palladium(II) complexes with β-diketone pyridine type ligands: from liquid crystals to metal fluorescence probes, J. Mater. Chem., 2011, 21, 1255–1263. DOI: 10.1039/ c0jm02698k
11. M. J. Mayoral, P. Cornago, R. M. Claramunt, M. Cano, Pyridyl and pyridiniumyl β-diketones as building blocks for palladium(II) and allyl-palladium(II) isomers. Multinuclear NMR structural elucidation and liquid crystal behaviour, New J. Chem., 2011, 35, 1020–1030. DOI: 10.1039/c0nj00938e
12. C. G. Bazuin, D. Guillon, A. Skoulios, J.-F. Nicoud, The thermotropic mesophase structure of two long-chain alkyl pyridinium halides, Liq. Cryst., 1986, 1, 181–188. DOI: 10.1080/02678298608086504
13. J. J. H. Nusselder, J. B. F. N. Engberts, H. A. Van Doren, Liquid crystalline and thermochromic behaviour of 4-substituted 1-methylpyridinium iodide surfactants, Liq. Cryst., 1993, 13, 213–225. DOI: 10.1080/02678299308026295
14. C. J. Bowlas, D. W. Bruce, K. R. Seddon, Liquid-crystalline ionic liquids, Chem. Commun., 1996, 1625–1626. DOI: 10.1039/CC9960001625
15. S. Ujiie, A. Mori, Cubic mesophase formed by thermotropic liquid crystalline ionic systems-effects of polymeric counter ion, Mol. Cryst. Liq. Cryst., 2005, 437, 1269–1275.
16. C. Cruz, B. Heinrich, A. C. Ribeiro, D. W. Bruce, D. Guillon, Structural study of smectic A phases in homologous series of N-alkylpyridinium alkylsulphates, Liq. Cryst., 2000, 27, 1625–1631. DOI: 10.1080/026782900750037185
17. D. W. Bruce, S. Estdale, D. Guillon, B. Heinrich, Mesomorphic N-alkylpyridinium dodecylsulphates, Liq. Cryst., 1995, 19, 301–305. DOI: 10.1080/02678299508031983
18. F. Neve, O. Francescangeli, A. Crispini, J. Charmant, A2[MX4] Copper(II) pyridinium salts. From ionic liquids to layered solids to liquid crystals, Chem. Mater., 2001, 13, 2032–2041. DOI: 10.1021/cm000804d
19. T. Mihelj, V. Tomašić, Thermal behavior of dodecylpyridinium-based surfactant salts with varied anionic constituent, J. Dispers. Sci. Technol., 2014, 35(4), 581–592. DOI: 10.1080/01932691.2013.811683
20. G. Kohnen, M. Tosoni, S. Tussetschläger, A. Baro, S. Laschat, Counterion Effects on the mesomorphic properties of chiral imidazolium and pyridinium ionic liquid, Eur. J. Org. Chem., 2009, 5601–5609. DOI: 10.1002/ejoc.200900730
21. M. Tosoni, S. Laschat, A. Baro synthesis of novel chiral ionic liquids and their phase behavior in mixtures with smectic and nematic liquid crystals, Helv. Chim. Acta, 2004, 87, 2742–2749. DOI: 10.1002/hlca.200490247
22. M. Tabrizian, A. Soldera, M. Couturier, C. G. Bazuin, Pyridinium salt liquid crystals effect of mesogen extension and alkyl chain length, Liq. Cryst., 1995, 18, 475–482. DOI: 10.1080/02678299508036647

23. D. Ster, U. Baumeister, J. Chao, C. Tschierske, G. Israel, Synthesis and mesophase behaviour of ionic liquid crystals. J. Mater. Chem., 2007, 17, 3393–3400. DOI: 10.1039/b705519f
24. C. M. Gordon, J. D. Holbrey, A. R. Kennedya, K. R. Seddon, Ionic liquid crystals: hexafluorophosphate salts, J. Mater. Chem., 1998, 8, 2627–2636. DOI: 10.1039/A806169F
25. Y. Haramoto, S. Ujiie, M. Nanasawa, New pyridinium type ionic liquid crystal compounds with a 1,3-dioxane ring in the principal structure, Liq. Cryst., 1996, 21, 923–925. DOI: 10.1080/02678299608032911
26. J. Baudoux, P. Judeinstein, D. Caharda, J.-C. Plaquevent, Design and synthesis of novel ionic liquid/liquid crystals (IL2Cs) with axial chirality, Tetrahedron Lett., 2005, 46, 1137–1140. DOI: 10.1016/j.tetlet.2004.12.097
27. Y. Z. Yousif, A. A. Othman, W. A. Al-masoudi, P. R. Alapati, Some novel cholesteric liquid crystals, Liq. Cryst., 1992, 12, 363–368. DOI: 10.1080/02678299208031053
28. J. Tao, J. Zhong, P. Liu, S. Daniels, Z. Zeng, Pyridinium-based ionic liquid crystals with terminal fluorinated pyrrolidine, J. Fluor. Chem., 2012, 144, 73–78. DOI: 10.1016/j.jfluchem.2012.07.009
29. N. G. Filippi, D. Zambelli Mezalira, S. Ovalle, E. Westphal, Study of the mesomorphic behaviour through the structure modification of azo and acetylene pyridinium and imidazolium-based ionic liquid crystals, Liq. Cryst., 2016. DOI: 10.1080/02678292.2016.1199817
30. Y. Kosaka, T. Kato, T. Uryu, Thermotropic liquid crystalline ionic stilbazoles and their miscible mixtures with non-ionic carbazolyl compounds, Liq. Cryst., 1995, 18, 693–698. DOI: 10.1080/02678299508036678
31. K. Binnemans, C. Bex, R. Van Deun, Ionic liquid crystals with hemicyanine chromophores, J. Incl. Phenom. Macrocycl. Chem., 1999, 35, 63. DOI: 10.1023/A:1008146431142
32. D. Haristoy, D. Tsiourvas, Novel ionic liquid-crystalline compounds bearing oxadiazole and pyridinium moieties as prospective materials for optoelectronic applications, Chem. Mater., 2003, 15, 2079–2083. DOI: 10.1021/cm021365g
33. D. Haristoy, D. Tsiourvas, Effect of counterions on the thermotropic and thermochromic properties of ionic liquid crystals, Liq. Cryst., 2004, 31, 697–703. DOI: 10.1080/02678290410001675110
34. E. Westphal, D. Henrique da Silva, F. Molin, H. Gallardo, Pyridinium and imidazolium 1,3,4-oxadiazole ionic liquid crystals: a thermal and photophysical systematic investigation, RSC Adv., 2013, 3, 6442–6454. DOI: 10.1039/C3RA23456H
35. J. A. Pedro, J. R. Mora, E. Westphal, H. Gallardo, H. D. Fiedler, F. Nome, Photophysical study and theoretical calculations of an ionic liquid crystal bearing oxadiazole, J. Mol. Struct., 2012, 1016, 76–81. DOI: 10.1016/j.molstruc.2012.02.046
36. F. Lo Celso, I. Pibiri, A. Triolo, R. Triolo, A. Pace, S. Buscemib, N. Vivona, Study on the thermotropic properties of highly fluorinated 1,2,4-oxadiazolylpyridinium salts and their perspective applications as ionic liquid crystals, J. Mater. Chem., 2007, 17, 1201–1208. DOI: 10.1039/b615190f
37. Pana, F. L. Badea, M. Ilis, T. Staicu, M. Micutz, I. Pasuk, V. Cîrcu, Effect of counterion on the mesomorphic behavior and optical properties of columnar pyridinium ionic liquid crystals derived from 4-hydroxypyridine, J. Mol. Struct., 2015, 1083, 245–251. DOI: 10.1016/j.molstruc.2014.11.059
38. S. Kumar, S. K. Pal, Ionic discotic liquid crystals: synthesis and characterization of pyridinium bromides containing a triphenylene core, Tetrahedron Lett., 2005, 46, 4127–4130. DOI: 10.1016/j.tetlet.2005.03.201
39. L. Cui, V. Sapagovas, G. Lattermann, Synthesis and thermal behaviour of liquid crystalline pyridinium bromides containing a biphenyl core, Liq. Cryst., 2002, 29, 1121–1132. DOI: 10.1080/02678290210155589
40. R. G. Santos-Martell, A. Ceniceros-Olguín, L. Larios-López, R. J. Rodríguez-González, D. Navarro- Rodríguez, B. Donnio, D. Guillon, Synthesis and thermotropic liquid-crystalline properties of N-alkylpyridinium bromides substituted with a

terphenylene moiety, Liq. Cryst., 2009, 36, 787–797. DOI: 10.1080/02678290903060618

41. M. R. Imam, M. Peterca, U. Edlund, V. S. K. Balagurusamy, V. Percec, Dendronized supramolecular polymers self-assembled from dendritic ionic liquids, J. Polym. Sci.: Part A: Polym. Chem., 2009, 47, 4165–4193. DOI: 10.1002/pola.23523
42. Pană, I. Pasuk, M. Micutz, V. Cîrcu, Nematic ionic liquid crystals based on pyridinium salts derived from 4-hydroxypyridine, CrystEngComm., 2016, 18, 5066–5069. DOI: 10.1039/C6CE00618C
43. W. Li, J. Zhang, B. Li, M. Zhang, L. Wu, Branched quaternary ammonium amphiphiles: nematic ionic liquid crystals near room temperature, Chem. Commun., 2009, 5269–5271. DOI: 10.1039/B909605A
44. L. Lu, N. Sharma, G.A.N. Gowda, C.L. Khetrapal, R.G. Weiss, Enantiotropic nematic phases of quaternary ammonium halide salts based on trioctadecylamine, Liq. Cryst., 1997, 22, 23–28. DOI: 10.1080/026782997209630
45. K. Goossens, P. Nockemann, K. Driesen, B. Goderis, C. Görller-Walrand, K. Van Hecke, L. Van Meervelt, E. Pouzet, K. Binnemans, T. Cardinaels, Imidazolium ionic liquid crystals with pendant mesogenic groups, Chem. Mater., 2008, 20, 157–168. DOI: 10.1021/cm702321c
46. Pană, M. Ilis, M. Micutz, F. Dumitrascu, I. Pasuk, V. Cîrcu, Liquid crystals based on silver carbene complexes derived from dimeric bis(imidazolium) bromide salts, RSC Adv., 2014, 4, 59491–59497. DOI: 10.1039/C4RA11023D
47. S. Ahn, S. Yamakawa, K. Akagi, Liquid crystallinity-embodied imidazolium-based ionic liquids and their chiral mesophases induced by axially chiral tetra-substituted binaphthyl derivatives, J. Mater. Chem. C, 2015, 3, 3960–3970. DOI: 10.1039/C4TC02968B
48. X. H. Cheng, X. Q. Bai, S. Jing, H. Ebert, M. Prehm, Self-assembly of imidazolium-based rodlike ionic liquid crystals: transition from lamellar to micellar organization, Chem. Eur. J., 2010, 16, 4588–4601. DOI: 10.1002/chem.200903210
49. X. Liu, J. L. Liu, B. Cai, X. M. Ren, A charge transfer salt consisted of bis(maleonitriledithiolato)zincate dianion and 1,1'-didecyl-4,4'-bipyridinium exhibiting uncommon nematic mesophase behavior, Inorg. Chem. Commun., 2011, 14, 1428–1431. DOI: 10.1016/j.inoche.2011.05.038
50. M. Ghedini, D. Pucci, Cyclopalladation of 5-(1-hexyl)-2{[4′-(1-undecyloxy)phenyl]}-pyrimidine. Synthesis and characterization of mononuclear complexes, J. Organomet. Chem., 1990, 395, 105–112. DOI: 10.1016/0022-328X(90)85265-Z
51. Liebmann, C. Mertesdorf, T. Plesnivy, H. Ringsdorf, J. H. Wendorff, Complexation of transition metal ions with substituted aza macrocycles: induction of columnar mesophases by molecular recognition, Angew. Chem. Int. Ed. Engl., 1991, 30, 1375–1377. DOI: 10.1002/anie.199113751
52. J. W. Goodby, G. H. Mehl, I. M. Saez, R. P. Tuffin, G. Mackenzie, R. Auzely-Velty, T. Benvegnu, D. Plusquellec, Liquid crystals with restricted molecular topologies: supermolecules and supramolecular assemblies, Chem. Commun., 1998, 2057–2070. DOI: 10.1039/A802762E
53. B. Ringstrand, A. Jankowiak, L. E. Johnson, P. Kaszynski, D. Pociecha, E. Gorecka, Anion-driven mesogenicity: a comparative study of ionic liquid crystals based on the [closo-1-CB9H10]– and [closo-1-CB11H12]–clusters, J. Mater. Chem., 2012, 22, 4874–4880. DOI: 10.1039/C2JM15448J
54. Jankowiak, J. Kanazawa, P. Kaszynski, R. Takita, M. Uchiyama, [closo-1-CB11H11-1-Ph]– as a structural element for ionic liquid crystals, J. Organomet. Chem., 2013, 747, 195–200. DOI: 10.1016/j.jorganchem.2013.05.034
55. V. Causin, G. Saielli, Effect of asymmetric substitution on the mesomorphic behaviour of low-melting viologen salts of bis(trifluoromethanesulfonyl)amide. J. Mater. Chem. 2009, 19, 9153–9162. DOI: 10.1039/b915559g
56. V. Causin, G. Saielli, Effect of a structural modification of the bipyridinium core on the phase behaviour of viologen-based bistriflimide salts. J. Mol. Liq. 2009, 145, 41–47. DOI: 10.1016/j.molliq.2008.11.013

57. G. Casella, V. Causin, F. Rastrelli, G. Saielli, Viologen-based ionic liquid crystals: induction of a smectic A phase by dimerisation, Phys. Chem. Chem. Phys., 2014, 16, 5048–5051. DOI: 10.1039/C3CP54628D
58. M. Bonchio, M. Carraro, G. Casella, V. Causin, F. Rastrelli, G. Saielli, Thermal behaviour and electrochemical properties of bis(trifluoromethanesulfonyl)amide and dodecatungstosilicate viologen dimers, Phys. Chem. Chem. Phys., 2012, 14, 2710–2717. DOI: 10.1039/C2CP23580C
59. G. Casella, V. Causin, F. Rastrelli, G. Saielli, Ionic liquid crystals based on viologen dimers: tuning the mesomorphism by varying the conformational freedom of the ionic layer, Liq. Cryst., 2016, 43, 1161–1173. DOI: 10.1080/ 02678292. 2016.1161852
60. J. C. Díaz-Cuadros, L. Larios-López, R. J. Rodríguez-González, B. Donnio, D. Guillon, D. Navarro-Rodríguez, Ionic liquid crystals bearing bipyridinium and pentaphenylene groups, J. Mol. Liq., 2010, 157, 133–141. DOI: 10.1016/j.molliq.2010.09.002
61. K. Tanabe, T. Yasuda, M. Yoshio, T. Kato, Viologen-based redox-active ionic liquid crystals forming columnar phases, Org. Lett., 2007, 9(21), 4271–4274. DOI: 10.1021/ol701741e.
62. Pană, M. Iliş, T. Staicu, I. Pasuk, V. Cîrcu, Columnar bis(pyridinium) ionic liquid crystals derived from 4-hydroxypyridine: synthesis, mesomorphism and emission properties. Liq. Cryst., 2016, 43, 381–392. DOI: 10.1080/02678292.2015.1116630
63. S. Kohmoto, T. Chuko, S. Hisamatsu, Y. Okuda, H. Masu, M. Takahashi, K. Kishikawa, Piezoluminescence and liquid crystallinity of 4,4′-(9,10-anthracenediyl) bispyridinium salts crystal. Growth Des., 2015, 15 (6), 2723–2731. DOI: 10.1021/acs.cgd.5b00028
64. K. Tanabe, Y. Suzui, M. Hasegawa, T. Kato, Full-color tunable photoluminescent ionic liquid crystals based on tripodal pyridinium, pyrimidinium, and quinolinium salts, J. Am. Chem. Soc. 2012, 134, 5652–5661. DOI: 10.1021/ja3001979
65. K. Tanabe, T. Yasuda, T. Kato, Luminescent ionic liquid crystals based on tripodal pyridinium salts, Chem. Lett., 2008, 37, 1208–1209. DOI: 10.1246/cl.2008.1208

CHAPTER 5

Zero Sound Propagation in Femto-Scale Quantum Liquids

Yoritaka Iwata

GSI Helmholtzzentrum für Schwerionenforschung, Darmstadt, Germany

ABSTRACT

Charge equilibration has been recognized as a dominant process at the early stage of low-energy heavy-ion reactions. The production of exotic nuclei is suppressed under the appearance of charge equilibration, in which the proton-neutron ratios of the final reaction products are inevitably averaged. Therefore charge equilibration plays one of the most crucial roles in the synthesis of chemical elements. Focusing on how and when the charge equilibration takes place, zero sound propagation in femto-scale quantum liquids is explained.

Keywords: Quantum liquids; charge equilibration

1. INTRODUCTION

This article is concerned with Fermionic property of heavy ions (many-nucleon systems with the size up to several 10-femtometers) colliding at the energy of a few MeV per nucleon. In the following, we refer to those collisions as low-energy heavy-ion collisions. Fusion is not necessarily achieved in the low-energy heavy-ion collisions, neither is fragmentation. Reaction dynamics and the resulting products can be drastically different depending on the impact parameter, the mass of colliding ions, and so on. Therefore the reaction mechanism of low-energy heavy-ion collisions is worth investigating to understand the production of chemical elements. This is deeply concerned with an open problem as for the existence and the origin of chemical elements including their production mechanism.

Charge equilibrium in heavy-ion reactions means the states with the proton-neutron ratio corresponding to the average of the colliding ions, and charge equilibration is the process leading towards charge equilibrium. The chemical property of final products can be different depending on whether charge equilibration appears or not. There is a relatively long research history for the charge equilibration. In fact many experiments of low-energy heavy-ion collisions were carried out in the 1960s. In those experiments most of the final products were reported to be in charge equilibrium, even when fragmentation takes place [1]. There should not be any restrictions to the proton-neutron ratio of the final products if the final product forms a kind of stable bound system, so that these experimental results cannot be trivially understood.

One of the most important features of charge equilibration is its rapidness taking as much as a few 10^{-22} s [1]. This time scale is actually short in the order of magnitude compared with the typical reaction time of low-energy heavy-ion collisions (1000 fm/c $\sim 10^{-20}$ s), so that charge equilibration has been recognized as an inevitable and dominant process in low-energy heavy-ion reactions. The relation between charge equilibration and the isovector giant dipole resonance has been studied relatively well because of the correspondence in their time scales, but no decisive conclusion has ever been obtained. Indeed, including the question of "when does charge equilibration take place ?", many things could not be explained merely by the isovector giant dipole resonance. For the theoretical research on the relation between charge equilibration and the isovector giant dipole resonance, the importance of dipole mode to charge distribution of a fissioning nucleus was pointed out by HillWheeler [2] in the 1950s. Research on charge equilibration using time-dependent mean field calculations were started in the 1970s (Bonche-Ngô [3]). Recently research based on three-dimensional time-dependent mean field calculations was carried out by Simenel-Chomaz-de France [4,5]. In this article, based on Ref. [6], the unknown relation between charge equilibration and zero sound propagation is presented. It leads to a rather universal recognition of zero sound propagation in femtoscale quantum liquids, which cannot necessarily be reduced to the giant dipole resonance.

2. THEORY OF MATTER WAVE PROPAGATION

2.1. Landau's Fermi Liquid Theory

Let us denote a many-particle wave function by $\Psi(t, x_1, \cdots, x_n)$. It is assumed to satisfy

$$i\hbar\frac{\partial\Psi(t,x_1,\cdots,x_n)}{\partial t}=H\Psi(t,x_1,\cdots,x_n),$$

where H denotes the Hamiltonian operator. The solution can be represented by $\Psi(t,x)=e^{-itH/\hbar}\Psi(0,x)$ under a suitable boundary condition, if H does not de- pend on $\Psi(t,x_1,\cdots,x_n)$. Let the corresponding probability density be $\rho(t,x)$, and begin with the classic theory of sound propagation inside gases. Readers may wonder why sound propagation is related to charge equilibration, it will be clarified step by step. Let the equilibrium probability density be $\rho_0(x)$. If the fluctuation is added to the equilibrium:

$$\rho(t,x)=\rho_0(x)+\delta\rho(t,x)$$

a force arises from the gradient pressure. Here the essential property of sound propagation is extracted from the simplified linearized analysis. The equation of motion is given by

$$\frac{\partial(\rho v)}{\partial t}\sim\rho_0\frac{\partial v}{\partial t}=-\nabla P, \tag{1}$$

where v and P denote the velocity and the pressure, respectively. The pressure P depends on both ρ and the entropy S. On the other hand, the equation of continuity is given by

$$\frac{\partial\delta\rho}{\partial t}=-div(\rho v).$$

The right hand side is approximated by $-\rho_0\, div\, v$ to the lowest order. After differentiating this equation by t, an equality is derived together with the divergence of Equation (1). If we further assume that P is expanded with respect to $\delta\rho$ at a fixed entropy, the lowest order contribution brings about

$$\frac{\partial^2\delta\rho}{\partial t^2}=\left(\frac{\partial P}{\partial\rho}\right)_S\Delta\delta\rho \tag{2}$$

This is a wave equation for $\delta\rho$, where $\sqrt{(\partial P/\partial\rho)_S}$ represents the propagation speed of the density. For example, $\sqrt{(\partial P/\partial\rho)_S}$ is given by $v_F/\sqrt{3}$ for the perfect Fermi gas in its ground state (S=0), where v_F means the Fermi velocity. This type of propagation is called first sound, which provides a picture for particles propagating with changing density.

On the other hand, we should pay attention to the propagation of particle without changing density. This type of propagation is called zero sound, which has a finite frequency even when the wave number is equal to zero (cf. zero point vibration). To understand zero sound, the linear response is considered for a given external field $H_{ex}(t)$:

$$\begin{cases} i\hbar \dfrac{\partial \Psi(t, x_1, \cdots, x_n)}{\partial t} = \left(H + H_{ex}(t)\right)\Psi(t, x_1, \cdots, x_n), \\ H_{ex}(t) = \int \mathrm{d}^3x\, n(t,x) U_{ex}(t,x), \end{cases}$$

where $n(t,x)$ denotes the number density operator. In particular, if we restrict ourselves to the impulsive perturbation $U_{ex}(t,x) = \bar{U}_{ex} e^{iqx} \delta(t)$, the linear response is

$$\delta\langle n(t,x)\rangle$$
$$= \bar{U}_{ex} e^{iq\omega} (2\pi)^{-1} \int \mathrm{d}\omega\, e^{-i\omega t} \bar{U}_{ex}^{-1}(q)\left(\left[\kappa^R(q,\omega)\right]^{-1} - 1\right),$$

where $\kappa^R(q,\omega)$ denotes the retarded generalized dielectric function. The resonance frequency can be calculated by the pole of the integrand of the right hand side. Eventually we assume the phonon dispersion relation ($\Omega_q = c_0 q$), which reproduces the frequency of propagating wave, and is nonzero even when the wave number is equal to zero:

$$1 = V(q)\Pi^{0R}\left(q, \Omega_q - i\gamma_q\right)$$

then the resonance frequency is obtained. Consequently the dispersion relation at long wave length becomes

$$\lim_{q\to 0} \frac{\pi^2\hbar^2}{mk_F V(q)} = \frac{1}{2} x \log\left(\frac{x+1}{x-1}\right) - 1,$$

where $x = m\omega/\hbar k_F q$; m and k_F denote the mass and the Fermi wave number, respectively. Note that this relation is the representation in momentum space. The lower-limit of the velocity for zero sound is the Fermi velocity, because the non-damping mode can only exist when $x > 1$ (see the denominator inside the logarithmic function). If $V(q) \to V(0)$ is assumed in the limit $q \to 0$, the left hand side becomes $\pi^2\hbar^2/(m_F V(0))$, and the propagation velocity of zero sound is represented using $V(0)$. For the weak-coupling limit ($V(0) \ll \hbar^2/mk_F$):

$$c_0 \sim v_F\left\{1 + 2\exp\left(-\frac{2\pi^2\hbar^2}{mk_F V(0)} - 2\right)\right\},$$

and for the strong-coupling limit ($V(0) \gg \hbar^2/mk_F$):

$$c_0 \sim v_F\left\{\frac{V(0)}{3\pi^2\left(\hbar^2/mk_F\right)}\right\}^{1/2}$$

This shows that the propagation velocity of zero sound is almost equal to v_F in the weak-coupling limit. It is just $\sqrt{3}$ times faster than the previously seen first sound velocity of the perfect Fermi gas. A rather general discussion shows that zero sound is faster than first sound [7,8]. Roughly speaking, the matter waves without changing density propagate more easily than those with changing density, because the change in density possibly leads to the appearance of larger restoring force. Zero sound provides a picture of particle exchange within a quite short time (but not instantaneous).

Two different sound propagations in Fermi liquids have been discussed in terms of whether they entail the density change or not. In particular there exists zero sound in Fermionic many-body system, which is different from and faster than the ordinary sound. Zero sound has been known to arise from the collective dynamics of the Fermionic many-body systems. All the collective dynamics is actually based on the propagation of zero sound, while there are various representations (various modes) for the collective dynamics. One distinct difference between zero and first sounds is their relation to collisions between particles. First sound appears when the states are in local thermal equilibrium. This corresponds to the situation when the mean interparticle collision time is sufficiently smaller than the oscillation period of the propagating wave. Meanwhile zero sound is associated with the collective excitation mode, which disappears when there are many collisions between particles. This corresponds to the situation when the mean interparticle collision time is larger than the oscillation period of the propagating wave. Although such a collisionless energy regime is expected to be realized due to the Pauli principle, its validity should be confirmed for each individual physical system. Consequently zero sound is expected to be important to low-energy phenomena, while first sound becomes effective at higher energies. For the usage of the terminology low and high energies, we had better note that both sounds are scaled by v_F (i.e. Fermi energy). For more details of sound propagation, see Section 5 of Ref. [9].

2.2. Nucleon Propagation in Heavy-Ion Collisions

As a theoretical research on zero sound in Fermionic many-body systems, Landau's Fermi liquid theory [10, 11] is well known, where zero sound propagation was actually seen in liquid ^{3}He [12,13]. However, there is no guarantee that such a sound plays a role in heavy-ion reactions. There are two essential differences between heavy-ion reactions and the ^{3}He case; the physical system consists of finite numbers of nucleons, and the event is accomplished within a finite time interval. That is to say, both size and time are highly restricted in heavy-ion reactions. Indeed, in the context of many-nucleon systems, the main interest of zero sound propagation was not in heavy-ion reactions but in nuclear vibrations (for example, see Ref. [14]). In the following the mechanism of charge equilibration is discussed with respect to whether it is achieved by nucleon propagation with or without changing density.

3. MECHANISM OF CHARGE EQUILIBRATION

3.1. Nucleon Propagation Realizing Charge Equilibration

As an example of Fermionic many-body systems, we consider many-nucleon system. There are two types of nucleons, that is, protons and neutrons. Charge equilibration is the mixing of protons and neutrons due to the time evolution, therefore it is a kind of chemical mixing. Apart from such a chemical equilibration, there are several kinds of equilibration in heavy-ion reactions, that is, mass equilibration, momentum equilibration, and thermal equilibration. Among them charge equilibration (chemical equilibration) has drawn special attention because of its crucial role in the synthesis of chemical elements.

Let us consider the collision of two ions. The existence of an upper energy limit for charge equilibration has been pointed out by Iwata-Otsuka-Maruhn-Itagaki [6] for the first time. The limit is presented by a formula explaining both experiments and numerical calculations based on microscopic three-dimensional time-dependent meanfield theory. Furthermore, the upper energy limit has been concluded to be determined by the Fermi energy.

$$\frac{E_{lab}}{A} = \frac{\hbar^2 \left(3\pi^2 \rho_{min}\right)^{2/3}}{2m} + \frac{e^2 Z_1 Z_2}{4\pi\varepsilon_0 r_0} \frac{A_1 + A_2}{A_1 A_2 \left(A_1^{1/3} + A_2^{1/3}\right)} \tag{3}$$

$$\rho_{min} = \min_i \left(\frac{N_i \left(\frac{4\pi r_0}{3} A_i^{1/3}\right)^{-1}}{\left(1-3\overline{\varepsilon}\right)\left(1+\overline{\delta}\right)}, \frac{Z_i \left(\frac{4\pi r_0}{3} A_i^{1/3}\right)^{-1}}{\left(1-3\overline{\varepsilon}\right)\left(1-\overline{\delta}\right)} \right) \tag{4}$$

where i=1,2 identifies the two colliding ions, and E_{lab} means the energy in the laboratory frame.A_1 and A_2, which satisfy $A = A_1 + A_2$, $A_1 = Z_1 + N_1$ and $A_2 = Z_2 + N_2$, denote the masses of the two colliding ions, where Z_1, Z_2, N_1 and N_2 denote the proton and the neutron numbers of each colliding ion (labeled by i), respectively. $\overline{\varepsilon}$ and $\overline{\delta}$ are parameters introduced based on Ref. [15,16]. This formula provides an upper energy limit of charge equilibration, which arises from the nucleons propagating at the Fermi velocity. In heavy-ion collisions, there are four different Fermi velocities (Fermi energies), because there are two different kinds of nucleons and two colliding ions. The first term on the right hand side of Equation (3) calculates those four Fermi velocities. If only the Fermi velocities (for neutrons and protons) of the total merged system are taken, experimental results cannot be explained because the charge equilibration is accomplished before the formation of a compound nucleus (see the experimental comparison shown in Ref. [6]). In particular, for the many-nucleon system, special structures such as skin and halo exist, and those structures change the Fermi energy. These effects are taken into account in $\overline{\varepsilon}$ and $\overline{\delta}$ in Equation (4). For the nucleon wave associated with each Fermi energy to propagate throughout both colliding ions after touching, the minimum is taken in Equation (4). This treatment does not seem to be so important at a glance, but it contributes to derive the fact that different reactions having exactly the same composite nucleus can result in a different upper energy limit (for the difference of upper energy limit, see Figure 2 of Ref. [6]). If it is not for such a treatment, we cannot explain both numerical calculations and experiments. Furthermore, for a correct comparison, it is necessary to estimate the relative velocity at contact time. The most considerable effect here is the deceleration due to the Coulomb repulsion, which becomes prominent for cases when the masses of the colliding ions are larger. This effect is considered in the second term on the right hand side of Equation (3). Consequently it was confirmed in Ref. [6] that the final fragments achieving charge equilibrium drastically decrease at an incident energy higher than the energy shown in the formula.

The nucleons propagating at the Fermi velocity, which is represented by Equations (3) and (4), correspond to the propagation of zero sound. Indeed, according to the calculation of the mean free path of nucleons inside nuclear matter [17], collisions between nucleons were shown to appear and increase rapidly if the incident energy of heavy-ion collisions becomes a few 10% larger than the Fermi energy (of the composite nucleus). In such situations, zero sound itself disappears. The possible propagation speed of zero sound is not not different from the amplitude of the Fermi velocity. This fact explains the reason why Equations (3) and (4) are related to zero sound propagation.

Figure 1 visualizes the propagation of charge equilibrating flow: neutron-rich flow from ^{52}Ca (N/Z = 32/20) to ^{36}Ca (N/Z = 16/20). Cases with two different impact parameters and two different energies are compared. Because this heavy-ion reaction is classified to class II of the classification shown in Figure 1 of Ref. [18] and the "N/Z = 1"-line is located between the initial points of ^{52}Ca and ^{36}Ca on the N-Z plane, its charge equilibration dynamics can be measured mostly by the flow of neutrons. The final products are different, where only the case with b=6.1 fm and E/A = 1.0 MeV results in fragmentation, while the other cases result in fusion. Despite such significant differences in final products, the propagation speed of charge equilibrating flow is almost the same. Indeed, the wave front of N/Z = 1.10 propagates 6.5 fm in 0.75 × 10^{-22} s after the neck formation (the neck is formed at 6.0 × 10^{-22} s for E/A = 1.0 MeV and 3.0 × 10^{-22} s for E/A = 2.0 MeV). It shows that the propagation speed of charge equilibrating flow is in dependent of the incident energy (the relative velocity of collision) and the impact parameter, while the total neutrons included in neutron-rich flow depends highly on the incident energy. In particular the propagation speed of charge equilibrating flow is faster than the relative velocity of collision (**Table 1**).

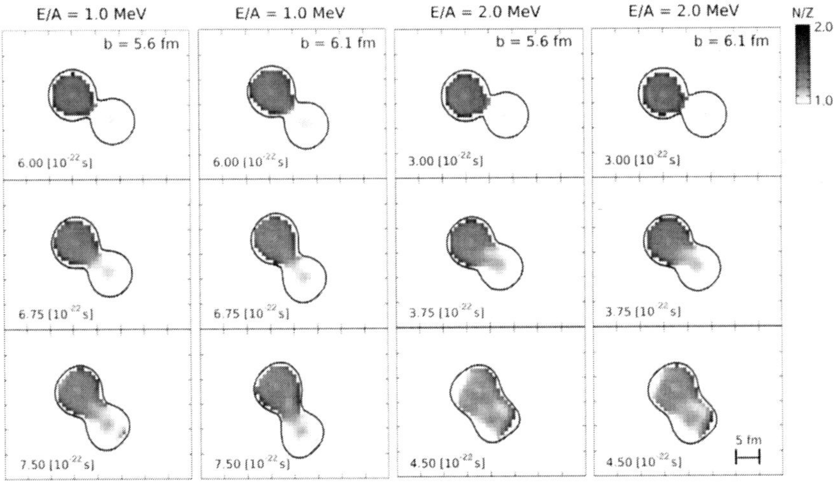

Figure 1. (Color online) Propagation of neutron-rich flow is depicted for the collision between ^{52}Ca and ^{36}Ca, where ^{52}Ca and ^{36}Ca correspond to the ions coming from the left and right hand sides, respectively. The colored parts correspond to the parts with $N/Z > 1$ (each frame is 40×30 fm^2), and the density contour equal to 0.02 fm^{-3} is shown by a thick black curve. Three-dimensional time-dependent mean-field calculations with a Skyrme interaction (SLy4d) are carried out; the singleparticle wave functions are represented on a Cartesian grid with the spacing of 0.8 fm, and the time unit of calculation is set to 1.5×10^{-24} s. The initial distance between the two colliding ions are set to 20 fm, then the relative velocity of collision is given.

Table 1. Comparison of speeds, where $|v_F|$ is fixed to 1/3 of the speed of light (corresponding to the nuclear standard value). The propagation speed of charge-equilibrating flow is calculated by the propagation speed of the wave front of N/Z = 1.10, where the propagation speed is the same for collisions with the two different relative velocities and the two different impact parameters (Figure 1). The relative velocity of collision at the contact is slower than that at the initial time, because of the deceleration due to the Coulomb repulsion.

Motion	Speed	Description
Propagation of charge-equilibrating flow	$0.90\,|v_F|$	$\sim 6.5/(0.75 \times 10^{-22})$ fm/s
Relative velocity for $E/A = 2.0$ MeV	$0.36\,|v_F|$	Speed given at the initial time
Relative velocity for $E/A = 1.0$ MeV	$0.23\,|v_F|$	Speed given at the initial time

With respect to zero sound, what was clarified in Ref. [6] can be summarized in the following three points. First, zero sound propagation plays a role in heavy-ion collisions. Second, the fast charge equilibration, which is achieved within the order of 10^{-22} s, is realized as nucleon propagation without changing density. Third, there exists an upper energy limit for the fast charge equilibration, which corresponds to the energy limit at which zero sound can survive. For example, as is shown in Ref. [6], the upper energy limit explains the experimental values: the appearance of charge equilibration for ^{40}Ar + ^{58}Ni at E_{lab}/A = 7.0 MeV and ^{56}Fe + ^{165}Ho at E_{lab}/A = 8.3 MeV, and the disappearance of charge equilibration for ^{112}Sn + ^{124}Sn at E_{lab}/A = 50 MeV. For the terminology of "fast" charge equilibration, it takes into account the existence of another kind of charge equilibration that has nothing to do with zero sound propagation. Such charge equilibration, which appears at higher energies, is more related to the first sound. Therefore its process is relatively slow compared to the fast charge equilibration, and insufficient to lead to fusion or to charge equilibrium for most fragments. Charge equilibration at higher energies was also studied well (for example, see references [8-10,20-23] of Ref. [6]).

As a remark, the previous research on charge equilibration with respect to the collective dynamics is mentioned. As is discussed, charge equilibration was studied in association with the isovector giant dipole resonance. Charge equilibration is sometimes related to the isovector dipole resonance, but not in all cases; i.e. the concept of resonance is too restrictive to explain charge equilibration. Nevertheless isovector giant dipole resonance means that the modes related to the composite nucleus play a role, it cannot explain the different upper energy limits for different reactions having exactly the same composite nucleus. In addition, an isovector mode different from the isovector giant dipole resonance sometimes appears (see Figure 4 of Ref. [6]). Although it is always true that the fast charge equilibration is achieved by the collective dynamics, the fast charge equilibration is not necessarily achieved by the isovector giant dipole resonance.

3.2. Origin of Charge Equilibration

Apart from how and when charge equilibration takes place, here we see the reason why charge equilibration takes place. First of all, the propagation of zero sound is expected to be efficient to any kind of nucleon propagations and vibrations. The answer is obtained by clarifying the origin of charge equilibration. When the two ions have a contact during the heavy-ion collisions, large fluctuations appear in the shape and the internal structure. Zero sound is expected to contribute to stabilization by changing both the shape and the internal structure, which can be understood by the contribution of each term included in the Bethe-Weizsäcker mass formula [19,20]:

$$B(A) = a_{vol} A + a_{surf} A^{2/3} + a_{coul} Z^2 A^{-1/3} + a_{sym} \frac{(N-Z)^2}{A},$$

where the coefficients are given by $a_{vol} \sim -16$ MeV, $a_{surf} \sim 20$ MeV, $a_{coul} \sim 0.751$ MeV and $a_{sym} \sim 21.4$ MeV, respectively [21]. After contact, the first and second terms contribute to stabilization by changing the shape such as the volume and surface area, while the third and forth terms contribute mainly to stabilization by changing the internal structure. It is reasonable that the shape change (including vibration) due to zero sound propagation leads to stabilization, and we do not go further into detail. Here we focus on stabilization by changing the internal structure. The effect due to the symmetry energy (the forth term) should play a principal role, because the Coulomb energy (the third term) is actually small except for the collisions between very heavy ions. This symmetry energy is the principal driving force of the stabilization by changing the internal structure, and its effect acting on zero sound propagation is fast charge equilibration. In this manner charge equi libration is closely related with zero sound propagation.

4. SUMMARY

The propagation of charge equilibrating flow has been visualized, and a physical interpretation has been given to the fast charge equilibration; i.e., the fast charge equilibration is realized by zero sound propagation in which their propagation speeds coincide with each other. This means that the collective dynamics of the many-nucleon system and thus its Fermionic statistical property are essential to the fast charge equilibration. In this context the upper energy limit for the fast charge equilibration corresponds to the upper-limit energy for zero sound propagation to be effective. Consequently charge equilibration should be regarded as the exchange of nucleons (charge exchange), because zero sound does not necessarily entail density change.

The correspondence of zero sound propagation in femto-scale systems is not only the giant dipole resonance, but also the flow propagating almost at the Fermi velocity. Indeed, the similarity between neutron-rich flow shown in **Figure 1** and the isovector giant dipole resonance is quite limited. Once the window between the two nuclei has opened, nucleons can cross relative freely into the other fragment, and this leads to an exchange of protons and neutrons with a speed governed by the Fermi velocity. After this rapid initial exchange the oscillation corresponding to giant dipole resonance appears. It is suggested that a prominent role of zero sound in femto-scale systems can be experimentally detected by the appearance of the fast charge equilibration; whether most of the final products of heavy-ion reactions are in charge equilibrium of not.

As is discussed, the fast charge equilibration is not a process appearing only at a certain energy. The fast charge equilibration universally appears in low-energy heavy-ion reactions at energies lower than the upper energy limit, instead. Note that the fast charge equilibration process becomes operational only if the two colliding ions interact by the nuclear force.

This work is based on the collaboration with Profs. T. Otsuka, J. A. Maruhn, and N. Itagaki, and supported by the Helmholtz alliance HA216/EMMI. The author thanks to Prof. J. A. Maruhn for reading the manuscript carefully. Encouragement from Dr. C. Simenel at the initial stage of this research is acknowledged.

REFERENCES

1. H. Freiesleben and J. V. Kratz, "N/Z-Equilibration and Nucleon Exchange in Dissipative Heavy-Ion Collisions," Physics Reports, Vol. 106, No. 1-2, 1984, pp. 1-120. doi:10.1016/0370-1573(84)90092-9
2. D. L. Hill and J. A. Wheeler, "Nuclear Constitution and the Interpretation of Fission Phenomena," Physical Review, Vol. 89, No. 5, 1953, pp. 1102-1145. doi:10.1103/PhysRev.89.1102
3. P. Bonche and N. Ngô, "Time-Dependent Hartree-Fock Description of Charge Equilibration in Heavy-Ion Collisions," Physics Letters, Vol. 105, No. 1, 1981, pp. 17-21.
4. C. Simenel, Ph. Chomaz and G. de France, "Quantum Calculation of the Dipole Excitation in Fusion Reactions," Physical Review Letters, Vol. 86, No. 14, 2001, pp. 2971-2974.doi:10.1103/PhysRevLett.86.2971
5. C. Simenel, Ph. Chomaz and G. de France, "Fusion Process Studied with Preequilibrium Giant Dipole Resonance in Time Dependent Hartree-Fock Theory," Physical Review C, Vol. 76, 2007, Article ID: 024609. doi:10.1103/PhysRevC.76.024609
6. Y. Iwata, T. Otsuka, J. A. Maruhn and N. Itagaki, "Suppression of Charge Equilibration Leading to the Synthesis of Exotic Nuclei," Physical Review Letters, Vol. 104, No. 25, 2010, Article ID: 252501. doi:10.1103/PhysRevLett.104.252501
7. A.A. Abrikosov and I. M. Khalatnikov, "The Theory of a Fermi Liquid (the Properties of Liquid ^{3}He at Low Temperatures)," Reports on Progress in Physics, Vol. 22, 1959, pp. 329-367. doi:10.1088/0034-4885/22/1/310
8. J. Wilks, "The Properties of Liquid and Solid Heilium," Oxford University Press, Oxford, 1967.
9. A.L. Fetter and J. D. Walecka, "Quantum Theory of Many-Particle Systems," Dover-Publications, Mineola, 2003.
10. L. D. Landau, "The theory of a Fermi liquid," Soviet Physics—JETP, Vol. 3, 1956, p. 920.

11. L. D. Landau, "Oscillation in a Fermi liquid," Soviet Physics—JETP, Vol. 5, 1957, p. 101.
12. A.J. Leggett, "A Theoretical Description of the New Phases of Liquid ^{3}He," Reviews of Modern Physics, Vol. 47, No. 2, 1975, pp. 331-414.
13. W. R. Abel, A. C. Anderson and J. C. Wheatley, "Propagation of Zero Sound in Liquid He^{3}at Low Temperatures," Physical Review Letters, Vol. 17, No. 2, 1966, pp. 74-78. doi:10.1103/PhysRevLett.17.74
14. P. Ring and P. Schuck, "The Nuclear Many-Body Problem," Springer-Verlag, New York, 1980.
15. W. D. Myers and W. J. Swiatecki, "Average Nuclear Properties," Annals of Physics, Vol. 55, No. 3, 1969, pp. 395-505.
16. W. D. Myers, "Droplet Model Isotope Shifts and the Neutron Skin," Physics Letters, Vol. 30, No. 7, 1969, pp. 451-454.
17. M. T. Collins and J. J. Griffin, "Mean Free Path of Nucleons in a Fermi Gas at Finite Temperature," Nuclear Physics A, Vol. 348, No. 1, 1980, pp. 63-74. doi:10.1016/0375-9474(80)90545-X
18. Y. Iwata, T. Otsuka, J. A. Maruhn and N.Itagaki, "Geometric Classification of Nucleon Transfer at Moderate Low-Energies," Nuclear Physics A, Vol. 836, No. 1-2, 2010, pp. 108-118. doi:10.1016/j.nuclphysa.2010.01.242
19. H. A. Bethe and F. Bacher, "Nuclear Physics I. Stationary States of Nuclei," Reviews of Modern Physics, Vol. 36, No. 2, 1936, pp. 82-229.
20. C. F. von Weizsäcker, "Zur Theorie der Kernmassen," Zeitschrift Für Physik A: Hadrons and Nuclei, Vol. 96, No. 7-8, 1935, pp. 431-458. doi:10.1007/BF01337700
21. W. Greiner and J. A. Maruhn, "Nuclear Models," SpringerVerlag, Berlin, 1996. doi:10.1007/978-3-642-60970-1

CHAPTER 6

The Dynamical Properties on Ionic Liquids: Insights from Molecular Dynamics Study

Tateki Ishida*

Department of Theoretical and Computational Molecular Science, Institute for Molecular Science, Japan

1. INTRODUCTION

Ionic liquids (ILs) have attracted many researchers in the areas including physics and chemistry because of their characteristics that are different from conventional molecular liquids and, today, ILs have been one of interesting subjects of scientific study. ILs are consisted of ions and liquids at or near room temperature, and show negligible vapor pressure, thermal and chemical properties, and so on [1-8]. Also, ILs have been widely used as solvents for organic reactions with the expectation of high yields [1]. Most interesting features of ILs can be attributed to remarkable interionic interactions, and these can be an important key factor to study the characteristics of ILs at molecular level. From the results of both experimental and theoretical investigations, it has been recognized that the interionic interaction of ILs could determine physical and chemical properties.

From an experimental side, both femtosecond optically heterodyne-detected Raman-induced Kerr effect spectroscopy (OHD-RIKES)[3, 9, 10] and THz time-domain spectroscopy (THz-TDS)[11, 12] have been applied to investigate the intermolecular vibrational dynamics in ILs. In particular, with OHD-RIKES studies, the possibility to control a property such as shear viscosity by substituting an atomic element in an ionic unit has been reported [13, 14]. On the other hand, from theoretical and computational viewpoints in recent years, ILs have been chosen to study static properties such as structural and thermophysical properties[15-18] and novel interionic dynamics under solvation dynamics [19-21], dynamical properties [22-26], and Kerr spectra analyses [14, 27]. It has been suggested from the simulation studies by Ishida et al. [14] that interionic properties in ILs could be effectively adjustable by substituting an atomic unit in an ion unit in addition to a combination of cations and anions. Also, it has been pointed out that the interplay of motions between cation and anion species could play an important role in specific interionic interactions of ILs.[28]

Now, we can consider two factors important to understand specific interionic interactions in ILs. One is the interionic interaction depending on specific correlations such as cross-correlation terms between cations and anions, and the other is polarization effects due to many-body interactions caused by cations and anions in ILs. While a large number of experimental approaches have been applied to investigate these subjects, molecular-level understanding of many specific properties of ILs has been left unresolved. Obviously, theoretical researches are suitable to tackle these problems to which experimental procedures are not accessible. Thus, it is expected that such computational method as molecular dynamics (MD) simulations enables us to obtain significant information of ILs, utilizing the force field with well-parameterized potential functions and partial charges [29].

With the MD simulation procedures, it is possible to study the effects of the cross-correlations on dynamic collective motions of ions in ILs which are considered to govern the strength and behavior of couplings between ionic motions through interionic interactions [28]. The computation of the time correlations of velocity and momentum between a tagged ion and other unlike ions at different distances provides cross-correlation and momentum correlation functions [28, 30, 31]. Utilizing these calculated functions, we can investigate not only interionic interactions at molecular level but also how collective motions in ILs can proceed accompanied by the momentum transfer between ions in the target IL system.

On the other hand, electrostatic interactions between ions in ILs could be modulated due to many-body interactions and, then, it could emerge as polarization effects caused by the distortion of electron densities under anisotropic environment in ILs. It has been pointed out that the inclusion of polarization effects is significant to investigate the characteristics of ILs in structural and dynamical properties [23, 24]. Therefore, it is required for us to carry out the MD simulation, introducing such procedures as a point dipole model and a polarization energy term into the total potential energy representation of the system [32]. In addition to electrostatic interactions, describing the variation and relaxation of the polarizability anisotropy of ILs is important to investigate dynamics in ILs. For achieving this, we need to compute time-dependent polarization effects on a target system due to environmental effects in ILs. Theoretically, when we would take polarizability anisotropy into account, to track the change of molecular polarizability tensor partly dependent on molecular orientations such as rotational motions would be required. With the calculation of the time correlation function (TCF) of off-diagonal elements of the total polarizability of the system, we can investigate collective properties with the result of the polarizability anisotropy relaxation of the system.

In this chapter, we choose a 1-butyl-3-methylimidazolium cation based ILs with hexafluorophosphate anion, $[BMIm][PF_6]$, and bis(trifluoromethylsulfonyl)amide anion, $[BMIm][NTf_2]$, as target systems. (See Figure 1 for all the molecular structures of $[BMIm]^+$, $[PF_6]^-$ and $[NTf_2]^-$.) Firstly, we focus on the collective properties of $[BMIm][PF_6]$ with the cross-correlation functions of ionic species in the IL and polarization effects on ionic motions. As a second, we investigate dynamical properties of $[BMIm][NTf_2]$. Below, we start from the explanation for the velocity cross-correlations of a central atom with neighboring atoms. Then, we show how to evaluate both cross-correlation and momentum correlation functions of the target IL. In addition, we describe the introduction of a polarization energy term and the procedure to calculate induced dipole moments and the polarization energy. Following those, the theoretical background of polarization TCFs is given. Computational details are also summarized. In later sections, we discuss polarization effects on interionic interactions and specific properties related to those based on the MD simulation results. We show computation results obtained from performing MD simulation and computing the polarization TCF with the dipole-dipole (DID) approximation [4,5]. Also, we examine the relation between the anisotropic polarizability relaxation and collective motions of ionic species. Lastly, we discuss relaxation processes of ILs including an explanation of important points in studying dynamical properties on ILs.

(a)

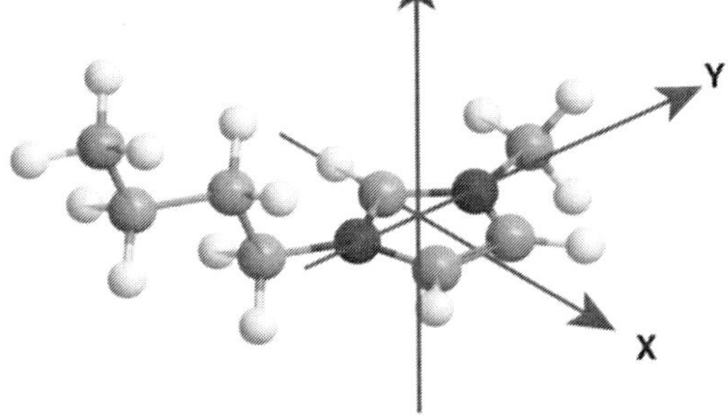

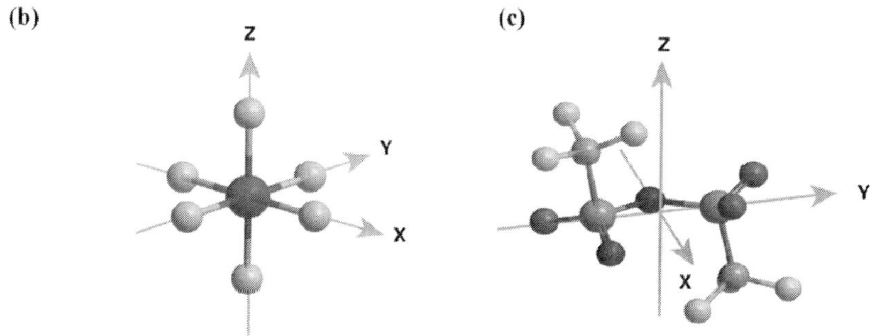

Figure 1. Molecular structures and definitions of body-fixed coordinate axes: (a) $[BMIm]^+$, (b) $[PF_6]^-$ and (c) $[NTf_2]^-$. (See text.) In $[BMIm]^+$, the *Y* direction is along the line connecting two nitrogen (blue colored) atoms in the ring, and the *Z* direction is set perpendicular to the ring plane and the *Y* direction axis. The *X* directionis set in the ring plane orthogonal to both the *Y* and the *Z* axes. In [PF6] $^-$, the *X*, *Y*, and *Z* direction axes are set equivalently. In $[NTf_2]^-$, the Y axis is along the line connecting two sulfurs (dark yellow colored), and the Z axis is set perpendicular the S-N-S plane and the X axis. The X axis is set in the S-N-S plane orthogonal to both the Y and Z axes.

2. TRACKING IONIC MOTIONS THROUGH INTERIONIC INTERACTIONS: CROSS-CORRELATION, POLARIZATION EFFECTS, AND DYNAMICAL PROPERTIES

As mentioned in the previous section, for many problems of ILs to which experimental procedures are not accessible, theoretical investigations with computer simulation procedures are promising and suitable to obtain detailed information at molecular level. In particular, specific correlations such as cross-correlations between cations and anions seem not to be feasible to detect experimentally, but the MD simulations enable us to evaluate those. Also, for tracing dynamics in ILs such as librational and reorientational dynamics of ionic species, cross-correlated ionic motions, and the influence of polarization coming from many-body effects caused by cations and anions in ILs from a microscopic point of view, the MD simulations are considered to be one of useful and powerful tools. In this chapter, we will show and explain the dynamical properties on ILs based on MD simulation results taking into consideration mainly following points:

1. how can cross-correlated ionic motions in ILs be modified by interionic dynamics and electronic polarizability effects ?
2. how can the collective dynamics through interionic interactions in ILs be tracked by computer simulation procedures ?
3. what kinds of properties with simulation data do we have to check and care in the investigation of dynamical properties of ILs ?
4. what kinds of subjects of ILs can be or should be investigated theoretically, considering important properties of ILs experimentally observed ?

The items 1 and 2 are main parts in this chapter, including the key results mentioned above. The item 3 is considered to be important for suggesting (or notifying) attention to researchers who are working on analyzing the dynamical properties on ILs. The item 4 includes future perspectives. In particular, we will show and discuss new aspects of ILs in each section related to the items 1, 2, and 3. Also, some of theoretical backgrounds and computational procedures for studying the dynamical properties of ILs are given in each related section, below.

At first, let us consider following points:

1. how can the collective dynamics through interionic interactions cause the unique physical and chemical properties of ILs ?
2. how can interionic dynamics be modified by electronic polarizability effects ?

The former includes the investigation of the contribution of ionic motions due to Coulombic interactions to velocity cross-correlation functions. In particular, through the analysis of the longitudinal and nonlongitudinal contributions to the velocity cross-correlation function in ILs, we will be able to investigate interionic interactions in detail. Also, important properties for physical and chemical interests such as case effects seems to be within the scope of unique collective dynamics in ILs. The later covers the relation between polarizability correlation functions and interionic interactions for ILs. To investigate these points, we can utilize useful information for static properties obtained from computer simulations, but those are not often enough to extract the details of specific interactions.

Here, we give an example that it is difficult for us to find the importance of the interactions between cation and anion species only from static properties. Figure 2 displays the computed radial distribution functions, g(r), comparing non-polarizable model with polarizable model [28]. As shown in the figure, we can observe only small difference in RDFs between two models, except that the g(r) of cation-anion is different from those of

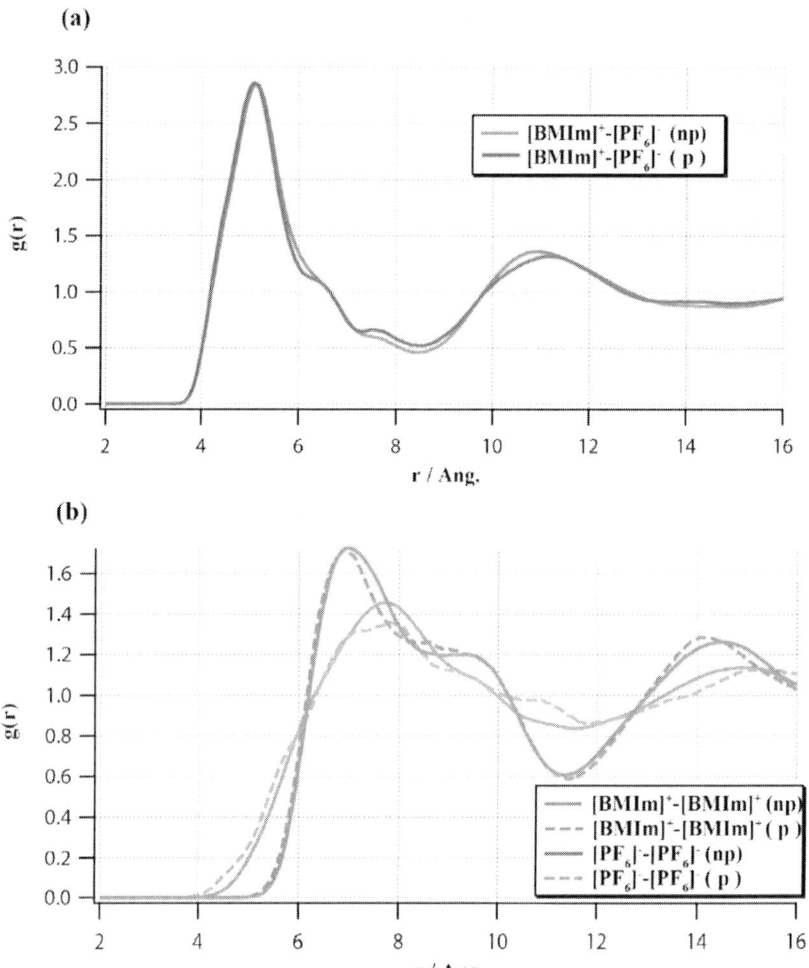

Figure 2. Radial distribution functions (rdfs) for the center of mass of [BMIm][PF_6] for both nonpolarizable (np) and polarizable (p) models: (a) $[BMIm]^+$ - $[PF_6]^-$ and (b) $[BMIm]^+$ - $[BMIm]^+$ and $[PF_6]^-$ - $[PF_6]^-$.

cation-cation and anion-anion as easily deduced. But, obviously, it is not feasible for us to observe information other than strong spatial correlations and sequential ordering of cation and anion pairs. In addition, the analysis of averaged static structures such as RDFs is not enough to investigate remarkable polarization effects such as screening effects influenced by polarization. Below, we give a schematic explanation and discuss that to consider interionic interactions with cross-correlation analyses is important to investigate the interplay between cation and anion species in ILs, and it is shown that the cross-correlation analyses provide contrastive features.

Different from a usual velocity autocorrelation functions (VACF), cross-correlation functions describe interactions between unlike (ionic) species (that is, between a cation and an anion in ILs) and show opposite features to VACFs. In Figure 3, these features are explained schematically. Here, it should be emphasized that cross-correlation functions provide more information on interionic interactions than that static properties such as RDFs include. As seen in Figure 3 ((a) and (b)), the variation of cross-correlation function correlates with the alteration of the VACF. In particular, the cross-correlation function shows the increasing toward the maximum peak where the VACF approaches a minimum point. These features indicate that it is possible for us to track the time evolution of cross-correlations (between cation and anion species in ILs) as collective dynamics. In addition, it is observed in Figure 3 ((c)) that a particle (ionic species) is bouncing back and forth between like and unlike particles (ions). This implies collective (ionic) motions between coordination shells, thus, it is indicated that not only interionic interactions but also momentum transferring among ionic species could be extracted by analyzing cross-correlation functions. Also, considering that these cross-correlation functions could be modulated by the strength of interionic interactions and the coupling between cation and anion motions, cross-correlation functions would be largely influenced by polarization effects.

On the other hand, as another type of the appearance of interplay between cross-correlation and polarization effects, we consider the polarizability anisotropy and its relaxation of an IL system. These correspond to the variation of the sum of molecular polarizability depending on time. Therefore, it is required to compute the change of molecular polarizablity on each molecule due to interionic interactions and interaction-induced multipole effects. Obviously, it is expected that molecular polarizabilites are influenced by interionic cross-correlations. In later sections, we introduce the theoretical background of polarizability TCF and its application to the study of ILs, and show how degree these are effective and discuss the importance of considering cross-correlations.

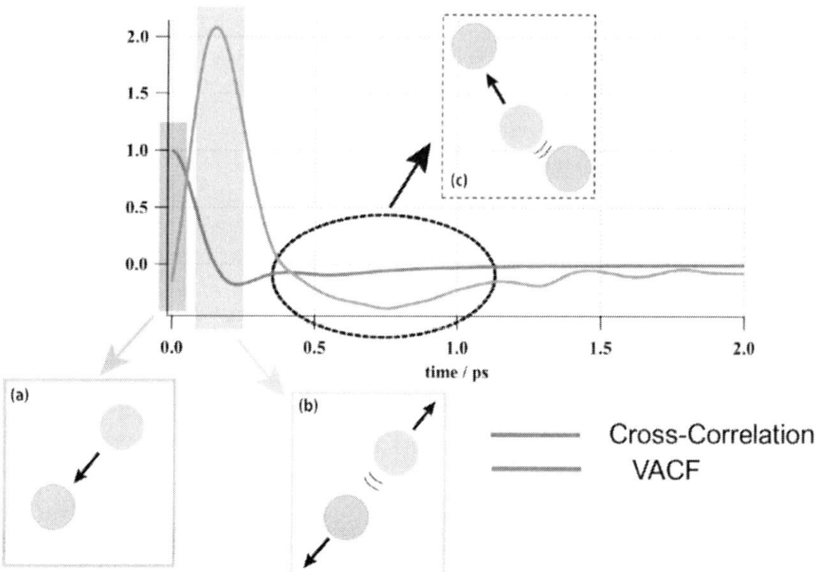

Figure.3. Schematic explanation for a cross-correlation and a VACF: (a) short-time region: an ion approaches a counter ion, and, at the same time, the VACF decreases, (b) the cross-correlation functions start decreasing after the maximum (see text) and (c) bounced ions again approaches counter ions.

3. THEORETICAL BACKGROUND

Here, we introduce cross-correlation functions, and then, give an explanation of a polarizable model and polarizability time correlation function. Computational details are also given.

3.1. Cross-Correlation Function

A velocity autocorrelation function (VACF) is defined for calculating the velocity correlation of a same particle as follows,

$$C(t) = \langle v_i(t) \cdot v_i(0) \rangle \langle v_i(0) \cdot v_i(0) \rangle^{-1} \tag{1}$$

where $v_i(t)$ is the velocity of i-th species. Different from the VACF, the time cross-correlation functions between the velocity of a central species i and the velocity of a neighboring species *j* different from the particle *i* is defined as follows [30, 31],

$$\mathrm{CR}_n^{ij}(t) = \langle \mathrm{v}_i(0)\cdot \mathrm{v}_j(t)\rangle_n \left(\langle \mathrm{v}_i^2\rangle\langle \mathrm{v}_j^2\rangle\right)^{-1/2} \tag{2}$$

where $\mathrm{v}_i(t)$ is the velocity of the species *i* and $\mathrm{v}_j(t)$ is the velocity of the species *j*. $\langle v_i^2\rangle$ and $\langle v_j^2\rangle$ represent the mean square velocities. Here, it should be noted that $\langle\rangle_n$ represents a restricted statistical average [30, 31],

$$\langle \mathrm{v}_i(0)\cdot \mathrm{v}_j(t)\rangle_n = \left\langle \mathrm{v}_i(0)\sum_j \mathrm{v}_j(0)\cdot\theta\left(r_{ij}(0)-a_n\right)\cdot\theta\left(b_n-r_{ij}(0)\right)\right\rangle / N_n \tag{3}$$

where θ represents the step function. a_n and b_n are set as the positions of the *n*th minima of the radial distribution function of the system, *g(r)*. The N_n in Eqs. (2) and (3) is the coordination number in the region between $r = a_n$ and b_n. as follows,

$$N_n = 4\pi\rho\int_{a_n}^{b_n} r^2 g(r)\,dr \tag{4}$$

where ρ is the number density of the system.

Here, we can define the cross-correlation in ILs as that between a centered cation and the total contribution of other anions [28, 30],

$$\mathrm{CR}_n^{\mathrm{CA}}(t) = N_n\langle \mathrm{v_C}(0)\cdot \mathrm{v_A}(t)\rangle_n \left(\langle \mathrm{v_C}^2\rangle\langle \mathrm{v_A}^2\rangle\right)^{-1/2} \tag{5}$$

where “C” and “A” in super- and subscript represent cation and anion, respectively. By interchanging “C” and “A” in super- and subscript in the above equation, we can easily derive the formulation for the cross-correlation between a centered anion and other cations.

Also, with the computation of the cross-correlation functions between the velocity of a central ion and velocities of neighboring distinct ions, we can analyze the momentum transfer between distinct ions in ILs. Introducing the momentum correlation function [28, 31], the transfer of the momentum of a cation to distinct cations and anions is defined as follows,

$$\begin{aligned} P_n^{\mathrm{C}(total)}(t) &= \left(N_n^{\mathrm{CC}}\langle p_+(0)\cdot p_+(t)\rangle_n + N_n^{\mathrm{CA}}\langle p_+(0)\cdot p_-(t)\rangle_n\right)\left(p_+^2\right)^{-1} \\ &= P_n^{cation\to cation(\mathrm{CC})}(t) + P_n^{cation\to anion(\mathrm{CA})}(t) \end{aligned} \tag{6}$$

where p_+ and p_- mean the momentums of cation and anion, respectively. The transference of momentum of an anion to distinct anions and cations, $P_n^{A(total)}(t)$, is also given as follows,

$$P_n^{A(total)}(t) = \left(N_n^{AA}\langle p_+(0)\cdot p_+(t)\rangle_n + N_n^{AC}\langle p_+(0)\cdot p_-(t)\rangle_n\right)\left(p_+^2\right)^{-1}$$
$$= P_n^{anion\rightarrow anion(AA)}(t) + P_n^{anion\rightarrow cation(AC)}(t) \quad (7)$$

3.2. Polarization Effects

3.2.1. A Polarizable Model

Here, we introduce a polarizable model considering induced dipole moments and explain the procedure for computing induced dipole moments and the many-body polarization energy [33-35].

The total potential energy of the system under the resulting polarizable force field is defined as follows,

$$V_{tot} = V_{bond} + V_{nonbond} + V_{pol} \quad (8)$$

where the terms V_{bond} and $V_{nonbond}$ are intra- and intermolecular interaction energies. The polarization energy, V_{pol}, is decomposed into the three terms as follows,

$$V_{pol} = V_{charge-dipole} + V_{dipole-dipole} + V_{self} \quad (9)$$

where the charge-dipole contribution, $V_{charge-dipole}$, the dipole-dipole contribution, $V_{dipole-dipole}$, and the self-polarizability term, V_{self}, are defined, respectively, as

$$V_{charge-dipole} = -\sum_i \mu_i \cdot E_i \quad (10)$$

$$V_{dipole-dipole} = \sum_{i>j} \mu_i \cdot T_{ij} \cdot \mu_j \quad (11)$$

$$V_{self} = \sum_i \frac{1}{2\alpha_i} \mu_i \cdot \mu_i \quad (12)$$

In the above equations, E_i is the electric field on atom *i*, produced by the partial charges of all other surrounding atoms, and μ_i represents the induced dipole moment on atom *i*. Also, αiαimeans the isotropic atomic polarizability of atom *i*. In Eq. (11), the T_{ij}, is the dipole field tensor element defined as follows,

$$T_{ij} = \frac{1}{r_{ij}^3}\left[1 - \frac{3\mathbf{r}_{ij}\mathbf{r}_{ij}}{r_{ij}^2}\right] \tag{13}$$

where $r_{ij} = r_i - r_j$. The induced dipole moment on atom *i* is given by

$$\mu_i = \alpha_i\left(E_i - \sum_{j\neq i} T_{ij}\mu_j\right) \tag{14}$$

With Eqs. (10), (11), and (12), the polarization energy, E_{pol}, is summarized as follows,

$$V_{pol} = -\frac{1}{2}\sum_i \mu_i \cdot E_i \tag{15}$$

3.2.2. Polarizability Time-Correlation Function (TCF)

The polarizability anisotropy of the system can be tractable by calculating the TCF of off-diagonal elements of the total polarizability. Here, the theoretical background of the polarizability TCF is summarized briefly.

At first, we define the total polarizability of the system, $\Pi(t)$, that is the sum of the molecular polarizability, $\Pi^M(t)$, and the interaction-induced polarizability, $\Pi^{II}(t)$, as follows,

$$\Pi(t) = \Pi^{\mathrm{M}}(t) + \Pi^{\mathrm{II}}(t) \tag{16}$$

where *t* represents the time dependence, and subscripts M and II mean the molecular part and the interaction-induced part, respectively. The molecular part is given by the sum of the polarizability tensors of isolated gas phase molecular polarizability in the laboratory frame,

$$\Pi^{\mathrm{M}}(t) = \sum_{i=1}^{N} \alpha_i(t) \tag{17}$$

where N is the number of molecules, and α_i is the polarizability tensor of molecule i. For the formulation of the interaction-induced part, we employ the dipole-induced-dipole (DID) model approximation [36, 37], which assumes that the molecular polarizabilities are modified due to a dipolar coupling with the influence of higher order unconsidered. The interaction-induced polarizability in the DID approximation is given as follows,

$$\Pi^{\mathrm{II}}(t) = \sum_{i=1}^{N} \sum_{j \neq i}^{N} \alpha_i(t) \cdot T_{ij}(t) \cdot \tilde{\alpha}_j(t) \tag{18}$$

where T_{ij} means the dipole interaction tensor between molecules i and j. $\tilde{\alpha}_i(t)$ is the effective polarizability for molecule i defined by the following equation including the interaction-induced effects,

$$\tilde{\alpha}_i(t) = \alpha_i(t) + \sum_{j \neq i}^{N} \alpha_i(t) \cdot T_{ij}(t) \cdot \tilde{\alpha}_j(t) \tag{19}$$

Equation (19) can be solved by the calculation procedure that is called the all-orders DID approximation [37]. It should be noted that the DID model employed here assumes a center-center DID model, where it recognizes the polarizability as concentrated in the center of mass of the molecule.

Here, we give the representation of the total system polarizability including the cationic and anionic components [14, 27],

$$\Pi(t) = \Pi^{\mathrm{C}}(t) + \Pi^{\mathrm{A}}(t) \tag{20}$$

where the superscripts C and A represent cation and anion species, respectively. Referring to Equations. (16), (17), and (18), the $\Pi^C(t)$ and $\Pi^A(t)$ defined as follows,

$$\Pi^{\mathrm{C}}(t) = \sum_{i=1} (\alpha_i^{\mathrm{C(M)}}(t) + \alpha_i^{\mathrm{C(II)}}(t)) \tag{21}$$

$$\Pi^{\mathrm{A}}(t) = \sum_{j=1} (\alpha_j^{\mathrm{A(M)}}(t) + \alpha_j^{\mathrm{A(II)}}(t)) \tag{22}$$

where the indices k and l go over all cations and anions, respectively. Then, the total polarizability of the system, $\Pi(t)$, can be rewritten, as follows,

$$\Pi(t) = \frac{1}{3}\mathrm{Tr}\left(\Pi^{C}(t) + \Pi^{A}(t)\right)\mathbf{I}$$
$$+\left\{\Pi^{C}(t) - \frac{1}{3}\mathrm{Tr}\left(\Pi^{C}(t)\right)\mathbf{I}\right\} + \left\{\Pi^{A}(t) - \frac{1}{3}\mathrm{Tr}\left(\Pi^{A}(t)\right)\mathbf{I}\right\}$$
$$= \frac{1}{3}\mathrm{Tr}\left(\Pi^{C}(t) + \Pi^{A}(t)\right)\mathbf{I} + \beta^{C}(t) + \beta^{A}(t) \tag{23}$$

Where $\beta^{C}(t)=\left\{\Pi^{C}(t)-\frac{1}{3}\mathrm{Tr}\left(\Pi^{C}(t)\right)I\right\}$ and $\beta^{A}(t)=\left\{\Pi^{A}(t)-\frac{1}{3}\mathrm{Tr}\left(\Pi^{A}(t)\right)I\right\}$ and Iis the unit tensor. Finally, the polarizability TCF, *ϕ(t)*, can be rewritten as follows,

$$\phi(t) = \left\langle \mathrm{Tr}\left[\left(\beta^{C}(0) + \beta^{A}(0)\right)\cdot\left(\beta^{C}(t) + \beta^{A}(t)\right)\right]\right\rangle$$
$$= \left\langle \mathrm{Tr}\left(\beta^{C}(0)\cdot\beta^{C}(t)\right)\right\rangle + \left\langle \mathrm{Tr}\left(\beta^{A}(0)\cdot\beta^{A}(t)\right)\right\rangle$$
$$+ \left\langle \mathrm{Tr}\left(\beta^{C}(0)\cdot\beta^{A}(t)\right)\right\rangle + \left\langle \mathrm{Tr}\left(\beta^{A}(0)\cdot\beta^{C}(t)\right)\right\rangle$$
$$= \phi^{C}(t) + \phi^{A}(t) + \phi^{C\text{-}A}(t) \tag{24}$$

where

$$\phi^{C}(t)=\left\langle \mathrm{Tr}\left(\beta^{C}(0)\cdot\beta^{C}(t)\right)\right\rangle$$
$$\phi^{A}(t)=\left\langle \mathrm{Tr}\left(\beta^{A}(0)\cdot\beta^{A}(t)\right)\right\rangle$$
$$\phi^{C-A}(t)=\left\langle \mathrm{Tr}\left(\beta^{C}(0)\cdot\beta^{A}(t)\right)\right\rangle + \left\langle \mathrm{Tr}\left(\beta^{A}(0)\cdot\beta^{C}(t)\right)\right\rangle$$

The nuclear response function, *R*(*t*), is represented as the time derivative of the polarizability TCF,

$$R(t) = -\frac{1}{k_B T}\frac{\partial}{\partial t}\phi(t) \tag{25}$$

3.3. Computational Details

Utilizing the sets of force field parameters for $[BMIm]^+$, $[PF_6]^-$ and $[NTf_2]^-$ [38-40], MD simulations have been performed with the DL_POLY molecular dynamics suite [41]. In all the simulations, all the stretching bonds were constrained with the SHAKE algorithm [42]. 125 ion pairs (4000 atoms for $[BMIm][PF_6]$ and 5000 for $[BMIm][NTf_2]$) were set in a cubic box under the periodic boundary condition. The lengths of cubic box size were set to be 35.47 and 38.46 Å to reproduce the experimental data of the densities of $[BMIm][PF_6]$ and $[BMIm][NTf_2]$ at 298 K [13, 43]. 12 Å was set as cutoff length. The time step was 2 fs. The long-range Coulomb and polarization terms were computed with the Ewald's summation technique [44]. Firstly, each system was equilibrated at 600 K for 15 ns in the NVE run and then successively cooled down to 298 K in several stages using velocity scaling. Then, 20 ns NVT run for equilibration at 298 K was carried out. After these equilibration runs, trajectories were recorded every 20 fs (50 fs for $[BMIm][NTf_2]$) and collected during 10 ns (20 ns for $[BMIm][NTf_2]$) production runs. For atomic polarizabilities, from the literature [45, 46], we adopted 1.152, 0.705 and 0.0885 Å^3 for the C, N, and H atoms of the cation, respectively, and 0.121 and 3.630 Å^3 for the F and P atoms of the anion, respectively. We considered the distance and vector between the atomic sites of distinct molecules in the dipole interaction tensor, $\mathrm{T}=(\boldsymbol{I}-3\hat{\boldsymbol{r}}\hat{\boldsymbol{r}})/r^3$, where $\hat{\boldsymbol{r}}=\boldsymbol{r}/r$ Then, the range of the attenuation of dipolar interactions at short distances, s, was evaluated with the Thole's definition [47], $s = 1.662(\alpha_i\alpha_j)^{1/6}$ with atomic polarizabilities, $\alpha_i$andα_j. Also, the dipole interaction tensor we used is given by [47]

$$\mathrm{T} = (\mathbf{I} - 3\hat{\mathbf{r}}\hat{\mathbf{r}}) / r^3 \quad (r > s) \tag{28}$$

$$\mathrm{T} = \left[(4\mathrm{a}^3 - 3\mathrm{a}^4)\mathbf{I} - 3\mathrm{a}^4\hat{\mathbf{r}}\hat{\mathbf{r}} \right] / r^3 \quad (r \leq s) \tag{29}$$

where a = r/s. Equation (14) is solved with an iterative procedure at each time step, and then, the criterion value of iterative solution for induced dipole moment was set to 0.001 D. For the computation of the polarizability TCF of $[BMIm][NTf_2]$, we used molecular polarizabilities of 14.372 Å^3 for $[BMIm]^+$ and 11.259 Å^3 for $[NTf_2]^-$.[48] The body-fixed coordinate axes set in the cation and the anion are shown in Figure 2. For more computational procedures in detail, interested readers should refer to references [14] and [28].\

4. CROSS-CORRELATION, MOMENTUM CORRELATION, AND POLARIZATION EFFECTS

The computed velocity cross-correlation functions [28] are shown in Figures 4 and 5. These figures show the comparison of velocity cross-correlation functions observed around the cation, $[BMIm]^+$, and the anion, $[PF_6]^-$, placed at the center, for both the nonpolarizable and polarizable models. From the result of the cation-anion RDF in Figure 2, we selected 3.5, 8.4, 8.4, and 14.75 Å for the value of a_1, b_1, a_2, and b_2 in Equation (4) for the nonpolarizable model, respectively, and 3.5, 8.5, 8.5, and 15.0 Å for the polarizable model, respectively. With these values, the first ($C_1(t)$) and second ($C_2(t)$) coordination shells were specified. It should be noted that the $C_1(t)$ between the center anion and distinct anions is not shown in Figure 5 because the $C_1(t)$ is almost zero corresponding to the result that the anion-anion RDF result is almost zero at the region of the specified first coordination shell (see RDFs in Figure 2(b)). Also, it should be noted that the initial values, $C_n(0)$, are negative since the system size is finite as have been pointed out [30]

As shown in Figures 4 and 5, the VACFs are also displayed for comparison. As briefly explained in Section 2 with Figure 3, the initial rise of the cross-correlation function, $C_1(t)$, appears toward the maximum for both models, corresponding to the decay of the VACF to the minimum (see the (a) and (b) in Figure 3). These results indicate that the initial momentum of the central ion is gained by neighboring ions immediately after $t = 0$. In addition, following decay profiles (see the (c) in Figure 3) are seen. These are ascribed to the spread of transferred momentum to the outer coordination shells. The peak height of the $C_1(t)$ in the polarizable model is larger than that in the nonpolarizable model. These significant results indicate that the momentum transfer can be intensified from the more distant coordination shells to the first ones due to both the charge-dipole and dipole-dipole interactions by polarization effects in addition to charge-charge Coulomb interactions. As the characteristics of the VACF profile, it is noted that the minima of the VACF are located at the positions later than those of the maxima of the velocity cross-correlation functions for the first coordination shell. These features imply that the momentum transmitted to the neighbors at the first shell is regained partly by the central cation or anion, bouncing back and forth for some time, as have been pointed out in the literature on the computer simulations of simple liquid binary mixtures [28, 31].

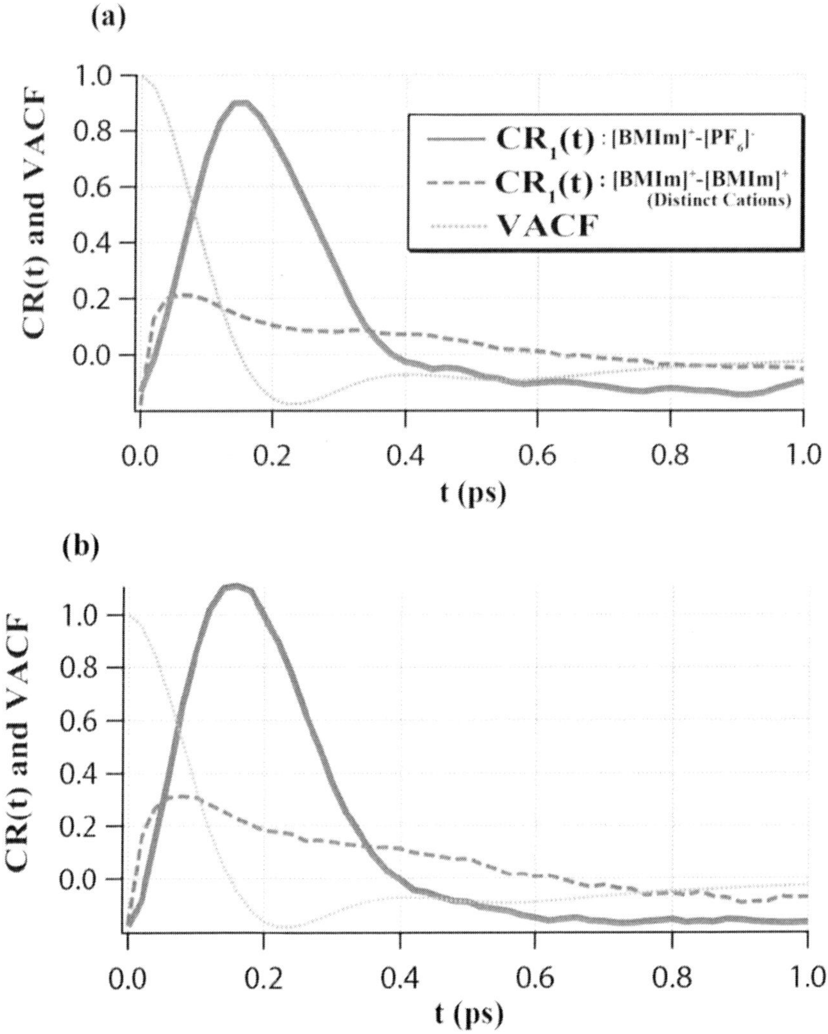

Figure 4. The Calculated velocity cross-correlation functions between the $[BMIm]^+$ set at the center in the first coordination shell and $[PF_6]^-$, and between the center cation and other cations, including VACF: (a) in the nonpolarizable model and (b) in the polarizable model.

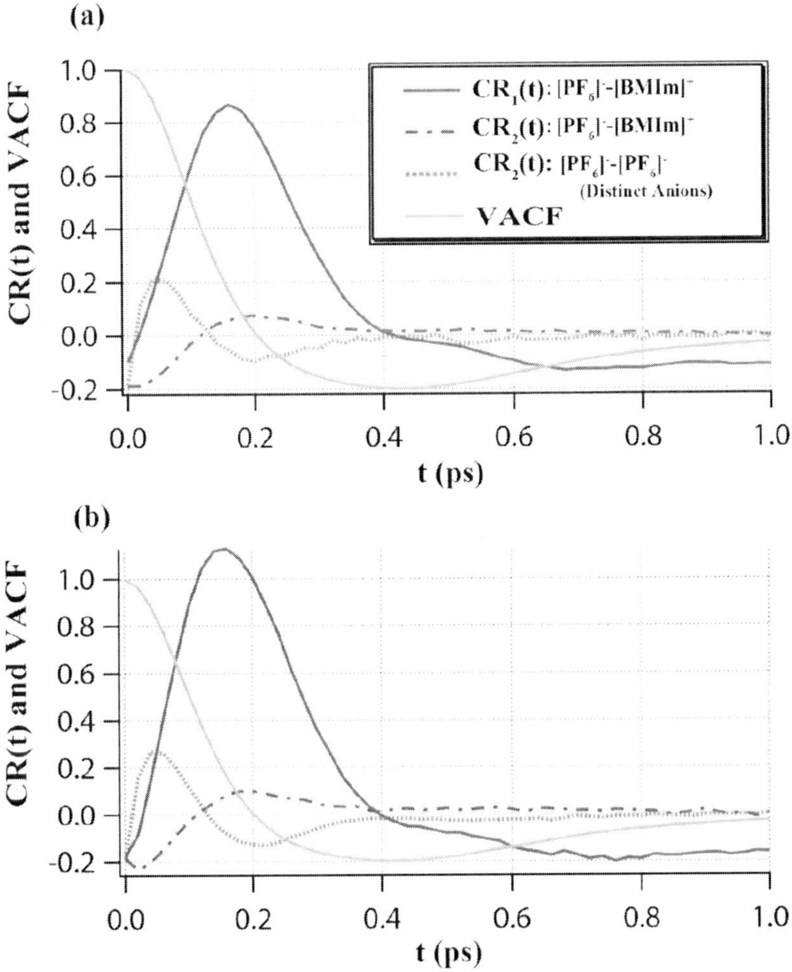

FIGURE 5.The Calculated velocity cross-correlation functions between the $[PF_6]^-$ set at the center in the first and second coordination shells and $[BMIm]^+$, and between the center anion and anions, including VACF: (a) in the nonpolarizable model and (b) in the polarizable model.

On the other hand, an interesting feature is seen in the $C_1(t)$ between distinct cations (or the $C_2(t)$ for distinct anions). As indicated in Figures 4 and 5, the cross-correlation function for distinct ions reaches the maximum point earlier than the $C_1(t)$ for the cation-anion cross-correlation, even though the peak height is smaller than that in the cation-anion cross-correlation function. On these results, it is considered that the alteration of cage effects increases the probability that a cation (or an anion) meets a distinct cation (or anion) at early time region.

Also, in Figures 4 and 5, it is clearly observed that the cross-correlation function between distinct like ions is influenced by the modulation of cage effects due to polarization effects and that its peak height is enhanced in the polarizable model [28]. As shown in Figures 4 and 5, when VACFs pass the value of zero to negative, the cross-correlation functions reach the maximum point. Thus, this implies that the central cation (or anion) is likely to lose its initial momentum. In addition, as seen in Figure 5, the peak position of $C_2(t)$ is shifted to the time region later than that of $C_1(t)$. This is consistent with the consideration that the momentum of the central ion is transferred from the first coordination shell to the second.

Here, we examine the characteristics of the cross-correlation functions. As shown in Figure 6, we consider the decomposition of a cross-correlation function into the longitudinal (denoted as L in the figure), $C_n{}^{L}(t)$, and nonlogitudinal (denoted as NL in the figure), $C_n{}^{NL}(t)$, contributions as follows [28, 31, 49],

$$CR_n(t) = CR_n^{L}(t) + CR_n^{NL}(t) \tag{30}$$

where $CR_n{}^{L}(t)$ is represented as the velocity cross-correlation along the direction designated by the center of masses of distinct ions at t = 0. We can compute $CR_n{}^{L}(t)$ with the following equation [28, 31, 49],

$$CR_n^{L}(t) = N_n \left\langle v_i{}^{L}(0) \cdot v_j{}^{L}(t) \right\rangle_n \left(\left\langle v_i{}^2 \right\rangle \left\langle v_j{}^2 \right\rangle \right)^{-1/2} \tag{31}$$

where $v_i{}^{L}(t) = v_i(t)[r_{ij}(0)/r_{ij}(0)]$ and $vi(t)$ is the velocity of the ionic species *i*. Also, $rij(0)$ means the direction vector between the center of masses of the distinct ions i and j. $\langle v_i{}^2\rangle$and $\langle v_j{}^2\rangle$ ⟨vj2⟩represent the mean square velocities. $CR_n{}^{NL}(t)$ can be computed with $CR_n(t)$ and $CR_n{}^{L}(t)$. Figure 7 shows the computed $C_n{}^{L}(t)$ and $C_n{}^{NL}(t)$ functions for a centered cation at the first coordination shell (n = 1) both for the nonpolarizable and polarizable models [28]. All the results indicate that the velocity cross-correlations at the short time region up to 0.4 ps are predominantly governed by the longitudinal function, $C_n{}^{L}(t)$.

These results clearly indicate that the Coulomb interactions between neighboring ions are mainly effective on the ionic motions as "driving forces" at the short time. In addition, the magnitude of the longitudinal correlation of ions in the polarizable model is larger than that in the nonpolarizable model. Therefore, it is considered that the longitudinal motions are strongly affected by polarization effects. On the decay behavior, similar features are obtained in both models. These indicate that polarization effects on the longitudinal motion complete mostly at the short time region.

Next, the computed results of momentum correlation functions [28] are displayed in Figures 8 and 9. As shown in Figures 8 and 9 in common, the initial momentum of the cation (or anion) is mainly transferred to the close anions (or cations) while the momentum correlation with distinct cations (or anions) is smaller. These results are consistent with the consideration that strong Coulomb interactions between the cations and the anions enhance the possibility of approaching or attracting each other. Also, while the contribution of the momentum correlation between distinct cations, $P_1^{CC}(t)$, is smaller than that between the cation and anion molecules, $P_1^{CA}(t)$, the peak height of the $P_1^{CA}(t)$ in the polarizable model is larger than that in the nonpolarizable model. These results indicate that the transference of the momentum between distinct cations could be intensified by both charge-dipole and dipole-dipole interactions coming from polarization effects. As we could observe in the cross-correlation functions (see Figures 4 and 5), the maxima of the momentum correlation functions between distinct cations emerge at earlier time region in comparison with those of the momentum correlations between cation and anion species.

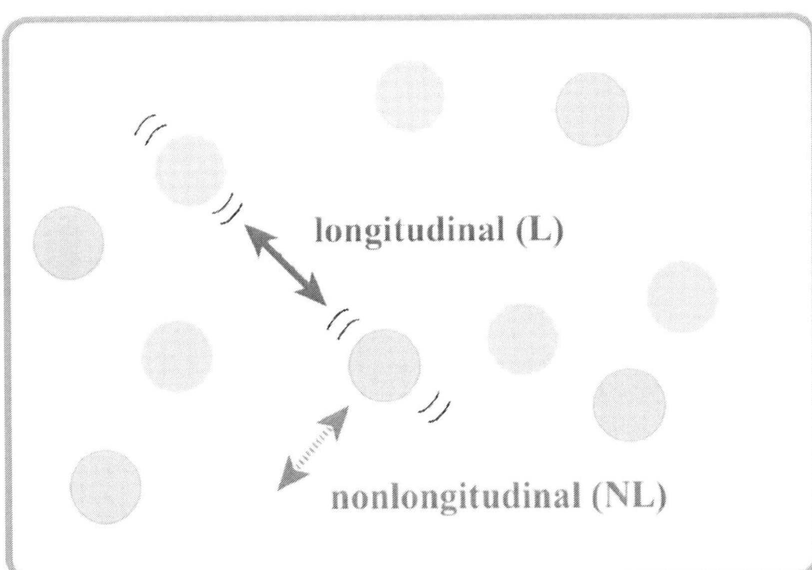

Figure 6. Schematic explanation of the longitudinal (L) and nonlongitudinal (NL) of a cross-correlation function.

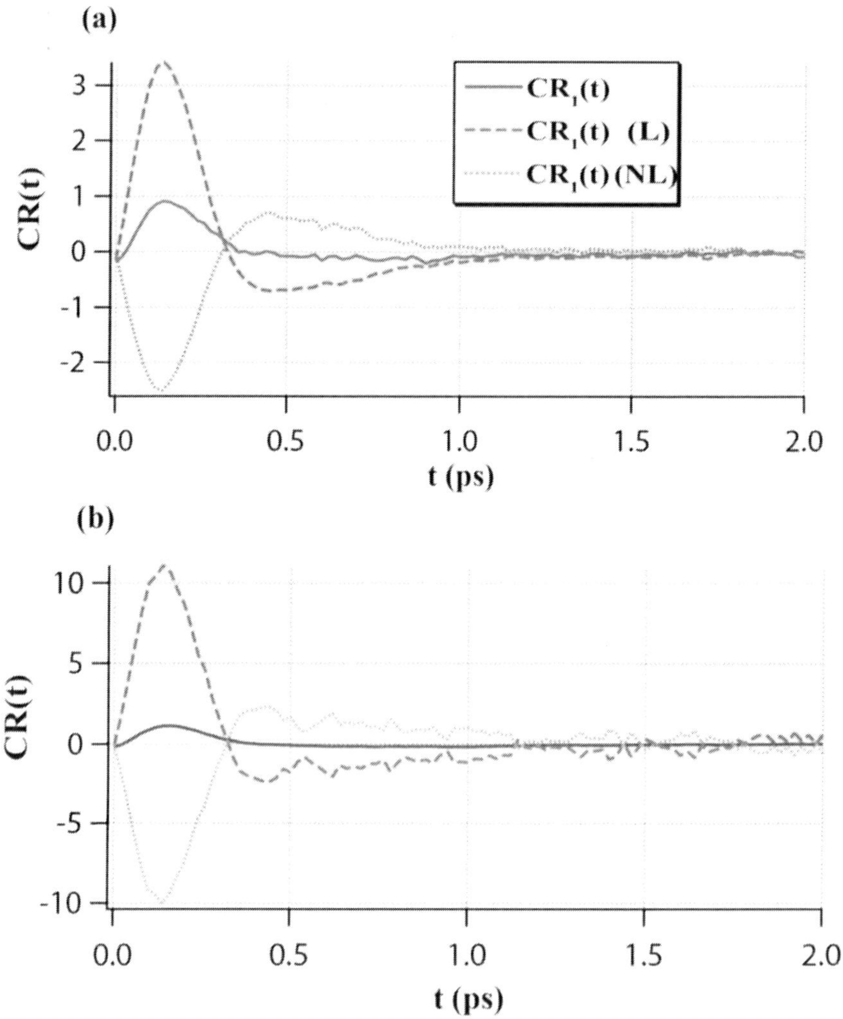

Figure 7. The simulated longitudinal and nonlongitudinal contributions to the velocity cross-correlation function: (a) in the nonpolarizable model and (b) in the polarizable model.

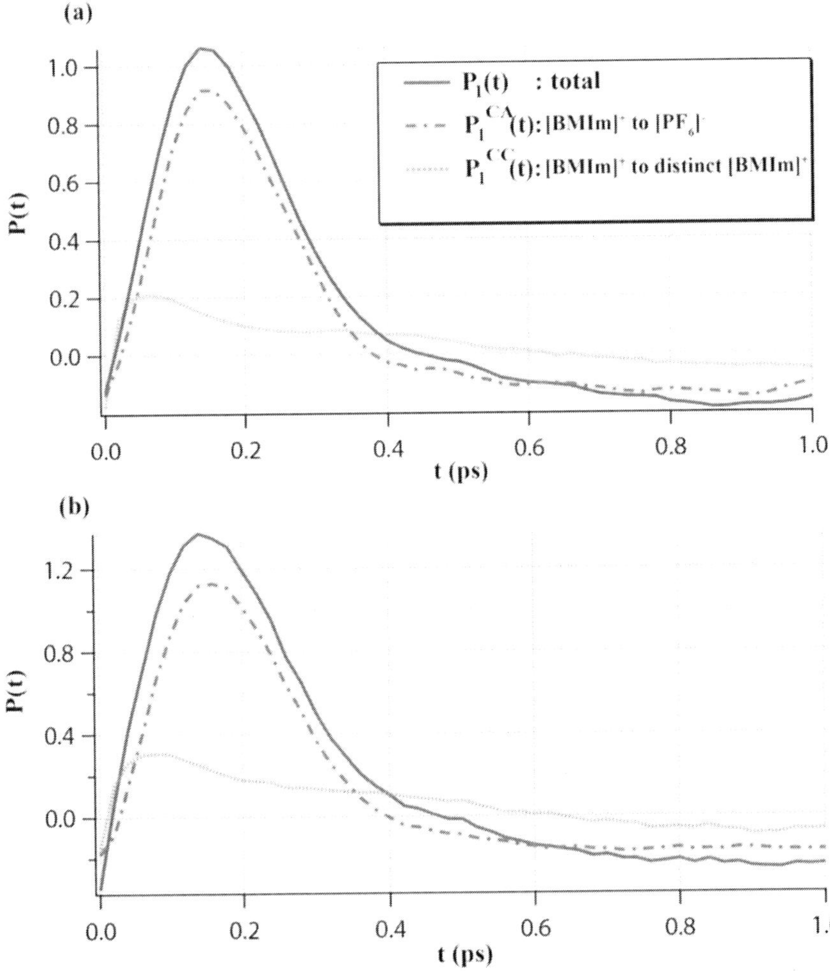

Figure 8. The calculated momentum correlation functions for the $[BMIm]^+$ at the center in the first coordination shell with the VACF: (a) in the nonpolarizable model and (b) in the polarizable model.

Based on these results, obviously, it is indicated that the variation of cage effects promotes the transfer of the initial momentum of the central cation to distinct cations rather than anions. Then, the cage effects is considered to be weakened by polarization effects, though the degree of this effect is likely to be relatively small as deduced from the figures.

On the other hand, the momentum correlations for the center anion in Figure 9 show distinct features from those for the center cation. The momentum correlation function between distinct anions, $P_1^{AA}(t)$, indicates much smaller contributions to the total momentum correlations, $P_1(t)$, in both the nonpolarizable and polarizable models, compared with the $P_1^{AC}(t)$. Therefore, these indicate that the initial momentum of the center anion is mainly transferred to neighboring cations, and the transference between distinct anions is not enhanced. These are consistent with the consideration that the cation-anion interactions promote the propagation of the momentum one after another, as mentioned in the cation case. In addition, Figure 9 shows more interesting features. Compared with the results for the central cation in Figure 8, while the $P_1^{AA}(t)$ does not contribute negligibly to the total momentum correlation in the nonpolarizable model, it indicates characteristic oscillating behavior in the short time region up to 0.4 ps in the polarizable model, and the latter has a relatively larger contribution to the $P_1(t)$ up to 0.1 ps than in the nonpolarizable model. This clearly indicates the decrease of cage effects through polarization effects and implies that the interionic interactions between distinct anions could become effective by the charge-charge and dipole-dipole interactions [28].

5. RELAXATION PROCESSES AND DYNAMICAL PROPERTIES

Firstly, the relaxation feature of IL system is examined with the computed polarizability TCF and its time derivative, and then, we discuss dynamical behavior of cations and anions in ILs, and consider what kinds of properties with simulation data we have to check and care in the study of dynamical properties of ILs.

Figure 10 shows the computed polarizability TCF for each component and the time derivative of the total polarizability TCF for [BMIm][NTf_2]. As seen in Figure 10(a), the relaxation of the polarizability TCFs seems to be slower than in usual liquids. In particular, the relaxation of the anion shows faster decay than that of the cation. Also, for each component, it takes about 2~3 ns to approach toward zero. Thus, this indicates that it is required for us to perform a production MD simulation run for, at least, a few ns to study the relaxation behavior of the polarizability TCF. Also, as seen in Figure 10(a), the cation-anion cross-correlation indicates similar variation to the total polarizability TCF. Therefore, the relaxation behavior of the system strongly correlates with the cation-anion cross-correlation, and it is emphasized that tracking cross-correlation terms is very important. On the other hand, corresponding to this result, the time derivative of the polarizability TCF shows long-decay feature extending up to the nanosecond times, as seen in Figure 10(b). Considering that the time derivative of the polarizability

TCF is directly related to optical Kerr effect (OKE) response,[14] these results are indicative of how long a MD simulation has to be carried out to examine the relaxation of the total system polarizability and the OKE response.

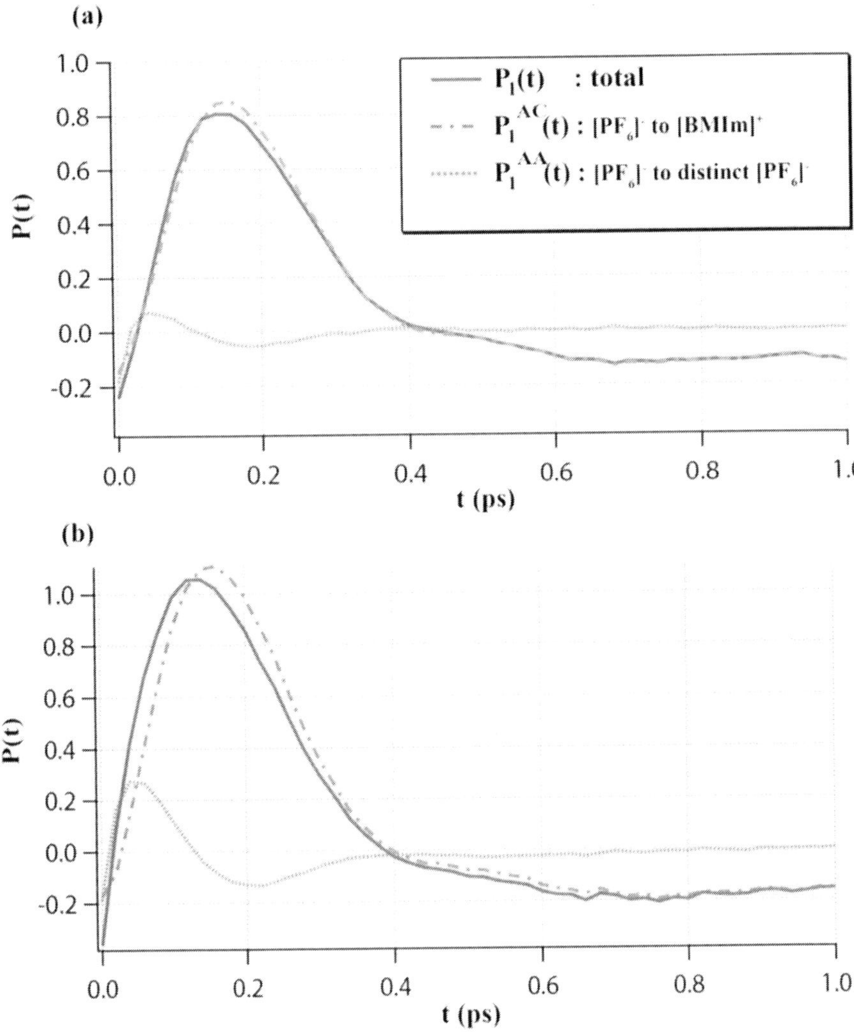

Figure 9. The calculated momentum correlation functions for the $[PF_6]^-$ at the center in the first coordination shell with the VACF: (a) in the nonpolarizable model and (b) in the polarizable model.

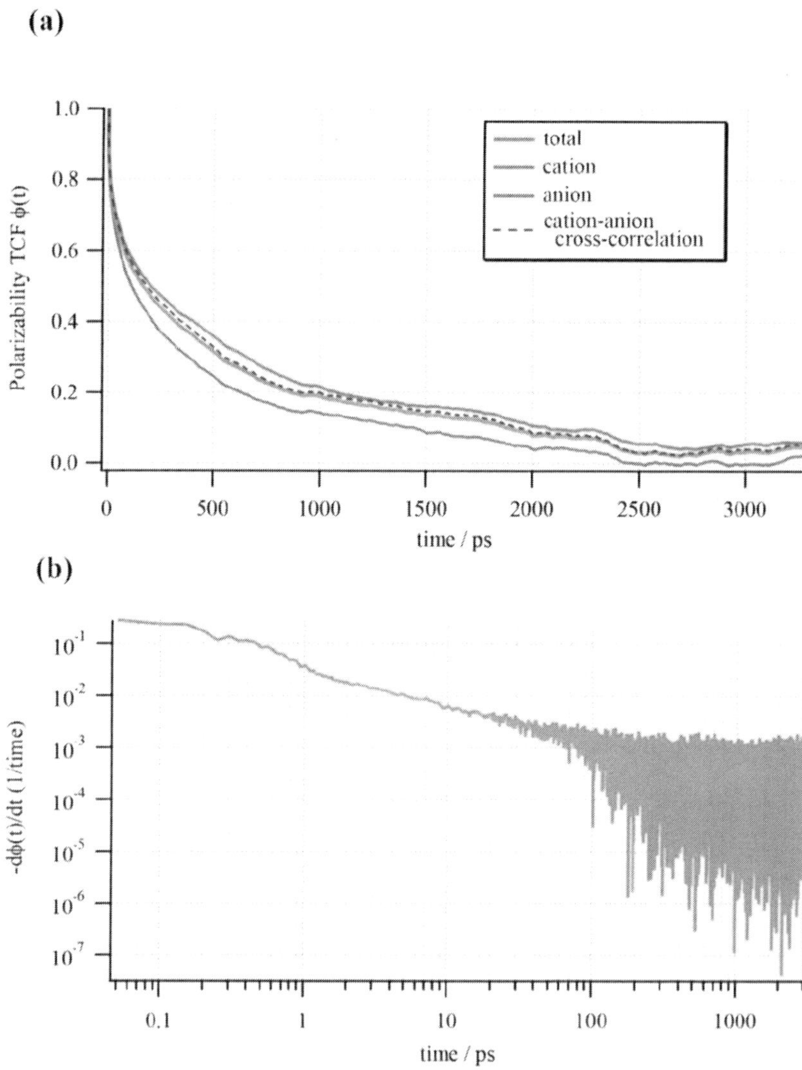

Figure 10. The computed polarizability TCF for each component (a) and the time derivative of the total polarizability TCF (b) for [BMIm][NTf2] up to about 3 ns.

The relaxation process of ILs implies long-time dynamics of ILs. In particular, it has been known that diffusive motion of each ion in ILs at room temperature is usually much slower than in usual liquids.[1, 16] Therefore, a MD simulation would need to be run for a few ns or more. Then, we are able to use a procedure to check whether a system is in the diffusive regime. The procedure is to compute β(t) = dlog(MSD)/dlog(t),[16] where MSD means the mean-squared displacement (MSD) of the center of mass of ions, $\left\langle \sum_{i=1}^{N} [r_i(t) - r(0)]^2 \right\rangle$. In the case that β(t) = 1, the system is in the diffusive regime, while when b(t) < 1, the system is in the sub-diffusive regime. In Figure 11, computed β(t)s for [BMIm][NTf_2] are shown. These results indicate that both the cation and the anion reach the β(t) region between 0.8 and 1 (the shadowed area in Figure 11) after about 16 ns. This β(t) region is considered to be almost in the diffusive regime. Therefore, Figure 11 obviously indicates that, for confirming reliable self-diffusivities carefully, we need to perform a longer MD simulation than about 15 ~ 20 ns.

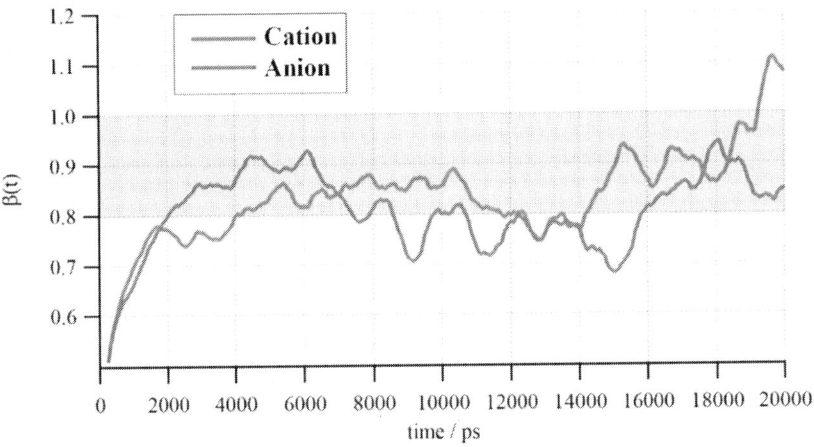

Figure 11. β(t) (see text) for $[BMIm]^+$ and $[NTf_2]^-$. Also, see text for the shadowed area.

Figure 12 displays the results calculated for the MSD and, in comparison, the non-Gaussian parameter, $\alpha(t)$, [50, 51]

$$\alpha(t) = 3\left\langle r^4(t)\right\rangle \Big/ 5\left\langle r^2(t)\right\rangle^2 - 1 \tag{32}$$

where $r(t)$ is the displacement of an ion at time t with respect to its position at t = 0. In Figure 12, the MSDs of $[PMIm]^+$ and $[NTf_2]^-$ indicate three typical dynamic ranges (regions), respectively. Also, the behavior of the short- and long-time regimes of MSD for both the cation and the anion is similar to each other.

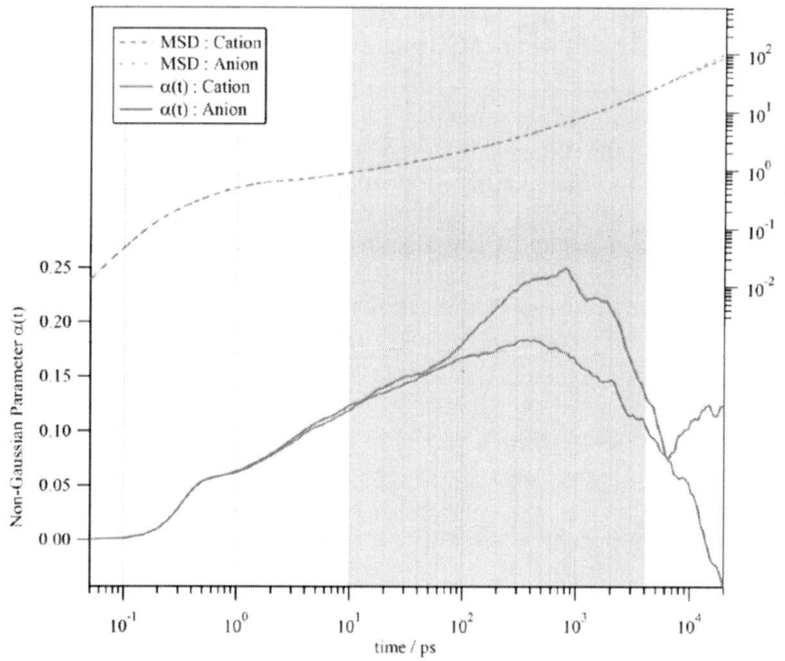

FIGURE 12. MSDs and non-Gaussian parameters for $[BMIm]^+$ and $[NTf_2]^-$. See text for the shadowed area.

For $[BMIm]^+$, first region is a microscopic regime until about 10 ps. Second is a crossover regime (the shadowed area in the Figure 12) until about 3-5 ns, and third regime is a sub-linear time dependence (sub-diffusive region) toward 20 ns. Also, has a double peak structure. One is at about 0.5 ps and the other at about 800 ps. In addition, a small shoulder structure emerges at between two peaks. As clearly seen in Figure 12, the short-time maximum peak corresponds to the microscopic region of the MSD, while the long-time maximum is located at around the center of the crossover regime of the MSD. On the other hand, for the anion, $[NTf_2]^-$, $\alpha(t)$ has a double peak structure similar to that of the cation. In particular, the long-time maximum peak is slightly shifted to about 300 ps, while the short-time maximum is locacted at the almost same time region as the cation. The time regime where the long-time maximum appear, is in good accord with those of $[BMIm]^+$, as shown in Figure 12. Therefore, considering that the magnitude of the deviation from the Gaussian behavior is remarkably larger for $[BMIm]^+$ than for $[NTf_2]^-$ as shown in Figure 12, it is expected that the relaxation behavior of the cation in perturbed $[BMIm][NTf_2]$ IL might be largely different from that of the anion. This consideration implies the possibility for us to find a distinct relaxation process in the OKE response (compare Figure 10 with Figure 12). Also, it should be noticed that the diffusive regime (linear time-dependence region) after about 10 ns in Figure 12 coincides with the diffusive region in Figure 11 (after about 16 ns) for both the cation and the anion.

In particular, the shadowed area in Figure 12 covers the time range of the full width at half maximum (FWHM) of the second peak of $\alpha(t)$. As pointed out previously,[52] this FWHM range is likely to include a slow decay region usually known as the α relaxation which is characteristic of glass-forming supercooled liquids in general. Also, recently, it has been reported that similar behavior could appear for ILs at room temperature.[51, 53, 54] Therefore, we had better study the α relaxation in ILs with the observations of the non-Gaussian parameter in addition to information on spatial relaxation such as intermediate scattering function. Also, it is suggested that, when we investigate dynamical properties of ILs, the crossover, the sub-diffusive, and the diffusive regimes have to be carefully examined, as one of criterions for reliable research. Thus, only with a long-time MD simulation, it is considered that we are able to investigate dynamical properties of ILs to which experimental procedures are not accessible.

6. CONCLUSION

In this chapter, we introduced cross-correlation function analyses with a polarizable model, polarizability TCFs for investigating the relaxation behavior of ILs, and some of procedures for studying dynamical properties. We showed the importance of considering cross-correlation functions and related properties, showing some of examples. Firstly, we employed the polarizable model based on point dipole treatment to the investigation of the polarization effect on the target IL, [BMIm][PF_6]. With the MD simulation data for both the nonpolarizable and polarizable models, velocity cross-correlation analyses were shown, and we presented the momentum correlation functions between the cation and anion species in the IL. Next, we computed polarizability TCFs of [BMIm][NTf_2] and discussed their relaxation behavior. In addition, we investigated dynamical properties of ILs such MSDs and non-Gaussian parameters and considered what kinds of properties with simulation data we have to check and care in the study of dynamical properties of ILs. This chapter is summarized as follows:

1. In the study and discussion of velocity cross-correlation functions, it was shown that polarization effects could enhance the cross-correlations between $[BMIm]^+$ and $[PF_6]^-$ in the polarizable model in comparison with that in the nonpolarizable model. These features are ascribed to interionic interactions through attractive

forces coming from the charge-dipole and dipole-dipole interactions caused by polarization effects in addition to charge-charge Coulomb interactions. Based on the results of computed cross-correlation between distinct cations (or anions), it was shown that, at early time region, the modulation of cage effects through polarization effects could improve the probability of approach between like ions. Also, by decomposing the cross-correlation function into the longitudinal and nonlongitudinal components, it was indicated that, at the short time region, the velocity cross-correlation is predominantly controlled by the longitudinal contribution. In addition, it was indicated that, compared with the longitudinal correlation in the nonpolarizable model, the longitudinal component is further modified in the polarizable model.

2. On the momentum correlation functions between $[BMIm]^+$ and $[PF_6]^-$, it was exhibited that the correlation between $[BMIm]^+$ and $[PF_6]^-$ plays a important role. Also, it was shown that the contribution of the cross-correlation between distinct anions could be enhanced in the polarizable model. This result indicates that cage effects could be diminished with polarization effects, implying that the interionic interactions between distinct anions could be intensified by the charge-charge and dipole-dipole interactions related to polarization effects. Therefore, as has been pointed out in the literature [23, 24, 29], it is considered that the cage effect in ILs could be reduced by many-body polarization effects.
3. Both the computed polarizability TCF for each component and the time derivative of the total polarizability TCF for $[BMIm][NTf_2]$ showed long-time decay behavior. It was indicated that those took about 2~3 ns to approach toward zero. Therefore, this indicates that it is required for us to perform a production MD simulation run. Also, the relaxation behavior of ILs was investigated with the calculation of $\beta(t) = d\log(MSD)/d\log(t)$ as an indicator of the diffusive region. Our results suggested that long-time dynamics of ILs has to be studied with a longer MD simulation than about 15 ~ 20 ns for confirming reliable self-diffusivities in ILs, carefully. Furthermore, we examined MSDs and non-Gaussian parameters for $[BMIm][NTf_2]$. From our studies, it was exhibited that the magnitude of the deviation from the Gaussian behavior is remarkably larger for $[BMIm]^+$ than for $[NTf_2]^-$, and that, comparing with corresponding MSDs, it is possible to study diffusive motion of the cation and the anion. In addition, it was suggested that these studies imply the possibility for us to find a distinct relaxation process in the OKE response. Therefore, it is suggested that we had better study the α relaxation in ILs with the observations of the non-Gaussian parameter in addition to

information on spatial relaxation such as intermediate scattering function. Lastly, it is concluded, only with a long-time MD simulation (> 15 ~ 20 ns), that we are able to investigate the dynamical properties of ILs to which experimental procedures are not accessible.

7. ACKNOWLEDGEMENTS

1. Wasserscheid, P., & Welton, T. (2008). Ionic Liquids in Synthesis, Weinheim, WileyVCH.
2. Wasserscheid, P., & Keim, W. (2000). Ionic liquids- New "solutions" for transition metal catalysis. Angew. Chem., Int. Ed., 39, 3773-3789.
3. Castner, E. W. Jr, Wishart, J. F., & Shirota, H. (2007). Intermolecular Dynamics, Interactions, and Solvation in Ionic Liquids. Acc. Chem. Res., 40, 1217-1227.
4. Weingaertner, H. (2007). Understanding ionic liquids at the molecular level: facts, problems, and controversies. Angew. Chem., Int. Ed., 47, 654-670.
5. Ohno, H. (2005). Electrochemical Aspects of Ionic Liquids, Hoboken, Wiley-Interscience.
6. Rogers, R. D., & Voth, G. A. (2007). Special Issue on Ionic Liquids. Special Issue on Ion- ic Liquids. Acc. Chem. Res., 40(11).
7. Wishart, J. F., & Castner, E. W. Jr. (2007). Special Issue on Physical Chemistry of Ionic Liquids.Special Issue on Physical Chemistry of Ionic Liquids. Special Issue on Physical Chemistry of Ionic Liquids. J. Phys. Chem. B, 111(18).
8. Ohno, H., & Fukumoto, K. (2008). Progress in ionic liquids for electrochemical reaction matrices. Electrochemistry, 76, 16-23.
9. Shirota, H., Wishart, J. F., Castner, E. W., & Jr , . (2007). Intermolecular Interactions and Dynamics of Room Temperature Ionic Liquids That Have Silyl- and Siloxy-Substituted Imidazolium Cations. J. Phys. Chem. B, 111, 4819-4829.
10. Xiao, D., Rajian, J. R., Hines, J., , L. G., Li, S., Bartsch, R. A., & Quitevis, E. L. (2008). Nanostructural Organization and anion effects in the optical Kerr effect spectra of bi- nary ionic liquids mixtures. J. Phys. Chem. B, 112, 13316-13325.
11. Koeberga, M., Wu, C.-C., Kimc, D., & Bonn, M. (2007). THz dielectric relaxation of ionic liquid:water mixtures. Chem. Phys. Lett., 439, 60-64.
12. Yamamoto, K., Tani, M., & Hangyo, M. (2007). Terahertz Time-Domain Spectroscopy of Imidazolium Ionic Liquids. J. Phys. Chem. B, 111, 4854-4859.
13. Shirota, H., Nishikawa, K., & Ishida, T. (2009). Atom substitution effects of [XF6]- in ionic liquids. 1. Experimental study. J. Phys. Chem. B, 113, 9831-9839.
14. Ishida, T., Nishikawa, K., & Shirota, H. (2009). Atom substitution effects of [XF6]- in ionic liquids. 2. Theoretical study. J. Phys. Chem. B, 113, 9840-9851.
15. Refer to the sec.4 in ref.1 and references therein.
Maginn, E. J. (2007). Atomistic Simulation of the Thermodynamic and Transport Properties of Ionic Liquids. Acc. Chem. Res., 40, 1200-1207.
16. Lopes, J. N. A. C., & Padua, A. A. H. (2006). Nanostructural Organization in Ionic Liquids. J. Phys. Chem. B, 110, 3330-3335.
17. Bhargava, B. L., & Balasubramanian, S. (2007). Refined potential model for atomistic simulations of ionic liquid [bmim][PF6]. J. Chem. Phys., 127(114510), 1-6.
18. Znamenskiy, V., & Kobrak, M. N. (2004). Molecular Dynamics Study of Polarity in Room-Temperature Ionic Liquids. J. Phys. Chem. B, 108, 1072-1079.
19. Shim, Y., Duan, J., Choi, M. Y., & Kim, H. J. (2003). Solvation in molecular ionic liquids. J. Chem. Phys., 119, 6411-6414.

20. Kobrak, M. N. (2006). Characterization of the solvation dynamics of an ionic liquid via molecular dynamics simulation. J. Chem. Phys., 125(064502), 1-11.
21. Hu, Z. H., & Margulis, C. J. (2006). Heterogeneity in a room-temperature ionic liquid: Persistent local environments and the red-edge effect. Proc. Natl. Acad. Sci., U.S.A., 103, 831-836.
22. Yan, T., Burnham, C. J., Del Popolo, M. G., & Voth, G. A. (2004). Molecular Dynamics Simulation of Ionic Liquids: The Effect of Electronic Polarizability. J. Phys. Chem. B, 108, 11877-11881.
23. Jiang, W., Yan, T., Wang, Y., & Voth, G. A. (2008). Molecular Dynamics Simulation of the Energetic Room-Temperature Ionic Liquid, 1-Hydroxyethyl-4-amino-1,2,4-triazo- lium Nitrate (HEATN). J. Phys. Chem. B, 112, 3121-3131.
24. Urahata, S. M., & Ribeiro, M. C. C. (2005). Single particle dynamics in ionic liquids of 1-alkyl-3-methylimidazolium cations. J. Chem. Phys., 122(024511), 1-9.
25. Urahata, S. M., & Ribeiro, M. C. C. (2006). Collective excitations in an ionic liquid. J. Chem. Phys., 124(074513), 1-8.
26. Hu, Z., Huang, X., Annapureddy, H. V. R., & Margulis, C. J. (2008). Molecular Dynamics Study of the Temperature-Dependent Optical Kerr Effect Spectra and Intermolecular Dynamics of Room Temperature Ionic Liquid 1-Methoxyethylpyridinium Dicyanoamide. J. Phys. Chem. B, 112, 7837-7849.
27. Ishida, T. (2011). Molecular dynamics study of the dynamical behavior in ionic liquids through interionic interactions. J. Non-Cryst. Solids, 357, 454-462.
28. Leach, A. R. (2001). Molecular Modeling, Principles and Applications, Harlow, Pearson Education.
29. Verdaguer, A., Padró, J. A., & Trullàs, J. (1998). Molecular dynamics study of the velocity cross-correlations in liquids. J. Chem. Phys., 109, 228-234.
Verdaguer, A., & Padró, J. A. (2001). Computer simulation study of the velocity cross correlations between neighboring atoms in simple liquid binary mixtures. J. Chem. Phys., 114, 2738-2744.
30. Frenkel, D., & Smit, B. (2002). Understanding Molecular Simulation, From Algorithms to Applications, London, Academic Press.
31. Bernardo, D. N., Ding, Y., Krogh-Jespersen, K., & Levy, R. M. (1994). An Anisotropic Polarizable Water Model: Incorporation of All-Atom Polarizabilities into Molecular Mechanics Force Fields. J. Phys. Chem., 98, 4180-4187.
32. Předota, M., Cummings, P. T., & Chialvo, A. A. (2002). Pair approximation for polarization interaction and adiabatic nuclear and electronic sampling method for fluids with dipole polarizability. Mol. Phys., 100, 2703-2717.
33. Ahlström, P., Wallqvist, A., Engström, S., & Jönsson, B. (1989). A molecular dynamics study of polarizable water. Mol. Phys., 68, 563-581.
34. Frenkel, D., & Mc Tague, J. P. (1980). Molecular dynamics studies of orientational and collision-induced light scattering in molecular fluids. J. Chem. Phys., 72, 2801-2818.
35. Geiger, L. C., & Ladanyi, B. M. (1987). Higher order interaction-induced effects on Rayleigh light scattering by molecular liquids. J. Chem. Phys., 87, 191-202.
36. Lopes, J. N. C., Deschamps, J, & Padua, A. A. H. (2004). Modeling Ionic Liquids Us- ing a Systematic All-Atom Force Field. J. Phys. Chem. B, 108, 2038-2047.
37. Lopes, J. N. C., Deschamps, J., & Padua, A. A. H. (2004). Additions and Corrections: Modeling Ionic Liquids Using a Systematic All-Atom Force Field. J. Phys. Chem. B, 108, 11250.
38. Köddermann, T, Paschek, D, & Ludwig, R. (2007). Molecular Dynamic Simulations of Ionic Liquids: A Reliable Description of Structure, Thermodynamics and Dynamics. ChemPhysChem, 8, 2464-2470.
39. Smith, W., & Forster, T. R. (2001). The DL_POLY_2 User Manual, Daresbury, United Kingdom, Daresbury Laboratory.
40. Allen, M. P., & Tildesley, D. J. (1987). Computer Simulation of Liquids, Clarendon, Oxford.

41. Shirota, H, Mandai, T, Fukazawa, H, & Kato, T. (2011). Comparison between Dica tionic and Monocationic Ionic Liquids: Liquid Density, Thermal Properties, Surface Tension, and Shear Viscosity. J. Chem. Eng. Data, 56, 2453-2459.
42. Nymand, T. M., & Linse, P. (2000). Ewald summation and reaction field methods for potentials with atomic charges, dipoles, and polarizabilities. J. Chem. Phys., 112, 6152-6160
43. Van Duijnen, P. T., & Swart, M. (1998). Molecular and Atomic Polarizabilities: Thole's Model Revisited. J. Phys. Chem. A, 102, 2399-2407.
44. Lide, D. R. (2006). CRC Handbook of Chemistry and Physics, Boca Raton, CRC Press.
45. Thole, B. T. (1981). Molecular Polarizabilities Calculated with a Modified Dipole Interaction. Chem. Phys., 59, 341-350.
46. Molecular polarizabilities of [BMIm]+ and [NTf2]- were calculated with the same method in [14]. For molecular polarizability tensor elements of [BMIm]+, the report- ed values were used [14]. With the same procedures as in [14], molecular polarizabil- ity tensor elements of [NTf2]- were calculated, and the values of 9.8855, 1.4440 x 10-3 ,13.577-5x 10^{-4},-1.0539, and 10.315 Å^3 were used for α_{XX}, α_{XY}, α_{YY}, α_{XZ}, α_{YZ}, and α_{ZZ}, re- spectively.
47. Verdaguer, A., & Padró, J. A. (2000). Velocity cross-correlations and atomic momentum transfer in simple liquids with different potential cores. Phys. Rev. E, 62, 532-537.
48. Rahman, A. (1964). Correlations in the Motion of Atoms in Liquid Argon. Phys. Rev., 136, A405-A411.
49. Del Pópolo, M. G., & Voth, G. A. (2004). On the Structure and Dynamics of Ionic Liquids. J. Phys. Chem. B, 108, 1744-1752.
50. Colmenero, J., Alvarez, F., & Arbe, A. (2002). Self-motion and the α relaxation in asimulated glass-forming polymer: Crossover from Gaussian to non-Gaussian dynam- ic behavior. Phys. Rev. E, 65(041804), 1-12.
51. Ngai, K. L. (2011). Relaxation and Diffusion in Complex Systems, New York, Springer.
52. Bhargava, B. L., Klein, M. L., & Balasubramanian, S. (2008). Structural Correlations and Charge Ordering in a Room-Temperature Ionic Liquid. ChemPhysChem, 9, 67-70.

CHAPTER 7

Modeling of Ionic Liquid Systems: Phase Equilibria and Physical Properties

Filipa M. Maia, Noelia Calvar, Emilio J. González, Aristides P. Carneiro, Oscar Rodriguez and Eugénia A. Macedo

LSRE-Laboratory of Separation and Reaction Engineering, Associate Laboratory LSRE/LCM, Department of Chemical Engineering, Faculty of Engineering, University of Porto, Rua Dr. Roberto Frias, Porto, Portugal

1. INTRODUCTION

Ionic liquids (ILs) are a class of salts with a melting temperature below 100 °C, and the study of these compounds is considered priority by the U.S. Environmental Protection Agency. Due to their specific properties, which can be adjusted by changing either the cation or the anion, ILs have received great attention by the scientific community as potential replacements for volatile organic solvents (VOCs), and nowadays, ILs are starting to leave academic labs and find their way into a wide variety of industrial applications [1]. For example, ILs are used for the dispersion of nano-materials at IOLITEC, Air Products uses ILs instead of pressurized cylinders as a transport medium for reactive gases, ION Engineering is commercializing technology using ILs and amines for CO_2 capture and natural gas sweetening, and many others.

In order to apply these new compounds in different processes, the study of their physical properties, pure or mixed with other solvents, and phase equilibria (vapor-liquid, liquid-liquid, and solid-liquid equilibria) is crucial from a technological point of view. For example, density is fundamental to develop equations of state and it is also required for the design of different equipments, while viscosity is necessary for the design of processing units and pumping systems, and to study heat and mass transfer processes [2]. On the other hand, the refractive index can be used as a measure of the electronic polarizability of a molecule and can provide useful information when studying the forces between molecules or their behavior in solution [3].

As known, ILs also show an interesting potential to be used in separation processes and extraction media. Therefore, the knowledge of the mutual solubilities of molecular solvents and ILs prior to their industrial applications is also of primary importance. Moreover, many factors that control the phase behavior of these ionic salts with molecular solvents may be described from the phase equilibrium data.

However, as the number of possible ILs is enormous, this cannot be accomplished via experimental determination. Thus, it is very important to obtain models or empirical equations able to describe satisfactorily the experimental data.

In this chapter, a revision of the different equations applied for the modeling of physical properties of pure ILs and their mixtures, and phase equilibria of binary and ternary mixtures containing ILs, is presented and discussed. Future trends regarding the use of new models, namely equations of state accounting for association effects, are also focused.

2. PHYSICAL PROPERTIES

2.1. Pure ionic liquids

Since ILs are relatively new compounds, experimental data on physical properties, such as density, viscosity, or refractive index of pure ILs and its mixtures with other solvents are required for the design of different equipment and processing units and very useful for developing accurate theoretical models.

Due to innumerable number of ILs that can be synthesized, experimental measurements are impractical for selection of a suitable IL for a specific application. Therefore, development of correlations and theoretical approaches allowing accurate modeling of IL-based systems is essential. This section shows the most common empirical equations used to correlate the temperature dependence of some of the physical properties of ILs.

For pure ILs, temperature dependence of physical properties such as density, speed of sound, or refractive index is very important for the successful and large-scale use of these compounds. Usually, this dependence is described using simple polynomial expressions, mainly equations of first, second and third order [4-6].

Several papers were also published concerning the experimental densities of pure ILs as a function of temperature and pressure [6-9]. The Tait equation [10] with four adjustable parameters is commonly used to fit these experimental data [6,8]. This equation is an integrated form of an empirical equation representative of the isothermal compressibility behavior versus pressure, and it can be expressed as:

$$\rho\left(T,\ p,\ C,\ B\right)=\frac{\rho(T,\ 0.1\ \mathrm{MPa})}{1-C\cdot Ln\left(\frac{B(T)+p}{B(T)+0.1\ \mathrm{Mpa}}\right)} \tag{1}$$

where ρ (T, 0.1 MPa) represents the temperature dependence of density at 0.1 MPa. C is an adjustable parameter, and $B(T)$ is commonly expressed as a second-order polynomial:

$$B(T)=\sum_{i=0}^{2} B_i\cdot T^{\,i} \tag{2}$$

Regarding the variation of viscosity with temperature for pure ILs, a large number of empirical equations for correlating this property of pure fluids and mixtures can be found in literature [4,5,11-14]. The most commonly used equation is an Arrhenius-like law:

$$\eta=\eta\cdot exp\left(-\frac{E_a}{RT}\right) \tag{3}$$

where the viscosity at infinite temperature (η_∞) and the activation energy (E_a) are characteristic parameters generally adjusted from experimental data.

According to Seddon *et al.* [15], the Arrhenius law can generally be applied when the cation presents only a limited symmetry. If it is not the case, and especially in the presence of small and symmetrical cations with low molar mass, other equations such as the Vogel–Fulcher–Tamman (VFT) equation [16-18], or the modified VFT (mVFT) equation are recommended. This kind of expressions includes an additional adjustable temperature parameter (T_0) to the exponential term:

$$\eta = A \cdot exp\left(\frac{B}{T - T_0}\right) \tag{4}$$

$$\eta = A \cdot T^{0.5} \cdot exp\left(\frac{B}{T - T_0}\right) \tag{5}$$

where *A* and *B* are also adjustable parameters.

Another empirical equation to correlate viscosity data with temperature was proposed by Litovitz [19]:

$$\eta = A \cdot exp\left(\frac{B}{RT^3}\right) \tag{6}$$

This equation is used at ambient pressures and has the advantage of containing only two fitting parameters. Comparing all these equations, in general, the best fits for the variation of viscosity with temperature for pure ILs are obtained with the VFT equation [5].

As reflected by Harris *et al.* [11], the general form of the pressure dependence of the shear viscosity of liquids is greater than exponential at moderate pressures and less than exponential at very high pressures. Taking this into account, these authors modified the VFT and Litovitz equations in order to include the temperature and pressure dependence of the viscosity for several pure imidazolium-based ILs. The new equations were defined as:

$$\eta = exp\left(A + Bp + \frac{(C + Dp + Ep^2)}{T^3}\right) \tag{7}$$

$$\eta = exp\left(A' + B' + \frac{(C' + D'p + E'p^2)}{T - T_0}\right) \tag{8}$$

As for the density, a Tait-form equation is also used to correlate the pressure dependence of viscosity demonstrating good correlations [12,13].

$$\frac{\eta(T,\ p,\ C,\ B)}{\eta(T,\ 0.1\ MPa)} = \exp\left(C \cdot Ln\left(\frac{B(T) + p}{B(T) + 0.1\ \text{Mpa}}\right)\right)$$

(9)

where $\eta(T, 0.1\ \text{MPa})$ represents the temperature dependence of viscosity at 0.1 MPa. C is an adjustable parameter, and $B(T)$ is also commonly expressed as a second-order polynomial.

A hybrid Tait–Litovitz equation at elevated pressures (up to 126 MPa) was also presented in the literature to correlate the viscosity data for a series of room-temperature ILs [14]. The Litovitz equation is firstly used to correlate the data at ambient pressure, and then the Tait parameters are fitted for the higher pressures. This equation has the advantages of containing fewer fitting parameters than other models and simplicity of data analysis. The results show a good fit between the experimental data and those predicted by this equation.

2.2. Binary And Ternary Mixtures

In order to better understand the nature of ILs and design any future technological processes, detailed knowledge on the physical properties of ILs mixed with other solvents is required. During the last few years, the number of studies on thermophysical and thermodynamic properties of pure ILs and their mixtures with molecular solvents has increased significantly [4,20-24].

As for pure ILs, the dependence of the physical properties with temperature and composition is also correlated using empirical equations. In general, the change of density, speed of sound, refractive index and viscosity with composition is typically fitted to a polynomial expression although other more specific equations can be also found in literature. As example, the Connors and Wright equation [25] is employed to describe the variation of density with composition:

$$\rho = \rho_1 - \left[1 + \frac{b x_1}{(1 - a x_1)}\right](1 - x_1)(\rho_1 - \rho_2)$$

(10)

where ρ_1 and ρ_2 are densities of pure compounds; x_i is the mole fraction of component i of the mixture; and a and b are fitting parameters. Although this equation was initially adopted only for the concentration dependencies of surface tension, Geppert-Rybczynska *et al.* [20] employed this equation for density getting a fit better than using any other simple polynomial.

As it is known, the above mentioned physical properties can be used to obtain the corresponding excess properties, which are generally fitted to a Redlich-Kister type equation [26]:

$$\delta Q = x_1(1 - x_1)\sum_{i=0}^{M} A_i(2x_1 - 1)^i \tag{11}$$

where δQ is the excess property, x is the mole fraction, A_i are the adjustable parameters and M is the degree of the polynomial expansion.

An extended version of the Redlich-Kister equation, which takes into account the dependence on composition and temperature simultaneously, is also used to fit the excess properties [22]:

$$\delta Q = x_1(1 - x_1)\frac{\sum_{i=0}^{M} A_i(2x_1 - 1)^i}{1 + \sum_{j=1}^{N} B_j(2x_1 - 1)^j} \tag{12}$$

In order to take into account the influence of temperature on the excess properties, all the coefficients A_i and B_i are usually expressed as a second-order polynomial.

Density, refractive index, and viscosity data for ternary mixtures containing ILs are also common in the literature [27-29] and they are usually fitted to polynomial expressions. In these cases, weight fractions are often used instead of mole fractions, due to the large difference of molar mass between ILs and most organic solvents.

For the modeling of the excess properties, such as excess molar volumes, viscosity deviations, refractive index deviations or excess free energies of activation of viscous flow, the use of empirical equations is commonly adopted. Although the Redlich-Kister equation is also applied to correlate the excess properties for ternary systems containing ILs [29], the equations more widely employed are those proposed by Cibulka [30], Singh *et al.* [31] and Nagata and Sakura [32]. All the correlative models are capable of representing the behavior of the ternary mixtures with a higher or lesser degree of accuracy, although that developed by Cibulka usually leads to a better agreement with experimental data [28].

Cibulka equation:

$$Q_{123}^E = Q_{12}^E + Q_{13}^E + Q_{23}^E + x_1x_2x_3(A + Bx_1 + Cx_2) \tag{13}$$

Singh et al. equation:

$$Q^E_{123} = Q^E_{12} + Q^E_{13} + Q^E_{23} + Ax_1x_2x_3 + Bx_1(x_2 - x_3) + Cx_1^2(x_2 - x_3)^2 \quad (14)$$

Nagata and Sakura equation:

$$Q^E_{123} = Q^E_{12} + Q^E_{13} + Q^E_{23} + x_1x_2x_3A \quad (15)$$

where *A*, *B* and *C* are fit parameters and Q^E_{ij} is the contribution to the excess property of the constituent binary mixtures evaluated by Redlich-Kister equation.

3. PHASE EQUILIBRIA

Despite the large number of published articles and the broad fields of applications, there has not been a model explicitly derived for phase equilibria of ILs. This lack of models intended for systems containing ILs has forced researchers to use the equations available. But these equations were intended for ionic solutions, or not intended for ions at all. Thus, a model to account for a medium which is composed of ions, without a molecular solvent, is still needed. Nevertheless, the phase equilibria of ILs and their mixtures are being modeled in the literature. The models used will be described ahead. In general, those based on the excess Gibbs energy such as Wilson, NRTL or UNIQUAC were the first to appear. Lately, models with modifications for association (UNIQUAC ASM, NRTL1) and for electrolytes (PDH, e-NRTL) were also applied for this kind of systems in order to improve the results obtained with the initial models. Besides, Equations of State (EoS) have also been applied, especially for mixtures with gases and for broad ranges of pressure.

3.1. G^e-Based Models

Most of the g^E models available in the literature are for non-electrolytes. Thus, many authors have been using these models, or models for electrolyte solutions, for the phase equilibria of systems containing ILs. For example, for binary systems containing ILs it is common to use the model developed by Debye-Hückel (which was derived for small salt concentrations), although it is not recommendable for solutions at high ionic concentration.

The models for the correlation of these experimental data can be split into two main groups: *i*) ion-interaction Pitzer models and *ii*) local composition models.

i. The model developed by Pitzer has created a new generation of theories which use multiparameter regression. In the ion-interaction Pitzer model [33] the ion-interaction parameters are dependent on temperature and pressure, and it takes into account the Debye-Hückel constant.

The three-parameter Pitzer-ion interaction model has been successfully used for modeling vapor-liquid data of mixtures of ILs with water [34-37] or with alcohol [38] and has the following form for a binary 1:1 electrolyte solution:

$$\phi - 1 = f^{\phi} + \mathrm{m}B^{\phi} + \mathrm{m}^2C^{\phi} \tag{16}$$

where

$$f^{\phi} = -A_{\phi}\, I^{1/2} / (1 + bI^{1/2}) \tag{17}$$

$$A_{\phi} = (1/3)(2\pi \mathrm{N_A} d_s)^{1/2}\, (\mathrm{e}^2 / 4\pi\varepsilon_0\varepsilon kT)^{3/2} \tag{18}$$

$$B^{\phi} = \beta^{(0)} + \beta^{(1)} \exp(-\alpha_1 I^{1/2}) \tag{19}$$

In these equations, $\beta^{(0)}$, $\beta^{(1)}$ and C^{ϕ} are ion-interaction parameters of the Pitzer model that are dependent on temperature and pressure, and A_{ϕ} is the Debye-Hückel constant for the osmotic coefficient on the molal scale. Also, N_A is the Avogadro number, e is the proton charge, ε_0 is the permittivity of vacuum, and k is the Boltzman constant. The term I is the ionic strength in molality, $I = 1/2\ \Sigma m_i z_i^2$, where m_i is the molality of ith ion and z_i is the absolute value for i^{th} ionic charge. The remaining symbols have their usual meanings. The values for constants b and α were $b = 1.2$ kg$^{1/2}$ mol$^{-1/2}$ and $\alpha = 2$ kg$^{1/2}$ mol$^{-1/2}$, respectively.

In the last years, the Extended Pitzer model of Archer [39,40], in which the third adjustable parameter in the Pitzer model is replaced by a two-parameter function depending on the ionic strength, has demonstrated its accuracy in modeling binary Vapor-Liquid Equilibria (VLE) of systems containing ILs. In this model, the equation for B^{ϕ} is extended to:

$$B^{\phi} = \beta^{(0)} + \beta^{(1)} \exp(-\alpha_1 I^{1/2}) + \beta^{(2)} \exp(-\alpha_2 I^{1/2}) \tag{20}$$

and a new equation is introduced:

$$C^{\phi} = C^{(0)} + C^{(1)} \exp(-\alpha_3 I^{1/2}) \tag{21}$$

In the previous equations, the ion interaction parameters of the extended Pitzer model of Archer are $\beta^{(0)}, \beta^{(1)}, \beta^{(2)}, C^{(0)}$ and $C^{(1)}$, dependent on temperature and pressure, and $\alpha_1, \alpha_2, \alpha_3$, and b can be adjustable parameters or kept fixed at constant values; they are usually fixed to the values: α_1 = 2 kg$^{1/2}$mol$^{-1/2}$, α_2 = 7 kg$^{1/2}$ mol$^{-1/2}$, α_3 = 1 kg$^{1/2}$ mol$^{-1/2}$ and b = 3.2 [41-44], although other values can be used [45]. This model is widely used in literature for binary mixtures containing ILs and water or alcohol or acetonitrile [46-55] with very satisfactory results.

ii. The local composition (LC) models give a better empirical description and have physical meaning for the correlation of osmotic and activity coefficients. There are several LC models reported in literature for the modeling of phase equilibria experimental data; such as UNIversal QUAsiChemical (UNIQUAC) [56], Non-Random Two Liquids (NRTL) [57], electrolyte NRTL (e-NRTL) [58], Non-Random Factor (NRF) [59], modified NRTL (MNRTL) [60], Mean Spherical Approximation NRTL (MSA-NRTL) [61] or Extended Wilson (EW) [62]. Furthermore, models with modifications for association such as UNIQUAC ASM or NRTL 1 are also applied for this kind of systems, although their use is less common [63,64].

In local composition models it is assumed that the activity coefficient is composed by two terms: a long-range contribution (LR) and a short-range contribution (SR):

$$\ln \gamma_i = \ln \gamma_i^{LR} + \ln \gamma_i^{SR} \tag{22}$$

The well-known models of UNIQUAC and NRTL calculate the activity coefficients as follows:

ii.i In the UNIQUAC model [56], the long-range contribution term is expressed as:

$$\ln \gamma_i^{LR,UNIQUAC} = \ln\left(\frac{\phi}{x_i}\right) + \frac{z}{2} q_i \ln\left(\frac{\theta_i}{\phi_i}\right) + l_i - \frac{\phi_i}{x_i}\sum_{j=1}^{N} x_j l_j \tag{23}$$

where the term l_i is defined as a function of the external surface of the molecule and the bond segments:

$$l_i = \frac{z}{2}(r_i - q_i) - (r_i - 1) \tag{24}$$

and the coordination number, z, is set to 10.

The short-range contribution in this model is expressed as:

$$\ln \gamma_i^{SR,UNIQUAC} = q_i \left[1 - \ln\left(\sum_j \theta_j \tau_{ji} \right) - \sum_{j=1}^{N} \frac{\theta_j \tau_{ij}}{\sum_k \theta_k \tau_{kj}} \right] \tag{25}$$

The expression for the energy parameter, τ_{ij}, the volume fraction, ϕ_i, and the surface fraction, θ_i, are given for the next equations:

$$\tau_{ji} = \exp\left(-\frac{u_{ji} - u_{ii}}{RT} \right) = \exp\left(\frac{\Delta u_{ij}}{RT} \right) \tag{26}$$

$$\theta_i = \frac{q_i x_i}{\sum_j q_j x_j} \quad \phi_i = \frac{r_i x_i}{\sum_j r_j x_j} \tag{27}$$

where u_{ij} and u_{ji} are the energetic parameters, and r_i and q_i are the structural parameters corresponding to relative volume and surface area of the component *i*, respectively.

This model has been used for VLE of binary systems containing ILs [65]. The equation has also been used successfully for Liquid-Liquid Equilibria (LLE) of ternary systems including an IL [66,67], and even for Solid-Liquid Equilibria (SLE) of binary systems [68-70]. The main problem using the UNIQUAC model is the need for structural parameters r_i and q_i, the volume and surface area of the component. Despite for molecular components these parameters can be obtained easily (even from group-contribution data), in the case of ILs values for both anion and cation are calculated and summed. Different procedures have been proposed [66,68] and used successfully.

ii.In the NRTL model [57], the activity coefficients are calculated as follows:

$$\ln \gamma_i^{NRTL} = \left(\frac{\sum_{j=1}^{n} \tau_{ji} x_j G_{ji}}{\sum_{k=1}^{n} x_k G_{ki}} \right) + \sum_{j=1}^{n} \left(\frac{x_j G_{ij}}{\sum_{k=1}^{n} x_k G_{kj}} \right) \left(\tau_{ij} - \frac{\sum_{m=1}^{n} \tau_{mi} x_m G_{mi}}{\sum_{k=1}^{n} x_k G_{kj}} \right) \tag{28}$$

where

$$G_{ij} = \exp(-\alpha_{ij} \tau_{ij}) \tag{29}$$

$$\tau_{ij} = \frac{g_{ij} - g_{jj}}{RT} = \frac{\Delta g_{ij}}{RT} \quad \tau_{ji} = \frac{g_{ji} - g_{ii}}{RT} = \frac{\Delta g_{ji}}{RT} \tag{30}$$

where x represents the mole fraction, g_{ij} is an energy parameter that characterizes the interaction of species i and j, R is the gas constant, T is the absolute temperature, and the parameter $\alpha_{ij} = \alpha_{ji}$ is related to the nonrandomness in the mixture. Although α_{ij} can be adjusted, it can also be considered fixed in a value, usually between 0.2 and 0.47.

This is the local composition model most widely used in literature for binary and ternary systems containing ILs, regarding VLE [71-79] with alcohols or water, LLE of binary [80,81] and ternary systems [67,82-85] with ethanol or hydrocarbons or water and also SLE [68-70].

For the explanation of the following models, the Pitzer-Debye-Hückel (PDH) equation [33] has been used as the long-range term on a mole fraction scale as proposed by Chen *et al.* [86], assuming that the studied mixtures involve completely dissociated electrolytes. For the solvent, the formulation of this equation is:

$$\ln \gamma_1^{PDH} = \frac{2 A_\phi I_x^{3/2}}{\sqrt{M_S} \left(1 + \rho I_x^{1/2} \right)} \tag{31}$$

in which ρ is the number density of ionic species and the term I is the ionic strength on a mole fraction basis, $I = 1/2 \Sigma x_i z_i^2$.

The short-range contribution calculated using different models, such as electrolyte NRTL (e-NRTL) [58], non-random factor (NRF) [59], modified NRTL (MNRTL) [60], mean spherical approximation NRTL (MSA-NRTL) [61] or Extended Wilson (EW) [62] are explained below.

ii. In the e-NRTL model [58], the short-range contribution for the activity coefficient is calculated as:

$$\ln \gamma_1^{e-NRTL} = 2X_c^2\left[\frac{2\tau_{ca,m}\exp\left(-2\alpha\tau_{ca,m}\right)}{\left(X_1+2X_c\exp\left(-\alpha\tau_{ca,m}\right)\right)^2}+\frac{\tau_{m,ca}\exp\left(-\alpha\tau_{m,ca}\right)}{\left(X_1\exp\left(-\alpha\tau_{m,ca}\right)+X_c\right)^2}\right] \tag{32}$$

where α is the nonrandomness factor (usually set to 0.2) and X_i is the effective mole fraction of the component i, calculated as $X_i = j_i x_i$ with $j_i = z_i$ for ions and $j_i = 1$ for solvent. In this model, $\tau_{ca,m}$ and $\tau_{m,ca}$ are the adjustable parameters.

Examples of the correlation of VLE for binary mixtures [38,73-75] and ternary mixtures [73-75] containing ILs and ethanol or water can be found in literature. Nevertheless, the literature is scarce in examples for SLE [70] and LLE [80,81].

ii.iv The NRF model [59] calculates the short-range contribution for the activity coefficient of the solvent as:

$$\ln \gamma_1^{NRF} = X_c\left\{\left(\frac{\nu X_c\lambda_E}{z_c\nu_c}\right)\left(1+\frac{\exp(-\lambda_E/Z)}{\left(X_c\exp(-\lambda_E/Z)+X_1\right)^2}\right) - \frac{X_1\lambda_S\left((\nu/z_c\nu_c)-2\exp(-\lambda_S/Z)\right)}{2X_c\exp(-\lambda_S/Z)+X_1}\cdot\left(2-X_1\left(1+\frac{1}{2X_c\exp(-\lambda_S/Z)+X_1}\right)\right)\right\} \tag{33}$$

where ν and ν_c are the total number of ions into which the salt dissociates and the number of cations in one mole of the salt, respectively, and Z is the coordination number (usually set to 8). In this model λ_E and λ_S are the adjustable parameters; its use is not common for describing the vapor-liquid equilibria of systems containing ILs [38,77].

ii.v The short-range contribution in the MNRTL model has been developed by Jaretum and Aly [60], and Sardroodi *et al.* [87] used it in the presented form for the first time:

$$ln\gamma_1^{MNRTL} = 4X_c^2[\tau_{ca,m}W_{ca,m} / (X_1 + 2W_{ca,m}X_c)^2 + \tau_{m,ca}(W_{m,ca} - 1) / (X_c + X_1W_{ca,m})^2] \tag{34}$$

where $\tau_{ca,m}$ and $\tau_{m,ca}$ are the parameters of the model, X_i is the effective mole fraction, and the next expression is assumed:

$$W_i = exp\ (-\alpha\tau_i + w_i) \tag{35}$$

where α is the nonrandomness factor (usually fixed to 0.2) and $\omega_{ca,m}$ and $\omega_{m,ca}$ are the adjustable parameters.

Among the local composition models, it is one of those giving lower deviations for the modeling of VLE of systems with ILs [34-37,47-54,77].

ii.vi The equation for calculating the short-range contribution for the activity coefficient of the solvent given by the MSA-NRTL model [61] is as follows:

$$\ln\gamma_1^{MSA-NRTL} = \frac{A}{B^2} + v_c v_a\left(C_c + C_a\right) + v\tau^2_{mc,ac}x_S^2 x_S^i\left(D_c + D_a\right) \tag{36}$$

taking into account that:

$$A = (v_c p_{cm}\tau_{cm} + v_a p_{am}\tau_{am})(v_c p_{cm} + v_a p_{am}) \tag{37}$$

$$B = v_c p_{cm} + v_a p_{am} + x_S^1 \tag{38}$$

$$C_c = p_{cm,ac}\tau_{mc,ac} / \left(v_a + x_S^i p_{mc,ac}\right)^2 \tag{39}$$

$$C_a = p_{ma,ca}\tau_{ma,ca} / \left(v_c + x_S^i p_{ma,ca}\right)^2 \tag{40}$$

$$D_c = \frac{\nu_c p_{mc,ac}}{\left(\nu_a + x_S^i p_{mc,ac}\right)^2 \left(1 - \frac{\nu_a \alpha \tau_{mc,ac}}{\nu_a + x_S^i p_{mc,ac}}\right)} \tag{41}$$

$$D_a = \frac{\nu_a p_{ma,ca}}{\left(\nu_c + x_S^i p_{ma,ca}\right)^2 \left(1 - \frac{\nu_c \alpha \tau_{ma,ca}}{\nu_c + x_S^i p_{ma,ca}}\right)} \tag{42}$$

$$\mathrm{p_i} = \exp\left(-\alpha \tau_{\mathrm{i}}\right) \tag{43}$$

$$\tau_{\mathrm{ma,ca}} = \tau_{\mathrm{mc,ac}} + \tau_{\mathrm{am}} - \tau_{\mathrm{cm}} \tag{44}$$

$$\tau_{\mathrm{mc,ac}} = \tau^1_{\mathrm{mc,ac}} + \tau^2_{\mathrm{mc,ac}} X_1 \tag{45}$$

$$x^i = x_1 / x_S = 1 / x_S - \nu \tag{46}$$

in which x_1 and x_S are the mole fraction of solvent and salt, respectively.

The correlation using this model it is not common for the treatment of the VLE data of systems containing ILs [77].

ii.vii The EW model was presented by Zhao *et al.* [62], and it describes the short-range contribution for the activity coefficient as:

$$\ln\gamma_1^{EW} = -C\left(\ln\left(\frac{X_1+2X_c\exp\left(-E_{e1}/CRT\right)}{X_1+2X_c}\right)+\right.$$

$$\left.2X_c\left(\frac{\left(1-\exp\left(-E_{e1}/CRT\right)\right)X_1}{\left(X_1+2X_c\exp\left(-E_{e1}/CRT\right)\right)\left(X_1+2X_c\right)}+\frac{X_c\left(\exp\left(-E_{1e}/CRT\right)-1\right)}{\left(X_1\exp\left(-E_{1e}/CRT\right)+X_c\right)\left(X_1+X_c\right)}\right)\right) \quad (47)$$

where C is the coordination number (set to 10 [77]). In this model, E_{e1} and E_{1e} are the adjustable parameters. This model has been checked for its use in VLE of aqueous systems with ILs [77].

TABLE 1. Literature examples for the modeling of phase equilibria in systems containing ILs

Model	References		
	VLE	LLE	SLE
Pitzer	[34-38]		
Extended Pitzer of Archer	[46-55]		
UNIQUAC	[65]	[66,67]	[68-70]
NRTL	[71-79]	[67,80-85]	[70-72]
e-NRTL	[38,73-75]	[80,81]	[70]
NRF	[38,77]		
MNRTL	[34-37,47-54,77]		
MSA-NRTL	[77]		
EW	[77]		

A list of representative examples found in literature for the modeling of phase equilibria in systems containing ILs with the above mentioned models is presented in Table 1.

Among these correlation models, those which have demonstrated to give the best results in VLE are the Extended Pitzer model of Archer, the NRTL and the MNRTL models. Regarding LLE and SLE, there are less examples available, specially comparing different models. Nevertheless, it is clear that

NRTL is, by far, the most used equation. It is also important to highlight that differences in performance among the models are small.

3.2. Equations Of State

Equations of state (EoS) are powerful tools, which can be used to describe the properties of pure fluids or their mixtures. In the last 10 years, this kind of models has been widely applied to describe the properties of pure ILs, as well as to model the phase equilibrium (VLE and LLE) of mixtures containing them.

3.2.1. Peng-Robinson

The Peng-Robinson EoS was developed in 1976 by Peng and Robinson [88] and can be expressed as:

$$P = \frac{RT}{V - b} - \frac{a}{V(V + b) + b(V - b)} \tag{48}$$

with:

$$b = 0.077796 \frac{RT_c}{P_c} \tag{49}$$

$$a = 0.457235 \frac{R^2 T_c^2}{P_c} \alpha \tag{50}$$

$$\alpha = \left[1 + m\left(1 - \sqrt{\frac{T}{T_c}}\right)\right]^2 \tag{51}$$

$$m = 0.37464 + 1.54226\omega - 0.26992\omega^2 \tag{52}$$

where T_c is the critical temperature, P_c is the critical pressure and ω is the acentric factor. For most ILs, the critical properties and the acentric factor are impossible to determine experimentally, because they start to decompose even before the temperature reaches the boiling point. For this reason, Valderrama and co-workers have applied an extended group contribution method, the modified Lydersen-Joback-Reid method, to determine the critical properties, boiling temperatures and acentric factors of several ILs [89-92]. This method only requires knowledge of the structure of the ILs and its molecular weight. Since there are no experimental data available for these properties, the accuracy of the method is verified by comparing calculated liquid densities of the ILs to experimental data available. The results show that the method is sufficiently accurate for several applications, with average absolute deviations (AAD) between calculated and experimental liquid densities in the range of 5 to 6%. The properties determined by this research group have been widely used by the scientific community when modeling the properties of ILs and its mixtures, using this or other equations.

For mixtures of fluids, mixing rules have to be applied to parameters *a* and *b*, which imply the use of one or more binary interaction parameters. For systems containing ILs, the Wong-Sandler mixing rules have been widely used [93-96], as well as the quadratic [97,98], van der Waals [94], [99,100] and Mathias-Klots-Prausnitz mixing rules [101,102].

The application of the Peng-Robinson EoS to systems with ILs has been mainly focusing on the VLE with CO_2 and other gases. For example, Shin *et al.* [103] modeled the high-pressure solubilities of CO_2in ILs of the family $[C_nmim][Tf_2N]$ (1-alkyl-3-methylimidazolium bis(trifluoromethylsulfonyl) imide) using this EoS and quadratic mixing rules with two temperature-dependent binary interaction parameters. They obtained AAD between calculated and experimental equilibrium pressures between 10 and 14%, for which they conclude that the model can satisfactorily predict the solubility of high-pressure CO_2 in this family of ILs over a wide range of pressures up to the supercritical region of CO_2. Also Álvarez and Aznar [93] used the Peng-Robinson EoS to model the VLE of binary systems composed of IL + supercritical CO_2 or CHF_3 and IL + hydrocarbons. In this work, the van der Waals and Wong-Sandler mixing rules were used, with UNIQUAC and NRTL models used as excess Gibbs energy models in the Wong-Sandler mixing rules. Their results present AAD in pressure between 2 and 24% for low pressures and between 7 and 52% for high pressures. The authors conclude that the EoS is not able to represent the data at high pressures; however, it performs well at low pressures. Later, Álvarez *et al.* [104] used the Peng-Robinson EoS with the Wong-Sandler mixing rules, once again with UNIQUAC and NRTL models for the excess Gibbs energy, but also using the COSMO-SAC model. They modeled the isobaric VLE of 1-ethyl-3-methylimidazolium ethylsulfate ($[C_2mim][EtSO_4]$) with propionaldehyde or valeraldehyde. When using UNIQUAC or NRTL models, they obtain AAD below 0.15% for both systems. Since the COSMO-SAC model is a predictive model, *i.e.*, it does not require any adjustable binary interaction parameters, the authors regard the application of this model with the Peng-Robinson

EoS and the Wong-Sandler mixing rules as pure predictive results. The AAD obtained in this way was 0.3 and 2% for each system, respectively. As a final example, Ren and Scurto [99] used the Peng-Robinson EoS with the van der Waals one fluid mixing rule to model the VLE and VLLE of imidazolium-based ILs and the refrigerant gas 1,1,1,2-tetrafluoroethane. The authors obtained AAD values between 0.4 and 7.4% for VLE and concluded that the model is able to represent the bubble-point data with excellent agreement. However, when using the interaction parameters regressed from the VLE data alone to predict compositions of the VLLE transition to LLE, only satisfactory results were obtained for all systems.

3.2.2. Soave-Redlich-Kwong

A Soave modification of the Redlich-Kwong EoS [105] has been frequently applied to systems with ILs. The Soave-Redlich-Kwogn (SRK) EoS was introduced in 1972 by Giorgio Soave, and can be expressed as:

$$P = \frac{RT}{V - b} - \frac{a(T)}{V(V + b)} \tag{53}$$

with:

$$a(T) = 0.427480 \frac{R^2 T_c^2}{P_c} \alpha(T) \tag{54}$$

$$b = 0.08664 \frac{RT_c}{P_c} \tag{55}$$

In the works involving systems with ILs, the temperature dependent part of the *a* parameter has been modeled by the empirical form:

$$\alpha(T) = \sum_{k=0}^{\leq 3} \beta_k \left(\frac{T_c}{T} - \frac{T}{T_c} \right)^k \tag{56}$$

Coefficients β_K are usually determined to reproduce the vapor pressure of the pure compound. However, since there are no available experimental vapor pressure data for most ILs, the coefficients are treated as adjustable fitting parameters in the calculations [106]. The authors have found that for ILs, only one adjustable parameter, β_1, is sufficient (with β_0=1 and β_2= β_3=0).

In the case of this EoS, modified van der Waals-Berthelot mixing rules have been used, with four binary interaction parameters.

Shiflett and Yokozeki [106] first used this EoS to correlate the solubility of CO_2 in ILs l-butyl-3-methylimidazolium hexafluorophosphate ($[C_4mim][PF_6]$) and l-butyl-3-methylimidazolium tetrafluoroborate ($[C_4mim][BF_4]$). The authors obtained an excellent fit between experimental and calculated solubility data, with standard deviations of 17 and 10.5 kPa for each system, respectively. They used the same model to correlate the VLE of CO_2 and other gases (like ammonia [107,108] or SO_2 [109]) in different ILs. Later, the same authors modeled the solubility of water in several different ILs using the same EoS [110]. The four binary interaction parameters were determined using binary VLE data from the literature, having obtained standard deviations lower than 0.6 kPa for all systems. The graphical results presented by the authors show a very good agreement between experimental and calculated solubilities of water in all ILs considered. The same research group has also used the SRK EoS to study the phase behavior of ternary mixtures $CO_2/H_2/[C_4mim][PF_6]$ [111], CO_2/SO_2/1-butyl-3-methylimidazolium methyl sulfate ($[C_4mim][MeSO4]$) [112], $CO_2/H_2S/[C_4mim][PF_6]$ [113] and $CO_2/H_2S/[C_4mim][MeSO_4]$ [114]. For all the previous studies the authors obtained good agreement between calculated and experimental data. Very recently, Shiflett *et al.* [115] have also modeled the ternary system N_2O/CO_2/1-butyl-3-methylimidazolium acetate ($[C_4mim][Ac]$) with the purpose of understanding the separation of N_2O and CO_2 using ILs. They determined the binary interaction parameters using VLE data (either their own or from the literature) for the pairs $N_2O/[C_4mim][Ac]$, $CO_2/[C_4mim][Ac]$ and N_2O/CO_2. Unlike what happened in the previously mentioned ternary systems, the phase behavior prediction of the ternary system of $N_2O/CO_2/[C_4mim][Ac]$ may not be guaranteed based on the binary interaction parameters alone. This happens because for systems containing mixtures with the chemical complex formation (hydrogen-bonding, charge-transfer complex, etc.) as in the case of the binary $CO_2/[C_4mim][Ac]$ here present, the third component (in this case, N_2O) may interfere with the binary interactions of $CO_2/[C_4mim][Ac]$, originating a decrease or an increase of the complex formation and changing the binary interaction parameters of this pair in the ternary system. Consequently, one of the binary interaction parameters of the pair $CO_2/[C_4mim][Ac]$ was modified so as to fit the experimental ternary VLE data. This way, the model provides an excellent agreement between experimental and calculated data for the ternary system.

3.2.3. Statistical Associating Fluid Theory

The original Statistical Associating Fluid Theory (SAFT) was developed in 1989 [116] based on Wertheim's first-order thermodynamic perturbation theory [117-120]. Its main advantage over the traditional cubic EoS, is that it takes into account the structure of the molecule, similarly to group contribution models. It regards the molecules as chains of hard-spheres, which contain multiple association sites. Several variations of the SAFT EoS have been used to model the phase behavior of systems containing ILs: tPC-PSAFT [121-125], soft-SAFT [126-128], hetero-SAFT [129], PCP-SAFT [130] and PC-SAFT [131].

Kroon *et al.* [121] applied the tPC-PSAFT model to the phase behavior of IL + CO_2 systems and later Karakatsani *et al.* [122] applied it to the correlation of the solubility of CO_2, CO, O_2 and CHF_3 in the same ILs. Pure component parameters for the ILs were estimated from ILs experimental thermodynamic data (density, enthalpy, and entropy of dissolution of CO_2) and physicochemical data for the constituent ions (size, polarizability, number of electrons). The cross-association parameters were estimated from enthalpy and entropy data for the dissolution of CO_2 in the ILs. A temperature-dependent binary interaction parameter *kij* was adjusted in order to fit the model to experimental VLE data. The results of the model were in good agreement with experimental data. Later, Karakatsani and Economou [123] used the same parameters to model the VLE of the ternary system CO_2/acetone/[C_4mim][PF_6], also obtaining satisfactory agreement between calculated and experimental data. Additionally, Economou *et al.* [124] applied the same model to describe the VLE of the binary [C_8mim][BF_4]/benzene and the ternary CO_2/H_2O/[C_4mim][NO_3] (1-butyl-3-methylimidazolium nitrate) systems. For binary mixture calculations, a binary interaction parameter was fitted to the experimental data and the model was capable of accurately correlating the phase equilibria of the binary and of the ternary IL mixtures with polar solvents. In a more recent work, Karakatsani *et al.* [125] proved that the model can accurately predict the phase equilibrium of non-polar solvent/IL mixtures, without the use of any adjustable binary interaction parameter, by applying it to binary and ternary mixtures of ILs with organic solvents and water. They also concluded that in the case of aqueous solvents, the dissociation of IL has to be incorporated explicitly into the model in order to obtain a good correlation with the experimental data.

Andreu and Vega [126] used the soft-SAFT EoS to describe the solubility of CO_2 in ILs. They modeled the families of ILs [$C_n$mim][BF_4] and [$C_n$mim][PF_6] as Lennard-Jones chains with one associating site in each molecule. The chain length, size and energy parameters of the ILs were obtained by fitting the model predictions to available density data, obtaining AAD values lower than 0.2%. For the association parameters of ILs, values previously used for alkanols were adopted. The authors found that the model correlations and experimental data for VLE are in good agreement. They later used the same model to describe the solubility of hydrogen, CO_2 and xenon in ILs of the family [$C_n$mim][Tf_2N] [127]. In this case, the ILs were modeled as Lennard-Jones chains with three associating sites in each molecule. The pure component parameters for ILs were obtained as in the previous work, and a good description of experimental solubilities was obtained. Recently, Llovell *et al.* [128] re-parameterized the model for the [$C_n$mim][Tf_2N] family of ILs and modeled the solubilities of methanol and ethanol. Good agreement was found between the predictions of the model and the experimental data, with AAD values below 5% for methanol and around 10% for ethanol. They also modeled the VLE of IL/water systems, with the model providing good agreement with the experimental data, with AAD values between 6 and 12%. As for the LLE of mixtures of ILs with water, the authors were not able to obtain reasonable predictions, having obtained significant deviations in the water-rich phase. Quantitative agreement was achieved by using two adjustable parameters, which were adjusted only to the water-rich phase of the aqueous [C_4mim][Tf_2N] mixture and used in a predictive manner for the IL-rich phase and for the aqueous mixtures with [C_2mim][Tf_2N] and [C_6mim][Tf_2N].

Ji and Adidharma [129] used the heterosegmented SAFT (hetero-SAFT) to describe the solubility of CO_2 in the families of ILs [$C_n$mim][BF_4], [$C_n$mim][PF_6] and [$C_n$mim][Tf_2N]. The molecules of ILs were divided into groups representing the alkyl chain, the cation head and the anion. To account for the electrostatic/polar interactions between cation and anion, the spherical segments representing the cation head and the anion were assumed to have one association site each, which can only associate to each other. The parameters for the alkyl chains were obtained from those of the corresponding n-alkanes and the parameters for groups representing the cation head and the anion, including the two association parameters, were fitted to experimental IL density data. The model was capable of satisfactorily describing the solubility of CO_2 in the ILs studied.

Finally, Paduszynski *et al.* [130] used the Perturbed-Chain Polar Statistical Associating Fluid Theory (PCP-SAFT) to model the LLE of the IL 1-methyl-1-propylpiperidinium bis(trifluoromethylsulfonyl)imide [C_3mpip][NTf_2] with several alkan-1-ols. They modeled the IL as strongly associating molecules, with symbol A_1 representing a positive site which corresponds to the nitrogen atom on the cation and its proximity, and symbol B_1 representing a negative site which corresponds to the delocalized charge due to the oxygen molecules on the anion. They defined each type of associating site in an identical way; however, they only allowed A_1B_1 interactions to take place. Additionally, they assumed that each molecule has 5 positive sites of type A_1 and 5 negative sites of type B_1. The pure component parameters of the IL were determined by fitting to liquid density and total solubility parameter data. Self-association of IL and alkan-1-ols, as well as cross-association, were accounted for and one linearly temperature-dependent binary interaction parameter was needed in order to obtain qualitative agreement between calculated and experimental data. AAD values obtained were between 0.6 and 5%. On a different work, Paduszynski and Domanska [131] modeled the LLE of systems composed by piperidinium-based ILs ([C_3mpip][NTf_2] and [C_4mpip][NTf_2]) and several aliphatic hydrocarbons using the Perturbed-Chain Statistical Associating Fluid Theory (PC-SAFT) model. Pure component parameters for ILs were obtained as in the previous work and they used activity coefficients at infinite dilution of hydrocarbons in ILs reported in the literature to optimize the binary interaction parameter, which was again considered to be linearly temperature-dependent, with two adjustable parameters. In the calculation of the LLE of the systems, they obtained AAD values between 1.5 and 8%. Additionally, the authors decided to test their approach for the cross-associating systems [C_3mpip][NTf_2] + 1-pentanol and [C_3mpip][NTf_2] + water. Regarding the LLE of [C_3mpip][NTf_2] + 1-pentanol, the authors claim that the experimental data is well described by the model, although deviations between calculated and experimental compositions increase as the temperature increases and the model ends up over predicting the upper critical solution temperature by about 15 K. The AAD values obtained for this system were 52% for the IL-rich phase and 1.7% for the alcohol-rich phase. As for the system [C_3mpip][NTf_2] + water, the authors conclude that the PC-SAFT model is surprisingly good at describing the experimental data, but only when the IL-rich phase is considered alone. For the water-rich phase, the model predicts solubilities of IL much lower than those observed experimentally by several orders of magnitude. The authors justify this fact because the molecular model chosen for the IL is not appropriate for dilute solutions of IL in water, in which case the cation and anion of the IL are probably dissociated, which results on increased solubility.

3.2.4. Other Eos

Several other EoS have been used to model systems with ILs. For example, Tsioptsias et al. [132] used the Non-Random Hydrogen-Bonding (NRHB) model [133] to describe the phase behavior of binary systems containing ILs of the family [$C_n$mim][Tf_2N], obtaining good agreement between model correlations and experimental data. Wang et al. [134] used the square well chain fluid (SWCF) EoS [135] to model the solubilities of gases such as CO_2, C_3H_6, C_3H_8 and C_4H_{10} in several ILs. Breure et al. [136] used a group contribution EoS to study the phase behavior of binary systems of ILs of the families [$C_n$mim][PF_6] and [$C_n$mim][BF_4] with CO_2, also obtaining good agreement between model predictions and experimental data. Very recently, Maia et al. [137] applied the Cubic Plus Association (CPA) EoS [138], which combines the SRK EoS with an advanced association term similar to that of the SAFT type models, to describe the VLE with CO_2 and the LLE with water of ILs [C_2mim][Tf_2N] and [C_4mim][Tf_2N]. Good agreement was obtained between calculated and experimental data, even though smaller AAD percentage values were obtained for the VLE than for the LLE.

4. CONCLUSIONS

Regarding the treatment of physical properties of pure ILs, temperature dependence of physical properties such as density, speed of sound, or refractive index is described using simple polynomial expressions, mainly equations of first, second and third order. For the viscosity, usually the VFT or mVFT equations are strongly recommended. For binary mixtures containing ILs, the dependence of the physical properties with temperature and composition is also correlated using empirical equations, and their excess properties are generally fitted to a Redlich-Kister type equation. For the fitting of the excess properties of ternary systems containing ILs, the most widely used equation is that proposed by Cibulka.

For the correlation of experimental data concerning phase equilibria of mixtures containing ILs, several g^E-based models have been applied in literature (Pitzer, Extended Pitzer model of Archer, UNIQUAC, NRTL, e-NRTL, NRF, MNRTL, MSA-NRTL, EW), being the NRTL model the one that unifies simplicity and satisfactory results for the treatment of vapor-liquid, liquid-liquid and solid-liquid equilibria.

The use of EoS for the modeling of phase equilibria involving ILs is frequent. Unlike what happens with g^E models, most of the literature with EoS involve VLE data, rather than LLE or SLE. Nevertheless, many authors have proved that excellent results can be obtained in data correlation or, for some cases, even prediction. The main difficulty with the application of EoS is the calculation of pure component parameters for ILs. Up to date, a general procedure has not yet been defined. Very recently, a complete review on the use of EoS with ILs, with special emphasis on the obtention of model parameters, has been published by Maia and co-workers [137]. The interested reader is directed to that work for further details.

REFERENCES

1. Short PL. Out of the ivory tower. Chemical and Engineering News 2006;84(17) 15-21
2. Aparicio S, Atilhan M, Karadas F. thermophysical properties of pure ionic liquids: review of present situation. Industrial and Engineering Chemistry Research. 2010;49(20)9580-9595

3. Tariq M, Forte PAS, Gomes MFC, Lopes JNC, Rebelo LPN. Densities and refractive indices of imidazolium and phosphonium based ionic liquids: effect of temperature, alkyl chain length, and anion. Journal of Chemical Thermodynamics. 2009;41(6) 790-798
4. Widegren JA, Magee JW. Density, viscosity, speed of sound, and electrolytic conductivity for the ionic liquid 1-hexyl-3-methylimidazolium bis(trifluorome thylsulfonyl) imide and its mixtures with water. Journal Chemical and Engineering Data2007;52(6) 2331-2338
5. Seoane RG, Corderí S, Gómez E, Calvar N, González EJ, Macedo EA, Domínguez A.Temperature dependence and structural influence on the thermophysical properties of eleven commercial ionic liquids. Industrial and Engineering Chemical Research 2012;51(5) 2492–2504
6. Gu Z, Brennecke JF. Volume expansivities and isothermal compressibilities of imidazoliumand pyridinium-based ionic liquids. Journal Chemical and Engineering Data 2002;47(2) 339-345
7. Paduszyński K, Domańska U, Chang H. Thermodynamic modeling of ionic liquidsystems: development and detailed overview of novel methodology based on the PC-SAFT. Journal of Physical Chemistry B - Condensed Phase 2012;116(16) 5002–5018
8. Tome LIN, Gardas RL, Carvalho PJ, Pastoriza-Gallego MJ, Piñeiro MM, Coutinho JAP. Measurements and correlation of high-pressure densities of phosphonium based ionic liquids. Journal Chemical and Engineering Data 2011;56(5) 2205–2217
9. Machida H, Sato Y, Smith RL. Pressure–volume–temperature (PVT) measurements ofionic liquids ([bmim+][PF6−], [bmim+][BF4−], [bmim+][OcSO4−]) and analysis with the Sanchez–Lacombe equation of state. Fluid Phase Equilibria 2008;264(1-2) 147–155
10. Dymond JH, Malhotra R. The Tait equation: 100 years on. International Journal of Thermophysics 1988;9(6) 941-951
11. Harris KR, Kanakubo M, Woolf LA. Temperature and pressure dependence of the viscosity of the ionic liquids 1-methyl-3-octylimidazolium hexafluorophosphate and 1-methyl-3-octylimidazolium tetrafluoroborate. Journal Chemical and Engineering Data 2006;51(3) 1161-1167
12. Tomida D, Kumagai A, Kenmochi S, Qiao K, Yokoyama C. Viscosity of 1-hexyl-3-methylimidazolium hexafluorophosphate and 1-octyl-3-methylimidazolium hexafluorophosphate at high pressure. Journal Chemical and Engineering Data 2007;52(2) 577-579
13. Bandres I, Alcalde R, Lafuente C, Atilhan M, Aparicio S. On the viscosity of pyridinium based ionic liquids: an experimental and computational study. Journal of Physical Chemistry B 2011;115(43) 12499–12513
14. Ahosseini A, Scurto AM. Viscosity of imidazolium-based ionic liquids at elevated pressures: cation and anion effects. International Journal of Thermophysics 2008;29(4) 1222-1243
15. Seddon KR, Starck AS, Torres, MJ. ACS Symposium Series 901: Washington DC, 2004.
16. Vogel H. The law of the relation between the viscosity of liquids and the temperature. Phys. Z. 1921;22, 645-646
17. Fulcher GS. Analysis of recent measurements of the viscosity of glasses. Journal of the American Ceramic Society 1925;8(6) 339-355
18. Tammann G, Hesse W. The dependence of viscosity upon the temperature of supercooled liquids. Zeitschrift für anorganische und allgemeine Chemie 1926;156(1) 245-257
19. Litovitz, TA. Temperature dependence of the viscosity of associated liquids. Journal of Chemical Physics 1952;20(7) 1088-1089
20. Geppert-Rybczyńska M, Heintz A, Lehmann JK, Golus A. Volumetric properties of binary mixtures containing ionic liquids and some aprotic solvents. Journal Chemical and Engineering Data 2010;55(9) 4114–4120

21. Heintz A, Klasen D, Lehmann JK, Wertz C. Excess molar volumes and liquid-liquid equilibria of the ionic liquid 1-methyl- 3-octyl-imidazolium tetrafluoroborate mixed with butan-1-ol and pentan-1-ol. Journal of Solution Chemistry 2005;34(10) 1135–1144
22. Vercher E, Llopis FJ, González-Alfaro MV, Martínez-Andreu A. Density, speed of sound, and refractive index of 1-ethyl-3-methylimidazolium triiifluoromet hanesulfonate with acetone, methyl acetate, and ethyl acetate at temperatures from (278.15 to 328.15) K. Journal Chemical and Engineering Data 2010;55 (3) 1377–1388
23. González EJ, Domínguez A, Macedo EA. Excess properties of binary mixtures containing 1-hexyl-3-methylimidazoliumbis(trifluoromethylsulfonyl)imide ionic liquid and polar organic compounds. Journal of Chemical Thermodynamics 2012;47(1) 300-311
24. García-Miaja G, Troncoso J, Romaní L. Excess enthalpy, density, and heat capacity for binary systems of alkylimidazolium-based ionic liquids + water. Journal of Chemical Thermodynamics 2009;41(2) 161–166
25. Connors KA, Wright JL. Dependence of surface tension on composition of binary aqueous-organic solutions. Analytical Chemistry 1989;61(3) 194-198
26. Redlich O, Kister AT. Algebraic representation of thermodynamic properties and the classification of solutions. Industrial and Engineering Chemistry 1948;40(2) 345–348
27. González B, Calvar N, González E, Domínguez A. Density and viscosity experimental data of the ternary mixtures 1-propanol or 2-propanol + water + 1-ethyl-3-methylimidazolium ethylsulfate. Correlation and prediction of physical properties of the ternary systems. Journal of Chemical and Engineering Data 2008;53(3) 881–887
28. Gómez E, González B, Calvar N, Domínguez A. Excess molar properties of ternary system (ethanol + water + 1,3-dimethylimidazolium methylsulphate) and its binary mixtures at several temperatures. Journal of Chemical Thermodynamics 2008;40(8) 1208–1216
29. Andreatta AE, Arce A, Rodil E, Soto A. Physico-chemical properties of binary andternary mixtures of ethyl acetate + ethanol +1-butyl-3-methyl-imidazolium b is(tri- fluoromethylsulfonyl) imide at 298.15 K an atmospheric pressure. Journal of Solution Chemistry 2010;39(3) 371–383
30. Cibulka I. Estimation of excess volume and density of ternary liquid mixtures of non electrolytes from binary data. Collection of Czechoslovak Chemical Communication 1982;47(5) 1414–1419
31. Singh P, Nigam R, Sharma S, Aggarwal S. Molar excess volumes of ternary mixtures of nonelectrolytes. Fluid Phase Equilibria 1984;18(3) 333–344
32. Nagata M, Sakura J. Refractive index and excess volume for binary liquid mixtures. Part 1. Analyses of new and old data for binary mixtures. Journal of the Chemical Society, Faraday Transactions 1 1987;83(8) 2449–2457
33. Pitzer KS. Activity Coefficients in Electrolyte Solutions. Ed. CRC Press: Boca Raton, FL; 1991
34. Shekaari H, Mousavi SS. Influence of alkyl chain on the thermodynamic properties of aqueous solutions of ionic liquids 1-alkyl-3-methylimidazolium bromide at different temperatures. The Journal of Chemical Thermodynamics 2009;41 90–96
35. Shekaari H, Mousavi SS, Mansoori Y. Thermophysical properties of ionic liquid, 1-pentyl-3-methylimidazolium chloride in water at different temperatures. International Journal of Thermophysics 2009;30 499–514
36. Shekaari H, Mousavi SS. Osmotic coefficients and refractive indices of aqueous solutions of ionic liquids containing 1-butyl-3-methylimidazolium halide at T = (298.15 to 328.15) K. Journal of Chemical and Engineering Data 2009;54 823–829
37. Shekaari H, Armanfar E. Physical properties of aqueous solutions of ionic liquid, 1-propyl-3-methylimidazolium methyl sulfate, at T = (298.15 to 328.15) K. Journal of Chemical and Engineering Data 2010;55 765–772

38. Sardroodi JJ, Atabay M, Azamat J. Isopiestic determination of the osmotic coefficient and vapour pressure of N-R-4-(N,N-dimethylamino)pyridinium tetrafluoroborate (R = C4H9, C5H11, C6H13) in the ethanol solution at T = 298.15 K. The Journal of Chemical Thermodynamics 2012;49 70–74
39. Archer DG. Thermodynamic properties of the NaBr+H2O System. Journal of Physical and Chemical Reference Data 1991;20 509–555
40. Archer DG. Thermodynamic properties of the NaCl+H2O system. II. Thermodynamic properties of NaCl(aq), NaClaq2H2(cr), and phase equilibria. Journal of Physical and Chemical Reference Data 1992;21 793–829
41. Shekaari H, Zafarani-Moattar MT. Osmotic coefficients of some imidazolium based ionic liquids in water and acetonitrile at temperature 318.15K. Fluid Phase Equilibria 2007;254 198–203
42. Nasirzadeh K, Neueder R, Kunz W. Vapor pressures, osmotic and activity coefficientsfor (LiBr + acetonitrile) between the temperatures (298.15 and 343.15) K. The Journal of Chemical Thermodynamics 2004;36 511–517
43. Nasirzadeh K, Neueder R, Kunz W. Vapor pressures, osmotic and activity coefficients of electrolytes in protic solvents at different temperatures. 2. Lithium bromide in ethanol. Journal of Solution Chemistry 2004;33 1429-1446
44. Nasirzadeh K, Neueder R, Kunz W. Vapor pressures, osmotic and activity coefficients of electrolytes in protic solvents at different temperatures. 3. Lithium bromide in 2-propanol. Journal of Solution Chemistry. 2005;34 9-24
45. Nasirzadeh K, Neueder R, Kunz W. Vapor pressures and osmotic coefficients of aqueous LiOH solutions at temperatures ranging from 298.15 to 363.15 K. Industrial and Engineering Chemistry Research 2005;44 3807-3814
46. Sadeghi R, Ebrahimi N. Ionic association and solvation of the ionic liquid 1-hexyl-3-methylimidazolium chloride in molecular solvents revealed by vapor pressure osmometry, conductometry, volumetry, and acoustic measurements. The Journal of Physical Chemistry B 2011;115 13227–13240
47. González B, Calvar N, Domínguez A, Macedo EA. Osmotic coefficients of aqueous solutions of four ionic liquids at T = (313.15 and 333.15) K. The Journal of Chemical Thermodynamics 2008;40 1346-1351
48. Calvar N, González B, Domínguez A, Macedo EA. Osmotic coefficients of binary mixtures of 1-butyl-3-methylimidazolium methylsulfate and 1,3-dimethyli midazolium methylsulfate with alcohols at T = 323.15 K. The Journal of Chemical Thermodynamics 2009;41 617-622
49. Calvar N, González B, Domínguez A, Macedo EA. Vapour pressures and osmotic coefficients of binary mixtures of 1-ethyl-3-methylimidazolium ethylsulfate and 1-ethyl- 3-methylpyridinium ethylsulfate with alcohols at T = 323.15 K. The Journal of Chemical Thermodynamics 2009;41 1439-1445
50. Calvar N, González B, Domínguez A, Macedo EA. Osmotic coefficients of binary mixtures of four ionic liquids with ethanol or water at T = (313.15 and 333.15) K. The Journal of Chemical Thermodynamics 2009;41 11-16
51. Calvar N, Gómez E, Domínguez A, Macedo EA. Vapour pressures, osmotic and activity coefficients for binary mixtures containing (1-ethylpyridinium ethylsulfate + several alcohols) at T = 323.15 K. The Journal of Chemical Thermodynamics 2010;42 625-630
52. Gómez E, Calvar N, Domínguez A, Macedo EA. Measurement and modeling of osmotic coefficients of binary mixtures (alcohol + 1,3-dimethylpyridinium methylsulfate) at T = 323.15 K. The Journal of Chemical Thermodynamics 2011;43 908-913
53. Calvar N, Gómez E, Domínguez A, Macedo EA. Determination and modelling of osmoticcoefficients and vapour pressures of binary systems 1- and 2-propanol with Modeling of Ionic Liquid Systems: Phase Equilibria and Physical Properties CnmimNTf2 ionic liquids (n = 2, 3, and 4) at T = 323.15 K. The Journal of Chemical Thermodynamics 2011;43 1256-1262

54. Calvar N, Gómez E, Domínguez A, Macedo EA. Study of the influence of the structureof the alcohol on vapor pressures and osmotic coefficients of binary mixtures alcohol + 1-hexyl-3-methylimidazolium bis(trifluoromethylsulfonyl)imide at T = 323.15 K. Fluid Phase Equilibria 2012;313 38-45
55. Calvar N, González EJ, Domínguez A, Macedo EA. Acoustic, volumetric and osmoticproperties of binary mixtures containing the ionic liquid 1-butyl-3-methylimidazolium dicyanamide mixed with primary and secondary alcohols. The Journal of Chemical Thermodynamics 2012;50 19-29
56. Abrams DS, Prausnitz, JM. Statistical thermodynamics of liquid mixtures: A new expression for the excess Gibbs energy of partly or completely miscible systems. AIChE Journal 1975;21 116-128
57. Renon, H.; Prausnitz, JM. Local compositions in thermodynamic excess functions for liquid mixtures. AIChE Journal 1968;14 135-144
58. Chen CC, Evans LB. A local composition model for the excess Gibbs energy of aqueous electrolyte systems. AIChE Journal 1986;32 444-454
59. Haghtalab A, Vera J. A nonrandom factor model for the excess Gibbs energy of electrolyte solutions. AIChE Journal 1988;34 803-812
60. Jaretum A, Aly G. New local composition model for electrolyte solutions: single solvent, single electrolyte systems. Fluid Phase Equilibria 1999;163 175-193
61. Papiconomou N, Simonin JP, Bernard O, Kunz W. MSA-NRTL model for the description of the thermodynamic properties of electrolyte solutions. Physical Chemistry Chemical Physics 2002;4 4435-4443.
62. Zhao E, Yu M, Sauve RE, Khoshkbarchi MK. Extension of the Wilson model to electrolyte solutions. Fluid Phase Equilibria 2000;173 161-175
63. Domanska U, Bogel-Łukasik E. Measurements and correlation of the (solid + liquid) equilibria of [1-decyl-3-methylimidazolium chloride + alcohols (C2-C12)]. Industrial and Engineering Chemistry Research 2003;42, 6986-6992
64. Domanska U, Bogel-Łukasik E, Bogel-Łukasik R. Solubility of 1-dodecyl-3-methylimidazoliumchloride in alcohols (C2-C12). Journal of Physical Chemistry B 2003;107 1858-1863
65. Kato R, Krummen M, Gmehling J. Measurement and correlation of vapor–liquid equilibria and excess enthalpies of binary systems containing ionic liquids and hydrocarbons. Fluid Phase Equilibria 2004;224 47-54
66. Banerjee T, Singh MK, Sahoo RK, Khanna A. Volume, surface and UNIQUAC interaction parameters for imidazolium based ionic liquids via Polarizable Continuum Model. Fluid Phase Equilibria 2005;234 64–76
67. Simoni LD, Lin Y, Brennecke JF, Stadtherr MA. Modeling liquid-liquid equilibrium of ionic liquid systems with NRTL, electrolyte-NRTL, and UNIQUAC. Industrial and Engineering Chemistry Research 2008;47 256-272
68. Domanska D, Mazurowska L. Solubility of 1,3-dialkylimidazolium chloride or hexafluorophosphate or methylsulfonate in organic solvents: effect of the anions on solubility. Fluid Phase Equilibria 2004;221 73–82
69. Carneiro AP, Rodríguez O, Macedo EA. Solubility of monosaccharides in ionic liquids – Experimental data and modeling. Fluid Phase Equilibria 2012;314 22– 28
70. Carneiro AP, Rodríguez O, Macedo EA. Solubility of xylitol and sorbitol in ionic liquids – Experimental data and modeling. The Journal of Chemical Thermodynamics 2012; http://dx.doi.org/10.1016/j.jct.2012.05.020
71. Calvar N, González B, Gómez E, Domínguez A. Vapor-liquid equilibria for the ternary system ethanol + water + 1-butyl-3-methylimidazolium chloride and the corresponding binary systems at 101.3 kPa. Journal of Chemical and Engineering Data 2006;51 2178-2181
72. Calvar N, González B, Gómez E, Domínguez A. Study of the behavior of the azeotropic mixture ethanol–water with imidazolium-based ionic liquids. Fluid Phase Equilibria 2007;259 51-56
73. Calvar N, González B, Gómez E, Domínguez A. Vapor–liquid equilibria for the ternary system ethanol + water + 1-ethyl-3-methylimidazolium ethylsulfate and the

corresponding binary systems containing the ionic liquid at 101.3 kPa. Journal of Chemical and Engineering Data 2008;53 820-825

74. Calvar N, González B, Gómez E, Domínguez A. Vapor-liquid equilibria for the ternary system ethanol + water + 1-butyl-3-methylimidazolium methylsulfate and the corresponding binary systems at 101.3 kPa. Journal of Chemical and Engineering Data 2009;54 1004-1008
75. Calvar N, Gómez E, González B, Domínguez A. Experimental vapor-liquid equilibria for the ternary system ethanol + water + 1-ethyl-3-methylpyridinium ethylsulfate and the corresponding binary systems at 101.3 kPa: Study of the Effect of the Cation. Journal of Chemical and Engineering Data 2010;55 2786-2791
76. Safarov J, Verevkin SP, Bich E, Heintz A. Vapor pressures and activity coefficients of n-alcohols and benzene in binary mixtures with 1-methyl-3-butylimidazolium octyl sulfate and 1-methyl-3-octylimidazolium tetrafluoroborate. Journal of Chemical and Engineering Data 2006;51 518-525
77. Shekaari H, Mousavi SS. Measurement and modeling of osmotic coefficients of aqueous solution of ionic liquids using vapor pressure osmometry method. Fluid Phase Equilibria 2009;279 73-79
78. Orchillés AV, Miguel PJ, González-Alfaro V, Vercher E, Martínez-Andreu A. 1-Ethyl- 3-methylimidazolium dicyanamide as a very efficient entrainer for the extractive distillation of the acetone + methanol system. Journal of Chemical and Engineering Data 2012;57 394-399
79. Jongmans MTG, Schuur B, de Haan AB. Binary and ternary LLE data of the system (ethylbenzene + styrene + 1-ethyl-3-methylimidazolium thiocyanate) and binary VLE data of the system (styrene + 1-ethyl-3-methylimidazolium thiocyanate). The Journal of Chemical Thermodynamics 2012;47 234-240
80. Domanska U, Marciniak A. Liquid phase behaviour of 1-butyl-3-methylimidazolium2-(2-methoxyethoxy)-ethylsulfate with organic solvents and water. Green Chemistry 2007;9 262–266
81. Maia FM, Rodríguez O, Macedo EA. LLE for (water + ionic liquid) binary systems using [Cxmim][BF4] (x = 6, 8) ionic liquids. Fluid Phase Equilibria 2010;296 184–191
82. Arce A, Rodríguez O, Soto A. Tert-Amyl Ethyl Ether separation from its mixtures with ethanol using the 1-butyl-3-methylimidazolium trifluoromethanesulfonate ionic liquid: Liquid-Liquid Equilibrium. Industrial and Engineering Chemistry Research 2004;43 8323-8327
83. Arce A, Rodríguez O, Soto A. A comparative study on solvents for separation of tertamyl ethyl ether and ethanol mixtures. New experimental data for 1-ethyl-3-methyl imidazolium ethyl sulfate ionic liquid", Chemical Engineering Science 2006;61 6929– 6935
84. Wytze Meindersma G, Podt AJG, de Haan AB. Ternary liquid–liquid equilibria for mixtures of toluene + n-heptane + an ionic liquid. Fluid Phase Equilibria 2006;247 158–168
85. Hu X, Yu J, Liu H. Liquid-Liquid Equilibria of the system 1-(2-Hydroxyethyl)-3-methylimidozolium tetrafluoroborate or 1-(2-hydroxyethyl)-2,3-dimethylimidozolium tetrafluoroborate + water + 1-butanol at 293.15 K. Journal of Chemical and Engineering Data 2006;51 691-695
86. Chen CC, Boston JF, Evans LB, Britt HI. Local composition model for excess Gibbs energy of electrolyte systems. Part I: Single solvent, single completely dissociated electrolyte systems. AICHE Journal 1982;28 588-596
87. Sardroodi JJ, Zafarani-Moattar MT. Vapor pressures and apparent molal volumes of the solutions of ZnCl2 in ethanol at 298.15 K. Fluid Phase Equilibria 2005;230 64-71
88. Peng DY, Robinson DB. A New Two-Constant Equation of State. Industrial & Engineering Chemistry Fundamentals 1976;15 59-64
89. Valderrama JO, Robles PA. Critical properties, normal boiling temperatures, and acentric factors of fifty ionic liquids. Industrial & Engineering Chemistry Research 2007;46 1338-1344

90. Valderrama JO, Sanga WW, Lazzu JA. Critical properties, normal boiling temperature, and acentric factor of another 200 ionic liquids. Industrial & Engineering Chemistry Research 2008;47 1318-1330
91. Valderrama JO, Rojas RE. Critical properties of ionic liquids. Revisited. Industrial & Engineering Chemistry Research 2009;48 6890-6900
92. Valderrama JO, Forero LA, Rojas RE, Serena L. Critical properties and normal boiling temperature of ionic liquids. Update and a new consistency test. Industrial & Engineering Chemistry Research 2012;51 7838-7844
93. Alvarez VH, Aznar M. Thermodynamic modeling of vapor–liquid equilibrium of binary systems ionic liquid+supercritical {CO2 or CHF3} and ionic liquid+hydrocarbons using Peng–Robinson equation of state. Journal of the Chinese Institute of Chemical Engineers 2008;39 353-360
94. Arce PF, Robles PA, Graber TA, Aznar M. Modeling of high-pressure vapor–liquid equilibrium in ionic liquids+gas systems using the PRSV equation of state. Fluid Phase Equilibria 2010;295 9-16
95. Yazdizadeh M, Rahmani F, Forghani AA. Thermodynamic modeling of CO2 solubility in ionic liquid ([Cnmim] [Tf2N]; n=2, 4, 6, 8) with using Wong-Sandler mixing rule, Peng-Rabinson equation of state (EOS) and differential evolution (DE) method. Korean Journal of Chemical Engineering 2011;28 246-251
96. Mattedi S, Carvalho PJ, Coutinho JAP, Alvarez VH, Iglesias M. High pressure CO2 solubility in N-methyl-2-hydroxyethylammonium protic ionic liquids. Journal of Supercritical Fluids 2011;56 224-230
97. Shariati A, Peters CJ. High-pressure phase behavior of systems with ionic liquids: measurements and modeling of the binary system fluoroform + 1-ethyl-3-methylimidazolium hexafluorophosphate. Journal of Supercritical Fluids 2003;25 109-117
98. Song HN, Lee BC, Lim JS. Measurement of CO2 solubility in ionic liquids: [BMP] [TfO] and [P14,6,6,6][Tf2N] by measuring bubble-point pressure. Journal of Chemical Engineering Data 2010;55 891-896
99. Ren W, Scurto AM. Phase equilibria of imidazolium ionic liquids and the refrigerant gas, 1,1,1,2-tetrafluoroethane (R-134a). Fluid Phase Equilibria 2009;286 1-7
100. Hwang S, Park Y, Park K. Measurement and prediction of phase behaviour for 1-alkyl- 3-methylimidazolium tetrafluoroborate and carbon dioxide: Effect of alkyl chain length in imidazolium cation. Journal of Chemical Thermodynamics 2011;43 339-343
101. Bogel-Łukasik R, Matkowska D, Bogel-Łukasik E, Hofman T. Isothermal vapour–liquid equilibria in the binary and ternary systems consisting of an ionic liquid, 1-propanol and CO2. Fluid Phase Equilibria 2010;293 168-174
102. Bogel-Łukasik R, Matkowska D, Zakrzewska ME, Bogel-Łukasik E, Hofman T. The phase envelopes of alternative solvents (ionic liquid, CO2) and building blocks of biomass origin (lactic acid, propionic acid). Fluid Phase Equilibria 2010;295 177-185
103. Shin EK, Lee BC, Lim JS. High-pressure solubilities of carbon dioxide in ionic liquids:1-Alkyl-3-methylimidazolium bis(trifluoromethylsulfonyl)imide. Journal of Supercritical Fluids 2008;45 282-292
104. Alvarez VH, Mattedi S, Aznar M. Isobaric (vapor+liquid) equilibria of 1-ethyl-3-methylimidazolium ethylsulfate plus (propionaldehyde or valeraldehyde): Experimental data and prediction. Journal of Chemical Thermodynamics 2011;43 895-900
105. Soave G. Equilibrium constants from a modified Redlich-Kwong equation of state. Chemical Engineering Science 1972;27 1197-1203
106. Shiflett MB, Yokozeki A. Solubilities and diffusivities of carbon dioxide in ionic liquids: [bmim][PF6] and [bmim][BF4]. Industrial & Engineering Chemistry Research 2005;44 4453-4464
107. Yokozeki A, Shiflett MB. Ammonia solubilities in room-temperature ionic liquids. Industrial & Engineering Chemistry Research 2007;46 1605-1610

108. Yokozeki A, Shiflett MB. Vapor–liquid equilibria of ammonia+ionic liquid mixtures. Applied Energy 2007;84 1258-1273
109. Shiflett MB, Yokozeki A. Chemical absorption of sulfur dioxide in room-temperature ionic liquids. Industrial & Engineering Chemistry Research 2010;49 1370-1377
110. Yokozeki A, Shiflett MB. Water solubility in ionic liquids and application to absorption cycles. Industrial & Engineering Chemistry Research 2010;49 9496-9503
111. Yokozeki A, Shiflett MB. Hydrogen purification using room-temperature ionic liquids. Applied Energy 2007;84 351-361
112. Shiflett MB, Yokozeki A. Separation of carbon dioxide and sulfur dioxide using room-temperature ionic liquid [bmim][MeSO4]. Energy Fuels 2010;24 1001-1008
113. Shiflett MB, Yokozeki A. Separation of CO2 and H2S using room-temperature ionic liquid [bmim][PF6]. Fluid Phase Equilibria 2010;294 105-113
114. Shiflett MB, Niehaus AMS, Yokozeki A. Separation of CO2 and H2S using room-temperature ionic liquid [bmim][MeSO4]. Journal of Chemical Engineering Data 2010;5 4785-4793
115. Shiflett MB, Elliott BA, Niehaus AMS, Yokozeki A. Separation of N2O and CO2 using room-temperature ionic liquid [bmim][Ac]. Separation Science and Technology 2012;47 411-421
116. Chapman WG, Gubbins KE, Jackson G, Radosz M. SAFT: Equation-of-state solution model for associating fluids. Fluid Phase Equilibria 1989;52 31-38
117. Wertheim MS. Fluids with highly directional attractive forces. I. Statistical thermodynamics. Journal of Statistical Physics 1984;35 19-34
118. Wertheim MS. Fluids with highly directional attractive forces. II. Thermodynamic perturbation theory and integral equations. Journal of Statistical Physics 1984;35 35-47
119. Wertheim MS. Fluids with highly directional attractive forces. III. Multiple attraction sites. Journal of Statistical Physics 1986;42 459-476
120. Wertheim MS. Fluids with highly directional attractive forces. IV. Equilibrium polymerization. Journal of Statistical Physics 1986;42 477-492
121. Kroon MC, Karakatsani EK, Economou IG, Witkamp GJ, Peters CJ. Modeling of the carbon dioxide solubility in imidazolium-based ionic liquids with the tPC-PSAFT equation of state. Journal of Physical Chemistry B 2006;110 9262-9269
122. Karakatsani EK, Economou IG, Kroon MC, Peters CJ, Witkamp GJ. tPC-PSAFT Modeling of gas solubility in imidazolium-based ionic liquids. Journal of Physical Chemistry C 2007;111 15487-15492
123. Karakatsani EK, Economou IG. Phase equilibrium calculations for multi-component polar fluid mixtures with tPC-PSAFT. Fluid Phase Equilibria 2007;261 265-271
124. Economou IG, Karakatsani EK, Logotheti G, Ramos J, Vanin, A.A. Multi-scale modeling of structure, dynamic and thermodynamic properties of imidazolium-based ionic liquids: Ab initio DFT calculations, molecular simulation and equation of state predictions. Oil & Gas Science and Technology 2008;63 283-293
125. Karakatsani EK, Economou IG, Kroon MC, Bermejo MD, Peters CJ, Witkamp GJ. Equation of state modeling of the phase equilibria of ionic liquid mixtures at low and high pressure. Physical Chemistry Chemical Physics 2008;10 6160-6168
126. Andreu JS, Vega LF. Capturing the solubility behavior of CO2 in ionic liquids by a simple model. Journal of Physical Chemistry C 2007;111 16028-16034
127. Andreu JS, Vega LF. Modeling the solubility behavior of CO(2), H(2), and Xe in [C(n)-mim][Tf(2)N] ionic liquids. Journal of Physical Chemistry B 2008;112 15398-15406
128. Llovell F, Valente E, Vilaseca O, Vega LF. Modeling complex associating mixtures with [Cn-mim][Tf2N] ionic liquids: predictions from the soft-SAFT equation. Journal of Physical Chemistry B 2011;115 4387-4398
129. Ji X, Adidharma H. Thermodynamic modeling of CO2 solubility in ionic liquid with heterosegmented statistical associating fluid theory. Fluid Phase Equilibria 2010;293 141-150

130. Paduszyński K, Chiyen J, Ramjugernath D, Letcher TM, Domańska U. Liquid–liquid phase equilibrium of (piperidinium-based ionic liquid+an alcohol) binary systems and modelling with NRHB and PCP-SAFT. Fluid Phase Equilibria 2011;305 43-52.
131. Paduszyński K, Domańska U. Solubility of aliphatic hydrocarbons in piperidinium ionic liquids: measurements and modeling in terms of perturbed-chain statistical associating fluid theory and nonrandom hydrogen-bonding theory. Journal of Physical Chemistry B 2011;115 12537-12548
132. Tsioptsias C, Tsivintzelis I, Panayiotou C. Equation-of-state modeling of mixtures with ionic liquids. Physical Chemistry Chemical Physics 2010;12 4843-4851.
133. Panayiotou C, Tsivintzelis I, Economou IG. Nonrandom hydrogen-bonding model of fluids and their mixtures. 2. Multicomponent mixtures. Industrial & Engineering Chemistry Research 2007;46 2628-2636
134. Wang T, Peng C, Liu H, Hu Y. Description of the pVT behavior of ionic liquids and the solubility of gases in ionic liquids using an equation of state. Fluid Phase Equilibria 2006;250 150-157
135. Hu Y, Liu H, Prausnitz, JM. Equation of state for fluids containing chainlike molecules. Journal of Chemical Physics 1996;104 396-404
136. Breure B, Bottini SB, Witkamp GJ, Peters CJ. Thermodynamic modeling of the phase behavior of binary systems of ionic liquids and carbon dioxide with the group contribution equation of state. Journal of Physical Chemistry B 2007;111 14265-14270
137. Maia FM, Tsivintzelis I, Rodrígez O, Macedo EA, Kontogeorgis GM. Equation of state modelling of systems with ionic liquids: Literature review and application with the Cubic Plus Association (CPA) model. Fluid Phase Equilibria; 332, 128–143
138. Kontogeorgis GM, Voutsas EC, Yakoumis IV, Tassios DP. An equation of state for associating fluids. Industrial & Engineering Chemistry Research 1996;35 4310-4318

CHAPTER 8

CO_2 Absorption by "Dry Ionic Liquids"

Mayuko Ishihara, Masaya Miyake, Mitsuru Satoh*

Department of Chemical Science and Engineering, Tokyo Institute of Technology, Tokyo, Japan

ABSTRACT

Dry Ionic Liquid (D-IL) never means some "water-free" ionic liquid but is a member of "Dry Matter (DM)". DM is a collective name for powdery substances that are composed of micro droplets as an inner core phase and surrounding hydrophobic silica nano particles as the shell part. When the core part is water, it is called Dry Water (DW), which is the first member of DM, while D-IL is the newest one. Because of the much larger surface area of DM compared with that of the inner phase in bulk state, this novel substances are expected to show excellent performance for any mass transfer through the gas-liquid interface. In the present study, we investigated CO_2 absorption by some D-ILs and a DM containing a polyamine in terms of the speed to the equilibrium and a mol-based absorption efficiency. Compared with the respective bulk systems, the D-IL and DM systems proved to be accelerated by ca.50 times without impairing the absorption efficiency.

Keywords: Dry Matter, CO_2 Absorption, Ionic Liquids, Silica Nanoparticles

1. INTRODUCTION

Dry water (DW) has been known since the mid-1960 s when a German patent reported its properties and manufacturing procedure with its name

[1]. DW may be taken as a kind of Pickering emulsion [2] , where air serves as the nonpolar continuous phase in place of common organic solvents. DW can be easily prepared by simply mixing water and hydrophobic fumed silica particles even in a home-use blending machine with a relatively high speed (typically >10^4 rpm) for several tens of seconds. The physicochemical properties of DW, however, has never been studied until two groups rediscovered DW; Binks and Murakami reported that the phase inversion of air-in-water type Pickering emulsion (foam) to water-in-air type one (DW) occurs by changing the particles' hydrophilicity (hydrophobicity) and/or the volume ratio of water to air [3]. Cooper et al. first reported DW as a promising material for methane storage in a form of gas hydrate [4].

Preparation conditions for DW have been rather extensively studied so far; Pezron and her coworkers investigated in their series of papers [5] - [7] successful conditions for DW and concluded that balance of the mechanical energy input and the interaction energy of water and the solid particles at the interface is the most critical factor. Namely, when vigorously mixed, only strongly hydrophobic particles produce DW, while with less hydrophobic particles one must charge a weaker mechanical power into the system, otherwise only a paste-like substance will be obtained.

The latter conclusion should also be valid for other members of "Dry Matter" (DM), which was first proposed by one of the present authors [8] as a collective name for a family of the powder like substances, e.g., DW, Dry polymer solution [9] , Dry Gel [10] and Dry Ionic Liquid (D-IL) [8]. Thus, when a highly hydrophobic particle is used, the more the hydrophobicity of the liquid (or the liquid is apt to wet the particles), the less mechanical energy should be applied. This in fact seems to be consistent with the much easier formation of so-called Liquid Marble (LM) [11] compared with that of DW. LM is typically prepared by simply rolling mm-size droplets on a solid particle bed [12]. Thus, the mechanical energy input during the preparation is negligibly small and the available range of the liquid properties such as the hydrophobicity or the surface tension for LM is rather large compared with that of DM, if those properties allow the particles wet but not immersion into the liquid [13] [14]. Thus, LM may be categorized as a sub-group of DM because of its similarity in the geometrical formation, while the preparation procedure is completely different from that of DM. And the very suitability of the latter's preparation procedure to mass production together with the much larger surface area than LM make DM much more promising functional materials than LM [15].

On the other hand, most members of DM have an inherent weak point; water easily evaporates through the rather porous shell made of silica particles. Although the evaporation may be virtually suppressed in a closed plastic (e.g., polypropylene) bottle, DW would gradually lose the inner water in an open air. In order to reduce this defect to some extent, Dry Gel

(DG) has been tested by Cooper et al. as a potential material for the methane hydrate container. However, studies on DG are scarcely available in literatures except for that prepared with gellan gum [10].

Dry Ionic Liquid (D-IL) is in principle free from the above disadvantage which originates from the escape of the inner liquid phase of DMs because IL has virtually no vapor pressure. Further, IL's stability at higher temperature may also extend its availability in versatile fields of application. With these unique properties, in fact, D-IL can be a promising material for CO_2 absorption; many ILs show high affinity to the gas and some specifically designed ILs to form a complex with CO_2 have also been reported [16]. Further, recovery of CO_2 and the absorption material may be easily performed by evacuation and/or moderate heating at high efficiency [17]. As a matter of fact, similar "high-efficiency" materials in which those outstanding properties of IL are utilized have been reported [18] [19] and especially some kinds of "supported ionic liquid phase (SILP) CO_2 absorbers have been devised. For example, Kolding et al. used an IL-loaded hydrophobic silica gel to confirm its high performance as an SILP CO_2 absorber [20]. However, many kinds of drawbacks (e.g., existence of open pores, low diffusivity or high tortuosity of gases in the pore, high cost for preparation of optimum SILPs, and so on) have been pointed out for SILPs as gas (e.g., CO_2) absorber or separator by G.E. Romanos et al. [17]. The authors have invented novel supported ionic liquid phase systems, which were named "inverse SILPs" and successfully avoided those limitations on the gas absorption and separation performance by utilizing the "inverse SLIPs". Structure of the inverse SLIPs is inherently the same as that of D-IL; powder like substance which is composed of ionic liquid micro droplets covered with silica nano particles. However, the former is largely different from the latter in some points. In the first place, the preparation process of the inverse SLIPs is much more complicated than the original D-IL preparation method; in the former case, an IL and silica nano particles are first mixed with ethanol for ca.2 hours to make a Pickering emulsion (the inner phase is IL and the continuum medium is ethanol) and then the ethanol must be removed in vacuum to obtain the final powder like substance, while in the latter it is enough to mix both materials (IL + silica) with air in a blender machine just for 90 s! Furthermore, the mixing ratio of the inverse SILPs was fixed at g (silica)/g (IL) = 1.5 (IL: 40 wt.%) [17] , while the original D-IL was prepared with a mixing ratio of ca.0.18 ([bmim][I]: 85 wt.%) [8]. Silica nano particles used were also different; The inverse SLIPs were prepared with hydrophilic and less hydrophobic types of fumed silica, i.e., HDK-T-30 and H 20. Free silanol group fractions on the surface are 100% and 50%, respectively, according to the manufacturer (Wacker), and the rest of the latter is methylated as Si-CH_3. The usage of such "hydrophilic" silica should be due to the hydrophilicity of ethanol which was used as a continuum medium of the Pickering emulsion. This means in turn that the inverse SLIPs thus prepared may be unstable in usual usage because air is most hydrophobic. On the other hand, the original D-IL is prepared with a more hydrophobic silica (HDK-H18 with 75% of silanol methylated) in air, which is the same environment as that for its practical applications. Thus, those cited examples for the differences between the inverse SLIPs and the original D-IL would suggest that the latter is more favorable for applications as gas absorber and separator materials.

In the present study, we attempted to prepare a D-IL with a CO_2-absorbing IL (1-butyl-3-methylimidazolium acetate ([bmim] [ace]) and checked the performance as a CO_2-absorption material in terms of speed and capacity. We selected this IL among so many candidates including amino-functionalized ILs [16] because it has already been reported by M. B. Shiflett and his coworkers that [bmim] [ace] is superior, even when used as a bulk liquid, to the traditional amine-based processes for CO_2 absorption or separation from the post combustion flue gas in economic terms [21]. Thus, if it is possible to prepare powder like D-IL with the relevant IL, the total performance for the CO_2 absorption should be more improved due to the enhanced gas absorption speed. Of course, the excellent performance of D-IL-like materials for the gas absorption has already been demonstrated, at least qualitatively, in the distinguished study by Romanos et al. [17]. Therefore we confined our primary goal to preparation of a powder like D-IL with [bmim] [ace] and confirmation of its feasibility as the gas absorption material. Further we also tested a potentiality of an IL + polyamine mixed system in a form of DM as CO_2 absorption material. Needless to say, one of the most distinguished and favorable properties of ILs must be their non-volatility. Polymers are also non-volatile and those containing amino groups in the monomer unit may be able to absorb more CO_2 than IL if the ability was compared per mass of the absorbents. Thus, if a polyamine could be incorporated into a D-IL, the resultant DM may be promising as a CO_2 absorption material even though compared with monoethanolamine that is now utilized in the industrial process [22]. In the present study, we for the first time explored these possibilities.

2. MATERIALS AND METHODS

2.1. Materials

1-butyl-3-methylimidazolium acetate ([bmim] [ace]) and 1-butyl-3-methylimidazolium iodide ([bmim] [I]) were purchased from Sigma Aldrich and used as received. Poly (allyl amine) hydrochloride (PAlAm·HCl) (MW: 15,000, Sigma Aldrich) was neutralized by NaOH and then filtrated through a Ultrafilter UP-10 (Advantec) for several times with pure water to remove inorganic ions and low molecular weight polymers. Concentration of PAlAm thus purified was estimated by conductometric titration to be ca.9.9%.

Hydrophobic fumed silica particle (HDK-H18, primary particle size: 5 - 30 nm), the surface OH groups of which are methylated with poly(dimethyl siloxane) by 75%, was purchased from Wacker Asahikasei Silicone Co. and used as received.

2.2. Methods

2.2.1. Preparation of D-IL

D-IL of [bmim] [ace] was prepared with a blender machine (Waring J-SPEC blender, container volume: 50 mL) by mixing 6.0 g of the IL or its aqueous solution and 1.05 g of the silica at a fixed speed (22500 rpm) for 90 s (30 s × 3 with an interval of 10 s each) at a room temperature. Partial remove of water from the prepared D-IL samples was performed with an infrared humidity meter (FD-720, Kett Co. Ltd.) at certain temperatures. D-IL containing PAlAm was also prepared with the same preparation procedure except for using a mixed solution of the IL and aq. PAlAm solution (9.9 wt.%) instead of the pure IL or the aq. solution.

2.2.2. CO_2 Absorption Measurements

Absorption performance of D-ILs or bulk liquids (IL, aq. solution of IL, aq. solution of PAlAm) for CO_2 was estimated by measuring pressure depression within connected glass cylinders due to the absorption of CO_2. The experimental set up which was assembled within a thermostat incubator (FMU-133I, Fukushima Kogyo Co. Ltd.) is schematically illustrated in Figure 1. First, CO_2 gas was introduced to Cylinder A (Hyper Glass Cylinder HPG-96-3 (90 mL), Taiatsu Garasu Kogyo Co. Ltd.) via Valve A at a partial pressure of ca.2 bar with Valve B closed. Then, a sample D-IL or a sample bulk liquid was set in Cylinder B with Valve B closed. After 30 min of incubation time, Valve B was opened and the pressure was recorded with a pressure gauge (KDM 30, Krone Co.). Since the pressure begins to decrease at the same time as the opening of Valve B, pressure at time 0 can't be directly measured. Thus, the initial partial pressure in Cylinder A + B ($p_{0(A+B)}$) was theoretically estimated by the following equation.

$$p_{0(A+B)} = p_{i(A)} V_A \Big/ \left(V_{A+B} - V_{sample} \right)$$

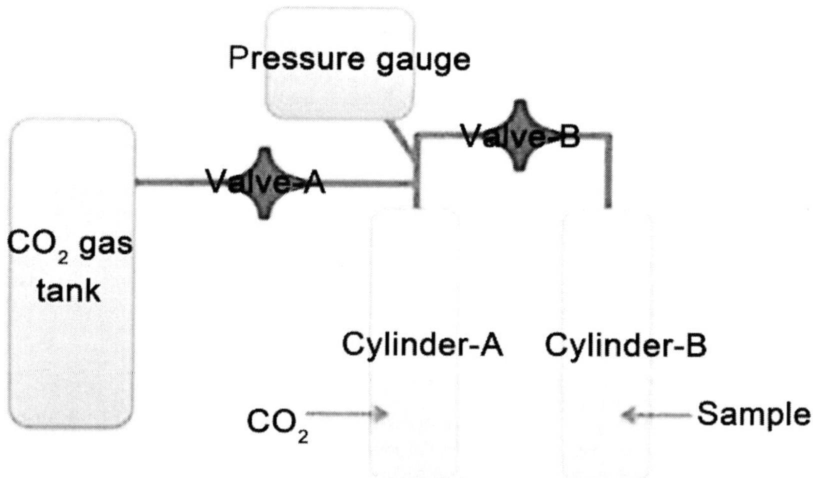

Figure 1. Diagrammatic illustration of the set up to measure the CO_2partial pressure.

where $p_{i(A)}$ is the initial partial pressure of CO_2 introduced into Cylinder A, V is the inner volume of the cylinders including the valve parts or the sample volume. $p_{i(A)}$ was set around at 2 bar. V_A and V_{A+B} were 95.8 cm^3 and 191.7 cm^3, respectively. Since V_{sample} (~2 cm^3) was rather small compared with V_{A+B}, the absorption experiments were carried out under an initial partial pressure of CO_2 ~1 bar. CO_2 absorption efficiency of a sample was estimated by a ratio of mol of absorbed CO_2 to mol of IL or PAlAm (mol/mol efficiency) or by mol/(mol × bar (equilibrium CO_2 pressure)) (mol/(mol bar) efficiency). According to a preliminary experiment, the efficiency turned out to be estimated within ±10% error.

3. RESULTS AND DISCUSSION

3.1. Preparation of D-IL

D-IL preparation was attempted by using [bmim] [ace] and the silica nanoparticles. Using the bulk liquid without added water, only a creamy or paste-like substance was obtained (Table 1). This failure in preparation of a powdery D-IL may be ascribed to relatively low surface tension (γ) of the pertinent IL; although γ of the IL is not available in literatures, γ of 1-ethyl-3-methylimidazolium acetate was reported to be ca.46 mN/m [23]. Thus, γ of [bmim] [ace] may be lower than that of [emim] [ace] because

Table 1. Preparation results of D-IL with [bmim] [ace].

Water content	0 wt%	10 wt%	20 wt%
Appearance of Product		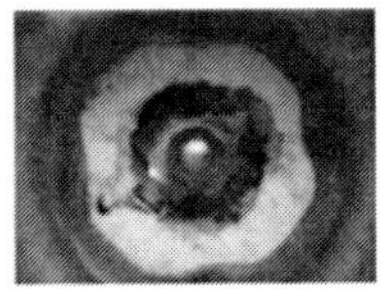	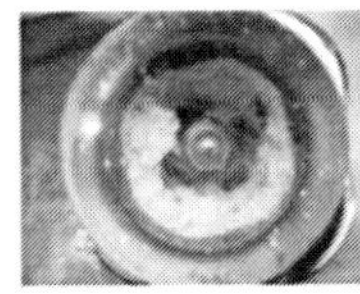
Water content	30 wt%	40 wt%	
Appearance of Product	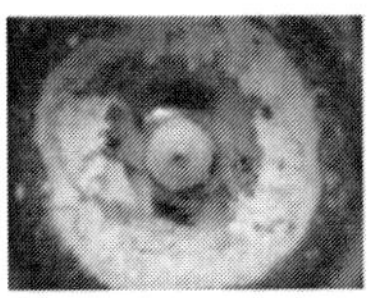	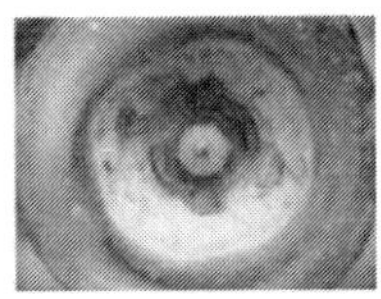	
Water content	50 wt%	60 wt%	
Appearance of Product	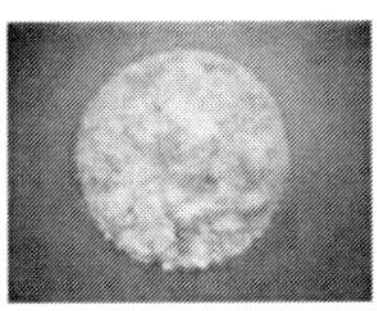	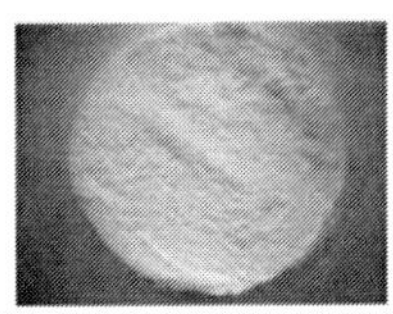	

the former alkyl group is larger and more hydrophobic than the latter. So far, only one successful preparation of powdery D-IL. Only with large amounts of water, it was successful; although the product prepared with the IL containing 50 wt% water was "souf- flé-like" to some extent, one with 60 wt% water was wholly powdery. Effects of the water addition and the resultant changes in the appearance of D-IL on the CO_2 absorption performance were examined as shown below.

3.2. CO_2 Absorption by D-IL Prepared with [bmim] [ace]

CO_2 absorption was monitored by measuring CO_2 partial pressure as a function of time for two kinds of bulk ILs; [bmim] [ace] and [bmim] [I]. The latter was used because it is the only IL that provides powdery D-IL without water addition [8]. If it were also a CO_2-absorbing IL, the D-IL would be promising from a view point of the expected high absorption speed. The experimental results are shown in Figure 2 and the experimental conditions and the respective CO_2 absorption abilities are given on Table 2. In the case of [bmim] [I], the decrement in the partial pressure was only slight and the absorption soon reached at the saturation.

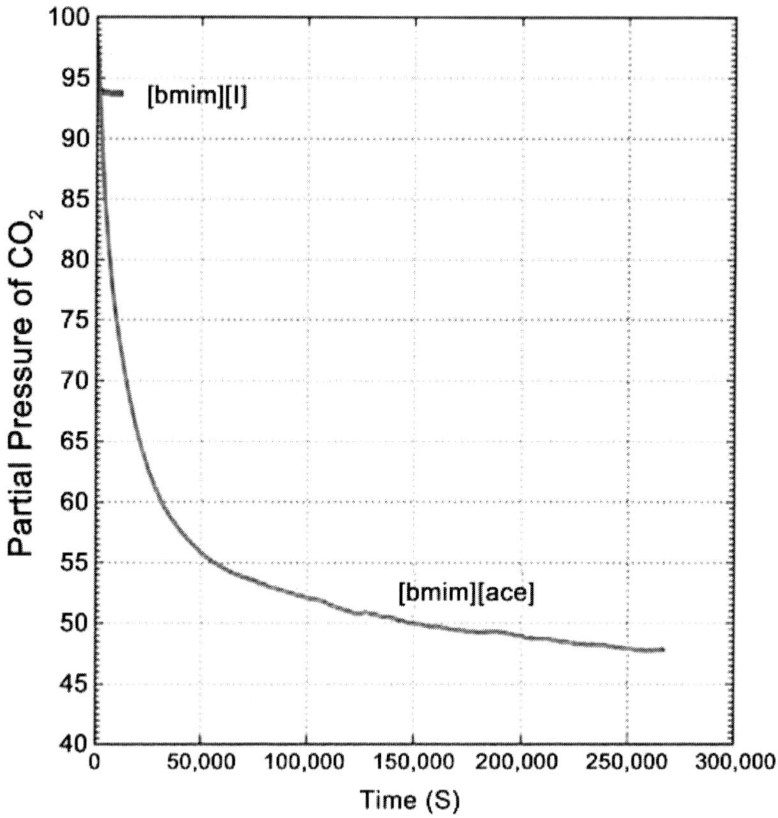

Figure 2. CO_2 absorption performance of [bmim] [I] and [bmim] [ace] observed as pressure profile.

On the other hand, CO_2 absorption by bulk [bmim] [ace] was significant even at low pressures as expected from literature data [24] [25] while it took so long time (ca.72 h) to reach the saturation. The respective performance on the basis of mol of CO_2 absorbed by mol of IL were 0.014 at the partial pressure $p \sim 0.94$ bar for [bmim] [I] and 0.32 at $p \sim 0.48$ bar for [bmim] [ace]. The latter value was comparable with the literature data by Yokozeki et al. [24] [25] ; 0.336 at $p = 0.502$ bar and $t = 298.1$ K. The smaller absorption (Figure 2), it may be due to the larger amount of IL contained in the D-IL compared with the bulk system (cf. Table 3). In fact, except for the initial stage of the absorption, those two profiles showed a similar tendency of pressure decrement. efficiency obtained in the present study may be safely ascribed to the lower pressure.

As a next step, CO_2 absorption performance by D-IL of [bmim] [ace] and those prepared by adding water to the IL was investigated in a similar way as the bulk IL cases. First of all, the creamy substance prepared without water addition was tested. The result was given in Figure 3 and on Table 3 together with the other ones. The creamy D-IL showed a similar performance with the bulk IL; although the initial speed of the CO_2 absorption, i.e., the initial decrement in the partial pressure was more significant than that of the bulk.

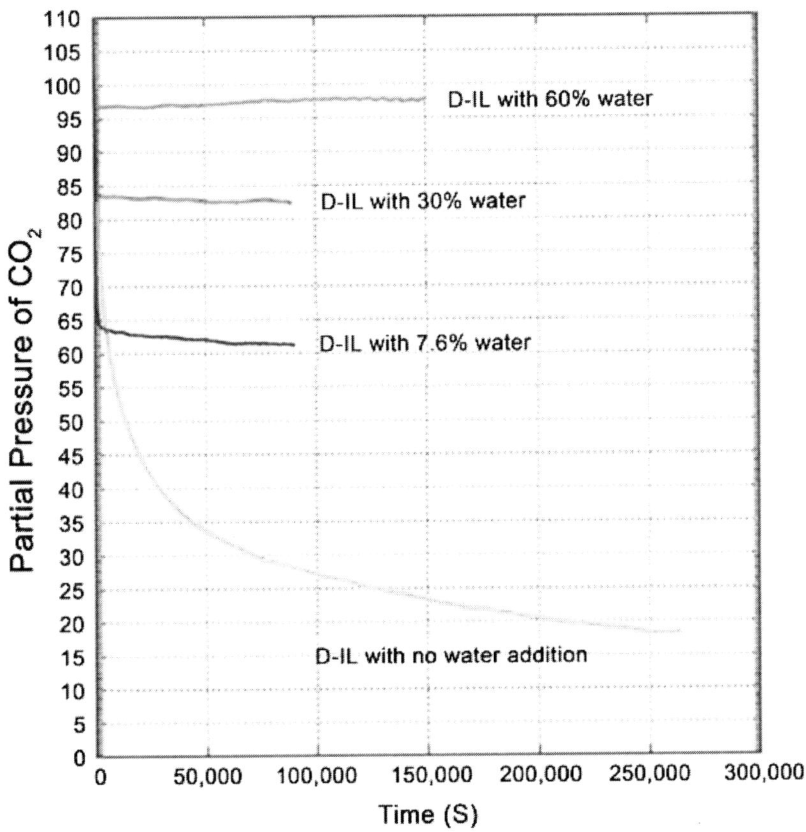

Figure 3. Effects of water addition on CO_2 absorption performance of [bmim] [ace]-based D-ILs.

Table 2. Experimental conditions and the results for CO_2 absorption by bulk [bmim] [I] and [bmim] [ace].

IL	IL amount (mol)	$p_{initial}$ (kPa)	p_{final} (kPa)	CO_2 absorbed (mol)	Absorption efficiency (mol/mol)
[bmim] [I]	0.013	95.8	93.7	1.6×10^{-4}	0.014
[bmim] [ace]	0.012	98.2	47.8	3.8×10^{-3}	0.32

Table 3. Experimental conditions and the results for CO_2 absorption by D-ILs with and without water addition.

Water content (%)	Sample appearance	IL amount (mol)	$p_{initial}$ (kPa)	p_{final} (kPa)	CO_2 absorbed (mol)	Absorption efficiency (mol/mol)
0	Creamy	0.025	104.9	18.4	6.5×10^{-3}	0.25
60	Powdery	0.012	103.8	96.7	5.3×10^{-4}	0.044
30	Powdery	0.011	99.6	82.4	1.3×10^{-3}	0.12
7.6	Souffle-like	0.011	98.8	61.2	2.9×10^{-3}	0.27

For example, at t = 50,000 s (~14 h), ca.83% of the initial amount of CO_2 was absorbed to the respective samples. The equilibrium states of CO_2 absorption seem to befter 2.5 - 2.6 × 10^5 s (69 - 72 h) for both systems. Thus, CO_2 absorption speed of the creamy D-IL proved to be no superior to that of the bulk IL sample. These results suggest that the IL in the creamy D-IL formed a continuous phase without increasing the effective surface area. As for the absorption ability, the molar ratio of CO_2 absorbed to the IL was estimated to be 0.25. This value, which was obtained at p = 0.18 bar, is comparable to literature values of the bulk IL at p = 0.1 bar; 0.19 - 0.23 [24] [25] , as a matter of course.

Next we tested powdery samples to confirm the expected enhancement in the absorption speed. The results are shown in Figure 3. As described in the previous section, a powdery D-IL of [bmim] [ace] was successfully prepared only when substantial amount of water was added as much as 60%. As seen from the figure and Table 3 the CO_2 absorption speed was so rapid that the equilibrium was established just within several minutes. The slight increase in the pressure after 5000 s may be due to evaporation of water. On the contrary to the advantage in the absorption speed, the 60% water containing D-IL showed a much lower CO_2 solubility than the bulk IL, just one eighth [26]. This deterioration in the absorption ability may be caused from the existence of water. According to a proposed mechanism for the chemical absorption of CO_2 to the relevant IL [26] [27] , C2 proton of the imidazolium ring is abstracted by an acetic acid counter anion to form a site reacting with CO_2. This mechanism has in fact been supported by some researches [28] [29] , one of which also suggested that water may inhibit the C2 proton abstraction reaction [28]. Thus, in order to obtain a D-IL material combining a high CO_2 solubility and an excellent absorption speed, it should be necessary to reduce the amount of water in the inner phase of the relevant D-IL samples as much as possible with a powdery state retained. Since preparations of D-IL with [bmim] [ace] containing 50% or less of water provide only creamy substances (Table 1), we attempted to remove water from the D-IL containing 60% of water with an IR humidity meter at 40°C and 80°C. With this treatment, water content of the D-IL sample was reduced to 30% and 7.6% at the respective temperatures. The absorption performance of these modified samples is given on Table 3 and Figure 3. With decreasing the water content, the CO_2 solubility into the samples increased as can be judged from the almost flat level of CO_2 pressure. The mol-based absorption efficiency of the 7.6% water sample was 0.27, which is still lower than 0.32. Since the equilibrium pressure at which the former value was obtained (61.2 kPa) was higher than the latter (47.8 kPa) by ca.30%, actual difference of the solubility should be more significant. This means that only a small amount of water may affect the CO_2 absorption to [bmim] [ace]. However, the absorption efficiency value, 0.27, is never low in light of such a relatively low equilibrium pressure (0.61 bar). Further, the absorption speed was also excellent; 90% of the saturation level was absorbed within ca.20 min. Although the 7.6% water sample was not powdery but Souffle-like, the speedy absorption suggests that the aqueous IL did not form a continuous phase but was finely divided by the silica particles with the surface wetted to some extent by the liquid. Thus, the "post-dehydration" procedure seems to be effective to prepare D-ILs as a high performance material for CO_2 absorption.

3.3. CO_2 Absorption by DM Containing [bmim] [ace] and PAlAm

Nonvolatile CO_2 absorption substances are not limited to ILs; polymers containing amino groups may also be utilized for such purpose. For a polyamine to be incorporated into a DM, the polymer must be solubilized in any liquid, ideally pure ILs. Thus, we attempted to dissolve PAlAm into [bmim] [ace] or [bmim] [I], while it proved to be in vain. Many neutral polymers are known to be soluble in ILs, while polyelectrolytes seem to be mostly insoluble in them [30]. Then, an aqueous solution of PAlAm was used to prepare a DM containing [bmim] [ace] and the polyamine. By setting the water content of the DM at 60 wt%, a Souffle-like DM was obtained.

Figure 4 and Table 4 show results of CO_2 absorption by the DM prepared above, together with those by the bulk aqueous PAlAm solution and the bulk IL + aq. PAlAm mixed solution (water content = 60 wt%). First of all, one should note the significant absorption ability of the bulk aq. solution of

PAlAm as large as 0.39, a mol-based value, which is comparable with 0.32 of bulk [bmim] [ace]; if those values are converted to the mol bar-based efficiency, the former becomes 0.61 and the latter 0.67. This relatively high CO_2 absorption ability of the pertinent polymer is still lower than the theoretical or ideal value, 0.5; in principle, two free amino groups of PAlAm may react with one CO_2molecule to form an ion-pair as the case of low molecular weight amines [31].

$$2R-NH_2 + CO_2 \rightarrow RNHCOO^- + RNH_3^+ \quad (1)$$

In the case of polyamines like PAlAm however the above ion-pair formation serves as a physical crosslinking of the polymer chains, leading to a gel formation [32]. In fact, in the above aqueous solution of PAlAm a phase separation was observed after CO_2

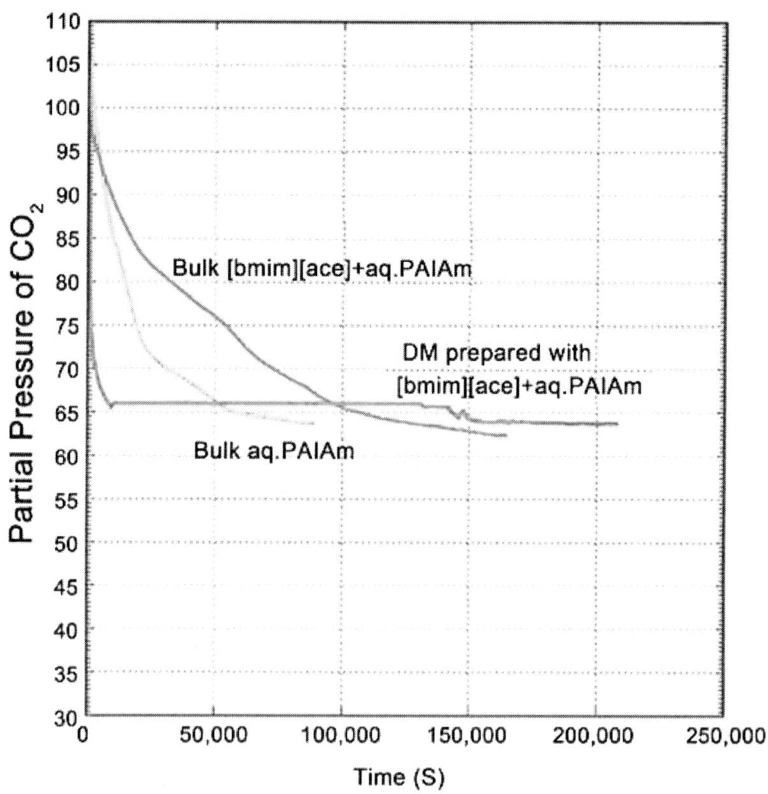

Figure 4. CO_2 absorption performance of samples containing aqueous PAlAm solution.

Table 4. Experimental conditions and the results for CO_2 absorption by samples containing aq. PAlAm.

Sample	Sample appearance	Amount of IL or amino group (mol)	$p_{initial}$ (kPa)	p_{final} (kPa)	CO_2 absorbed (mol)	Absorption efficiency (mol/mol)
Bulk aqueous PAlAm (9.9%)	Clear solution	8.4×10^{-3}	106.8	63.7	3.3×10^{-3}	0.39
Bulk [bmim] [ace] + aq. PAlAm	Clear solution	$8.8 \times 10^{-3} + 7.6 \times 10^{-3}$	100.0	62.4	2.8×10^{-3}	0.37*
DM with [bmim] [ace] + aq. PAlAm	Souffle-like	$6.8 \times 10^{-3} + 5.0 \times 10^{-3}$	102.7	63.7	2.9×10^{-3}	0.58*

*Values estimated on the basis of PAlAm mol.

absorption measurements. The post-measurement sample consisted of a turbid lower layer and a clear upper layer, which was separated by a white thin layer. By shaking the sample tube, the white layer was easily destroyed and a totally turbid liquid was obtained. If the white layer was a gel-like phase, it would retard the CO_2 absorption to some extent. In fact, after leaving the sample tube for two days, the whole lower layer became a soft gel.

A similar phase separation was also confirmed for the bulk [bmim] [ace] + aq. PAlAm system. In this case, however, the lower layer was already gel-like. Although the difference in the gel formation behavior between these two systems is not known at present, the slower CO_2 absorption of the bulk IL + PAlAm system may be ascribed to the more significant gel formation. As for the absorption efficiency, it became slightly worse than that of the bulk aq. PAlAm; if [bmim] [ace] in the mixed solution sample absorbed CO_2 to the same extent as the 60% water-containing sample (Table 3) in the mol bar-based absorption efficiency (0.044/0.967 = 0.046), the mol-based efficiency of the IL in the mixed solution sample may be estimated as 0.046 × 0.624 = 0.029. Then, a contribution from the IL to the total amount of CO_2 absorbed is estimated with this efficiency to be 2.6×10^{-4}mol. This means in turn that the contribution from PAlAm in the total absorption amount is ~2.5×10^{-3} mol of CO_2 and the efficiency becomes ca.0.33 mol/mol. This slightly lower efficiency, compared with 0.39 of the bulk PAlAm solution, may be ascribed to the more significant gel formation. Namely, the gel network might obstruct CO_2 to access to free amino groups. Thus, for the pertinent pressure profile given in Figure 4 to reach the true equilibrium, much more time would be necessary.

In the case of DM prepared with the IL + aq. PAlAm mixed solution, the CO_2 absorption speed was distinctly higher than those bulk systems. For example, it took ca.90 min to absorb 90% of the saturation, while it was ca.1600 min for the bulk mixed solution. The former elapsed time is, however, still much longer than that of the bulk IL system containing only 7.6% of water but no PAlAm, i.e., ca.20 min (Figure 3). This slowing down in the absorption speed, which may be due to the presence of PAlAm, may also be attributed to the gel formation as compared in the bulk systems. As for the absorption efficiency, however, the gel formation in the DM does not seem obstruct the CO_2 absorption; the mol-based absorption efficiency, 0.58, though it was estimated as a value per amino group of PAlAm, is higher than 0.5, the limiting value for the CO_2 absorption by PAlAm. This means that the IL also contributed to the significant CO_2 absorption to a substantial degree, although the respective contributions may not be separately estimated in the present study. Here we should note why [bmim] [ace] could absorb CO_2 to a significant degree irrespective of the coexistence of water. As discussed previously, the abstraction of the C2 proton of the IL cation by the counter anion (CH_3COO^-) should be interfered by water. The observed high level of CO_2 absorption suggests that this interference mechanism by water is not well working. As a probable mechanism for that, we postulate that amino groups of PAlAm are involved in the C2 proton abstraction. In fact, basicity of a primary aminc is much stronger than that of -COO^-; basicity measured by pK_b values of ethyl amine is larger than that of acetate anion by a factor of ca.10^6. Then, a next question is why the abstraction by the polymer in the bulk mixed solution system did not appear to contribute to an appreciable degree. This may be safely ascribed to the observed macroscopic phase separation. Because the polymer was mostly confined to the lower layer of the sample, the amino groups would be hardly available for the IL left in the upper layer. On the other hand, in the case of the DM system each DM particle contains only a tiny amount of liquid and has a relatively large surface area. For a typical DM having a diameter of 100 μm [8] , the surface/volume ratio is as large as ca.10^6 times of the bulk. Thus, it may be reasonable to assume that the gelation in the DM system prevailed through the tiny sample without leaving a separate PAlAm layer. A non-monotonic change in the CO_2 pressure observed for the DM sample around at $t = 10^4$ s may be corresponding to this rather rapid development of gelation through each droplet. A "transition" to a second "stationary" state found around at $t = 1.3 \times 10^5$ s (36 h) indicates that the CO_2 absorption process was not completed but just temporarily stopped by the gel formation. As stated above, PAlAm gel formed by ionic crosslinking according to Equation (1), where Rs should be replaced by polymer chains, seemed to be never robust. Thus, the gel layer, which must be much thinner than that formed for the bulk system, would be easily, at least partly, broken under a certain CO_2 pressure.

4. CONCLUSIONS

In the present study, we have shown a promising potential of [bmim] [ace] in a form of D-IL as CO_2absorption material. By the "post-dehydration" of the D-IL, the mol-based efficiency for the CO_2absorption was greatly improved without impairing the absorption speed. Further, addition of PAlAm to the D-IL as another CO_2 absorption component proved to be effective; the amino group seemed to improve the CO_2 absorption by the IL in the presence of water.

A main problem to be overcome for the D-IL to be employed as a competitive CO_2 absorption material was also evident; water, which impairs the CO_2 absorption of [bmim] [ace], was necessary to prepare a powdery or Souffle-like D-IL. One solution would be to employ an amine-containing IL. For example, a group of ILs consisting of a tetraalkylammonium cation and an amino acid anion seems to be promising [17] [33]. In fact, we have plenty of choice for ILs having high CO_2absorption ability [34]. Attempts have just started to endow such ILs with excellent absorption speed by forming them into D-ILs.

ACKNOWLEDGEMENTS

This work was partly supported by JSPS KAKENHI Grant Number JP15K05584.

REFERENCES

1. Brunner, H., Shutte, D. and Schmitz, F.-T. (1964) German Patent DE1467023.
2. Pickering, S.U. (1907) Emulsions. Journal of the Chemical Society, 91, 2001-2021. http://dx.doi.org/10.1039/CT9079102001
3. Binks, B.P. and Murakami, R. (2006) Phase Inversion of Particle-Stabilized Materials from Foams to Dry Water. Nature Materials, 5, 865-869. http://dx.doi.org /10.1038/nmat1757
4. Wang, W., Bray, C.L., Adams, D.J. and Cooper, A.I. (2008) Methane Storage in Dry Water Gas Hydrates. Journal of the American Chemical Society, 130, 11608-11609. http://dx.doi.org/10.1021/ja8048173
5. Forny, L., Perzon, I., Saleh, K., Guigon, P. and Komunjer, L. (2007) Storing Water in Powder Form by Self-Assembling Hydrophobic Silica Nanoparticles. Powder Technologies, 171, 15-24. http://dx.doi.org/10.1016/j.powtec.2006.09.006
6. Forny, L., Saleh, K., Pezron, I., Komunjer, L. and Guigon, P. (2009) Influence of Mixing Characteristics for Water Encapsulation by Self-Assembling Hydrophobic Silica Nanoparticles. Powder Technologies, 189, 263-269. http://dx.doi.org/ 10.1016/j.powtec.2008.04.030
7. Saleh, K., Forny, L., Guigon, P. and Pezron, I. (2011) Dry Water: From Physic-Chemical Aspects to Process-Related Parameters. Chemical Engineering Research and Design, 89, 537-544. http://dx.doi.org/10.1016/j.cherd.2010.06.005
8. Shirato, K. and Satoh, M. (2011) "Dry Ionic Liquid" as a New Comer to "Dry Matter". Soft Matter, 7, 7191-7193. http://dx.doi.org/10.1039/c1sm05999h
9. Miyake, M. and Satoh, M. (2015) Preparation Conditions of Dry Matter. Kobunshi Ronbunshu, 72, 617-623. http://dx.doi.org/10.1295/koron.2015-0026
10. Carter, B.O., Wang, W., Adams, D.J. and Cooper, A.I. (2010) Gas Storage in "Dry Water" and "Dry Gel" Clathrates. Langmuir, 26, 3186-3193. http://dx.doi. org/10.1021/la903120p

11. Aussillous, P. and Quéré, D. (2001) Liquid Marbles. Nature, 411, 924-927. http://dx.doi.org/10.1038/35082026
12. McEleney, P., Walker, G.M., Larmour, I.A. and Bell, S.E.J. (2009) Liquid Marble Formation Using Hydrophobic Powders. Chemical Engineering Journal, 147, 373-382. http://dx.doi.org/10.1016/j.cej.2008.11.026
13. Binks, B.P. and Tyowua, A.T. (2013) Influence of the Degree of Fluorination on the Behaviour of Silica Particles at Air-Oil Surfaces. Soft Matter, 9, 834-845. http://dx.doi.org/10.1039/C2SM27395K
14. Tan, T.T.Y., Ahsan, A., Reithofer, M.R., Tay, S.W., Tan, S.Y., Hor, T.S.A., Chin, J.M., Chew, B.K.J. and Wang, X. (2014) Photoresponsive Liquid Marbles and Dry Water. Langmuir, 30, 3448-3454. http://dx.doi.org/10.1021/la500646r
15. Carter, B.O., Adams, D.J. and Cooper, A.I. (2010) Pausing a Stir: Heterogeneous Catalysis in "Dry Water". Green Chemistry, 12, 783-785. http://dx.doi.org/10.1039/b922508k
16. Yang, Z.-Z., Zhao, Y.-N. and He, L.-N. (2011) CO2 Chemistry: Task-Specific Ionic Liquids for CO2 Capture/Activation and Subsequent Conversion. RSC Advances, 1, 545-567. http://dx.doi.org/10.1039/c1ra00307k
17. Romanos, G.E., Shulz, P.S., Bahlmann, M., Wasserscheid, P., Sapalidis, A., Katsaros, F.K., Athanasekou, C.P., Beltsios, K. and Kanellopoulos, N.K. (2014) CO2 Capture by Novel Supported Ionic Liquid Phase Systems Consisting of Silica Nanoparticles Encapsulating Amine-Functionalized Ionic Liquids. Journal of Physical Chemistry C, 118, 24437-24451. http://dx.doi.org/10.1021/jp5062946
18. Thomassen, P.L., Kunov-Kruse, A.J., Mossin, S.L., Kolding, H., Kegnas, S., Riisager, A. and Fehrmann, R. (2013) Separation of Flue Gas Components by SILP (Supported Ionic Liquid-Phase) Absorbers. ECS Transactions, 50, 433-442. http://dx.doi.org/10.1149/05011.0433ecst
19. Nurul, S.M., Bustam, M.A., Yunus, N.M. and Man, Z. (2014) Solid Supported [hmim][Tf2N] for CO2 Adsorption. Advanced Materials Research, 879, 149-154. http://dx.doi.org/10.4028/www.scientific.net/AMR.879.149
20. Kolding, H., Fehrmann, R. and Riisager, A. (2012) CO2 Capture Technologies: Current Status and New Directions Using Supported Ionic Liquid Phase (SILP) Absorbers. Science China Chemistry, 55, 1648-1656. http://dx.doi.org/10.1007/s11426-012-4683-x
21. Shiflett, M.B., Drew, D.W., Cantini, R.A. and Yokozeki, A. (2010) Carbon Dioxide Capture Using Ionic Liquid 1-Butyl-3-Methylimidazolium Acetate. Energy and Fuels, 24, 5781-5789. http://dx.doi.org/10.1021/ef100868a
22. Rochelle, G.T. (2009) Amine Scrubbing for CO2 Capture. Science, 325, 1652-1654. http://dx.doi.org/10.1126/science.1176731
23. 23. Schuermann, J., Huber, T., LeCorre, D., Mortha, G., Sellier, M., Duchemin, B. and Staiger, M.O. (2016) Surface Tension of Concentrated Cellulose Solutions In 1-Ethyl-3-Methyli-midazolium Acetate. Cellulose, 23, 1043-1050. http://dx.doi.org/10.1007/s10570-015-0850-5
24. Shiflett, M.B., Kasprzak, D.J., Junk, C.P. and Yokozeki, A. (2008) Phase Behavior of {Carbon Dioxide + [bmim][Ac]} Mixtures. Journal of Chemical Thermodynamics, 40, 25-31. http://dx.doi.org/10.1016/j.jct.2007.06.003
25. Yokozeki, A., Shiflett, M.B., Junk, C.P., Grieco, L.M. and Foo, T. (2008) Physical and Chemical Absorption of Carbon Dioxide in Room-Temperature Ionic Liquids. Journal of Physical Chemistry B, 112, 16654-16663. http://dx.doi.org/10.1021/jp805784u
26. Maginn, E.J. (2005) Design and Evaluation of Ionic Liquids as Novel CO2 Absorbents. Quarterly Technical Reports to DOE, 1-12.
27. Cadena, C., Anthony, J.L., Shah, J.K., Morrow, T.I., Brennecke, J.F. and Maginn, E.J. (2004) Why Is CO2 So Soluble in Imidazolium-Based Ionic Liquids? Journal of the American Chemical Society, 126, 5300-5308. http://dx.doi.org/10.1021/ja039615x
28. Gurau, G., Rodriguez, H., Kelly, S.P., Janiczek, P., Kalb, R.S. and Rogers, R.D. (2011) Demonstration of Chemisorption of Carbon Dioxide in 1,3-Dialkylimidazolium

Acetate Ionic Liquids. Angewandte Chemie, 123, 12230-12232. http://dx.doi.org/10.1002/ange.201105198

29. Cabaco, M.I., Besnard, M., Danten, Y. and Coutinho, J.A.P. (2012) Carbon Dioxide in 1-Butyl-3-Methyl Imidazolium Acetate. I. Unusual Solubility Investigated by Raman Spectroscopy and DFT Calculations. Journal of Physical Chemistry A, 116, 1605-1620. http://dx.doi.org/10.1021/jp211211n
30. Chen, Y., Zhang, Y., Ke, F., Zhou, J., Wang, H. and Liang, D. (2011) Solubility of Neutral and Charged Polymers in Ionic Liquids Studied by Laser Light Scattering. Polymer, 52, 481-488. http://dx.doi.org/10.1016/j.polymer.2010.11.034
31. Reynolds, A.J., Verheyen, T.V., Adeloju, S.B., Meuleman, E. and Feron P. (2012) Towards Commercial Scale Postcombustion Capture of CO2 with Monoethanolamine Solvent: Key Considerations for Solvent Management and Environmental Impacts. Environmental Science & Technology, 46, 3643-3654. http://dx.doi.org/10.1021/es204051s
32. Carretti, E., Dei, L., Baglioni, P. and Weiss, R.G. (2003) Synthesis and Characterization of Gels from Polyallylamine and Carbon Dioxide as Gallant. Journal of the American Chemical Society, 125, 5121-5129. http://dx.doi.org/10.1021/ja034399d
33. Yu, H., Wu, Y.T., Jiang, Y.-Y., Zhou, Z.Z. and Zhang, Z.-B. (2009) Low Viscosity Amino Acid Liquids with Asymmetric Tetraalkylammonium Cations for Fast Absorption of CO2. New Journal of Chemistry, 33, 2385-2390. http://dx.doi.org /10.1039/b9nj00330d
34. Zhang, X., Zhang, X., Dong, H., Zhao, Z., Zhang, S. and Huang, Y. (2012) Carbon Capture with Ionic Liquids: Overview and Progress. Energy & Environmental Science, 5, 6668-6681. http://dx.doi.org/10.1039/c2ee21152a

CHAPTER 9

Rapid Carbonation for Calcite from a Solid-Liquid-Gas System with an Imidazolium-Based Ionic Liquid

Abdul-Rauf Ibrahim †, Jean Bosco Vuningoma †, Yan Huang, Hongtao Wang and Jun Li *

Department of Chemical and Biochemical Engineering, College of Chemistry and Chemical Engineering, National Engineering Laboratory for Green Chemical Productions of Alcohols, Ethers and Esters, Xiamen University, Xiamen 361005, China

ABSTRACT

Aqueous carbonation of $Ca(OH)_2$ is a complex process that produces calcite with scalenohedral calcite phases and characterized by inadequate carbonate species for effective carbonation due to the poor dissolution of CO_2 in water. Consequently, we report a solid-liquid-gas carbonation system with an ionic liquid (IL), 1-butyl-3-methylimidazolium bromide, in view of enhancing the reaction of CO_2 with $Ca(OH)_2$. The use of the IL increased the solubility of CO_2 in the aqueous environment and enhanced the transport of the reactive species (Ca^{2+} and CO_3^{2-}) and products. The presence of the IL also avoided the formation of the $CaCO_3$ protective and passivation layer and ensured high carbonation yields, as well as the production of stoichiometric rhombohedral calcite phases in a short time.

Keywords: calcite; carbon dioxide; carbonation; solid-liquid-gas; ionic liquid

1. INTRODUCTION

Calcium carbonate ($CaCO_3$) is an inorganic chemical that occurs in different crystalline polymorphs at ambient pressure. These are anhydrous phases of vaterite, calcite and aragonite and hydrated phases of monohydrocalcite and hexahydrocalcite (ikaite) [1]. Anhydrous $CaCO_3$ is generally classified as rhombic calcite, needle-like aragonite or spherical vaterite, among which calcite is the most stable phase under normal atmospheric conditions [2]. The formation of any of the polymorphs is strictly guided by the temperature, supersaturation and pH of reaction solution [3]. However, chameleonic phase transformation was found between calcite, aragonite and vaterite, normal to abnormal and *vice versa*, in a CO_2expanded ethanol-water solution system applied for the solid-liquid-gas carbonation (SLGC) of $Ca(OH)_2$ [4]. $CaCO_3$ is used for several industrial applications and by living organisms for bone development [5]. For instance, it is used as an extender and to meliorate stability and exposure to resistance in paints [6,7]. As such, global consumption and demand for $CaCO_3$continues to increase [8,9]. Industrially, calcite is usually produced by the SLGC route, where CO_2 is bubbled through a $Ca(OH)_2$ slurry or slake lime [10,11,12,13]. Although the mechanism for the precipitation of $CaCO_3$ from SLGC is debated, many investigators have shown that the absorption of CO_2 is the rate determining step in the SLGC with a $Ca(OH)_2$suspension [10,11]. Current synthesis procedures are slow, energy intensive and less efficient, leading to low conversion yields. Several methods for synthesizing $CaCO_3$ effectively and with various properties using additives and surfactants [14,15], block copolymers [16,17] and acids [18,19] have been explored.

The excellent role of ionic liquids (ILs) as reaction media and catalysts is common knowledge, among which, imidazolium-based ILs are the most researched group. Studies have shown that the solubility of CO_2 in this group of ILs increases with increased pressure [20,21]. For example, Sudha and Khanna [22] in a molecular modeling of the solubility of CO_2 in several ILs reported the solubility of CO_2 in 1-butyl-3-methylimidazolium bromide, Bmim[Br], to be a 0.31 molar fraction. Once the absorption of CO_2 is a rate determining step in the SLGC with the $Ca(OH)_2$ suspension and CO_2 is very soluble in ILs, coupling an SLGC system with an IL will enhance the absorption of CO_2 and lead to higher conversion yields.

Previously, we conducted systematic research on a solid-gas carbonation (SGC) involving a dry $Ca(OH)_2$ system with a solid ionic liquid (SIL) and high-pressure/supercritical CO_2 and achieved the complete conversion of $Ca(OH)_2$ to $CaCO_3$ in 5 min [23,24]. The SIL served as the catalyst carrier and facilitated the effective dissolution of CO_2, leading to the syntheses of stoichiometric rhombohedral calcite phases [23,24]. In the current study, we report an SLGC involving a $Ca(OH)_2$ slurry system using an imidazolium-based IL in view of enhancing carbonation efficiency. The study searches for rapid carbonation for the production of nano-$CaCO_3$ and the capture of CO_2, since CO_2 capture is an important industrial application [23]. The IL was used in the current work as a reaction media to increase the solubility of CO_2 and to enhance the kinetics of the reaction in the liquid phase. Note that the IL is very miscible with water and can easily be recovered.

2. RESULTS AND DISCUSSION

Precipitation of $CaCO_3$ from aqueous carbonation of $Ca(OH)_2$ is an exothermic and complex process with an overall reaction equation as [5,7]:

$$Ca(OH)_{2(aq)} + CO_{2(aq)} \rightarrow CaCO_{3(s)} \downarrow + H_2O_{(l/v)} \quad (1)$$

Traditionally, SLGC produces calcite with a scalenohedral lattice {2 1 −1} [25,26] and is characterized by excess calcium species (Ca^{2+}) as a result of the limited solubility of CO_2 in water [27,28,29], leading to inadequate carbonate species (CO_3^{2-}) necessary for producing the stoichiometric rhombohedral phase [24,27]. Thus, any process that enhances the solubility of CO_2 and the dissolution of $Ca(OH)_2$ can positively affect the reaction rate and result in higher conversion yields. Therefore, increasing the solubility of CO_2 using an IL to facilitate the effective dissolution of $Ca(OH)_2$ will not only enhance carbonation yields, but also facilitate and produce the stoichiometric rhombohedral calcite phase [24,27].

2.1. Effect of Temperature

The effect of temperature on SLGC is well-known [7], and lower temperature is commonly adopted to increase the solubility of CO_2 in water. We investigated the effect of temperature (25.0, 30.0, 40.0, 50.0 and 60.0 °C) at atmospheric pressure (0.1 MPa) in 5 min with 0.5 g Bmim[Br]/5.0 g $Ca(OH)_2$ and 10 mL water. X-ray diffractometry (XRD) patterns of the produced samples and the corresponding conversions of $Ca(OH)_2$ at the different temperatures are shown in Figure 1. As can be seen, the conversion increased from 44.1% at 25.0 °C to only 45.9% at 40.0 °C and then decreased slightly to 37.8% at 60.0 °C, revealing that temperature did not have an obvious effect on the system and that the IL took effect. The highest conversion (47.7%) for the 0.1 MPa reactions, although not complete, is about the best value reported for identical systems with shorter reaction times in the literature, indicating the availability of the IL.

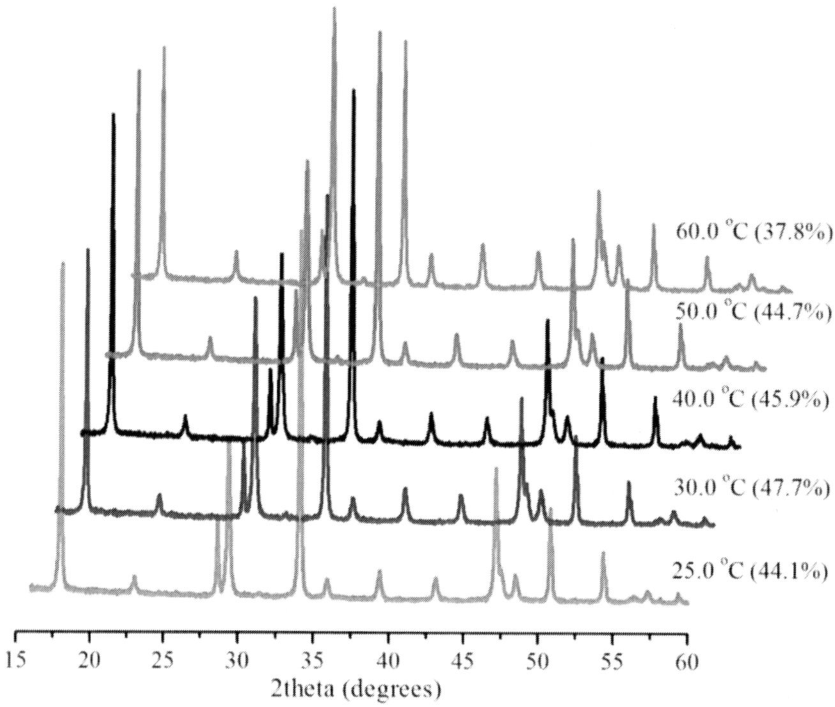

Figure 1. XRD (X-ray diffractometry) patterns and corresponding conversions (in brackets) for samples produced with 1-butyl-3-methylimidazolium bromide (Bmim[Br]) at various temperatures with 0.1 MPa in 5 min.

The samples were also characterized by scanning electron microscope (SEM) to investigate their morphology (typically shown in Figure 2). The SEM images revealed that irregular rhombohedral calcite (ICCD-PDF-2: 01-086-2334) particles were produced at the conditions investigated. Additionally, there was no unique and distinct difference between the particles produced.

Figure 2. SEM (scanning electron microscope) images of samples produced at (**a**) 40.0 °C and (**b**) 60.0 °C with 0.1 MPa in 5 min.

2.2. Effect of Pressure

Figure 3 shows the effect of pressure on the conversion at 30.0 °C in 5 min with 0.5 g Bmim[Br]/5.0 g $Ca(OH)_2$ and 10 mL water. At 0.1 MPa, there was incomplete conversion, because a 47.7% conversion was realized. This was evidenced by the presence of the characteristic peaks for both the starting material (at 18.0° and 34.2°) and the product (at 29.4°). A similar phenomenon was witnessed for the reaction at 1.0 MPa, resulting in a 64.7% conversion. However, at 5.0 MPa, the major peaks of the starting material almost disappeared, leading to pronounced major peaks of the product and substantial conversion (99.1%). Yet, a further increase of the pressure reduced the conversion slightly (Figure 3). The presence of the Bmim[Br] increased the conversions at 0.1 and 1.0 MPa (Figure 3). The quantity of dissolved CO_2 needed to produce CO_3^{2-}-species for carbonation is restricted by the low solubility of CO_2 in water at low pressures [5], but the Bmim[Br] enhanced the availability of the CO_3^{2-} species, due to the high solubility of CO_2 in the IL [22] and led to the high conversion yields.

Figure 4 shows typical SEM images of the samples produced at various pressures. The images confirmed the rhombohedral, but agglomerated, nature of the particles produced. A comparison of the particles produced at 5.0 MPa with Bmim[Br] (Figure 4c) and without Bmim[Br] (Figure 4d) revealed that the former sample consisted of the majority of less agglomerated and irregular, but smaller (≈60 nm) particles with a minority of bigger (≈300 nm) rhombohedral ones, while the latter sample had almost uniform, but highly agglomerated and bigger (≈100 nm) rhombohedral particles. Note that the conversion without the IL at 5.0 MPa and 30.0 °C in 5 min was 52.7%.

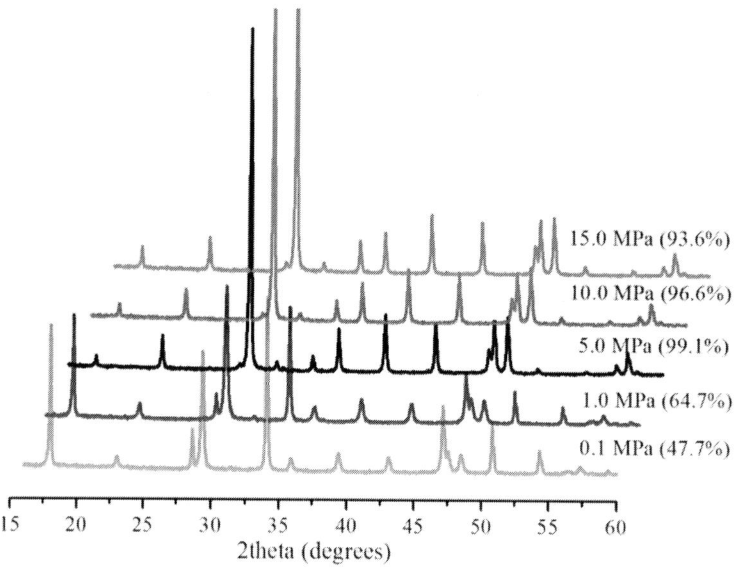

Figure 3. XRD patterns and corresponding conversions (in brackets) for samples produced with Bmim[Br] at various pressures at 30.0 °C in 5 min.

Figure 4. SEM images of samples produced with Bmim[Br] at (**a**) 0.1 MPa; (**b**) 1.0 MPa; (**c**) 5.0 MPa and without Bmim[Br] at (**d**) 5.0 MPa in 5 min at 30.0 °C.

2.3. Effect of Reaction Time

The extent of carbonation reaction is a function of time; hence, the effect of reaction time on the rate of conversion was investigated. Figure 5 shows the effect of reaction time on the conversion of $Ca(OH)_2$ with 0.5 g Bmim[Br]/5.0 g $Ca(OH)_2$and 10 mL water at 30.0 °C for 0.1 and 5.0 MPa. At 0.1 MPa, the slow reaction led to low conversion (26.3%) after 5 min without Bmim[Br]. The conversion increased to 34.6% in 30 min and 65.3% in 60 min without the IL. However, it was 47.7% in 5 min when the IL was introduced, but increased to 93.9% in 30 min and then 99.7% after 60 min. Obviously, the effect of the IL on the conversion at this low pressure was significant especially, within 30 min. The conversion increased from 34.6% in 30 min without the IL to 93.9% (≈59.3% increment) when the IL was added.

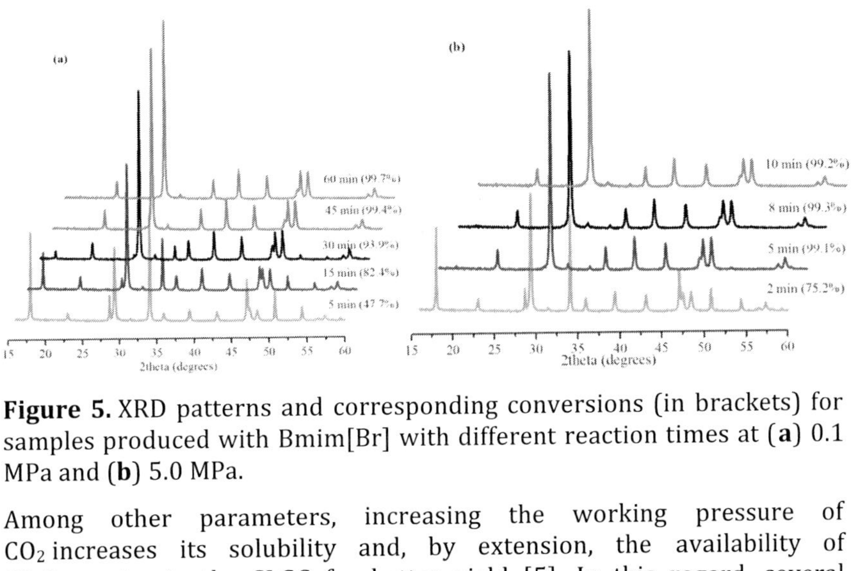

Figure 5. XRD patterns and corresponding conversions (in brackets) for samples produced with Bmim[Br] with different reaction times at (**a**) 0.1 MPa and (**b**) 5.0 MPa.

Among other parameters, increasing the working pressure of CO_2 increases its solubility and, by extension, the availability of CO_3^{2-} species in the SLGC for better yield [5]. In this regard, several experiments were carried out at 5.0 MPa with the Bmim[Br] to investigate the effect of time. For these cases, the reactions were very fast (almost complete conversions) within 5 min. Even so, when the reaction was monitored over a very short time (2 min), the positive influence of the IL was evident. The reaction resulted in a 75.2% conversion demonstrating the ability of the IL to facilitate faster reaction in the initial stages of the carbonation at medium pressure (5.0 MPa). Furthermore, a 47.7% conversion was obtained at 0.1 MPa in 5 min, whereas the conversion was 99.3% at 5.0 MPa, confirming the fast nature of the reaction in the presence of the IL at the medium pressure.

Representative SEM images for the samples produced at selected times are shown in Figure 6. Generally, and as expected, the samples produced with the Bmim[Br] exhibited well-defined particles (Figure 6a,c,e) compared to those produced without the IL (Figure 6b,d,f). Additionally, the effect of time on morphology was clearly visible. For instance, whereas the sample produced with the IL in 5 min consisted of a mixture of flower-like calcite (≈50 nm) and bundle-like hexagonal (inset) $Ca(OH)_2$ particles (≈50 × 900 nm; Figure 6a), those produced in 30 min showed a perfect mixture of smaller (≈50 nm) and larger (≈400 nm) rhombohedral calcites (Figure 6c). However, the SEM images (Figure 6) indicated the obvious increase of particle size following the reaction time. Note that Figure 4a and Figure 6a are different images of the same sample; Figure 6a was deliberately taken to show both $Ca(OH)_2$ and calcite particles, whereas Figure 4a showed only calcite particles.

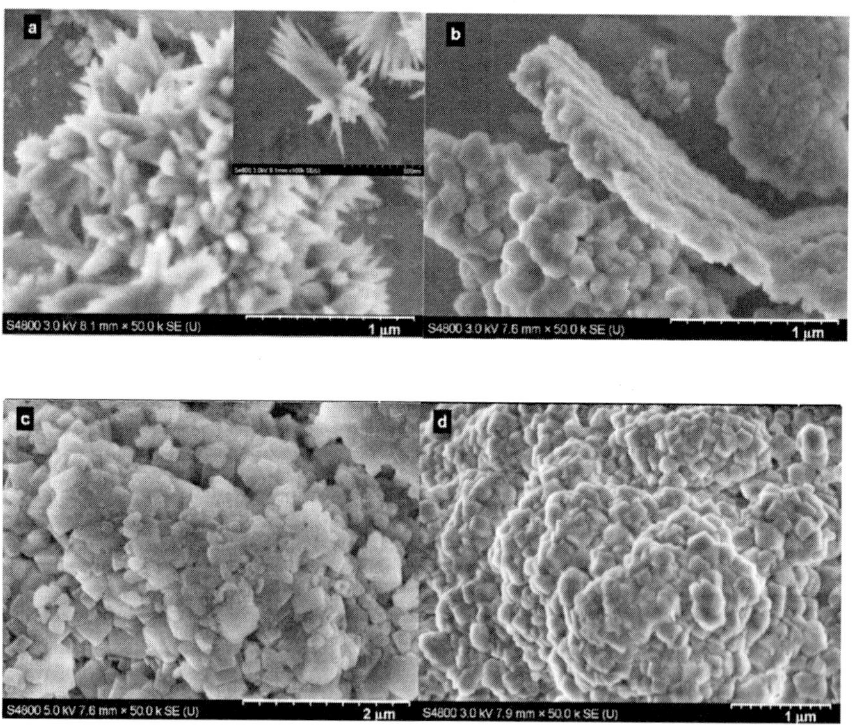

Figure 6. SEM images of the samples produced at 30.0 °C and 0.1 MPa. (**a**) with Bmim[Br] and (**b**) without Bmim[Br] in 5 min; (**c**) with Bmim[Br] and (**d**) without Bmim[Br] in 30 min; (**e**) with Bmim[Br] and (**f**) without Bmim[Br] in 60 min.

2.4. Effect of Amount of Water

Water is very vital for carbonation reactions, and the role of water has been reported accordingly [5,23]. The influence of the amount of water on conversion was thus investigated at 0.1 MPa and 30.0 °C in 5 min with 0.5 g Bmim[Br]/5.0 g $Ca(OH)_2$. For the zero-milliliter (0 mL) water, the IL was dried under vacuum for 36 h to remove traces of adsorbed water, and no additional water was added into the reactor during the reaction; so it was a dry carbonation (SGC) system [23]. The conversion results are shown in Figure 7, revealing increased conversion with an increasing amount of water. A conversion of 40.0% was obtained using 5.0 mL as opposed to 1.3% for the reaction without water (0 mL), but the conversion increased to 63.6% with 20.0 mL of water. Note that these were slurry systems and, as such, suspensions; hence the mentioning of the presence of water instead of concentrations or solutions.

2.5. Effect of Amount of Bmim[Br]

The presence of Bmim[Br] was shown to have a significant effect on the conversion at the low (0.1 MPa) and medium (5.0 MPa) pressures in 5 min (see Section 2.2 and Section 2.3). Therefore, we investigated the effect of the mass ratio of the IL to $Ca(OH)_2$ on the conversion at 0.1 MPa within 5 min at 30.0 °C. Figure 8 shows the conversion results of the samples produced, revealing increased conversion from 28.2% at 0.06 g IL/g $Ca(OH)_2$ to 94.4% at 1.0 g IL/g $Ca(OH)_2$. We are mindful of the comparatively expensive nature of ILs. However, about a 30% conversion of $Ca(OH)_2$ at 0.1 MPa with just 6% (0.06 g IL/1 g $Ca(OH)_2$) of the IL and a

99.1% conversion at 5.0 MPa with 10% (0.5 g IL/5 g $Ca(OH)_2$) of the IL in 5 min, perhaps, justify the use of the IL in terms of cost.

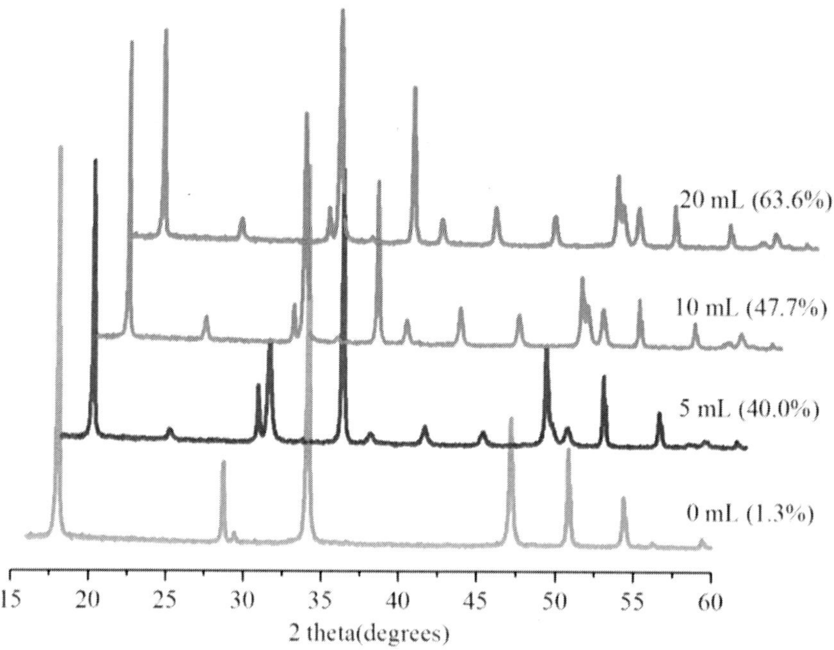

Figure 7. XRD patterns and corresponding conversions (in brackets) for samples produced with the Bmim[Br] and various amounts of water at 0.1 MPa and 30.0 °C in 5 min.

2.6. Significance of Bmim[Br] and the Mechanism

Enhanced dissolution of CO_2 is among the key parameters that control carbonation reactions, and for SLGC, enhancing the solubility of CO_2 in water is particularly important. As such, the use of the IL was expected to enhance the production of the CO_3^{2-} from the effective dissolution of CO_2 in the IL [22] and affect conversion yields. The influence of the IL on conversion was evident at lower pressure. For example, with 0.1 MPa at 30.0 °C in 5 min, the reaction without the IL gave a conversion of 26.3% against about 50% with the IL (Figure 8). Furthermore, the effect of the IL on the particle size and degree of agglomeration at the medium pressure (5.0 MPa) was noticeable. The sample produced with the IL was a mixture of smaller (≈60 nm) and bigger (≈300 nm) rhombohedral particles with less agglomeration (Figure 4c), whereas the particles synthesized without the IL were bigger rhombohedrals (≈100 nm) with more agglomeration

(Figure 4f). The IL also had a positive influence on the conversion (99.1%) in 5 min at the medium pressure, however; its availability was prominent when the reaction was monitored for 2 min (75.0% conversion), indicating that it induced a faster reaction within the initial stage of the carbonation. On the other hand, reaction with 0.10 g IL/g $Ca(OH)_2$ in 5 min resulted in a conversion of 47.7%. A decreased amount of the IL (0.06 g IL/g $Ca(OH)_2$ in 5 min), as expected, only worsened the conversion yield (28.2%), while an increased amount of the IL (1.0 g IL/g $Ca(OH)_2$ in 5 min) gave a conversion (94.4%) greater than 90.0%. Considering the cost implications of ILs, 1.0 g IL/g $Ca(OH)_2$ in 5 min for a 94.4% conversion may not be industrially prudent, although the IL may be recovered. Therefore, we suggest an optimum of 0.10 g IL/g $Ca(OH)_2$ (10% IL), but in 30 min, because a conversion of 93.9% was obtained in this case, even though a 99.7% conversion was realized in 60 min.

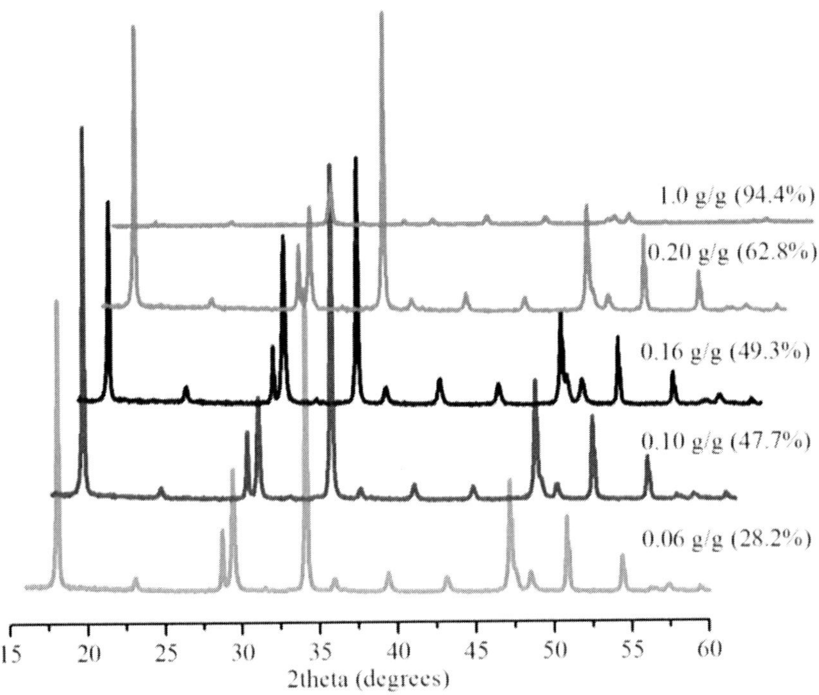

Figure 8. XRD patterns and corresponding conversions (in brackets) for samples produced with 10 mL water and various amounts of Bmim[Br] in 5 min.

The mechanism of the current SLGC system with Bmim[Br] is therefore similar to that of the SGC system with the solid ionic liquid previously reported [23,24]. Briefly, the Bmim[Br] absorbs water and CO_2 to initiate the reaction by forming hydrated layers. These layers promote the dissolution of $Ca(OH)_2$ to produce Ca^{2+} species and the conversion of CO_2 to CO_3^{2-} species. Because of the relatively high solubility of CO_2 in Bmim[Br], there is enough CO_3^{2-} avoiding the formation of the usual protective and passivation $CaCO_3$ layer, and this led to the formation of stoichiometric rhombohedral calcite particles.

2.7. Brunner-Emmet-Teller (BET) Surface Area and Thermogravimetric Analysis (TGA)

BET analysis was done to study the specific surface area and pore structure of the sample synthesized with the IL at 5.0 MPa and 30 °C in 5 min, since this gave almost complete conversion (99.1%). A surface area of 13.0 m^2/g was obtained. This result is comparable with values reported for SLGCs with similar conditions [4,30], but better than values reported for some SGCs [27]. The highest surface area for a similar system, but with a rotating disc reactor unit at ambient conditions, was reported to be between 20.4 and 23.3 m^2/g; however, the products were filtered through a 0.1-μm membrane [31].

The thermal stability of the samples produced at the lower and medium pressures was studied with a thermogravimetric analyzer. As shown in Figure 9, the total weight loss was 47.7% for the sample produced at 5.0 MPa, whilst a total weight loss of 34.6% was noticed for the sample produced at 0.1 MPa. The TGA results enabled the validation of the Rietveld whole pattern refinement of the XRD analysis (see Table 1). For this case, the weight loss due to the release of physically adsorbed water, the release of water from $Ca(OH)_2$ and the release of CO_2 from $CaCO_3$ as a result of thermal decomposition were recorded, and the weight percent of $CaCO_3$ in the sample was calculated with Equation (2).

$$CaCO_3\left[\text{wt }\%\right] = \frac{W_\text{L}}{W_\text{T}} \times \frac{M_{CaCO_3}}{n \times M_{CO_2}} \times 100 \qquad (2)$$

where W_T denotes the total weight of the sample (g), W_L denotes the weight loss (g), n denotes the number of moles of CO_2eliminated and M denotes molecular weight (g/mol).

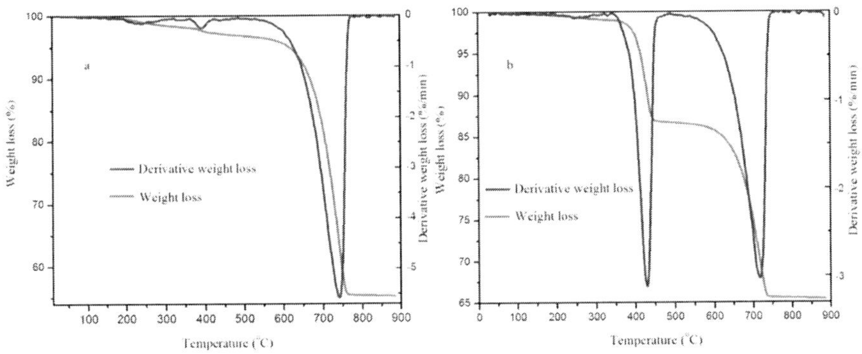

Figure 9. Thermogravimetric curves for the $CaCO_3$ produced: (**a**) 5.0 MPa and (**b**) 0.1 MPa in 5 min at 30.0 °C.

Table 1. Validation of XRD Rietveld refinements by TGA.

Sample	Rietveld (wt %)	TGA (wt %)	Difference (wt %)
5.0 MPa	99.1	96.2	2.9
0.1 MPa	47.7	49.7	2.0

The endothermic peak at about 740.6 °C for the sample produced at 5.0 MPa led to about 96.2% (mass) $CaCO_3$. Similarly, the endothermic peaks at about 428.6 °C attributed to decomposition of $Ca(OH)_2$ and at about 716.6 °C assigned to the decomposition of $CaCO_3$ for the sample produced at 0.1 MPa correspond to about 49.7% $CaCO_3$. These percentage values are in agreement with the results from the Rietveld refinements for the XRD patterns (Table 1).

3. EXPERIMENTAL SECTION

3.1. Materials

Analytical grade $Ca(OH)_2$ with 95% purity was purchased from Sinopharm Chemical and Reagents Co., Ltd, Shanghai, China. Carbon dioxide (99.99% purity) was sourced from Linde Gas, Xiamen Corporation Ltd., Xiamen, China, and 1-butyl-3-methylimidazolium bromide, Bmim[Br] with 99% purity was purchased from Shanghai Cheng Jie Chemical Co., Ltd., Shanghai, China.

3.2. Procedure

The setup for the process is shown in Figure 10, consisting of a CO_2 compressor (G447-400, Beijing HuiZhi M&E Facilities Co., Ltd., Beijing, China), a self-modified back pressure regulator (±0.1 MPa), a preheater, a check valve and a high pressure reactor (100 mL) with a maximum internal working pressure of 30 MPa, a temperature of 300 °C and outlet and inlet valves arranged, such that CO_2 could be bubbled straight into the bulk of the slurry and discharged immediately after the reaction. The reactor is equipped with a mechanical stirrer (about 500 rpm), thermocouple, a pressure meter and a temperature controller (±0.1 °C).

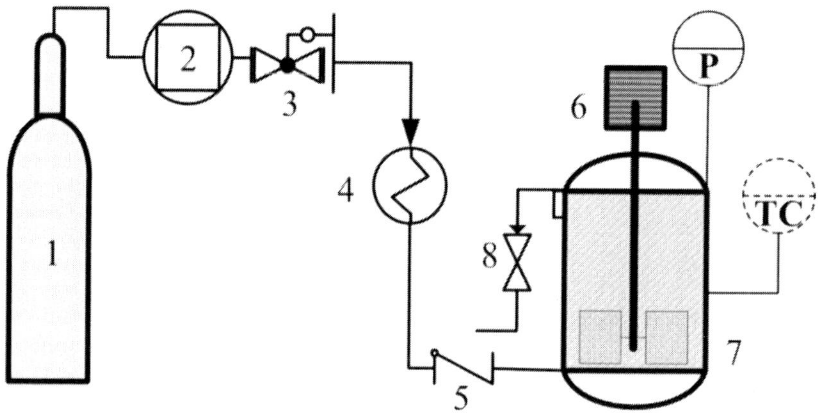

Figure 10. Illustration of the experimental setup: **1**, CO_2 cylinder; **2**, compressor; **3**, back pressure regulator; **4**, preheater; **5**, check valve; **6**, mechanical stirrer; **7**, reactor; **8**, outlet valve; TC, temperature controller; P, pressure indicator.

Typically, an amount of distilled water (mL) was added to 5.0 g of $Ca(OH)_2$ to produce a slurry. To this, Bmim[Br] of mass ratios of 0.5 g/5 g $Ca(OH)_2$ was loaded into the reaction vessel and heated under stirring (500 rpm) until thermal equilibrium (°C). CO_2 was fed into the vessel, brought to the desired pressure (MPa) and reacted for a set time (min). $CO_2$was deliberately released after the reaction; the product was collected, washed to recover the ionic liquid and dried in an oven (Jinhong, DZF-6020, Shanghai, China) at 100 °C for 12 h.

3.3. Characterization

The phase purity and quantity of the samples were analyzed with X-ray diffractometry (XRD), and the crystallite size was determined using Jade

5.0 from XRD patterns. Measurements of the XRD were carried out using X'Pert Pro (PANalytical B.V, Ea Almelo, The Netherlands) with a Cu anode operated at 40 kV, 30 mA and a wavelength (λ) of 1.540598 A. A continuous scan from 16° to 60° with a 2-theta step scan of 0.016711° and a step length of 10 s was employed. Calcite has a characteristic major peak at 29.41° corresponding to the {1 0 4} reflection plane, while $Ca(OH)_2$ has a major peak at 34.11°, matching the {0 0 1} reflection plane, followed by the peak at 18.08° [24]. In our case, however, the $Ca(OH)_2$ peak at 18.08° was the highest. Phase analysis and quantification was achieved by Rietveld whole pattern refinements using X'Pert HighScore plus (2.0) and the ICDD-PDF-2 database [24]. Specific surface area and pore size were determined by BET (Micrometric TRISTAR-3000, Micromeritics Instrument Corporation, Norcross, GA, USA), and isotherms were obtained by performing isothermal physical sorption with N_2 gas at 77.3 K. SEM (S4800, Hitachi, Tokyo, Japan) was employed to study the morphology of the samples by dispersing samples ultrasonically in absolute ethanol for 30 min and depositing on a silicon wafer. Thermal stability was studied by thermogravimetric analysis (NETZSCH TG 209 F1, Selb, Germany) under N_2 from 30 to 900 °C with a heating rate of 10 °C/min.

4. CONCLUSIONS

We investigated the effect of several variables on an SLG carbonation system using $Ca(OH)_2$ slurry with an imidazolium-based IL, Bmim[Br]. Based on the results, we conclude that: (1) the use of the IL positively influenced the carbonation reaction at relatively low pressure (0.1 MPa), such that with 1 g IL/1 g $Ca(OH)_2$, a conversion higher than 94.4% was achieved within 5 min; (2) the optimum condition (0.10 g IL/1 g $Ca(OH)_2$ in 30 min) at the lower pressure gave a conversion of 93.9%; (3) medium pressure (5.0 MPa) led to almost complete conversion (99.6%) in 5 min; (4) 5.0 mL of water or more was necessary for the rapid conversion of 5.0 g $Ca(OH)_2$, yet the removal of water from the system gave very low conversion; (5) the samples exhibited rhombohedral calcite structures with a high specific surface area; and (6) the amount of $CaCO_3$ obtained from the Rietveld refinement of XRD patterns was very close to that obtained from TG analysis.

ACKNOWLEDGMENTS

This work was supported by National Nature Science Foundation of China (NSFC) (No. 21276212) and Specialized Research Fund for the Doctoral Program of Higher Education (SRFDP) (No. 20100121110009).

REFERENCES

1. Gomez-Villalba, L.S.; Lopez-Arce, P.; de Buergo, M.A.; Fort, R. Atomic defects and their relationship to aragonite-calcite transformation in portlandite nanocrystal carbonation. *Cryst. Growth Des.* **2012**, *12*, 4844–4852.
2. Knez, S.; Klinar, D.; Golob, J. Stabilization of PCC dispersions prepared directly in the mother-liquid after synthesis through the carbonation of (hydrated) lime. *Chem. Eng. Sci.* **2006**, *61*, 5867–5880.
3. Montes-Hernandez, G.; Daval, D.; Chiriac, R.; Renard, F. Growth of nanosized calcite through gas-solid carbonation of nanosized portlandite under anisobaric conditions. *Cryst. Growth Des.* **2010**, *10*, 4823–4830.
4. Ibrahim, A.R.; Zhang, X.L.; Hong, Y.Z.; Wang, H.T.; Li, J. Instantaneous solid-liquid-gas carbonation of $Ca(OH)_2$ and chameleonic phase transformation in CO_2-expanded solution. *Cryst. Growth Des.* **2014**, *14*, 2733–2741.
5. Domingo, C.; Loste, E.; Gomez-Morales, J.; Garcia-Carmona, J.; Fraile, J. Calcite precipitation by a high-pressure CO_2carbonation route. *J. Supercrit. Fluids* **2006**, *36*, 202–215.
6. Ciullo, P.A. *Industrial Minerals and Their Uses: A Handbook and Formulary*; Noyes Publications: Westwood, NJ, USA, 1996; pp. 35–37.
7. Montes-Hernandez, G.; Renard, F.; Geoffroy, N.; Charlet, G.; Pironon, J. Calcite precipitation from CO_2–H_2O–$Ca(OH)_2$ slurry under high pressure of CO_2. *J. Cryst. Growth* **2007**, *308*, 228–236.
8. Harja, M.; Cimpeanu, C.; Bucur, R.D. The influence of hydrodynamic conditions on the synthesis of ultra-thin calcium carbonate. *J. Food Agric. Environ.* **2012**, *10*, 1191–1195.
9. Juvekar, V.A.; Sharma, M.M. Absorption of CO_2 in a suspension of lime. *Chem. Eng. Sci.* **1973**, *28*, 825–837.
10. Sada, E.; Kumazawa, H.; Lee, C.; Fujiwara, N. Gas-liquid mass transfer characteristics in a bubble column with suspended sparingly soluble fine particles. *Ind. Eng. Process Des. Dev.* **1985**, *24*, 255–261.
11. Xiang, L.; Xiang, Y.; Wen, Y.; Wei, F. Formation of $CaCO_3$ nanoparticles in the presence of terpineol. *Mater. Lett.* **2004**, *58*, 959–965.
12. Yagi, H.; Iwazawa, A.; Sonobe, R.; Matsubara, T.; Hikita, H. Crystallization of calcium carbonate accompanying chemical absorption. *Ind. Eng. Chem. Fundam.* **1984**, *23*, 153–158.
13. Meldrum, F.C.; Hyde, S.T. Morphological influence of magnesium and organic additives on the precipitation of calcite. *J. Cryst. Growth* **2001**, *231*, 544–558.
14. Qi, L.; Li, J.; Ma, J. Biomimetic morphogenesis of calcium carbonate in mixed solutions of surfactants and double-hydrophilic block copolymers. *Adv. Mater.* **2002**, *14*, 300–303.
15. Colfen, H.; Qi, L. A systematic examination of the morphogenesis of calcium carbonate in the presence of a double-hydrophilic block copolymer. *Chem. A Eur. J.* **2001**, *7*, 106–116.

16. Rudloff, J.; Antonietti, M.; Colfen, H.; Pretula, J.; Kaluzynski, K.; Penczek, S. Double-hydrophilic block copolymers with monophosphate ester moieties as crystal growth modifiers of $CaCO_3$. *Macromol. Chem. Phys.* **2002**, *203*, 627–635.
17. Mann, S.; Heywood, B.R.; Rajam, S.; Birchall, J.D. Controlled crystallisation of $CaCO_3$ under stearic acid monolayers. *Nature* **1988**, *334*, 692–965.
18. Naka, K.; Huang, S.C.; Chujo, Y. Formation of stable vaterite with poly(acrylic acid) by the delayed addition method. *Langmuir* **2006**, *22*, 7760–7767.
19. Aki, S.N.V.K.; Mellein, B.R.; Saurer, E.M.; Brennecke, J.F. High-pressure phase behavior of carbon dioxide with imidazolium-based ionic liquids. *J. Phys. Chem. B* **2004**, *108*, 20355–20365.
20. Anthony, J.L.; Maginn, E.J.; Brennecke, J.F. Solution thermodynamics of imidazolium-based ionic liquids and water. *J. Phys. Chem. B* **2001**, *105*, 10942–10949.
21. Cadena, C.; Anthony, J.L.; Shah, J.K.; Morrow, T.I.; Brennecke, J.F.; Maginn, E.J. Why is CO_2 so soluble in imidazolium-based ionic liquids? *J. Am. Chem. Soc.* **2004**, *126*, 5300–5308.
22. Sudha, S.Y.; Khanna, A. Evaluation of CO_2-ionic liquid systems through molecular modeling. *World Acad. Sci. Eng. Technol.* **2009**, *57*, 539–542.
23. Ibrahim, A.-R.; Vuningoma, J.B.; Hu, X.H.; Gong, Y.; Dan, H.; Hong, Y.Z.; Wang, H.T.; Li, J. High-pressure gas-solid carbonation route coupled with a solid ionic liquid for rapid synthesis of rhombohedral calcite. *J. Supercrit. Fluids***2012**, *72*, 78–83.
24. Ibrahim, A.-R.; Gong, Y.; Hu, X.H.; Hong, Y.Z.; Su, Y.Z.; Wang, H.T.; Li, J. Solid-gas carbonation coupled with solid ionic liquid for synthesis of $CaCO_3$: Performance, polymorphic control and self-catalytic kinetics. *Ind. Eng. Chem. Res.***2013**, *52*, 9515–9524.
25. Lopez-Periago, A.M.; Pacciani, R.; Garcia-Gonzalez, C.; Vega, L.F.; Domingo, C. A breakthrough technique for the preparation of high-yield precipitated calcium carbonate. *J. Supercrit. Fluids* **2010**, *52*, 298–305.
26. Chen, P.C.; Tai, C.Y.; Lee, K.C. Morphology and growth rate of calcium carbonate crystals in a gas-liquid-solid reactive crystallizer. *Chem. Eng. Sci.* **1997**, *52*, 4171–4177.
27. Montes-Hernandez, G.; Pommerol, A.; Renard, F.; Beck, P.; Quirico, E.; Brissaud, O. *In situ* kinetic measurements of gas-solid carbonation of $Ca(OH)_2$ by using an infrared microscope coupled to a reaction cell. *Chem. Eng. J.* **2010**, *161*, 250–256.
28. Beruto, D.T.; Botter, R. Liquid-like H_2O adsorption layers to catalyse the $Ca(OH)_2/CO_2$ solid-gas reaction and to form a non-protective solid product layer at 20 °C. *J. Eur. Ceram. Soc.* **2000**, *20*, 497–503.
29. Dickinson, S.R.; Henderson, G.E.; McGrath, K.M. Controlling the kinetic *versus* thermodynamic crystallisation of calcium carbonate. *J. Cryst. Growth* **2002**, *244*, 369–378.
30. Lopez-Periago, A.M.; Pacciani, R.; Vega, L.F.; Domingo, C. Monitoring the effect of mineral precursor, fluid phase CO_2-H_2O composition, and stirring on $CaCO_3$ crystallization in a supercritical-ultrasound carbonation proces. *Cryst. Growth Des.* **2011**, *11*, 5324–5332.
31. Kedra-Krolik, K.; Gierycz, P. Precipitation of nano-structured calcite in a controlled multiphase process. *J. Cryst. Growth* **2009**, *311*, 3674–3681.

CHAPTER 10

Tribological Properties of Ionic Liquids

Yuriko Kondo, Tahahiro Koyama and Shinya Sasaki

[1]*Department of Mechanical Engineering, Tokyo University of Science, Japan*

1. INTRODUCTION

An ionic liquid is a liquid salt consisting of anions and cations. Since Wilkes [1] synthesized a low melting point ionic liquid that is stable in air in 1992, there has been active fundamental research and engineering development directed at applications. In the field of tribology, where application is focused toward lubricants, the number of related papers [3]-[28] has increased every year since Liu et al. [2] published their research results in 2001. As a lubricant, ionic liquids are characterized by an extremely low vapor pressure, high thermal stability, and high ion conductivity. Based on these features, there are high expectations for the development of new ionic liquid applications as lubricants [29]-[37] in extreme environments, such as high temperatures [7]-[9] and vacuum [10]-[12], where the use of conventional lubricants is limited.

A lubricant must be used under various sliding conditions. Therefore, choosing an ionic liquid suitable for its application is necessary. Ionic liquids are also called designer's liquids because various characteristics can be created by different combinations of anions and cations. Owing to this wide variety of choices, it is important to understand the characteristics of the ionic liquid targeted for use as a lubricant. However, at present details of the lubrication mechanism are still not clearly understood, so a trial-and-error approach is inevitable for selecting ionic liquids.

This article introduces the advantages and issues related to the basic characteristics of ionic liquids as a lubricant, and it describes their future prospects, mainly for applications in vacuum.

2. IONIC LIQUID AS A LUBRICANT

2.1. Required Characteristics Of A Lubricant

In general, the coefficient of friction is highest under dry friction conditions, where no lubricant is used. The use of a lubricant controls this type of friction conditions and may suppress damage to surfaces. Fig. 1 shows the Stribeck curve used in tribology. This figure shows the transition of the lubricating conditions of the sliding surfaces: the friction coefficient is the vertical axis, and the bearing characteristic number (fluid viscosity $\eta \times$ sliding speed V/average surface pressure P) is the horizontal axis. As the bearing characteristic number increases, the sliding condition transits from boundary lubrication to mixed lubrication to hydrodynamic lubrication. In dry friction, solid surfaces in the real contact area are in direct contact and adhesion; a high frictional force occurs from the shearing resistance of the adhesive parts. In boundary lubrication, an absorbed layer or tribo-chemical reaction layer lies between surfaces in the real contact area to control the adhesion between solid interfaces and to reduce the shearing resistance; this in turn decreases the coefficient of friction relative to dry friction. In mixed lubrication, the coefficient of friction further decreases because the fluid film of small shearing resistance bears a portion of the load, and the real contact area are decreased. In hydrodynamic lubrication, the coefficient of friction is at its minimum because the real contact area disappears owing to the fluid film bearing the entire load, and the frictional force is due to only the viscous resistance of the fluid. However, when the fluid viscosity and sliding speed further increase the bearing characteristic number, the coefficient of friction rises owing to the increase in viscous resistance.

Because most ionic liquids are in liquid state near room temperature, they can be used as a base oil for the lubricant. In addition, their application as a grease [38] by mixing the lubricant with additives [4][5][16][22] and thickeners has also been examined. The required properties of the ionic liquid differ depending upon the lubricating state. Specifically, in boundary and mixed lubrication, a firm adsorption layer must be formed on the sliding surfaces to control adhesion in the real contact area. On the other hand, in hydrodynamic lubrication, the formation of a thick liquid film must cause load burden even in the region of high surface pressure and low speed; as a result, the viscosity characteristic becomes an important factor. However, because an excessive tribo-chemical reaction can increase wear and cause corrosion—thus increasing friction coefficient—the lubricating efficiency balance must always be considered.

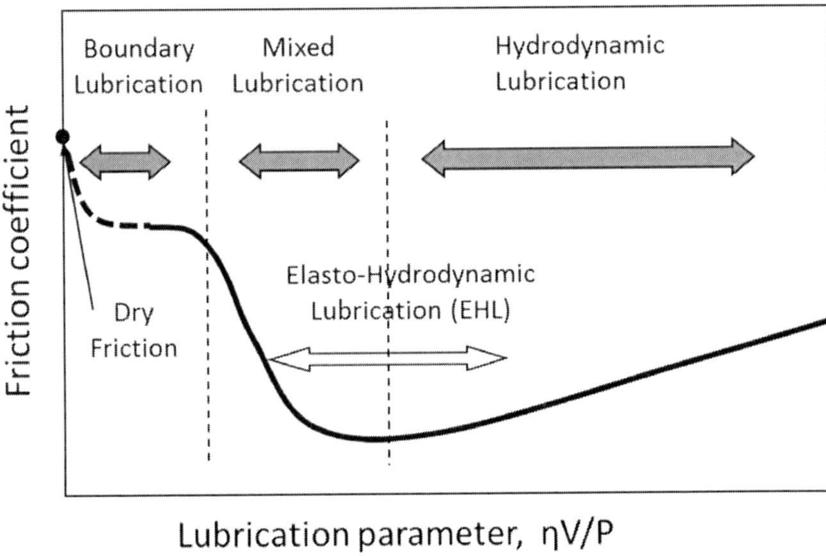

FIGURE 1. Schematic of the stribeck curve; the friction coefficient as a function of the lubrication parameter: $\eta V/P$. In this formula, η is the fluid viscosity, V is the relative speed of the surfaces, and P is the load on the interface per unit bearing width.

2.2. Hydrodynamic Lubrication Capacity Of Ionic Liquid

To decrease friction and control damage to the sliding surfaces, hydrodynamic lubrication without solid contact is desired. For hydrodynamic lubrication, viscosity bears an important role with regard to the lubricant quality. An ionic liquid is in liquid state at room temperature, and its viscosity largely depends on the combination of anions and cations; it can be as thick as malt syrup or as thin as water. The viscosity of liquids decreases when the temperature rises; ionic liquids are no exception. For lubricants, the relationship between the temperature and viscosity is expressed by the viscosity index (VI). In general, this index should show a small variation in viscosity. The VI of ionic liquid is higher than that of general minerals and synthetic oils (poly-α-olefin, PAO), as shown in Table 1. Fig. 2 shows the measured results for the temperature–viscosity relation of two ionic liquid mixtures made from the same cation ([BMIM][TFSI] and [BMIM][PF6]); this verified that different mixing ratios correspond to different viscosities [39]. This indicates the possibility of arbitrarily adjusting the viscosity by mixing several types of ionic liquids. Because of the advantages of low vapor pressure and high thermal stability, ionic liquids are especially suitable for hydrodynamic lubrication applications in special environments, such as high temperatures and vacuums. Applications in bearings under dynamic and static pressure are being examined [40].

For lubricant and grease used in the elasto-hydrodynamic lubrication (EHL) state, such as for rotating bearings and gears, the dominant physical property influencing lubrication capability is the viscosity under high pressure, which can be up to several gigapascals. Ohno et al. [41][42] measured the high-pressure viscosity of a methylimidazole-type ionic liquid and reported the crystallization behavior of molecular characteristic crystals to be similar to liquid crystal under high pressure. Regarding the physical properties, such as the high-pressure viscosity index of ionic liquid, more data are expected to be reported in the future. However, behavior that causes hindrances in the EHL state, such as that reported by Ohno et al. [43], has not been reported.

2.3. Boundary Lubrication Capacity Of Ionic Liquid

Most studies on the special tribological qualities of ionic liquids have been on the boundary lubricating capacity. The ionic liquids examined in these studies mainly use an imidazole derivative as the cation and fluorine as the halogen element, such as tetrafluoroborate [BF4], hexafluorophosphate [PF6], and bis(trifluoromethanesulphonyl)imide [TFSI], for the anion. Fig. 3 shows the results of an investigation into the lubrication capability of different ionic liquids. Bearing steel balls and disks were used as sliding materials. The conditions for friction evaluation were as follows: temperature = 50°C, load = 50 N, reciprocating frequency = 50 Hz, amplitude = 1 mm, and friction time = 60 min. Two types of ionic liquid ([BMIM][TCC] and [EMIM][DCN]) do not include halogen. On the other hand, ionic liquids containing halogen exhibit low friction and wear and show good boundary lubrication properties. When ionic liquid containing halogen is used for lubrication, metal fluoride forms on friction surfaces by a tribochemical reaction; because this reaction product operates as a boundary lubricating layer, satisfactory lubricity is shown.

On the other hand, ionic liquid containing a halogen such as fluorine has been known to cause corrosion in steel [16]-[22][44][45][46], aluminum alloy [17][21][22][45], bronze [45][46], and titanium alloy [9] sliding materials. The cause of corrosion has been reported to be the formation of hydrogen fluoride due to the decomposition of the ionic liquid; this is largely due to water being mixed into the ionic liquid as an impurity and participating in the reaction [21]. Decomposition and corrosion reactions of ionic liquid happen even in a static environment. However, it is more marked in a sliding environment. Because water from the surrounding atmosphere is mixed into the ionic liquid owing to enhancement by friction, metal fluoride formed on the friction surfaces is believed to further react with water by tribo-chemical reactions to generate hydrogen fluoride. Thus, corrosion occurs after friction. Fig. 4 shows the change in appearance of the friction surface for steel bearings after reciprocating sliding between the balls and disks, by using the hydrophobic ionic liquid [PP13][TFSI] as the lubricant in air at 50% relative humidity. Immediately after the rubbing

TABLE 1 Viscosity properties of ionic liquids and base oils *1: N-Methyl-N-propylpiperidinium bis(trifluoromethanesulfonyl)imide *2: Trihexyl (tetradecyl)phosphonium bis(trifluoromethylsolfonyl)imide *3: 1-Butyl-3-methylimidazolium tetrafluoroborate *4: 1-Butyl-3-methylimidazolium hexaflu orophosphate *5: 1-Butyl-3-methylimidazolium bis(trifluor omethanesulfonyl) imide *6: 1-Butyl-3-methylimidazolium iodide *7: 1-Ethyl-3- methylimidazolium dicyanamide *8: 1-Butyl-3-methylimidazolium tricyanomethane

Ionic liquids & Base oils		Kinematic viscosity [mm^2/s]		Viscosity Index VI
		40°C	100°C	
[PP13][TFSI] *1	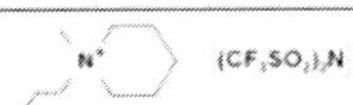	52.5	10.0	181
[P(h3)t][TFSI] *2		118	15.4	137
[BMIM][BF4] *3		44.5	9.01	189
[BMIM][PF6] *4		81.0	13.6	172
[BMIM][TFSI] *5	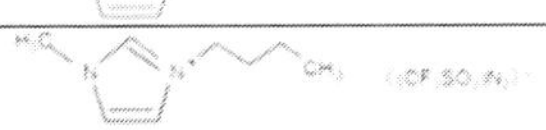	29.0	6.88	211
[BMIM][I] *6		203	21.0	122
[EMIM][DCN] *7	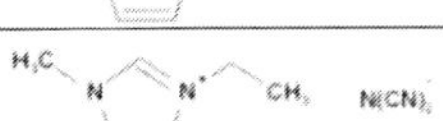	8.6	3.01	252
[BMIM][TCC] *8	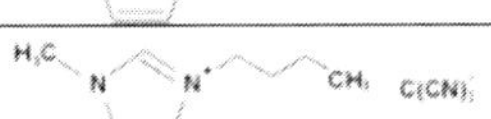	14.6	3.7	146
Synthetic oil (PAO 4)		16.8	3.90	129
Synthetic oil (PAO 6)		30.5	5.79	135
Mineral base oil		30.1	5.28	106

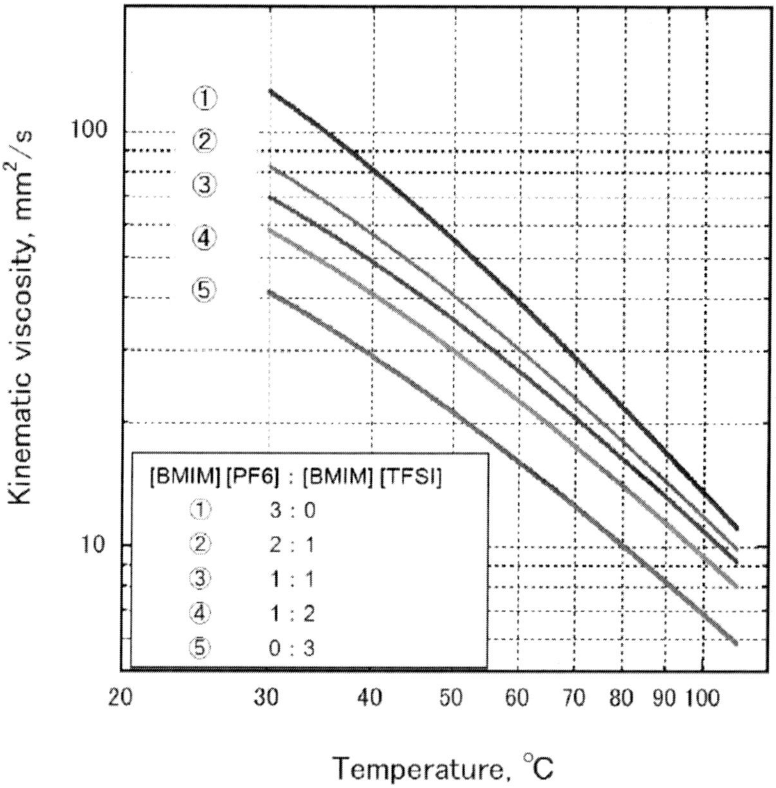

FIGURE 2. Viscosity property of mixed ionic liquids as a function of temperature

test, no remarkable corrosion was seen. However, after exposure to air for 24 h, a color change was observed for all parts touched by the ionic liquid. SEM-EDX analysis verified the composition of the corrosion product, containing mainly fluorine and oxygen, in the surface marked with pit-shaped corrosion [47]. Even after the [PP13][TFSI] was applied to the bearing steel surface exposed to air for 1 week, the occurrence of corrosion could not be verified. Therefore, friction is thought to promote the corrosion reaction and the decomposition of ionic liquid.

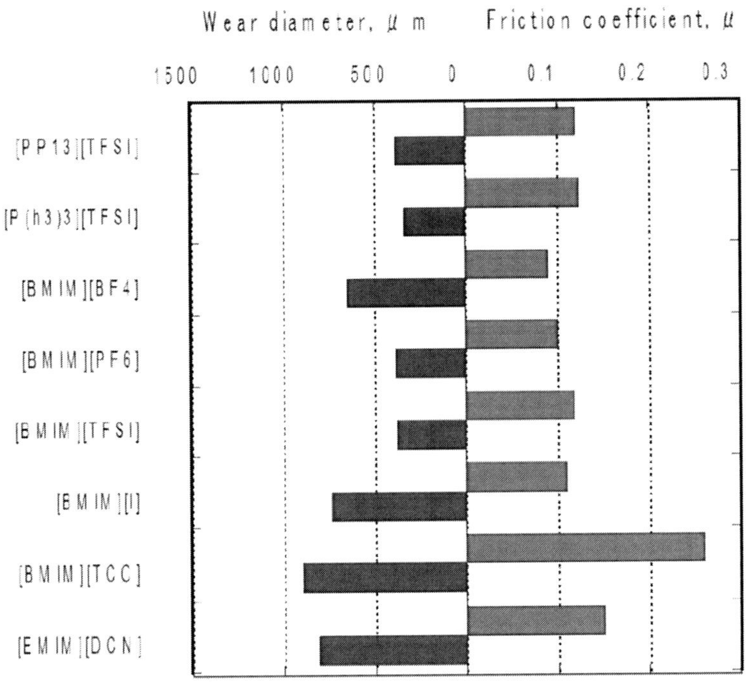

FIGURE 3. Lubricity of each ionic liquid for steel/steel sliding

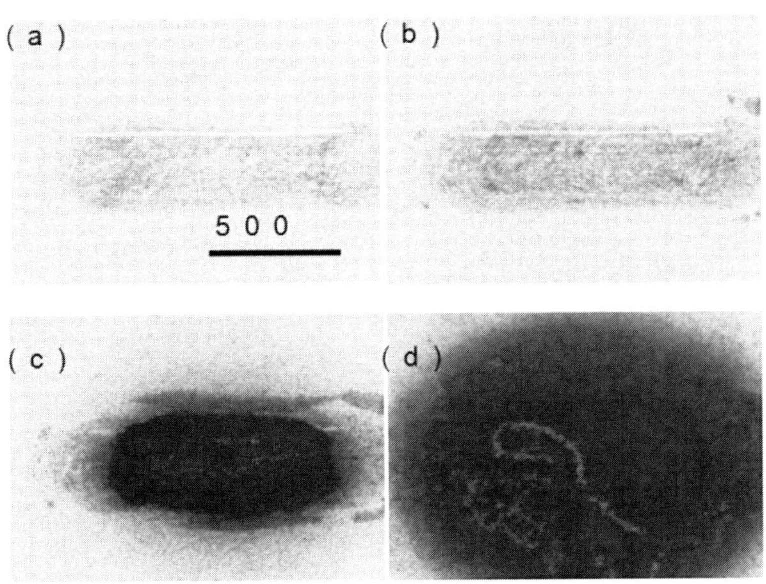

FIGURE 4. of corrosion on disk specimen after sliding test with [PP13][TFSI] at 50°C in air:(a)0.1 h, (b)1.0 h, (c)8.0h,(d) 24 h

2.4. Solutions To Corrosion Of Ionic Liquid

Although ionic liquids containing halogen have superior capacity in terms of boundary lubrication, they also have the problem of corrosion. There are three solutions to prevent corrosion.

2.4.1. Control Of Atmospheric Air

Based on the discovery of the relation between corrosion and water contamination, if an ionic liquid is used in an environment where it is hard to mix water with a hydrophobic ionic liquid having a low impurity concentration, the corrosion reaction can be controlled. Fig. 5 shows the friction and wear characteristics of the halogen-containing hydrophobic ionic liquid [PP13][TFSI] in air and in dry nitrogen [47]. In a dry nitrogen atmosphere, the coefficient of friction was stable and low even at 200°C, and the wear in this atmosphere was less in comparison with that in air. In addition, further corrosion was not observed on the specimen rubbed in dry nitrogen and left in air after the rubbing test, as shown in Fig. 6. From this, it is possible to prevent the corrosion in environments such as a vacuum, where there are almost no occurrences of mixture with water.

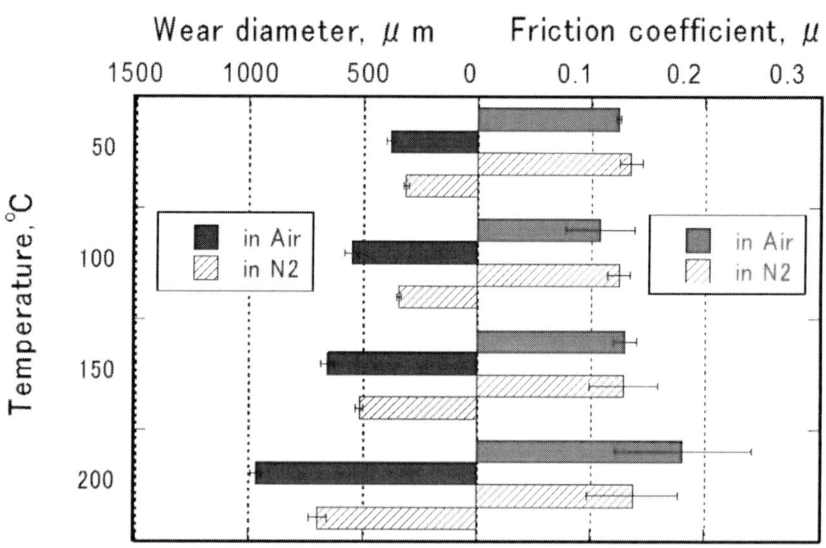

FIGURE 5. Comparison of friction and wear behavior of [PP13][TFSI] in air and in dry-nitrogen atmosphere

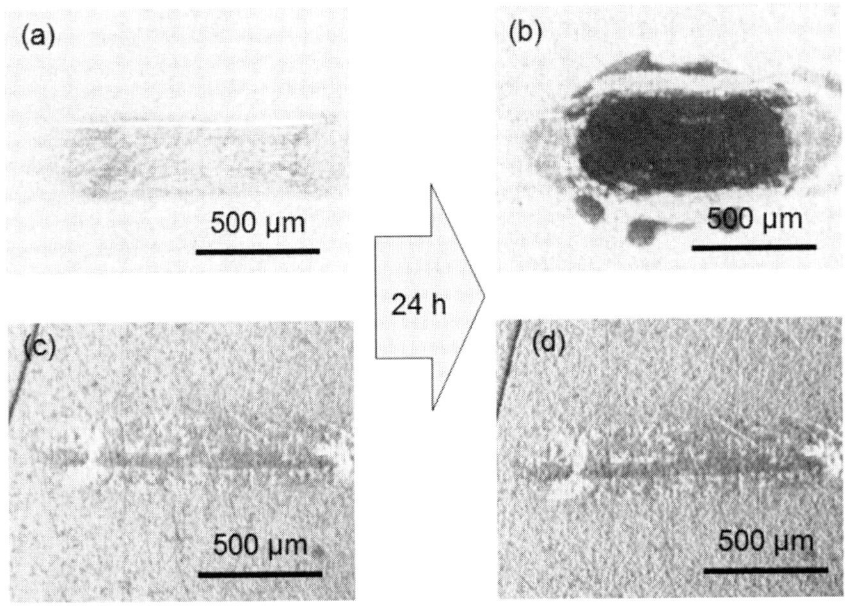

FIGURE 6. Optical micro-images of worn disk surfaces under lubrication with [PP13][TFSI] in air ((a),(b)) and dry-nitrogen ((c),(d)).(a) after 0.1 h, (b) after 24 h, (c) after 0.1 h and (d) after 24 h.

2.4.2. Control Of Metal Fluoride Formation Reaction

To control the occurrence of hydrogen halide, which causes corrosion, methods to form a protective film over the frictional surface to prevent excessive reaction between the halogen and metal from taking place have been considered. Fig. 3 compares the effect of the presence of phosphorus on friction and wear. [BMIM][PF6] and [BMIM][BF4] have the same cation but different anions; the former has phosphorus, whereas the latter does not. [PP13][TFSI] and [P(h3)3][TFSI] have the same anion but different cations: the former has phosphorus and the latter does not. Although the ionic liquids containing phosphorus have a somewhat higher coefficient of friction, they have substantially lower wear. Phosphorus in the ionic liquid is suspected to react with the frictional surface to form a phosphoric acid compound layer that is superior in wear resistance [19][23]-[25]; also, the formation of metal fluoride, which causes the occurrence of hydrogen fluoride, was controlled. For an ionic liquid containing phosphorus as a lubricant, because the advance of corrosion in a specimen left in air after rubbing was not observed, the suppression effect is believed to be a post-rubbing phenomenon.

2.4.3. Halogen-Free Ionic Liquid

To completely remove the corrosion reaction that originates from halogen, halogen-free ionic liquids should be selected. However, as shown in Fig. 3, their boundary lubrication ability is generally inferior to that of halogen-containing ionic liquids. However, when comparing halogen-free ionic liquids [BMIM][TCC] and [BMIM][BCN], the difference in anion results in a difference in lubricity. The discovery of a halogen-free ionic liquid with good boundary lubrication ability that does not depend on the formation of a metal halide layer is possible [48].

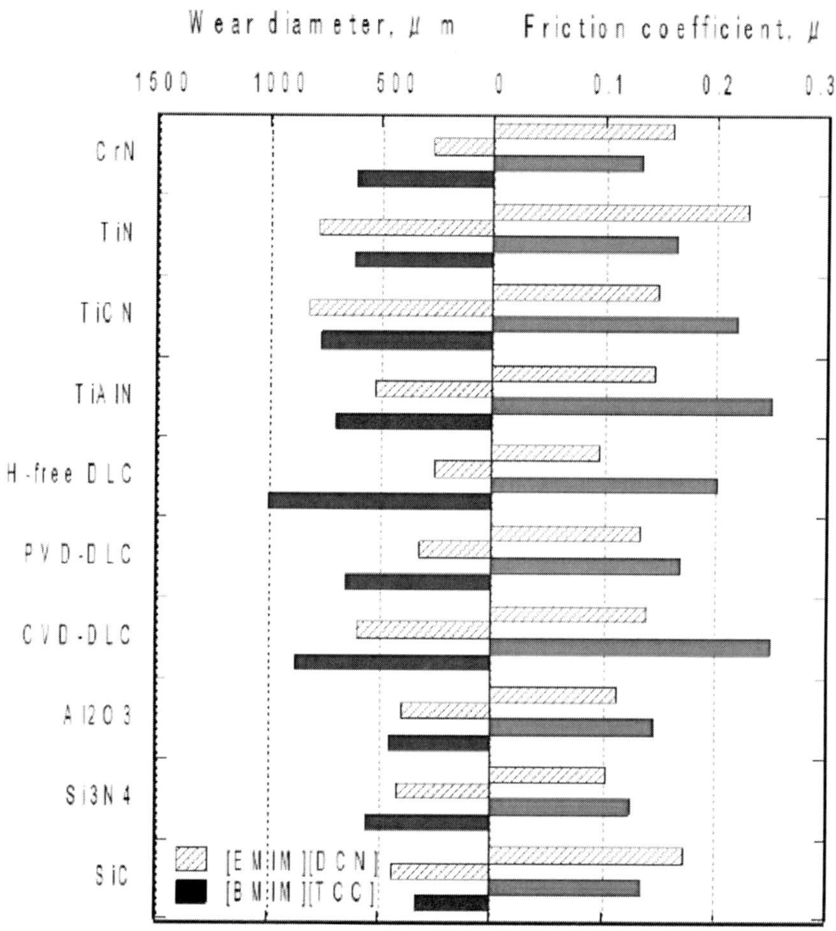

FIGURE 7. Friction and wear properties of hard-coatings and sintered-ceramics under lubrication with [EMIM][DCN] and [BMIM][TCC].

Fig. 7 shows the boundary lubricity of halogen-free ionic liquids corresponding to various wear resisting materials [49]. Although the lubrication ability of the two types of ionic liquid varied depending on the sliding materials, some samples showed good lubrication properties such as the combination of hydrogen-free diamond-like carbon (H-free DLC) and [BMIM][DCN]. Details on the lubricating mechanism of halogen-free ionic liquid are a future topic, but the application areas of halogen-free ionic liquid are expected to be broadened by selecting the combination of hydrogen-free diamond-like carbon with appropriate sliding materials.

3. APPLICATION AS LUBRICANT IN A VACUUM

3.1. Lubrication Properties In A Vacuum

In a vacuum, especially for machines in outer space, a lubricant with low vapor pressure and temperature stability is required owing to exposure to a severe temperature environment [38][43][50]. At present, perfluoropolyether (PFPE) and multiple-alkylated cycropentane (MAC) or grease that designates these as the base oil are used as the liquid lubricants in outer space because of their low vapor pressure. The lubricity of these lubricants is known to be improved by adding extreme pressure agents and the like, but issues such as evaporation of the additive and a decrease in the permanent viscosity of the base oil [43] remain to be resolved. As a non-additive base oil, ionic liquid has been reported to be superior to PFPE and MAC in terms of lubricity. In addition, as a grease, ionic-liquid-type grease has superior boundary lubrication capacity in a vacuum compared to systems of PFPE and MAC [38][50]. Regarding research on radiation for utilization in space, ionic liquid has been verified to have radiation resistance [38].

On the other hand, in industrial vacuum equipment, inevitable atmospheric release exists on sliding surfaces in most cases. Thus, extra attention must be paid to corrosion when using a halogen-containing ionic liquid. Furthermore, regarding rust prevention due to exposure to atmosphere, although the adsorption-type rust-preventive agent has been reported to be effective, these results were collected under static conditions, and the effect under sliding is unclear.

3.2. Occurrence Of Out-Gassing

With regard to out-gassing in a vacuum, the occurrence condition, gas type, and allowed quantity differ depending on the application of the vacuum. In

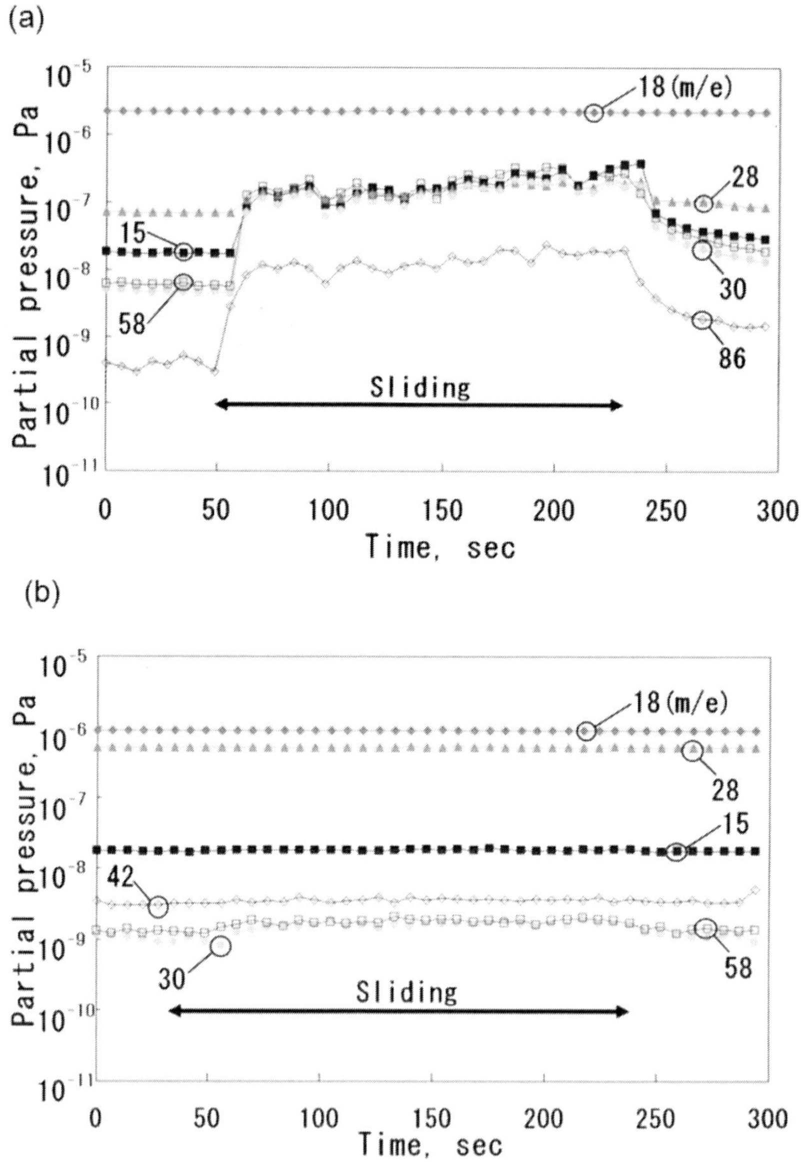

FIGURE 8. Partial pressure changes of generated gas species from ionic liquid during sliding under vacuum condition. (a) Titanium disk, (b) Steel disk

the case of materials for outer space equipment, measurement of the outgas (ASTM E595-93) requires a vacuum below 7 × 10-3 Pa for the sample and collector plate, a sample temperature of 125°C, and a collector temperature of 25°C (kept for 24 h) to calculate the loss mass ratio (mass change before and after test) (TML, total mass loss) and reagglutination material ratio of the collector plate (CVCM, collected volatile condensable materials) [51]. Ionic liquid based greases were developed and verified that the TML and CVCM are below 1.0% and 0.1% respectively [38]; this satisfies NASA's recommended values. In addition, the quantity of outgas that occurs during rubbing in an ionic liquid has been reported to be less than that in PFPE and MAC [50]. Thus, the outgas quality of an ionic liquid as a lubricant for space equipment applications can be concluded to be superior.

On the other hand, in some semiconductor processes, outgas is a contamination that can barely be tolerated. In such cases, attention must be paid to selecting the ionic liquid and sliding material. Fig. 8 shows the change in partial pressure in a vacuum during a sliding test between a SiC pin and a titanium or iron disk using N,N,N-trimethyl-N-propylammonium bis(trifluoromethanesulfonyl)imide (TMPA TFSI) as the lubricant [11]. When the iron was rubbed, a slight partial pressure rise at just m/e = 30 and 58 was verified. However, in the case of titanium, a clear rise in partial pressure was seen at m/e = 15, 30, 58, and 86 upon friction. Similar out-gassing results were seen in friction tests using aluminum. During the tribo-chemical reaction between the metal sliding surface and ionic liquid to form a metal fluoride, part of the ionic liquid decomposition is believed to release outgas into the vacuum. To suppress friction associated with out-gassing, a combination with a sliding surface having low chemical activity, such as DLC, is effective [52].

4. CONCLUSION

Ionic liquids have various superior qualities that traditional lubricants do not have. The development of new lubricating systems that use these features is expected. However, there are still remaining issues that must be overcome for the wide application of ionic liquids, such as disintegration and corrosion problems related to the stability and durability of ionic liquids and guidelines on optimizing the combination with sliding materials. However, these are fundamental issues that belong to the understanding of fundamental mechanisms of tribology rather than problems specific to ionic liquids. Thus, an attempt has recently been made to use an ionic liquid as a model chemical compound for understanding the action mechanism of lubricant additives. By developing an understanding of this type of fundamental lubricating mechanism, progress toward the use of ionic liquid lubricants can be expected.

REFERENCES

1. Wilkes, J. S., & Zaworokto, M. J. (1992). Air and water stable 1-ethyl-3-methylimidazolium based ionic liquids. *Journal of the Chemical Society, Chemical Communications*, 13, 965-967.
2. Ye, C., Liu, W., Chen, Y., & Yu, L. (2001). Room-temperature ionic liquids: a novel versatile lubricant. *Chemical Communications*, 21, 2244-2245.
3. Liu, W., Ye, C., Gong, Q., Wang, H., & Wang, P. (2002). Tribological performance of room-temperature ionic liquids as lubricant. *Tribology Letters*, 13, 81-85.
4. Kamimura, H., Kubo, T., Minami, I., & Mori, S. (2007). Effect and mechanism of additives for ionic liquids as new lubricants. *Tribology International*, 40, 620-625. nTribological Properties of Ionic Liquids http://dx.doi.org/10.5772/ 52595 137
5. Yu, B., Zhou, F., Pang, C. J., Wang, B., Liang, Y. M., & Liu, W. M. (2008). Tribological evaluation of α, ώ-diimidazoliumalkylene hexafluorophosphate ionic liquid and benzotriazole as additive. *Tribology International*, 41(8), 797-801.
6. Coerr, N., Gebseshuber, I. C., Holzer, D., Wanzenboek, H. D., Ecker, A., Pauschitz, A., & Granek, F. (2010). Evaluation of ionic liquids as lubricants. *Journal of Microengineering and Nanoelectronics*, 1, 29-34.
7. Canter, N. (2007). Using dicationic liquids as high temperature lubricants. *Tribology & Lubrication Technology*, 63(5), 12-13.
8. Phillips, B. S., John, G., & Zabinski, J. S. (2007). Surface chemistry of fluorine containing ionic liquids on steel substrates at elevated temperature using Mössbauer spectroscopy. *Tribology Letters*, 26, 85-91.
9. Jimenez, A. E., & Bermudez, M. D. (2010). Ionic liquids as lubricants of titanium-steel contact part 2: friction, wear and surface interactions at high temperature. *Tribology Letters*, 37, 431-443.
10. Suzuki, A., Shinka, Y., & Masuko, M. (2007). Tribological characteristics of imidazolium- based room temperature ionic liquids under high vacuum. *Tribology Letters*, 24, 307-313.
11. Yagi, T., Sasaki, S., Mano, H., Miyake, K., Nakano, M., & Ishida, T. (2009). Lubricity and chemical reactivity of ionic liquid used for sliding metals under high-vacuum conditions. *Proceedings of the Institution of Mechanical Engineers, Part J: Journal of Engineering Tribology*, 223, 1083-1090.

12. Iijima, S., Mazuko, M., Suzuki, A., Nogi, T., & Obara, S. (2011). Effect of oxide layer of metal surface on lubrication performance of liquid lubricants in high vacuum. *J. of Japanese Soc. of Tribologist*, 56(5), 320-330, (in Japanese).
13. Kamimura, H., Chiba, T., Watanabe, N., Kubo, T., Nanao, H., Minami, I., & Mori, S. (2006). Effects of carboxylic acids on wear and friction reducing properties for alkylimidazolium- derived ionic liquids. *Tribology Online*, 1, 40-43.
14. Xia, Y., Sasaki, S., Murakami, T., Nakano, M., Shi, L., & Wang, H. (2007). Ionic liquid lubrication of electrodeposited nickel Si3N4 composite coatings. *Wear*, 262, 765-771.
15. Lu, Q., Wang, H., Ye, C., Liu, W., & Xue, Q. (2004). Room temperature ionic liquid 1- ethyl-3-hexylimidazoliumbis (trifluoromethylsulfonyl)-imide as lubricant for steel/ steel contact. *Tribology International*, 37, 547-552.
16. Qu, J., Truhan, J. J., & Dai, S. (2006). Ionic liquids with ammonium cations as lubricants or additives. *Tribology Letters*, 22, 207-214.
17. Qu, J. (2009). Tribological characteristics of aluminum alloys sliding against steel lubricated by ammonium and imidazolium ionic liquids. *Wear*, 267, 1226-1231.
18. Liu, X., & Zhou, F. (2006). Benzotriazole as the additive for ionic liquid lubricant: one pathway towards actual application of ionic liquids. *Tribology Letters*, 23, 191-196. 138 Ionic Liquids - New Aspects for the Future
19. Zhang, L., Feng, D., & Xu, B. (2009). Tribological characteristics of alkylimidazolium diethyl phosphates ionic liquids as lubricants for steel-steel contact. *Tribology Letters*, 34, 95-101.
20. Zhao, Z., Shao, Y. W., Wang, T. M., Feng, D. P., & Liu, W. M. (2011). Study on corrosion property of a series of hexafluorophosphate ionic liquids on steel surface. *Corrosion Engineering Science and Technology*, 46(4), 330-333.
21. Bermudez, M. D., Jimenez, A. E., & Martinez-Nicolas, G. (2007). Study of surface interactions of ionic liquids with aluminium alloys in corrosion and erosion-corrosion processes. *Applied Surface Science*, 253(17), 7295-7302.
22. Jimenez, A. E., Bermudez, M. D., Carrion, F. J., & Martinez-Nicolas, G. (2006). Room temperature ionic liquids as lubricant additives in steel-aluminium contacts: influence of sliding velocity, normal load and temperature. *Wear*, 261, 347-359.
23. Minami, I., Inada, T., Sasaki, R., & Nanao, H. (2010). Tribo-chemistry of phosphonium- derived ionic liquids. *Tribology Letters*, 40, 225-235.
24. Fox, M. F., & Priest, M. (2008). Tribological properties of ionic liquids as lubricants and additives. Part 1: synergistic tribofilm

formation between ionic liquids and tricresyl phosphate. *Proceedings of the Institution of Mechanical Engineers, Part J: Journal of Engineering Tribology*, 222, 291-303.

25. Minami, I., Kita, M., Kubo, T., Nanao, H., & Mori, S. (2008). The tribological properties of trifluorotris (pentafluoroethyl) phosphate derived ionic liquids. *Tribology Letters*, 30, 215-223.
26. Shah, F. U., Glavatskih, S., Mac Farlane, D. R., Somers, A., Maria, Forsyth. M., & Antzutkin, O. N. (2011). Novel halogen-free chelated orthoborate-phosphonium ionic liquids: synthesis and tribophysical properties. *Physical Chemistry Chemical. Physics*, 13, 12865-12873.
27. Choa, S. H., Ludema, K. C., Potter, G. E., De Koven, B. M., Morgan, T. A., & Kar, K. K. (1995). A model for the boundary film formation and tribological behavior of a phosphazene lubricant on steel. *Tribology Transaction*, 38, 757-768.
28. Zhu, J., Chu, R., & Meng, X. (2009). Chemical structure of phosphazenes in relation to the tribological properties of a steel-on-steel system. *Lubrication Science*, 21, 103-109.
29. Bermudez, M. D., Jiménez, A. E., Sanes, J., & Carrion, F. J. (2009). Ionic Liquids as advanced lubricant fluids". *Molecules*, 14, 2888-2908.
30. Zhou, F., Liang, Y., & Liu, W. (2009). Ionic liquid lubricants: designed chemistry for engineering applications. *Chemical Society Reviews*, 38, 2590-2599.
31. Minami, I. (2009). Ionic liquids in tribology. *Molecules*, 14, 2286-2305.
32. Palacio, M., & Bhushan, B. (2010). A review of ionic liquids for green molecular lubrication in nanotechnology. *Tribology Letters*, 40(2), 247-268. Tribological Properties of Ionic Liquids http://dx.doi.org/10.5772/ 52595 139
33. Schlücker, E., & Waserscheid, P. (2011). Ionic liquids in mechanical engineering. *Chemie Ingeniuer Technik*, 83(9), 1476-1484.
34. Mori, S. (2005). Ionic liquids II. *Edited by Oono H.(CMC,Tokyo,)277 (in Japanese).*
35. Mori, S. (2009). Ionic liquids as a candidate for lubricants. *The journal of the Surface Finishing Society of Japan*, 60(8), 502-507, in Japanese).
36. Minami, I., & Mori, S. (2007). Tribology of ionic liquids. *Journal of The Surface Science Society of Japan*, 28(6), 311-317, (in Japanese).
37. Minami, I. (2012). Ionic liquid lubricants. *Jyunkatsukeizai*, 2, 2-9, (in Japanese).
38. Hayama, M., & Sasaki, S. (2012). Application of ionic liquids as space lubricants. *Proc. f JAST Tribology Conference, Tokyo*, 387-388, (in Japanese).

39. Koyama, T., Tsuboi, R., & Sasaki, S. (2011). Research on lubricity of ionic liquid for metal alloys. Proc. of JAST Tribology Conference, Tokyo (in Japanese), 47-48.
40. Mihashi, k., Yoshimoto, S., & Miyatake, M. (2008). Development of a small-sized hydrodynamic bearing. *Proc. of JAST Tribology Conference, Nagoya*, 491-492, (in Japanese).
41. Tomozawa, K., & Ohno, N. (2008). Construction of phase diagram of ionic liquid and liquid crystal by diamond anvil cell. Proc. of JAST Tribology Conference, Nagoya (in Japanese), 259-260.
42. Mia, S., Tomozawa, K., Morita, S., & Ohno, N. . (2009). High-pressure rheology of ionic liquid. *Proceedings of World Tribology Congress IV (Kyoto,)*, 572, 978-4-99001-399-8.
43. Ohno, N. (2012). Rheological and tribological properties of space lubricant under high pressure. *J. of Japanese Soc. of Tribologist*, 57(2), 103-111, (in Japanese).
44. Koyama, T., Kondo, Y., Tsuboi, R., & Sasaki, S. (2012). Research on lubricity of ionic liquid on bearing steel. Proc. of JAST Tribology Conference, Tokyo (in Japanese) , 365-366.
45. Uerdingen, M., Treber, C., Balser, M., Schmittc, G., & Wernerb, C. (2005). Corrosion behaviour of ionic liquids. *Green Chemistry*, 7, 321-325.
46. Gabler, C., Tomastik, C., Brenner, J., Pisarova, L., Doerra, N., & Allmaierb, G. (2011). Corrosion properties of ammonium based ionic liquids evaluated by SEM-EDX, XPS and ICP-OES. *Green Chemistry*, 13, 2869-2877.
47. Kondo, Y., Yagi, S., Koyama, T., Tsuboi, R., & Sasaki, S. Lubricity and corrosiveness of ionic liquids for steel-on-steel sliding contacts. *Proceedings of the Institution of Mechanical Engineers, Part J: Journal of Engineering Tribology (in press)*.
48. Sasaki, S., Kondo, Y., Koyama, T., & Tsuboi, R. (2012). Lubricity of halogen-free ionic liquids for hard coatings. *Books of synopses, 18th International Colloquium Tribology*, (Esslingen, Germany,), 161, 3-92481-397-3. 140 Ionic Liquids - New Aspects for the Future
49. Kondo, Y., Koyama, T., Tsuboi, R., Nakano, M., Miyake, K., & Sasaki, S. (2012). Study on tribological properties if halogen-free ionic liquids for hard materials. *Proc. Of JAST Tribology Conference,Tokyo*, 373-374.
50. Nogi, T. (2011). Challenge to the ultimate lubricants for space mechanisms. *J. of Japanese Soc. of Tribologist*, 56(9), 561-566, (in Japanese). http://matdb.jaxa.jp/Outgas/Outgasdoc1j.html.
51. Sasaki, S., Miyake, K., Nakano, M., & Ishida, T.. (2010). Lubricity of ionic liquids for hard coatings under high vacuum condition. *Books of synopses, 17th International Colloquium Tribology*, (Esslingen, Germany,), 202, 3-92481-380-9.

CHAPTER 11

A Rotating Annulus Driven by Localized Convective Forcing: A New Atmosphere-Like Experiment

Hélène Scolan[1], Peter L. Read[1]

[1] *Department of Physics, Atmospheric, Oceanic and Planetary Physics, University of Oxford, Oxford, UK*

ABSTRACT

We present an experimental study of flows in a cylindrical rotating annulus convectively forced by local heating in an annular ring at the bottom near the external wall and via a cooled circular disk near the axis at the top surface of the annulus. This new configuration is distinct from the classical thermally driven annulus analogue of the atmosphere circulation, in which thermal forcing is applied uniformly on the sidewalls, but with a similar aim to investigate the baroclinic instability of a rotating, stratified flow subjected to zonally symmetric forcing. Two vertically and horizontally displaced heat sources/sinks are arranged, so that in the absence of background rotation, statically unstable Rayleigh–Bénard convection would be induced above the source and beneath the sink, thereby relaxing strong constraints placed on background temperature gradients in previous experimental configurations based on the conventional rotating annulus. This better emulates local vigorous convection in the tropics and polar regions of the atmosphere while also allowing stably-stratified baroclinic motion in the central zone of the annulus, as in mid-latitude regions in the Earth's atmosphere. Regimes of flow are identified, depending mainly upon control parameters that in turn depend on rotation rate and the strength of differential heating. Several regimes exhibit baroclinically unstable flows which are qualitatively similar to those previously observed in the classical thermally driven annulus. However, in contrast to the classical configuration, they typically exhibit more spatio-temporal complexity. Thus, several regimes of flow demonstrate the equilibrated co-existence of, and interaction between, free convection and baroclinic wave modes. These new features were not previously observed in the classical annulus and validate the new setup as a tool for exploring fundamental atmosphere-like dynamics in a more realistic framework. Thermal structure in the fluid is investigated and found to be qualitatively consistent with previous numerical results, with nearly isothermal conditions, respectively, above and below the heat source and sink, and stably-stratified, sloping isotherms in the near-adiabatic interior.

1 INTRODUCTION

The circulation of the atmosphere is the result of several physical forcing processes: gravity, rotation, radiative exchanges (incoming sunlight, outgoing thermal infrared radiation), clouds and moisture processes, chemical reactions, interaction with surface topography, boundary processes, etc. Although numerical global circulation models solving equations of dynamics, thermodynamics, and continuity in 3D on a sphere are capable of reproducing and even predicting many features of the observed atmosphere, their complexity and the need to represent unresolved processes by semi-empirical parameterizations mean that insights and understanding of how the real atmosphere works may still be elusive.

In this context, laboratory analogues of the dynamics of the atmosphere continue to help scientists understand fundamental dynamical processes in the atmosphere by providing the possibility of precise, reproducible experiments to test ideas and hypotheses.

In the spirit of this approach, the idealization of the mid-latitude atmosphere in the laboratory as a differentially-heated, rotating, cylindrical annulus (see Fig. 1a) originally proposed by Hide, has proved to be a prolific and productive analogue of many features of the atmospheric circulation (Hide 1958; Fowlis and Hide 1965; Hide et al. 1977; Read et al. 2014). By combining only the three essential physical ingredients forcing the atmosphere, i.e., gravity, rotation, and differential heating, the 'classical' thermally driven rotating annulus setup provides an experimental configuration that enables both large-scale overturning circulations and baroclinic instabilities to be studied and explored. It consists of an assembly on a rotating platform with a fluid contained between two upright, coaxial cylinders that are maintained at different temperatures, with thermally insulating, horizontal endwalls (see Fig. 1b) and reproduces fluid motions and baroclinic instabilities that transport heat from the equator, where solar heating of the surface is most intense, towards the cooler polar regions in the atmosphere. In addition, the latitudinal dependence of the Coriolis force (the so-called 'beta-effect') can also be reproduced by simply adding a conical topography at the bottom inside the annulus (e.g., Bastin and Read 1997).

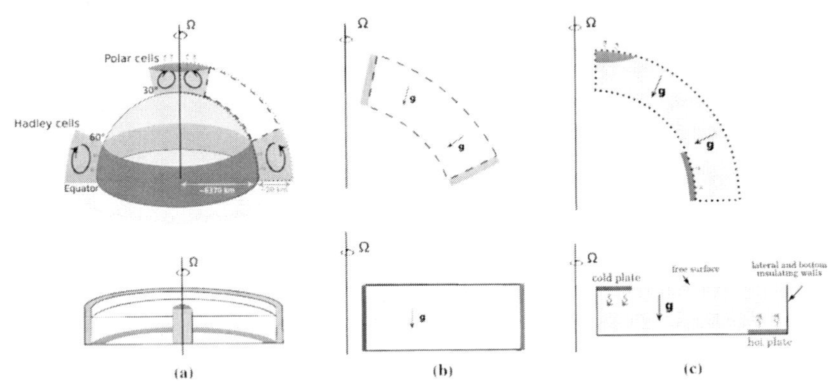

Figure. 1 a Sketch of a simplified structure of the global atmosphere and location of the dominant global thermal heat sinks and sources and its analogue in the laboratory in cylindrical configuration. **b**, **c** Schematic views of a section of mid-latitude atmosphere, as represented **b** in the classical annulus and **c** in the new annulus analogue, taking into account the heat sink in the upper atmosphere in the polar region and the heat source close to the ground in tropical regions. In each configuration sketched in 2b or 2c, the unwrapped rectangular shape obtained corresponds to one half *r-z* section of the experimental annular configuration. Axisymmetry with vertical axis is implicit and assumed near the inner vertical boundary near the heat sink region, as illustrated in **a**

Thus, this setup has proven to be a simple enough system to capture the fundamental physics and with enough complexity to represent nonlinear features and transition sequences to chaos similar to those of atmospheric dynamics (Hignett et al. 1985; Früh and Read 1997; Bastin and Read 1998; Read 2003; Von Larcher and Egbers 2005; Wordsworth et al. 2008; Vincze et al. 2014). It has also proved valuable as a tractable 'test bed' within which to test numerical codes and methods (Harlander et al. 2011; Vincze et al. 2015) as well as to benchmark statistical–dynamical analysis methods in widespread use in meteorology, such as data assimilation (Young and Read 2013). Moreover, it is even inspiring new experiments for studying non-terrestrial planetary atmosphere dynamics in the laboratory (Read et al. 2015; Yadav et al. 2016).

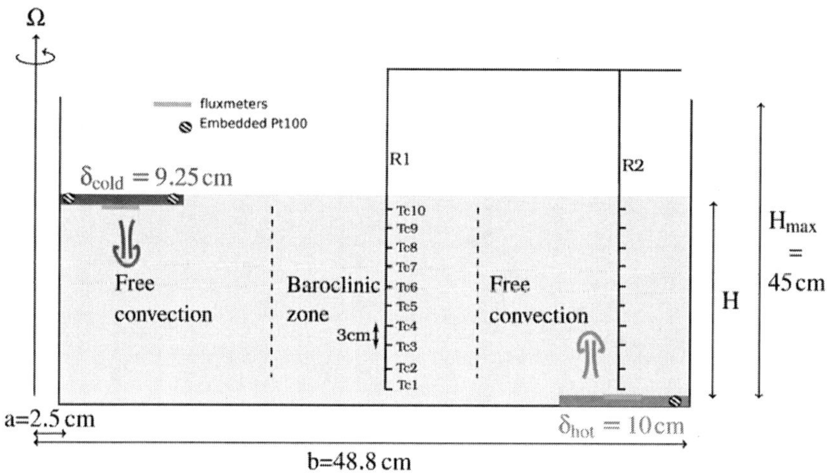

Figure. 2 Schematic cross section of half of the new annular experimental setup with local thermal heat source and heat sink horizontally and vertically displaced. 3 mm diameter rods instrumented with ten type T microthermocouples are introduced in the fluid to measure thermal stratification above the statically stable zone at mid-radius and centered above the heated external zone

The classical annulus configuration, however, has some important limitations as an analogue of the mid-latitude atmosphere, an important aspect of which derives from its use of isothermal vertical boundaries to provide heating and cooling. This leads to an intense, boundary layer dominated, overturning circulation which imposes a strong constraint on the background stratification in the working fluid, and which, therefore, limits the possibility for internal instabilities to change this, even at large equilibrated amplitudes. In contrast, in the real atmosphere (see Fig. 1a, c), the effective heat source in the tropics is located mostly near the ground (through absorption of infrared radiation re-radiated from the solar-heated surface), while the heat sink is predominantly located in the upper troposphere in the mid-latitudes and polar regions[1] (e.g., see Chan and Nigam 2009). The resulting circulation spontaneously partitions itself into a convectively unstable/neutral region in the tropics that interacts with a statically stable, baroclinic region at mid-latitudes that, despite being cooled from above, is statically stable.

Precisely *how* is heat passed from the convectively turbulent region in the tropics into the stably-stratified sub-tropics and mid-latitudes is still not well understood quantitatively. In particular, it is not clear what determines the characteristics of the observed vertical stratification in the mid-latitude atmosphere (its static stability, surface thermal contrast, and tropopause height) and more generally what are the dominant mechanisms for nonlinear equilibration of baroclinic and convective instabilities in the atmosphere (e.g., see Schneider 2006; Stone 2008 for reviews).

Diverse theoretical and numerical studies of stratified macroturbulence and equilibration processes have recently explored the possible role of baroclinic instabilities in acting to stabilise their own thermal environment (Zurita-Gotor and Lindzen 2007; Zurita-Gotor and Vallis 2009, 2010; Schneider and Walker 2006; Schneider 2006). The idea of a feedback between the baroclinic instability and the thermal stratification was originally formulated by Stone (1978). In his baroclinic adjustment theory, based on the two-layer quasi-geostrophic model (Phillips 1951), a rotating, stratified fluid may adjust the mean thermal structure to maintain a stably-stratified environment that is close to a marginally critical state for baroclinic instability (Stone 1978) with a criticality parameter,ξξ, that equilibrates to ξ≈1ξ≈1. ξξ is a measure of the isentropic slope and defined as (Zurita-Gotor and Vallis 2010),

$$\xi = -\frac{f}{\beta H}\frac{\partial_y \bar{\theta}}{\partial_z \bar{\theta}} \tag{1}$$

with $\bar{\theta}$ the mean potential temperature, $f \equiv 2\Omega \sin\phi$ (where ϕ is latitude) and β quantifies the (linearised) dependence of the coriolis force on latitude so that $f \simeq f_0 + \beta y$ with f_0 a constant. Recent studies, however, have disputed this analysis, since numerical simulations have shown that the two-layer model does not necessarily equilibrate to a value of $\xi \approx 1$ (e.g., Salmon 1978, 1980; Vallis 1988).

Nonlinear wave–wave interactions may play a significant role in determining whether the flow equilibrates to a marginally critical ($\xi \lesssim 1$) or supercritical ($\xi > 1$) state (Vallis 1988; Schneider and Walker 2006; Schneider 2006). Some evidence further suggests that wave–wave interactions may become spontaneously suppressed under some conditions (Schneider and Walker 2006; Schneider 2006), leading to marginally critical conditions and the absence of a large-scale turbulent energy cascade, although the detailed structure of the heat sources and sinks may also be important in determining ξ (Zurita-Gotor and Vallis 2009, 2010). Schneider (2004) further highlights that the inhibition of nonlinear eddy–eddy interactions offers an explanation for the historic success of linear and weakly nonlinear models of large-scale extratropical dynamics, but also rests on the postulate that the kinematic mixing properties of baroclinic eddies exhibit no essential vertical structure. The key point here is that this problem remains highly active and controversial, and laboratory experiments of the kind described here may provide some valuable new insights.

Table 1 Range of parameters of presented experiments in the new locally forced thermally driven rotating annulus setup

Parameter	**Symbol**	**Present range**	**Units**
Rotation rate	ΩΩ	0.01–1.1 (up to 3.5)	rad s^{-1}
Temperature difference	ΔTΔT	≈8–10≈8–10	K
Fluid properties at 20 °C (in the case of 17% glycerol solution)			
Density	ρρ	1041–1045	kg m^{-3}
Thermal expansion coefficient	αα	2.75×10^{-4}	K^{-1}
Kinematic viscosity	νν	1.71×10^{-6}	$m^2\ s^{-1}$
Thermal diffusivity	κκ	1.28×10^{-7}	$m^2\ s^{-1}$
Geometry			
Inner radius	*a*	0.025	m
Outer radius	*b*	0.488	m
Horizontal scale	L≡b−aL≡b−a	0.463	m
Cold plate radial width	δcoldδcold	0.0925	m
Heat annular radial width	δhotδhot	0.010	m
Mean fluid depth	*H*	0.30	m
Non-dimensional parameters			
Thermal Rossby	ΘΘ	2×10^{-2}–100	

number (Eq. (3))			
Ekman number (Eq. (9))	E	5×10^{-6}–3×10^{-4}	
Taylor number (Eq. (5))	Ta	10^{8}–10^{12}	
Rayleigh number (Eq. (7))	Ra	3.4×10^{-9}	
Boundary layer thicknesses ratio (8))	P	[0.1–1]	
Prandtl number (Eq. (11))	σσ	≈7≈7 (water only) or 13.4 (water–glycerol solution)	
Aspect ratio (Eq. (10))	ΓΓ	1.54	

In this study, to explore the impact of changes to the forcing on such baroclinic equilibration processes and the criticality parameter, we have constructed a new laboratory analogue of atmospheric circulation with a rotating annulus that is convectively forced by *local* differential heating on *horizontal* boundaries. We aim, thereby, to relax some of the thermal constraints on background temperature gradients in the 'classical' annulus experiments to better represent qualitatively how the thermal structure of the atmosphere itself is maintained, with particular reference to the equilibration mechanisms of convective and baroclinic instabilities. Section 2describes the characteristics of the new experimental apparatus. Validation of the setup and some preliminary results are presented in Sect. 3 and briefly discussed in Sect. 4.

2 EXPERIMENTAL SETUP

The new experimental configuration is inspired by the schematic model, as illustrated in Figs. 1c and 2. By unwrapping the spherical domain, this leads to the usual atmosphere-like cylindrical configuration, but where local differential heating at the horizontal boundaries is used to emulate the radiative heat sink in the upper atmosphere in the polar regions and the heat source near the ground in tropical regions. This allows the possibility of the formation of a statically stable (though baroclinically unstable) zone, sandwiched between convectively unstable regions over/underlying the heated or cooled boundaries (Wright et al. 2017) and permits a possible feedback of the resulting baroclinic instability onto the background stratification. Figure 2 illustrates the experimental configuration, with the fluid contained within an annular channel between two upright, rigid, coaxial, thermally insulating cylinders and a flat (or conical), horizontal base, within which the outermost δhot=10 cm in radius is maintained with a constant heat flux and equilibrates at a corresponding temperature T_b, while for $r<b-10$ cm, the lower boundary is thermally insulating (perspex material). The upper surface is free-slip except for the innermost $r\leq\delta$cold=9.25cm, for which the temperature T_a is fixed, thanks to circulating water within the cold central plate and such that $T_b>T_a$. The whole system rotates uniformly about the axis of symmetry in an anticlockwise direction at angular velocity Ω up to 3.5 rad s^{-1}.

Figure. 3 a Global view of the experimental configuration. The chiller used for the water circulation in the central cold plate can be seen on the left below the turntable control unit in *gray*. The two vertical rods inside the tank and their corresponding light-brown electrical wires connected to the ten thermocouples can be distinguished on the left-hand side of the tank. **b** 3D zoomed view of the design of the new setup with the electrically heated annular ring at the bottom of the tank shown in *red* and the cooling imposed through a circular disk at the centre of the tank at the upper surface in *blue* (with cold water circulation). **c** Top-view of the convection chamber showing the cold plate and aluminium annular ring instrumented with platinum resistance sensors and two heat fluxmeters (*orange squares* at the *bottom*) fixed onto the outer heated ring

The global design and a view of the whole assembly of the new experimental setup are presented in Fig. 3. This shows the perspex tank of internal radius b=48.8cm, filled with water and including the aluminium circular ring of radial width 9.25 cm attached to a central 5 cm diameter cylindrical post (i.e., with inner annulus radius a=2.5cm) cooled with water circulated from an external water bath in the lab frame (see Fig. 3a) via a DSTITMTM rotary union. Top-view cameras can be seen attached to the embedded rotating aluminium framework. The 10 cm wide, electrically heated, annular ring is composed of two aluminium plates, onto the lower face of each of which two flexible, plastic-coated Clarian™ heaters (each resistance ≈18±1Ω) are attached. All four heaters are connected in a parallel configuration to a 10A-35V TTI Thurlby Thandar TSX3510P power supply. Both the cold plate and the electrically heated annular plate are equipped with Captec™ heat fluxmeters and thermal probes, platinum resistance sensors and type T thermocouples, with watertight wire connections at the bottom of the tank (see Fig. 3c).

To monitor the temperature within the fluid and measure the thermal stratification, 3 mm diameter rods instrumented with ten type T microthermocouples have been introduced in a flexible way above the central (baroclinic) zone at mid-radius and above the heated external (convective) zone, as sketched in Fig. 2 and visible in Fig. 3a inside the tank. Calibration of the thermocouples was carried out independently by immersing the rods in the temperature-controlled reservoir of the chiller and using its built-in platinum resistance probe as a reference. As visible in Fig. 3a–c, an annular silver-coloured ring is located around the tank and consists of a series of white light LED arrays, collimated between two thin, aluminium plates to generate a 5 to 10 mm-thick horizontal white light sheet. On the shelf underneath the main turntable are located a remotely controlled computer, connected by a GPIB connection to the acquisition system for the thermal probes (a computer-controlled Agilent 34970A data logger unit with a 34901A card for measurements using 4-wire platinum resistance sensors and acquisition of voltages from fluxmeters, and a 34908A card for type T thermocouple measurements), and power supplies for the lights and annular heaters (see Fig. 3a). The data logger is set to acquire temperature measurements every 1 or 2 min in the general monitoring configuration and, depending on acquisitions, every 5 or 10 s in equilibrium phases.

The principal dimensionless parameters expected to govern the behaviour of the system are in common with the conventional rotating annulus experiment. Thus, the thermal Rossby number is a stability parameter comparing the characteristic velocity U_T of the thermally driven flow, consistent with the geostrophic thermal wind relation

$$T \approx \frac{g\alpha H \Delta T}{\Omega L}, \tag{2}$$

with the velocity scale ΩL based on the rotation rate and the typical horizontal size of the tank $L=b-a$:

$$\Theta = \frac{U_T}{\Omega L} = \frac{g\alpha H \Delta T}{\Omega^2 L^2} \tag{3}$$

with α the thermal expansion coefficient and ΔT the temperature difference between the plates (see Table 1).

The thermal Rossby number Θ can be related to the Burger number:

$$\mathrm{Bu} = \left(\frac{NH}{2\Omega L}\right)^2 = \left(\frac{R_\mathrm{d}}{L}\right)^2, \tag{4}$$

where N is the buoyancy frequency $N^2=-(g/\rho)\partial\rho/\partial z$, via a factor proportional to the mean slope of the isotherms (and isopycnals) $\sim\Delta T_{horiz}/\Delta T_{\mathrm{vert}}\sim$. Linked to the vertical static stability, Bu represent the ratio of $R_d=NH/(2\Omega)$, the Rossby deformation radius (a typical length scale of baroclinic motion) to the scale of the domain L, and is a key parameter determining the onset of baroclinic instability (Hide and Mason 1975).

The importance of viscous forces compared to Coriolis effects is described by the Taylor number in the form deduced by Fowlis and Hide (1965):

$$\mathrm{Ta} = \frac{4\Omega^2 L^5}{H\nu^2} \approx \left(\frac{\text{Coriolis acceleration}}{\text{friction force/unit mass}}\right)^2 \tag{5}$$

where symbols are as defined in Table 1.

Following Hignett et al. (1981), Read (1986), King et al. (2009) and Read et al. (2014), we anticipate that a significant parameter associated with the thermal structure of the system will be the squared ratio of the characteristic length scales of the buoyancy-driven thermal boundary layer (without rotation) and the Ekman layers. Assuming that the thermal boundary layer length scale $\ell_T=H/(2\mathrm{Nu})\sim HRa^{-\gamma}$, where Nu is the Nusselt number characterising the efficiency of the heat transfer in the fluid in comparison with pure conduction

$$\mathrm{Nu} = \frac{Q_{\mathrm{tot}}}{Q_{\mathrm{conductive}}}, \tag{6}$$

Ra is the Rayleigh number

$$\mathrm{Ra} = \frac{g\alpha \Delta T H^3}{\nu\kappa}, \tag{7}$$

and γ is an exponent that depends on the convective regime (e.g., see Chillà and Schumacher 2012), the boundary layer ratio parameter may be defined as

$$P = \left(\frac{\ell_T}{\ell_E}\right)^2 \propto \mathrm{Ra}^{-2\gamma} E^{-1}, \tag{8}$$

where $\ell_E = HE^{1/2}$ is the Ekman layer thickness with E the Ekman number:

$$E = \frac{\nu}{\Omega H^2}; \tag{9}$$

P is, therefore, directly proportional to Ω.

Other intrinsic parameters are determined once geometry and working fluid are chosen: the aspect ratio:

$$\Gamma = \frac{L}{H}, \tag{10}$$

the Prandtl number:

$$\sigma = \frac{\nu}{\kappa}, \tag{11}$$

which compares viscous and thermal diffusivities.

Experiments presented in this paper have been carried out in a flat-bottom configuration and with a 30 cm depth of either pure water or, to increase the density, a 17% solution by volume of glycerol and water (which also changed the viscosity and Prandtl number).

The two heaters were typically energised with an electrical power of 206W for the annulus, while the command temperature of the bath for the water circulation was set to 14 °C and air-conditioning was set to 21 °C in the room. This led to a maximum effective $\Delta T \approx 8$ to 10 °C between the cold plate and the aluminium annular ring at equilibrium, as illustrated in Fig. 4. This was associated with a measured total power from the heated ring and into the cold plate of around 150 and 50 W, respectively (Fig. 4b). The 100 W loss is consistent with the average $T_{bulk} \approx 23–25\,°C$ being larger than the temperature of the air-conditioned room (21 °C). Although the total power from the cold plate was smaller than the one from the heated annulus, the flux (in W/m^2) through the cold plate is found, as expected, to be larger than through the annulus, since the whole annulus surface area ($S \approx 0.27 m^2$) is larger than that of the cold plate ($S \approx 0.041 m^2$).

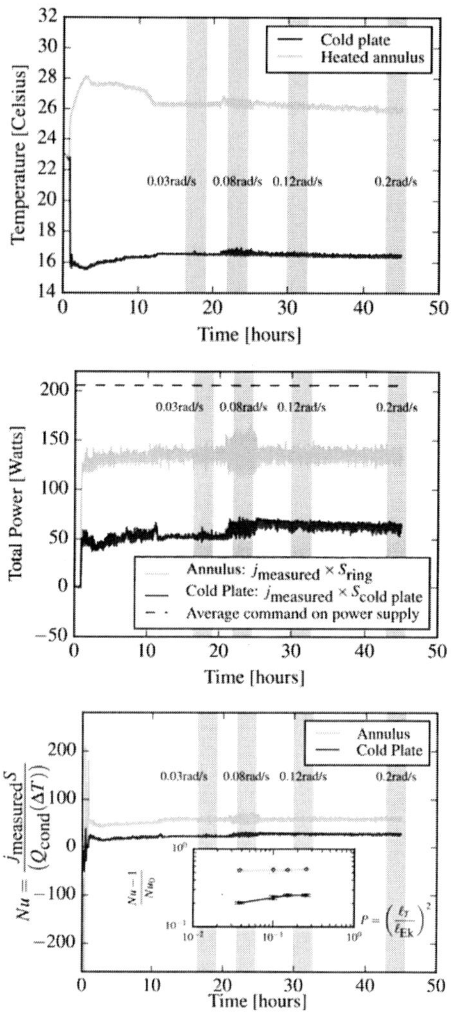

Figure. 4 Evolution of temperature measured with Pt100 platinum resistance sensors **a**, total power **b**, and Nusselt number through plates **c** for the glycerol and particles experiment. The *dashed line* in **b** shows the electrical input power. *Inset* in **c** corresponding evolution of the normalized Péclet number $(Nu-1)/Nu_0$ with *P*, squared ratio of non-rotating thermal boundary layer thickness and Ekman boundary layer thickness. Normalization is performed with the non-rotating Nusselt number Nu_0 estimated with a 2/7 scaling law following Wright et al. (2017)'s results, [see also Shraiman and Siggia (1990) and review by Chillà and Schumacher (2012)]. *Errors bars* shown are illustrating standard deviation on the averaging period at a given rotating rate

The whole system rotated uniformly about the axis of symmetry at angular velocity Ω = 0.03–1.1 rad s^{-1} in the current series of experiments. A summary of the values and ranges of the main parameters for the current experiments is listed in Table 1. After the first day or two of equilibration, changes in rotation rate were carried out and the system was allowed to relax to a new equilibrated dynamical regime for 1 h or more and thermally monitored before image acquisition.

In experiments using pure water, an exploration of the different regimes was performed by injecting a concentrated solution of fluoresceine dye to highlight vortices and other flow features, while in the water–glycerol experiments, 355–500 μm Pliolite particles matching the density of the solution (≈1.043×103 kg m^{-3}) were suspended in the fluid and illuminated by the LED light sheet to trace the flow. Images were acquired with a DFK 31BF03 *Imaging Source* firewire camera (resolution 1024 × 768) and an ethernet AVT Manta 609B camera (resolution 2752 × 2206) within the rotating frame of reference at a frame rate of 1 fps. In experiments with particles, streaklines were produced by superposing 20 or 50 images together in a given equilibrated regime. In addition, an FLIR i50 thermal imaging camera (resolution 240 × 320) was used to image the temperature variations across the surface of the water.

3 RESULTS

3.1 Thermal structure

As explained in introduction and sketched in Fig. 2a and as shown in numerical simulations of axisymmetric flow 2D regimes of the same system (see Fig. 5 and Wright et al. 2017), the formation of a statically stable though baroclinically unstable zone in the central area is expected at equilibrium, sandwiched between convectively unstable regions over/underlying the heated or cooled boundaries.

When background rotation is significant, two regimes were identified as a function of the rotation rate and classified as a function of *P*, the ratio of (non-rotating) thermal boundary layer to Ekman boundary layer thicknesses (Read 1986; Read et al. 2014; Wright et al. 2017). An (r, z) cross section of the temperature field from axisymmetric numerical simulations by Wright et al. (2017) of the similar configuration as the current experiment is illustrated in Fig. 5for the range *P* [0.1–1]. A secondary density current that flows beneath the heat sink, down the side

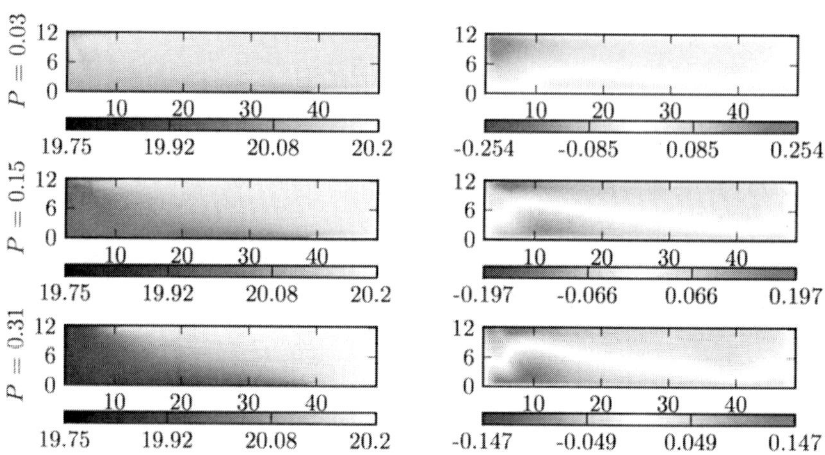

Figure. 5 Temperature (*left*) and azimuthal velocity (*right*) fields of 2D axisymmetric flows simulations (Wright et al. 2017) for *P* ratio in the range [0.1–1] [extract reproduced from data of Wright et al. (2017)]

of the inner cylinder and along the base towards the heat source, was found under very weak rotation conditions (Wright et al. 2017). For weak rotation (defined as $0.05 \lesssim P \lesssim 0.2$), this is replaced by a more uniform thermal gradient, with almost vertical isotherms close to the inner and outer cylinders, confined to Stewartson layers on the vertical boundaries, in association with a developing vertical shear in the azimuthal velocity (see Fig. 5 and Wright et al. 2017).

For moderate rotation rates with *P* of order or above unity ($0.2 \lesssim P \lesssim 2$), the thicknesses of the thermal boundary layer and the Ekman layer are comparable. Free convection then results in well-mixed, approximately isothermal, regions above and below the heat source and sink, respectively. Sandwiched between these two convective zones there emerges a stably-stratified baroclinic region with approximately uniformly sloping isotherms. In association with the horizontal temperature gradient obtained within the tank, a strong vertical shear of the azimuthal velocity is now seen to have developed, as consistent with the geostrophic thermal wind balance.

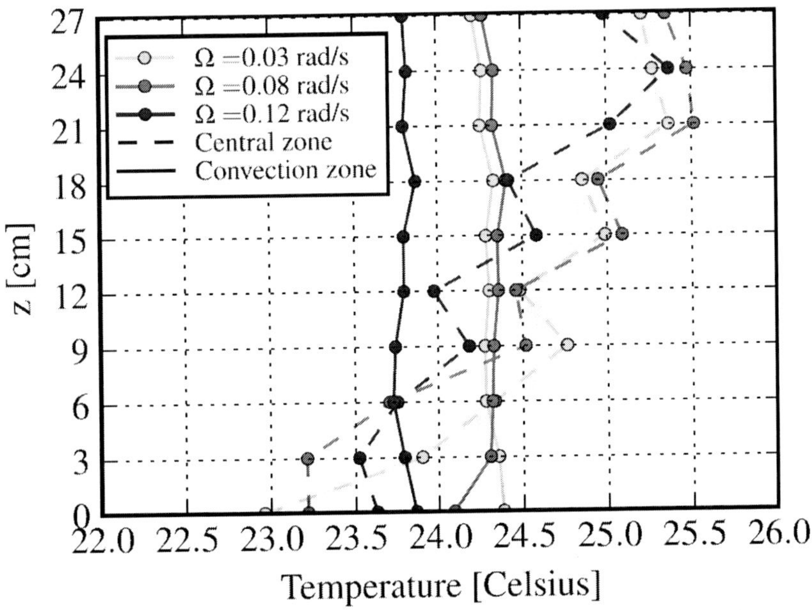

Figure. 6 Measurements of vertical temperature profiles for the mid-radius zone and the zone above the heated ring averaged for 2.5 h in the glycerol-solution experiment showing, respectively, statically stable baroclinic zone and isothermal features. *Gray* area (shown only on one profile in the central zone for clarity) is the estimated ±0.1 °C uncertainty. The measurement of the global stratification within the central zone (avoiding the two topmost temperature points) allows an estimation of the associated Burger numbers (defined in Eq. 4) of Bu≈[3.6,0.65,0.15]±0.1 for the corresponding Ω[rads−1]=[0.03,0.08,0.12]

This thermal structure found in the 2D simulations (Wright et al. 2017) was approximately validated in experiments using a water–glycerol solution, as shown in the time-averaged temperature profiles in Fig. 6, which were taken for different rotation rates at equilibrium (indicated by the shaded zones in the time series shown in Fig. 11): an overall stable stratification tendency is found in the central region, while an almost isothermal profile is indeed measured above the heated annular ring.

During the transition towards thermal equilibrium (before 15 h, not shown), the temperature measured by thermocouples at the bottom of probe R2 (very close to the heated ring near the external walls) rapidly rises as expected, while thermocouples at the top first of R2 become colder and then equilibrate to an almost isothermal configuration after 15 h. In the mean time, thermocouples on probe R1 in the central zone at mid-radius reveal an equilibration towards a final statically stable configuration, as confirmed by time-averaged temperature profiles in Fig. 6. As the squared ratio of thermal boundary layer to Ekman boundary layer thicknesses, *P*, is estimated to be below or around unity for the current experiments (see also values in Fig. 4c, inset), these thermometric results are consistent with the global picture of two almost well-mixed near-isothermal convection zones separated by a statically stable baroclinic zone as seen in simulations over the same range of *P* ratio (see Fig. 5).

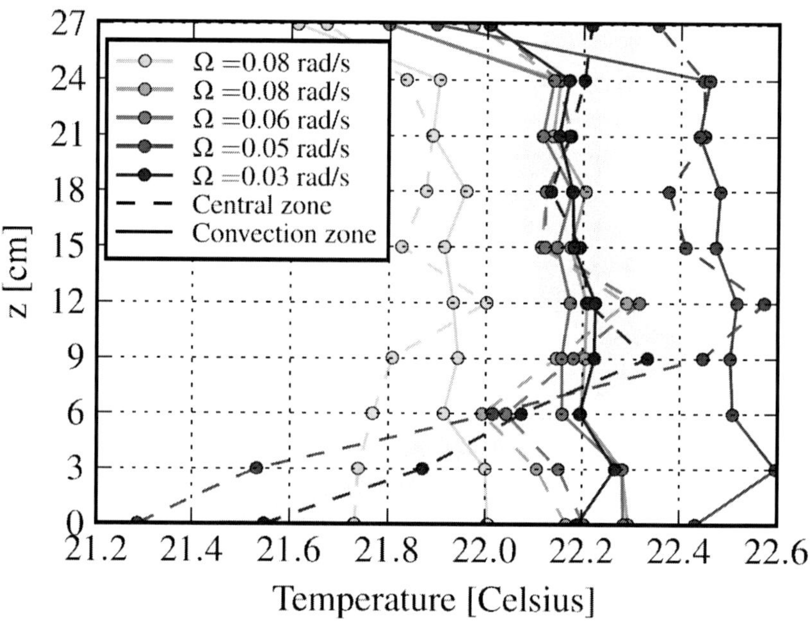

Figure. 7 Measurements of vertical temperature profiles for the mid-radius zone and the zone above the heated ring averaged for 1 h in the water-only experiment with Tcommand cold water bath=14Tcommand cold water bath=14 °C and input power in heated ring = 206 W. *Gray* area (shown only on one profile in the central zone for clarity) is the estimated ±0.1 °C incertainty

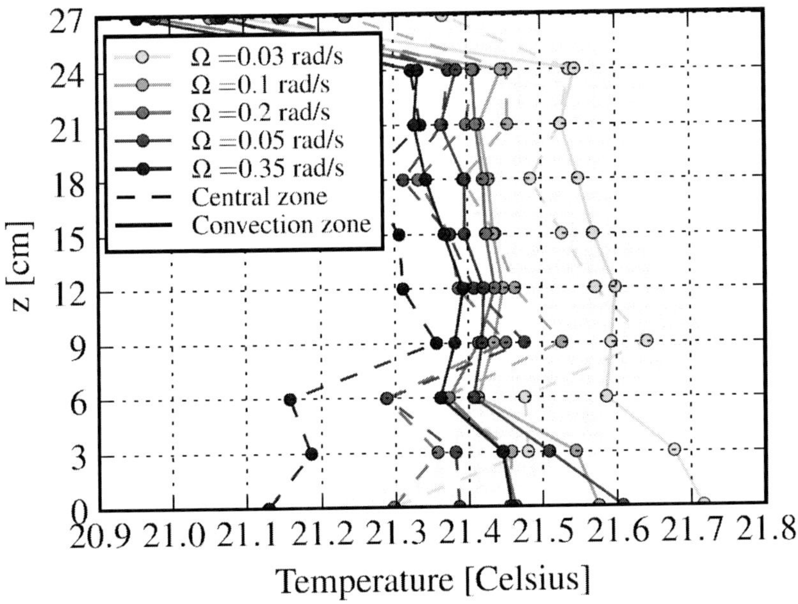

Figure. 8 Measurements of vertical temperature profiles for the mid-radius zone and the zone above the heated ring averaged for 20 min in the water-only experiment with Tcommand cold water bath=18Tcommand cold water bath=18 °C and input power in heated ring = 97.5 W. *Gray* area (shown only on one profile in the central zone for clarity) is the estimated ±0.1 °C incertainty

The thermal structure differs somewhat in the case of experiments using a water-only solution and dye visualisation (see Figs. 7, 8), but a statically stable interval is still present in the lower part of the profiles, merging into a more isothermal section in the upper part of the profile. This may be due to convection processes still at play as the flow gradually evolves towards equilibrium (given a shorter dwell time at each rotating rate) or alternatively may have been the result of evaporation and wind stress processes at the surface due to there having been no enclosure around the tank at that time (in contrast to the water–glycerol experiments conducted afterwards). We estimated the power lost at the free surface due to evaporation ($P_{evapo}=L_{hp}dV/dt$, , where L_h is the water latent heat and dV/dt the volumetric debit) by performing a rough monitoring of the decrease in time of the water level along the depth of the cold plate in water-only experiments. The power lost via evaporation process was found to be in the range 50–80 W, so not negligible compared to the power through the cold plate. This effect is likely to be still present in the glycerol case and temperature profiles in Fig. 6 in the mid-radius central zone also suggest the formation of a mixed layer in the uppermost few centimeters beneath the surface, in contrast to what is found in numerical simulations, where a rigid-lid boundary was used. However, this effect seems somewhat reduced, partly due to the physical properties of the mixture, or more likely mainly due to the enclosure of the whole setup with thick black fabric. Fluctuations can be seen in the interior of the same profiles within this central zone, however, which should be investigated in further experiments.

Thus, at least some of these observed differences between pure water and glycerol mixture experiments in temperature profiles are likely duc to evaporation effects. This illustrates the well-known imperative in rotating experiments with a free surface to reduce air exchange between air above the tank with the air in the room using thick curtains or panels or the use of rigid-lid experimental configurations, to prevent potentially non-negligible evaporation and wind stress effects. In the present case, the flux (in W/m^2) through the cold plate is much larger than through the free surface. Since the free surface is 17 times larger than the cold plate area, however, convective plumes can be expected to be more intense underneath the cold plate than the free surface.

3.2 Heat transport

The use of fluxmeters on the cold plate and on the heater ring allows us to estimate the Nusselt number, as illustrated in Fig. 4c, for the glycerol experiments. For estimating the conductive power Q_{cond} in the denominator of the measured Nusselt number in in Fig. 4c, the coefficient of proportionality of Q_{cond} with ΔT was retrieved from numerical calculation solving the steady conduction equation for the same control and geometrical parameters. Then, the heat transport efficiency is written in the form of normalized Péclet value $(\mathrm{Nu}-1)/\mathrm{Nu}_0$. The Péclet number associated with the annular heated ring is found fixed, while on the cold plate, it is found that it rises with the P ratio, i.e.,with Ω (Fig. 4c). Such two Nusselt number measurements associated with fluxmeters signals on each plate normally give two estimates of the same quantity. However, in reality, the power budget is distributed as $P_{\text{hot plate}}=P_{\text{cold plate}}+P_{\text{free surface}}+P_{\text{walls}}$, with lateral conduction through perspex walls P_{walls} likely to be negligible compared to the others, given the small difference between the bulk temperature and the room temperature. $P_{\text{free surface}}$ is likely dominated by evaporation and whose value estimated above is consistent with this global budget given the values of power measured through the heated annulus and cold plate. Therefore, the Nusselt number retrieved from the cold plate seems the more relevant estimate to better reflect the transverse thermal efficiency and the different flow dynamical regimes depending on rotating rate than the Nusselt based on the heated annulus measurements. Even if the configuration is fully 3D here and with the presence of baroclinic instabilities, in contrast to numerical axisymmetric 2D simulations, it turns out that a similar increase of normalized Péclet number from the cold plate as a function of *P* was found by Wright et al. (2017) in the [0.04–0.2] range.

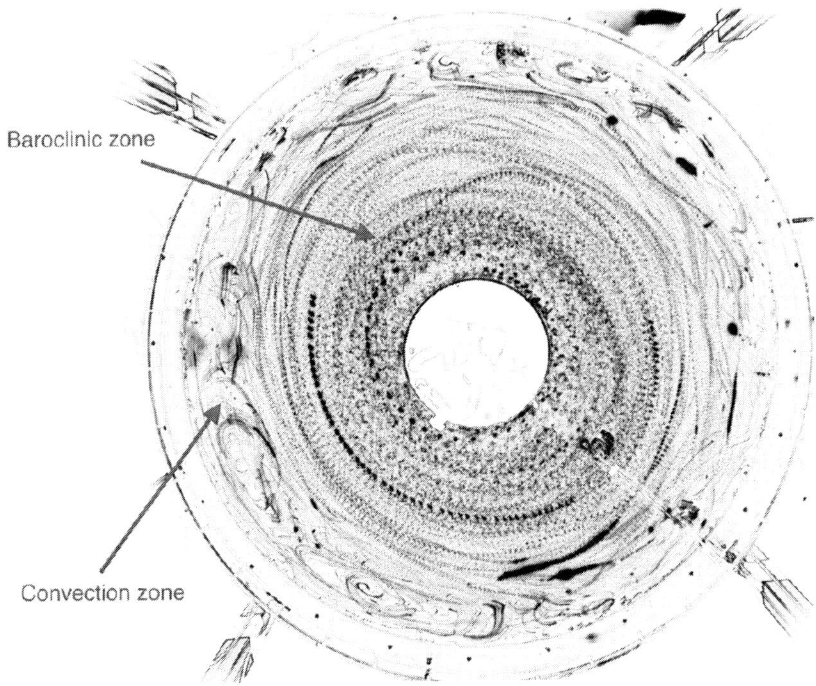

Figure. 9 Dynamical regime for Ω=0.08rads^{-1} in case of mixed glycerol–water solution experiment illustrated with streak lines of particles and showing the 'sandwich' configuration of low-order wavenumber baroclinic zone with convection zone (see also supplementary material Movie 3)

3.3 Horizontal flow

The flow configuration anticipated from the numerical simulations of Wright et al. (2017), of a baroclinic zone sandwiched between two convection zones, is further validated in horizontal streakline features, as illustrated in Figs. 9 and 10 (and the Online Resource video, for example, in the 0.08 and 0.12 rad s^{-1} cases; Movies 3 and 4). A clear separation can be noticed between the mid-radius zone, with the large amplitude baroclinic wave (Fig. 9 reveals a non-axisymmetric pattern consistent with baroclinic mode 1 or even an elliptic pattern corresponding to baroclinic mode 2), and a zone in an outer annular strip near the external wall presenting small-scale features likely to be remnants of the Rayleigh–Bénard convection plumes in that zone. Even in snapshots exhibiting a larger amplitude of baroclinic wave at Ω=0.12rads^{-1}, the meandering jet does not seem to reach the outer cylinder. This new experimental setup, therefore, offers the capability of exploring the properties of fully developed baroclinic instabilities with minimal interference from sidewall boundaries.

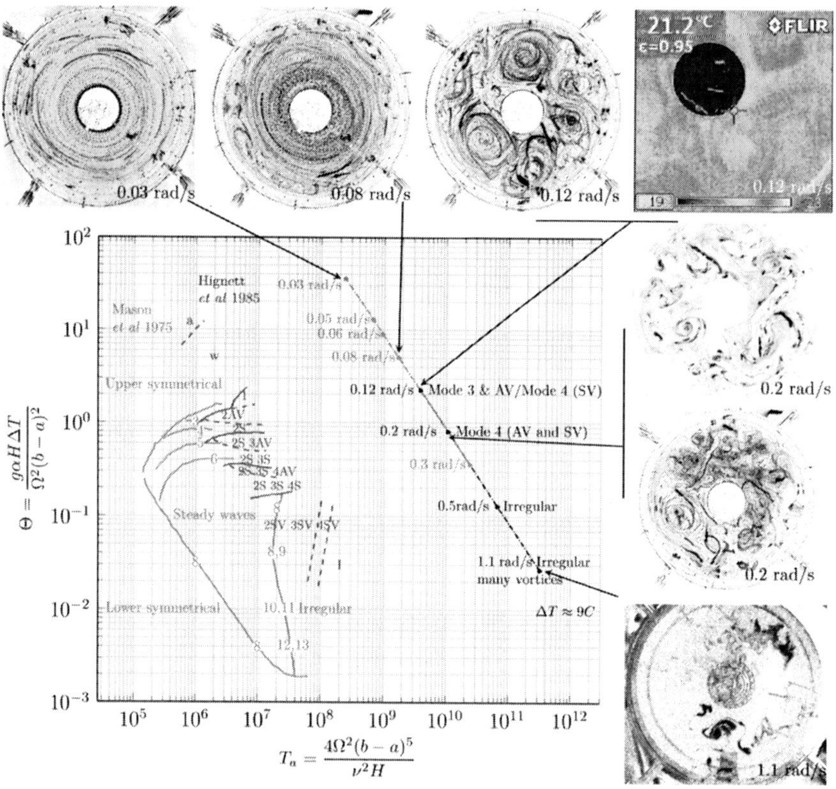

Figure. 10 Regime diagram as a function of Taylor number and thermal Rossby number. *Black* and *magenta points* correspond to two separate experimental runs and show the experimental dynamical regimes explored. Previous results from Mason (1975) and Hignett et al. (1985) in red and *blue lines* have been plotted for comparison. *Numbers* represent mode wavenumber and added symbols that follow original annotations in Hignett et al. (1985) with a, axisymmetric; *w* weak waves, *S* steady waves, *AV* amplitude vacillation, *SV* shape vacillation, *I* irregular. See also movies in supplementary material

At low rotation rates, the experimental flows observed are consistent with the structure of the azimuthal velocity predicted by 2D axisymmetric simulations (see cases $P = 0.03$ and $P = 0.15$ in Fig. 5 and Wright et al. 2017). The occurrence of higher values of azimuthal velocity underneath and near the inner cold plate than at larger radii is confirmed for the Ω=0.03 and 0.08rads^{-1} cases, as illustrated in the video of a close-up view of dye experiments (e.g., Movie 1 in the supplementary Online Resource) and in streakline movies (Movies 2–3 in supplementary Online Resource). This occurrence of an intense 'polar vortex' is particularly interesting as this characteristic has been suggested as a significant dynamical feature in slowly rotating planetary atmospheres such as on Titan (Lebonnois et al. 2012) and Venus.

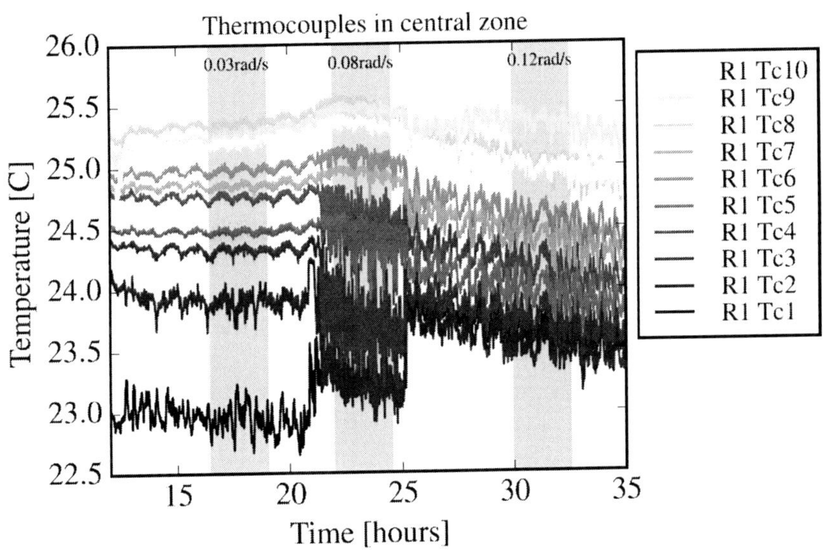

Figure. 11 Time series of thermocouples probes in the central zone at mid-radius for equilibrated state for the different rotating rates

3.4 Baroclinic instability

In the range of rotation rates and temperature differences used, it is expected that isopycnals in the statically stable region near mid-radius would be quite strongly sloped, allowing baroclinic instabilities to be able to grow. Figure 10 shows examples of typical observations of flow patterns found in the fluid, illustrated with passive dye tracer images (after

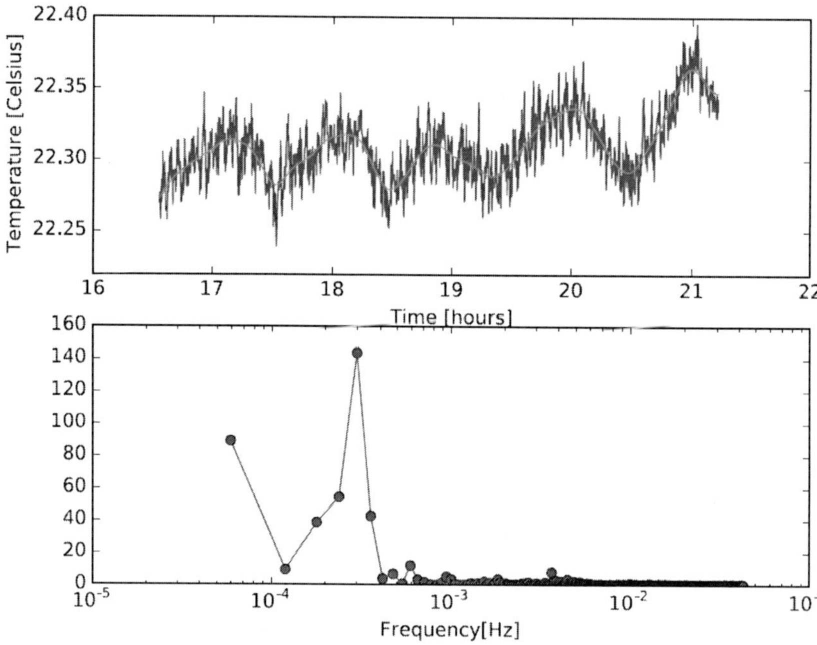

Figure. 12 **a** Zoomed view of time series of one of the thermocouple located in the mid-radius statically stable baroclinic zone for a rotating rate of 0.03 rad s^{-1} with the same high-frequency filtered signal superimposed. **b** Fourier plot associated with the visible $T = 55$ min drift

substraction and/or inversion of the image), particle streaklines and infrared thermographic camera images. Indeed, several modes of baroclinic instability were observed, from low-order, regular, azimuthal wave modes at small angular velocities to irregular and turbulent motions with small, chaotic vortices at higher rotation rates, as illustrated in Fig. 10 and in movies 2–5 in the supplementary Online Resource. For Ω=0.12rads^{-1}, the baroclinic wave is particularly highly developed and presents an almost steady mode 3 associated with a measured Burger number Bu = 0.15. Thus, the onset of strong baroclinic waves is found just above 0.08 rad s^{-1}which corresponds to Bu = 0.65 (see Figs. 6, 10), which is approximately consistent with the Eady criterion (see Hide and Mason (1975), with an instability threshold at Bu = 0.581). The evidence of weak instability for 0.03 rad s^{-1} up to 0.08 rad s^{-1} (Bu = 3.6 and 0.65, respectively) suggests the presence of instability beyond the Eady cutoff. This is also consistent with the zone of weak waves found by Hignett et al. (1985) (see symbol w in Fig. 10) suggesting a possible vertical confinement or surface intensification of baroclinic waves (Hide and

Mason 1978; Jonas 1980). This will be investigated in future work. Nonlinear amplitude vacillation features, similar to those found in classical rotating annulus experiments (e.g., Früh and Read 1997), were also observed. Thus, it was observed, e.g., at rotation rate 0.12 rad s^{-1}, that a baroclinic mode with the initial wavenumber 3 was found to evolve to a mode 4 (i.e., suggesting a form of nonlinear wavenumber vacillation). Other realizations at a similar rotation rate revealed an evolution of a mode 3 with varying amplitude (i.e., amplitude vacillation), as illustrated in supplementary material Movie 4.

In the same way, with a zoomed-in view of temperature time series from the thermocouple probes in the central zone at mid-radius, Fig. 11 also shows oscillatory patterns that reveal a slow drift of the lowest wavenumber components around the tank (τdrift≈55min) for Ω=0.03rads^{-1} (see also Fig. 12 for a closer view and Fourier analysis of it) and more complex, nonlinear baroclinic oscillations at higher rotation rates. A deeper analysis of these features is beyond the scope of this paper and will necessitate longer duration of data acquisition at a given equilibrium state for each rotation rate in future experiments; this will be presented elsewhere.

Finally, we compare our results with previous observations of baroclinic modes with the classical rotating annulus configuration. Figure 10 shows the location of the different observed modes in a Ta-Θ regime diagram on which the locations of various dynamical regimes from previous thermally driven annulus experiments using the classical configuration (Mason 1975; Hignett et al. 1985) are superimposed. This indicates a reasonably good agreement with our experiments in the regions of different wave regimes, between the new locally forced atmosphere-like configuration compared to the classic rotating annulus case.

4 CONCLUSION

A new atmosphere-like experiment has been designed in the form of a local, convectively forced, thermally driven, wide rotating annulus to emulate more effectively than in previous experimental studies the distribution of local heat sinks in the upper troposphere of the Earth in the polar regions and a heat source near the ground associated with strong convection in tropical regions. The first experiments with this new "baroclinic sandwich" setup exhibit the thermal structure in the (r, z) cross plane anticipated from the 2D numerical simulations of Wright et al. (2017), as well as a fast 'polar vortex' and the confirmation of coexisting baroclinic and convection zones in the flow structure. This validates this new experimental setup as a tool to study equilibration processes with coexisting baroclinic and convective instabilities in a form highly relevant to what occurs in a planetary atmosphere such as the Earth's.

Baroclinic instability wave modes are found to grow and equilibrate, in broad agreement with previous results in the more classical annular configuration, with similar zonal wavenumbers and vacillation behaviour. Future work will include the addition of a sloping bottom to take into account in the laboratory effects equivalent to lateral variations of the Coriolis parameter. Particle image velocimetry experiments will also then allow us to gain quantitative insights into the nonlinear baroclinic equilibration processes on the beta plane, as well as into the geostrophic turbulence regime, to help understand the peculiar characteristics and dynamics of a mid-latitude Earth-like atmosphere.

FOOTNOTES

[1] The situation in the polar atmosphere is complicated by the non-gray nature of atmospheric radiative transfer, which also (amongst other things) allows the surface to cool directly by radiating to space as well as the atmosphere.

ACKNOWLEDGEMENTS

The authors acknowledge support for this work by UK Engineering and Physical Sciences research Council with EPSRC grant EP/K029428/1. The authors are grateful to Bob Watkins and Andy Clack from Atmospheric, Oceanic & Planetary Physics, Oxford for their support during the experimental design. We credit Sylvie Su with the initial version of sketches of Fig. 1.

REFERENCES

1. Bastin M, Read P (1998) Experiments on the structure of baroclinic waves and zonal jets in an internally heated, rotating, cylinder of fluid. Phys Fluids (1994-present) 10(2):374–389. doi:10.1063/1.869530
2. Bastin ME, Read PL (1997) A laboratory study of baroclinic waves and turbulence in an internally heated rotating fluid annulus with sloping endwalls. J Fluid Mech 339:173–198. doi:10.1017/S0022112097005259
3. Chan SC, Nigam S (2009) Residual diagnosis of diabatic heating from era-40 and ncep reanalyses: Intercomparisons with trmm. J Clim 22(2):414–428. doi:10.1175/2008JCLI2417.1

4. Chillà F, Schumacher J (2012) New perspectives in turbulent Rayleigh–Bénard convection. Eur Phys J E 35(7):1–25. doi:10.1140/epje/i2012-12058-1
5. Fowlis W, Hide R (1965) Thermal convection in a rotating annulus of liquid: effect of viscosity on the transition between axisymmetric and non-axisymmetric flow regimes. J Atmos Sci 22(5):541–558. doi:10.1175/1520-0469(1965) 022<0541:TCIARA>2.0.CO;2
6. Früh WG, Read P (1997) Wave interactions and the transition to chaos of baroclinic waves in a thermally driven rotating annulus. Philos Trans R Soc Lond A Math Phys Eng Sci 355(1722):101–153. doi:10.1098/rsta.1997.0003MathSciNetMATH
7. Harlander U, von Larcher T, Wang Y, Egbers C (2011) Piv-and ldv-measurements of baroclinic wave interactions in a thermally driven rotating annulus. Exp Fluids 51(1):37–49. doi:10.1007/s00348-009-0792-5
8. Hide R (1958) An experimental study of thermal convection in a rotating liquid. Philos Trans R Soc Lond A Math Phys Eng Sci 250(983):441–478. doi:10.1098/rsta.1958.0004
9. Hide R, Mason P (1975) Sloping convection in a rotating fluid. Adv Phys 24(1):47–100
10. Hide R, Mason P (1978) On the transition between axisymmetric and non-axisymmetric flow in a rotating liquid annulus subject to a horizontal temperature gradient: Hysteresis effects at moderate taylor number and baroclinic waves beyond the eady cut-off at high taylor number. Geophys Astrophys Fluid Dyn 10(1):121–156
11. Hide R, Mason P, Plumb R (1977) Thermal convection in a rotating fluid subject to a horizontal temperature gradient: spatial and temporal characteristics of fully developed baroclinic waves. J Atmos Sci 34(6):930–950. doi:10.1175/1520-0469(1977) 034<0930:TCIARF>2.0.CO;2
12. Hignett BP, White A, Carter R, Jackson W, Small R (1985) A comparison of laboratory measurements and numerical simulations of baroclinic wave flows in a rotating cylindrical annulus. Q J R Meteorol Soc 111(467):131–154. doi:10.1002/qj.49711146705
13. Hignett P, Ibbetson A, Killworth PD (1981) On rotating thermal convection driven by non-uniform heating from below. J Fluid Mech 109:161–187. doi:10.1017/S0022112081000992
14. Jonas P (1980) Laboratory experiments and numerical calculations of baroclinic waves resulting from potential vorticity gradients at low taylor number. Geophys Astrophys Fluid Dyn 15(1):297–315
15. King EM, Stellmach S, Noir J, Hansen U, Aurnou JM (2009) Boundary layer control of rotating convection systems. Nature 457(7227):301–304. doi:10.1038/nature07647
16. Lebonnois S, Burgalat J, Rannou P, Charnay B (2012) Titan global climate model: A new 3-dimensional version of the ipsl titan gcm. Icarus 218(1):707–722
17. Mason P (1975) Baroclinic waves in a container with sloping end walls. Philos Trans R Soc Lond A Math Phys Eng Sci 278(1284):397–445. doi:10.1098/rsta.1975.0032
18. Phillips NA (1951) A simple three-dimensional model for the study of large-scale extratropical flow patterns. J Meteorol 8(6):381–394. doi:10.1175/1520-0469(1951) 008<0381:ASTDMF>2.0.CO;2
19. Read P, Jacoby T, Rogberg P, Wordsworth R, Yamazaki Y, Miki-Yamazaki K, Young R, Sommeria J, Didelle H, Viboud S (2015) An experimental study of multiple zonal jet formation in rotating, thermally driven convective flows on a topographic beta-plane. Phys Fluids (1994-present) 27(8):085–111
20. Read PL (1986) Regimes of axisymmetric flow in an internally heated rotating fluid. J Fluid Mech 168:255–289. doi:10.1017/S002211208600037XMATH
21. Read PL (2003) A combined laboratory and numerical study of heat transport by baroclinic eddies and axisymmetric flows. J Fluid Mech 489:301–323. doi:10.1017/S002211200300524XMATH
22. Read PL, Pérez EP, Moroz IM, Young RMB (2014) General circulation of planetary atmospheres. John Wiley & Sons, Inc, pp 7–44. doi:10.1002/9781118856024.ch1

23. Salmon R (1978) Two-layer quasi-geostrophic turbulence in a simple special case. Geophys Astrophys Fluid Dyn 10(1):25–52. doi:10.1080/ 0309192780 8242628MATH
24. Salmon R (1980) Baroclinic instability and geostrophic turbulence. Geophys Astrophys Fluid Dyn 15(1):167–211. doi:10.1080/03091928008241178
25. Schneider T (2004) The tropopause and the thermal stratification in the extratropics of a dry atmosphere. J Atmos Sci 61(12):1317–1340. doi:10.1175 /1520-0469(2004) 061<1317:TTATTS>2.0.CO;2
26. Schneider T (2006) The general circulation of the atmosphere. Annu Rev Earth Planet Sci 34:655–688. doi:10.1146/annurev.earth.34.031405.125144
27. Schneider T, Walker CC (2006) Self-organization of atmospheric macroturbulence into critical states of weak nonlinear eddy–eddy interactions. J Atmos Sci 63(6):1569–1586. doi:10.1175/JAS3699.1
28. Shraiman BI, Siggia ED (1990) Heat transport in high-rayleigh-number convection. Phys Rev A 42:3650–3653
29. Stone PH (1978) Baroclinic adjustment. J Atmos Sci 35(4):561–571. doi:10.1175/1520-0469(1978) 035<0561:BA>2.0.CO;2
30. Stone PH (2008) The atmospheric general circulation: some unresolved issues. Dyn Atmos Oceans 44(3):244–250. doi:10.1016/j.dynatmoce.2007.03.002
31. Vallis G (1988) Numerical studies of eddy transport properties in eddy-resolving and parametrized models. Q J R Meteorol Soc 114(479):183–204. doi:10.1002/qj.49711447910
32. Vincze M, Harlander U, von Larcher T, Egbers C (2014) An experimental study of regime transitions in a differentially heated baroclinic annulus with flat and sloping bottom topographies. Nonlinear Process Geophys 21(1):237–250. doi:10.5194/npg-21-237-2014
33. Vincze M, Borchert S, Achatz U, von Larcher T, Baumann M, Liersch C, Remmler S, Beck T, Alexandrov KD, Egbers C, Fröhlich J, Heuveline V, Hickel S, Harlander U (2015) Benchmarking in a rotating annulus: a comparative experimental and numerical study of baroclinic wave dynamics. Meteorologische Zeitschrift 23(6):611–635. doi:10.1127/metz/2014/0600
34. Von Larcher T, Egbers C (2005) Experiments on transitions of baroclinic waves in a differentially heated rotating annulus. Nonlinear Process Geophys 12(6):1033–1041. doi:10.5194/npg-12-1033-2005
35. Wordsworth R, Read P, Yamazaki Y (2008) Turbulence, waves, and jets in a differentially heated rotating annulus experiment. Phys Fluids (1994-present) 20(12):126–602. doi:10.1063/1.2990042MATH
36. Wright S, Su S, Scolan H, Young R, Read P (2017) Regimes of axisymmetric flow and scaling laws in a rotating annulus with local convective forcing. Fluids **(submitted)**
37. Yadav S, Banerjee A, Dey T, Balasubramanian S (2016) Velocity measurements using udv in rotating annulus subjected to quasi 2d temperature variations. Int J Curr Eng Technol. doi:10.14741/Ijcet/22774106/spl.5.6.2016.40
38. Young R, Read P (2013) Data assimilation in the laboratory using a rotating annulus experiment. Q J R Meteorol Soc 139(675):1488–1504. doi:10.1002/qj.2061
39. Zurita-Gotor P, Lindzen RS (2007) Theories of baroclinic adjustment and eddy equilibration. In: Schneider T, Sobel AH (eds) The global circulation of the atmosphere, chap 2. Princeton University Press, pp 22–24. ISBN:9780691121819
40. Zurita-Gotor P, Vallis GK (2009) Equilibration of baroclinic turbulence in primitive equations and quasigeostrophic models. J Atmos Sci 66(4):837–863. doi:10.1175/2008JAS2848.1
41. Zurita-Gotor P, Vallis GK (2010) Circulation sensitivity to heating in a simple model of baroclinic turbulence. J Atmos Sci 67(5):1543–1558. doi:10.1175/2009JAS3314.1

CHAPTER 12

Geminal Brønsted Acid Ionic Liquids as Catalysts for the Mannich Reaction in Water

Leqin He [1,2,†], Shenjun Qin [2,†], Tao Chang [2,*], Yuzhuang Sun [2] and Jiquan Zhao [1,*]

[1] *School of Chemical Engineering, Hebei University of Technology, Tianjin 300130, China*

[2] *Key Laboratory for Resource Exploration Research of Hebei Province, Hebei University of Engineering, Handan 056038, China*

ABSTRACT

Quaternary ammonium geminal Brønsted acid ionic liquids (GBAILs) based on zwitterionic 1,2-bis[*N*-methyl-*N*-(3-sulfopropyl)-alkylammonium]ethane (where the carbon number of the alkyl chain is 4, 8, 10, 12, 14, 16, or 18) and *p*-toluenesulfonic acid monohydrate were synthesized. The catalytic ionic liquids were applied in three-component Mannich reactions with an aldehyde, ketone, and amine at 25 °C in water. The effects of the type and amount of catalyst and reaction time as well as the scope of the reaction were investigated. Results showed that GBAIL-C_{14} has excellent catalytic activity and fair reusability. The catalytic procedure was simple, and the catalyst could be recycled seven times via a simple separation process without noticeable decreases in catalytic activity.

Keywords: Mannich reaction; geminal ionic liquid; water; catalyst; emulsion

1. INTRODUCTION

The Mannich reaction is one of the most important carbon–carbon bond-forming reactions in organic synthesis because of its atom economy and potential application in the synthesis of biologically active molecules [1–3]. In this reaction, an amine, two carbonyl compounds, and acid (or base) catalysts are used to produce β-amino carbonyl compounds, which constitute various pharmaceuticals, natural products, and versatile synthetic intermediates [4,5]. Conventional catalysts for the classic Mannich reaction involve inorganic and organic acids like HCl [6], proline [7], *p*-dodecylbenzenesulfonic acid [8], and several Lewis acids [9]. Reactions using these catalysts, however, often suffer from drawbacks including long reaction times, harsh reaction conditions, and difficult product separation. Utilization of solid catalysts may solve problems associated with current catalysts [10–12]. However, rapid deactivation and decreases in active sites per area limit the applications of solid catalysts to some degree. Therefore, new green catalysts are of great interest in this field.

Considering the focus on green synthesis in recent years, ionic liquid (ILs), including functionalized ones, have attracted considerable research attention [13–17]. ILs have been referred to as "designer solvents" because their physical and chemical properties can be adjusted by varying the cation and anion. These useful materials not only dissolve many organic and inorganic substances but are also readily recycled. Moreover, IL properties can be tuned to implement specific chemical tasks. Scientists have performed the Mannich reaction using various types of ILs. In some instances, catalysts and substrates may be dissolved into nearly neutral ILs, such as [BMIM][PF_6] [18], [emim][OTf] [19], and [Hmim][PF_6] [20] to promote the one-pot Mannich reaction. However, in most cases, acidic ILs are used both as the reaction medium and catalyst. Li *et al.* [21] reported a Mannich reaction catalyzed by a carboxyl-functionalized acidic IL [CMMIM][BF_4] in aqueous media. The use of the IL 2-pyrrolidinecarboxylic acid as a highly efficient organocatalyst for the asymmetric one-pot Mannich reaction has also been reported by Wang *et al.* [22]. Other researchers have reported that the Mannich reaction could be conducted using Brønsted acid ILs as catalysts and solvents [23–25]. Although extensive work has been conducted in this area, the disadvantages of the aforementioned catalytic systems, including the large amount of catalyst required and the necessity of an organic co-solvent, are evident. Moreover, the systems typically contain halogen and metal atoms, which, in some ways, limits their "greenness". Thus, synthesizing halogen-free and water-soluble ILs that can be used in a precise manner is necessary.

Organic reactions in water without the use of harmful organic solvents have recently become the focus of considerable research because water is a cheap, safe, and environmentally benign solvent. Some catalytic systems have been successfully applied in three-component Mannich-type reactions with aldehydes, amines, and ketones in water [26–29]. However, a major disadvantage in the use of an aqueous solvent is that most organic compounds are insoluble in water. To circumvent this disadvantage, surfactant-type catalysts [30,31], which solubilize organic materials or form micellar dispersions, have been utilized.

ILs with the molecular structure of gemini surfactants (geminal ILs), which are well known to show high surface activities, have gained significant research interest [32–34]. A number of reports have described the aggregation of these surface-active ILs in water. Therefore, geminal ILs may have potential applications in many areas.

In the present work, novel quaternary ammonium geminal Brønsted acid ionic liquids (GBAILs) were synthesized (see Scheme 1) and applied to catalyze three-component Mannich reactions in water. The ILs were characterized using IR, ^{1}H-NMR, and ^{13}C-NMR. Factors affecting the reaction, such as the type and amount of catalyst used and the reaction time, were investigated, and microscopic observations of colloidal particles created using GBAIL and organic substrates are presented an increase in the hydrocarbon chain length of the catalyst resulted in increases in product yield. To the best of our knowledge, Mannich reactions catalyzed by geminal ILs have yet to be reported.

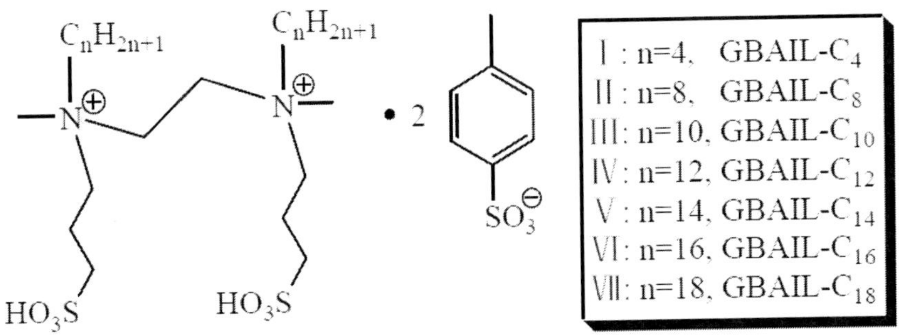

Scheme 1. Structures of the geminal Brønsted acid ionic liquids (GBAILs).

2. RESULTS AND DISCUSSION

2.1. Catalyst Screening for the Mannich Reaction

Mannich reactions of benzaldehyde, aniline, and cyclohexanone were chosen as model reactions to test the catalytic activities of GBAILs. The results are summarized in Table 1.

Table 1. Influence of the catalytic system on the Mannich reaction in water a.

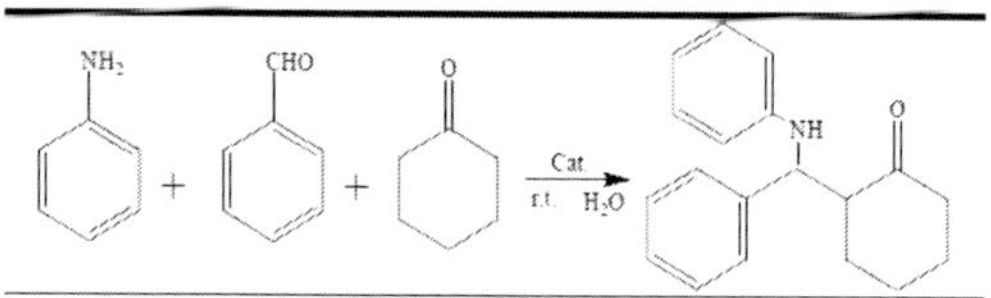

Entry	Catalyst	Yield/% [b]
1	GBAIL-C_4	64
2	GBAIL-C_8	70
3	GBAIL-C_{10}	76
4	GBAIL-C_{12}	82
5	GBAIL-C_{14}	88
6	GBAIL-C_{16}	87
7	GBAIL-C_{18}	85

[a] Reaction conditions: 1.0 mmol benzaldehyde, 1.0 mmol aniline, 1.0 mmol cyclohexanone, 0.050 mmol catalyst, 1.5 mL H_2O, 4 h, 25 °C; [b] Yield of the isolated product after purification by recrystallization.

All reactions produce moderate to high yields, which suggests that the GBAILs behave as Brønsted acids, surfactants, and effective catalysts for Mannich-type reactions in water. An increase in the length of the hydrocarbon chain of the catalyst resulted in an increased product yield (entries 1–5). The product yield was stable when the number of carbon atoms in the hydrophobic chain exceeded 14 (entries 6 and 7). These results show that long alkyl chains are crucial for efficient catalysis.

As shown in Figure 1a, the seven reaction tubes were stirred for 30 min at 25 °C. After standing for 1 min, turbid emulsions induced by GBAILs with short alkyl chains (less than C_{10}) formed two immiscible layers. By contrast, reaction mixtures with GBAILs with modest alkyl chain lengths (C_{12} or C_{14}) presented white turbid reaction mixtures after standing for 1 h. No phase separation was observed in the turbid emulsions induced by GBAIL-C_{16} or GBAIL-C_{18} after standing for 10 h. After completion (Figure 1b), waxy solids were observed in reactions catalyzed by GBAIL-C_4, GBAIL-C_8, and GBAIL-C_{10}, which may be due to low reaction rates in these systems. When the number of carbons in the hydrophobic chain exceeded

12, the particles became smaller with increasing hydrocarbon chain length. This result shows that long alkyl chains are necessary to form the colloidal dispersions assumed to promote efficient catalysis. Thus, considering the reaction rate and catalyst costs, GBAIL-C_{14} was selected as the most efficient catalyst for the Mannich reaction.

(a)

(b)

Figure 1. Different stages of the Mannich reaction catalyzed by GBAILs. (**a**) During the reaction; (**b**) at the end of reaction. The order of ionic liquids (ILs) from left to right is GBAIL-C_4, GBAIL-C_8, GBAIL-C_{10}, GBAIL-C_{12}, GBAIL-C_{14}, GBAIL-C_{16}, and GBAIL-C_{18}.

2.2. Reaction Conditions of the Mannich Reaction

Reactions of benzaldehyde, cyclohexanone, and aniline were used as model reactions to screen the effect of the amount of catalytic GBAIL-C_{14} (Table 2) on product yield. When the catalyst loading was increased from 1% to 5%, the product yield increased significantly from 43% to 88% (entries 1–5). The optimal amount of IL catalyst was 0.050 mmol (5 mol % based on the substrate), and further increases in catalyst amount did not substantially improve yields (entries 6 and 7). Therefore, we conclude that a sufficient number of active sites is available for the reaction to occur when the catalyst loading is 0.050 mmol. The effect of reaction time on the yield was also investigated. An increase in conversion was observed as the reaction time increased (entries 8–11). Further increases in reaction time only increased the product yield slightly (entries 12–14). This finding can be explained by the fact that the

Mannich reaction reaches equilibrium in approximately 5 h. As such, 5 h was considered the ideal reaction time.

Table 2. Effects of the amount of catalyst and reaction time on the Mannich reaction a.

Entry	Amount of catalyst/mmol	Time/h	Yield/% [b]
1	0.01	4	43
2	0.02	4	65
3	0.03	4	72
4	0.04	4	79
5	0.05	4	88
6	0.06	4	88
7	0.07	4	89
8	0.05	1	43
9	0.05	2	69
10	0.05	3	83
11	0.05	5	91
12	0.05	6	92
13	0.05	7	92
14	0.05	8	93

[a] Reaction conditions: 1.0 mmol benzaldehyde, 1.0 mmol aniline, 1.0 mmol cyclohexanone, 1.5 mL H_2O, 25 °C; [b] Yield of the isolated product after purification by recrystallization.

The reaction mixtures became turbid with increasing reaction time, which signifies the formation of colloidal dispersions. We observed the GBAIL-C_{14}-induced colloid via optical microscopy of the colloidal dispersions formed from GBAIL-C_{14} (5 mol %), benzaldehyde (1.0 mol), and aniline (1.0 mol) in water (1.5 mL) and found that spherical particles were formed (Figure 2). Most of the substrate and catalyst are presumably concentrated in the spherical particles, which function as hydrophobic reaction sites and enable rapid reaction in water.

2.3. Reusability of GBAIL-C_{14} in the Mannich Reaction

The reusability of the GBAIL-C_{14} catalyst in a model Mannich-type reaction was also investigated. After reaction, the products were isolated from the catalytic system by extraction with ethyl acetate. The combined aqueous extract (aqueous media containing the catalyst) was reused in subsequent runs without further purification. As shown in Figure 3, the catalyst can be reused at least seven times without any noticeable decrease in yield or reaction rate. The yields ranged from 91% to 85%. Compared with the reusability of traditional solvents and catalysts, the facile reusability of GBAIL-C_{14} is also an interesting property for environmental protection and economic reasons.

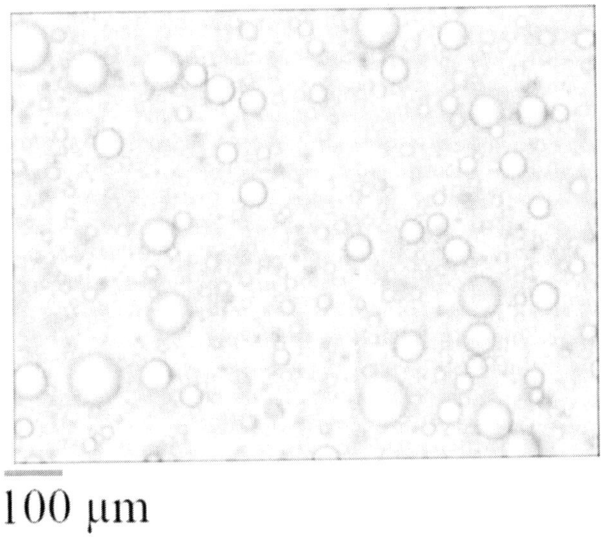

Figure 2. Optical micrograph of the reaction mixture.

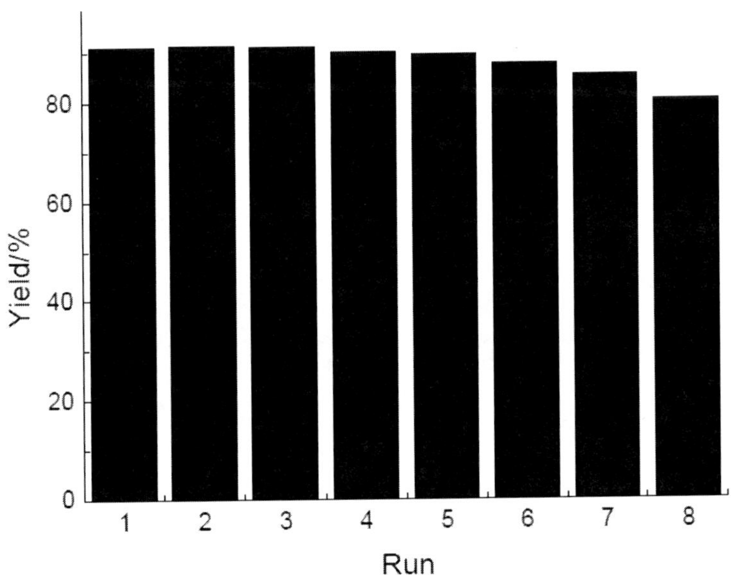

Figure 3. Reusability of the GBAIL-C_{14} catalyst. Reaction conditions: 1.0 mmol benzaldehyde, 1.0 mmol aniline, 1.0 mmol cyclohexanone, 0.050 mmol catalyst, 1.5 mL H_2O, 5 h, 25 °C.

2.4. Catalytic Performance of GBAIL-C_{14} in Mannich Reactions in Water

To generalize our protocol, we investigated the reactions of various aromatic aldehydes, aromatic amines, and cyclohexanone in the presence of GBAIL-C_{14} in water at room temperature. The results are summarized in Table 3. In general, the reaction proceeded smoothly to produce the corresponding products in reasonable to good yields ranging from 87% to 94%. The order of reactivity of the amines was *p*-chloroaniline > aniline > *p*-anisidine > *o*-anisidine (entries 1–4), which indicates the importance of the electronic and steric nature of the amines. Aromatic aldehydes and anilines with either electron-withdrawing or electron-donating substituents could undergo the reaction, as shown previously in [29] (entries 5–8). Ten millimolar-scale reactions were also conducted without any difficulties. For example, the reaction of benzaldehyde (10 mmol), aniline (10 mmol), and cyclohexanone (10 mmol) in the presence of 5 mol % of the catalyst produced the desired product with 87% yield (entry 9).

Table 3. Results of one-pot, three-component Mannich reactions with different substrates catalyzed by GBAIL-C_{14}a.

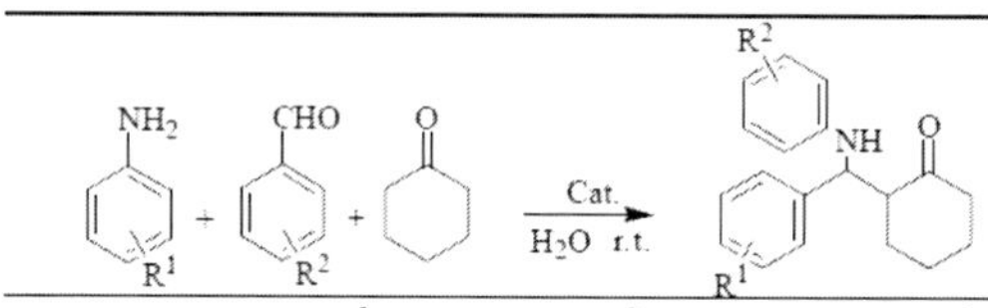

Entry	R^1	R^2	Yield/% [b]
1	H	H	91
2	*p*-Cl	H	94
3	*p*-OMe	H	89
4	*o*-OMe	H	85
5	H	*o*-NO_2	91
6	H	*p*-NO_2	91
7	H	*o*-Cl	92
8	H	*p*-OMe	90
9	H	H	87 [c]

[a] Reaction conditions: 1.0 mmol aromatic aldehyde, 1.0 mmol aromatic amine, 1.0 mmol cyclohexanone, 0.050 mmol catalyst, 1.5 mL H_2O, 5 h, 25 °C; [b] Yield of the isolated product after purification by recrystallization; [c] Reaction conditions: 10 mmol benzaldehyde, 10 mmol aniline, 10 mmol cyclohexanone, 0.50 mmol catalyst, 15 mL H_2O, 5 h, 25 °C.

3. EXPERIMENTAL SECTION

3.1. General

N,N-Dimethylethylenediamine was purchased from J&K Chemicals, Beijing, China. *n*-Butyl bromide, *n*-octyl bromide, *n*-decyl bromide, *n*-dodecyl bromide, *n*-tetradecyl bromide, and *p*-toluenesulfonic acid monohydrate were obtained from Aladdin Reagent Co., Shanghai, China and used without further purification. 1,3-Propanesulfonate (99%) was purchased from Energy Chemicals (Shanghai, China). *n*-Hexadecyl bromide and *n*-stearyl bromide were obtained from Chemxyz Reagent Co., Shanghai, China. Acetone, diethyl ether, ethanol, hexane, methanol, and sodium hydroxide were obtained from Tianjin Kemio Fine Chemical Institute, Tianjin, China and used directly.

FT-IR spectra were recorded using KBr tablets by a Bruker (Fällanden, Switzerland) Vector 22 spectrophotometer in the range of 400–4000 $cm^{-1.}$ ^{1}H- and ^{13}C-NMR spectra were obtained on a Bruker (Fällanden, Switzerland) AC-P400 spectrometer in D_2O and $CDCl_3$. Melting points were determined using an XT-4 (Keyi Electric Light Instrument Co., Ltd., Beijing, China) melting point apparatus and are reported uncorrected.

3.2. Synthesis of GBAILs

3.2.1. Synthesis of Geminal Sulfobetaine-Type Zwitterion

Geminal sulfobetaine-type zwitterionic 1,2-bis[*N*-methyl-*N*-(3-sulfopropyl)-alkylammonium]ethane (where the carbon number of the alkyl was 4, 6, 8, 10, 12, 14, 16, or 18) was synthesized by reaction of *N,N*-dimethylethylenediamine with *n*-alkyl bromide, followed by 1,3-propanesulfonate based on the literature [35] with some modifications.

The procedure was based on the synthesis of 1,2-bis[*N*-methyl-*N*-(3-sulfopropyl)- tetradecylammonium]ethane. *n*-Tetradecyl bromide (0.042 mol, 11.65 g) was added to a stirred solution of *N,N*-dimethylethylenediamine (0.020 mol, 1.76 g) in ethanol containing NaOH (0.042 mol, 1.68 g), and the mixture was refluxed for over 40 h. The formed inorganic salt was removed and hydrochloric acid gas was injected into the residue dissolved in acetone, thereby producing *N,N'*-dimethyl-*N,N'*-ditetradecylethylenediamine hydrochloric acid salts as white solids. The obtained solid was washed with acetone and neutralized with 10% NaOH aqueous solution, and the product was extracted with hexane. White waxy solids were obtained after solvent evaporation.

A fourfold molar excess of 1,3-propanesulfonate was added to a stirred solution of *N,N′*-dimethyl-*N,N′*-ditetradecylethylenediamine in acetone, and the mixture was refluxed for 24 h. After the mixture had cooled to room temperature, the obtained precipitate was filtered, washed with acetone, recrystallized from mixtures of acetone and methanol, and dried under reduced pressure to produce 1,2-bis[*N*-methyl-*N*-(3-sulfopropyl)-tetradecylammonium]ethane as a white solid.

3.2.2. GBAIL Preparation

Zwitterion acidification was accomplished by mixing the zwitterions with *p*-toluenesulfonic acid monohydrate based on a previously described method [29,36] with some modifications. Typically, the zwitterion (1 equiv.) and *p*-toluenesulfonic acid monohydrate (2 equiv.) were dissolved in 1 mL of ethanol. The mixture was magnetically stirred and heated to 80 °C in an open single-neck round-bottom flask until the solvent had evaporated. The mixture was then stirred for 5 h at 110 °C. After completion of the reaction, the viscous residue was dried under vacuum at 100 °C for 2 h. After cooling, a white solid product was obtained. The structures of intermediate materials and GBAILs were identified by IR, ^{1}H-NMR, and ^{13}C-NMR, and selected spectral data are listed in the supplementary information.

3.3. General Procedure for Mannich Reactions Catalyzed by GBAILs

In a typical procedure, a reaction tube was charged with the catalyst (1 to 7 mol %) in 1.5 mL of water. Aldehyde (1.0 mmol), ketone (1.0 mmol), and aniline (1.0 mmol) were added successively with stirring. The mixture was then stirred at 25 °C for the period of time listed in Table 2. After completion of the reaction (monitored by TLC), the mixture was extracted with ethyl acetate (2 mL) and the upper layer was removed via a pipette. Extraction was repeated three times, and the combined organic layers were concentrated and recrystallized from ethanol–acetone (v/v = 1:1) to produce the desired product. The reaction was repeated twice to obtain the average yield. The aqueous extract containing IL catalysts could be reused directly in subsequent runs without further purification.

4. CONCLUSIONS

In conclusion, several three-component Mannich-type reactions using aldehydes, amines, and cyclohexanone were efficiently catalyzed by GBAILs in water under mild conditions. The results showed that GBAIL-

C_{14} provides excellent catalytic activity because of its ability to induce emulsion formation. The IL formed stable colloidal particles in the presence of the substrates in water; this colloid formation has an essential function in accelerating the reaction. The catalytic procedure was simple and the catalyst could be recycled seven times via a simple separation process without noticeable decreases in catalytic activity. Therefore, we believe that the work reported in this paper will have potential applications in green chemistry.

ACKNOWLEDGMENTS

We acknowledge financial support by the National Natural Science Foundation of China (No. 21276061), the Science Foundation of Hebei province (No. B2012402001, D2014402046), program for One Hundred Innovative Talents in Universities of Hebei Province (No. BR2-204) and the Science and Technology Foundation of Universities of Hebei Province (No. Q2012030).

REFERENCES

1. Michael, A.; Bernhard, W.; Nikolaus, R. Modern variants of the mannich reaction. *Angew. Chem. Int. Ed* **1998**, *37*, 1044–1070.
2. Nuno, R.C.; Francesco, M.; Pedro, M.S.D.C.; Pedro, M.P.G. Boronic acids and esters in the petasis-borono mannich multicomponent reaction. *Chem. Rev* **2010**, *110*, 6169–6193.
3. Subramaniapillai, S.G. Mannich reaction: A versatile and convenient approach to bioactive skeletons. *J. Chem. Sci* **2013**, *125*, 467–482.
4. List, B. The direct catalytic asymmetric three-component Mannich reaction. *J. Am. Chem. Soc* **2000**, *122*, 9336–9337.
5. Bekircan, O.; Bektas, H. Synthesis of schiff and mannich bases of isatin derivatives with 4-amino-4,5-dihydro-1H-1,2,4-triazole-5-ones. *Molecules* **2008**, *13*, 2126–2135.
6. Blatt, A.H.; Gross, N. The addition of ketones to schiff bases. *J. Org. Chem* **1964**, *29*, 3306–3311.
7. List, B.; Pojarliev, P.; Biller, W.T.; Martin, H.J. The proline-catalyzed direct asymmetric three-component mannich reaction: Scope, optimization, and application to the highly enantioselective synthesis of 1,2-amino alcohols. *J. Am. Chem. Soc* **2002**, *124*, 827–833.
8. Manabe, K.; Mori, Y.; Kobayashi, S. Three-component carbon-carbon bond-forming reactions catalyzed by a Brønsted acid-surfactant-combined catalyst in water. *Tetrahedron* **2001**, *57*, 2537–2544.
9. Kobayashi, S.; Hamada, T.; Manabe, K. The catalytic asymmetric Mannich-type reactions in aqueous Media. *J. Am. Chem. Soc* **2002**, *124*, 5640–5641.
10. Azizi, N.; Torkiyan, L.; Saidi, M.R. Highly efficient one-pot three-component mannich reaction in water catalyzed by heteropoly acids. *Org. Lett* **2006**, *8*, 2079–2082.

11. Iimura, S.; Nobutou, D.; Manabe, K.; Kobayashi, S. Mannich-type reactions in water using a hydrophobic polymer-supported sulfonic acid catalyst. *Chem. Commun* **2003**, 1644–1645.
12. Nitabaru, T.; Kumagai, N.; Shibasaki, M. Catalytic asymmetric nitro-mannich reactions with a Yb/K heterobimetallic catalyst. *Molecules* **2010**, *15*, 1280–1290.
13. Welton, T. Room-temperature ionic liquids. Solvents for synthesis and catalysis. *Chem. Rev* **1999**, *99*, 2071–2083.
14. Chaturvedi, D. Recent developments on task specific ionic liquids. *Curr. Org. Chem* **2011**, *15*, 1236–1248.
15. Stepnowski, P. Application of chromatographic and electrophoretic methods for the analysis of imidazolium and pyridinium cations as used in ionic liquids. *Int. J. Mol. Sci* **2006**, *7*, 497–509.
16. Pârvulescu, V.I.; Hardacre, C. Catalysis in ionic liquids. *Chem. Rev* **2007**, *107*, 2615–2665.
17. Hallett, J.P.; Welton, T. Room-temperature ionic liquids: Solvents for synthesis and catalysis. *Chem. Rev* **2011**, *111*, 3508–3576.
18. Yang, X.; Wang, M.; Varma, R.; Li, C. Ruthenium-catalyzed tandem olefin migration—Aldol- and mannich-type reactions in ionic liquid. *J. Mol. Catal. A* **2004**, *214*, 147–154.
19. Akiyama, T.; Suzuki, A.; Fuchibe, K. Mannich-type reaction promoted by an ionic liquid. *Synlett* **2005**, *6*, 1024–1026.
20. Yong, F.F.; Teo, Y.C. Recyclable siloxy serine organocatalyst for the direct asymmetric Mannich reaction in ionic liquids. *Synth. Commun* **2011**, *41*, 1293–1300.
21. Li, J.; Peng, Y.; Song, G. Mannich reaction catalyzed by carboxyl-functionalized ionic liquid in aqueous media. *Catal. Lett* **2005**, *102*, 159–162.
22. Zheng, X.; Qian, Y.B.; Wang, Y. 2-Pyrrolidinecarboxylic acid ionic liquid as a highly efficient organocatalyst for the asymmetric one-pot Mannich reaction. *Eur. J. Org. Chem* **2010**, *2010*, 515–522.
23. Zhao, G.; Jiang, T.; Gao, H.; Han, B.; Huang, J.; Sun, D. Mannich reaction using acidic ionic liquids as catalysts and solvents. *Green Chem* **2004**, *6*, 75–77.
24. Sahoo, S.; Joseph, T.; Halligudi, S.B. Mannich reaction in brønsted acidic ionic liquid: A facile synthesis of β-amino carbonyl compounds. *J. Mol. Catal. A* **2006**, *244*, 179–182.
25. Dong, F.; Zhenghao, F.; Zuliang, L. Functionalized ionic liquid as the recyclable catalyst for mannich-type reaction in aqueous media. *Catal. Commun* **2009**, *10*, 1267–1270.
26. Manabe, K.; Kobayashi, S. Mannich-type reactions of aldehydes, amines, and ketones in a colloidal dispersion system created by a brønsted acid-surfactant-combined catalyst in Water. *Org. Lett* **1999**, *1*, 1965–1967.
27. Fang, D.; Luo, J.; Zhou, X.; Lui, Z. Mannich reaction in water using acidic ionic liquid as recoverable and reusable catalyst. *Catal. Lett* **2007**, *116*, 76–80.
28. Shimizu, S.; Shimada, N.; Sasaki, Y. Mannich-type reactions in water using anionic water-soluble calixarenes as recoverable and reusable catalysts. *Green Chem* **2006**, *8*, 608–614.
29. Chang, T.; He, L.; Bian, L.; Han, H.; Yuan, M.; Gao, X. Brønsted acid-surfactant-combined catalyst for the mannich reaction in water. *RSC Adv* **2014**, *4*, 727–731.
30. Wu, C.; Fu, X.; Ma, X.; Li, S.; Li, C. Threonine-surfactant organocatalysts for the highly diastereo- and enantioselective direct anti-Mannich reactions of hydroxyacetone. *Tetrahedron Lett* **2010**, *51*, 5775–5777.
31. Wu, C.; Fu, X.; Li, S. Simple and inexpensive threonine-based organocatalysts for the highly diastereo- and enantioselective direct large-scale syn-aldol and anti-Mannich reactions of α-hydroxyacetone. *Tetrahedron: Asymmetry***2011**, *22*, 1063–1073.
32. Ding, Y.; Zha, M.; Zhang, J.; Wang, S. Synthesis, characterization and properties of geminal imidazolium ionic liquids. *Colloids Surf. A* **2007**, *298*, 201–205.

33. Ao, M.; Xu, G.; Kang, W.; Meng, L.; Gong, H.; Zhou, T. Surface rheological behavior of gelatin/ionic liquid-type imidazolium gemini surfactant mixed systems. *Soft Matter* **2011**, *7*, 1199–1206.

34. Zhang, S.; Yan, H.; Zhao, M.; Zheng, L. Aggregation behavior of gemini pyrrolidine-based ionic liquids 1,1′-(butane-1,4-diyl)bis(1-alkylpyrrolidinium) bromide ([$C_n$py-4-$C_n$py][Br_2]) in aqueous solution. *J. Colloid Interface Sci* **2012**, *372*, 52–57.

35. Yoshimura, T.; Ichinokawa, T.; Kaji, M.; Esumi, K. Synthesis and surface-active properties of sulfobetaine-type zwitterionic gemini surfactants. *Colloids Surf. A* **2006**, *273*, 208–212.

36. He, L.; Qin, S.; Chang, T.; Sun, Y.; Gao, X. Biodiesel synthesis from the esterification of free fatty acids and alcohol catalyzed by long-chain Brønsted acid ionic liquid. *Catal. Sci. Technol* **2013**, *3*, 1102–1107.

CHAPTER 13

Phase-Averaged Flow Statistics in Compressors Using a Rotated Hot-Wire Technique

T. O. Jelly[1] · I. J. Day[1] · L. di Mare[2]

[1] *Whittle Laboratory, Engineering Department, University of Cambridge, Cambridge, UK*

[2] *Mechanical Engineering Department, Imperial College London, London, UK*

ABSTRACT

A technique based on a rotated hot wire has been developed to characterise the unsteady, three-dimensional flow field between compressor blade rows. Data are acquired from a slanted hot wire rotated through a number of orientations at each measurement point. Phase-averaged velocity statistics are obtained by solving a set of sensor response equations using a weighted, non-linear regression algorithm. The accuracy and robustness of the method were verified a priori by conducting a series of tests using synthetic data. The method is demonstrated by acquiring a full set of phase-averaged flow statistics in the wake of a compressor stator blade row. The technique allows three components of phase-averaged velocity, six components of phase-averaged deterministic stress, and six components of phase-averaged Reynolds stress to be recovered using a single rotated hot-wire probe.

Keywords: Hot-wire anemometry · Turbulence · Turbomachinery

1 INTRODUCTION

The fluid motion that develops between compressor blade rows is exceptionally complex. The relative motion of alternate rotating and non-rotating blade rows (see Fig. 1) produces a flow field that is inherently unsteady, three-dimensional and multi-scale.

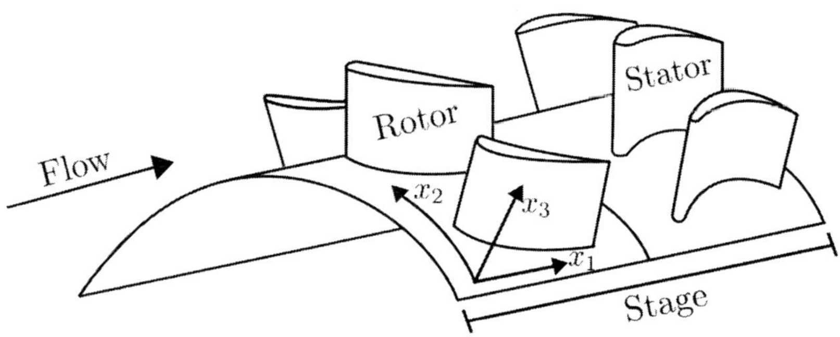

Figure 1 Sketch of a compressor stage showing the axial (x_1), circumferential (x_2), and radial (x_3) directions. Note that the outer casing of the compressor has been removed. This figure has been adapted from Bradshaw et al. (1981)

It is customary to divide the unsteadiness in compressors into two categories: (1) deterministic fluctuations related to the shaft and blade frequencies and (2) broadband stochastic fluctuations related to naturally occurring turbulence. The mean square of (1) and (2) are manifest as deterministic stresses and Reynolds stresses, respectively. Furthermore, statistical moments may exhibit periodicities in both space and time. The space–time dependence of flow statistics significantly complicates the closure of the phase-averaged Navier–Stokes equations, relative to the traditional Reynolds-averaged formulation (Reynolds and Hussain 1972).

The numerical prediction of compressor performance and stability rank amongst the greatest achievements of modern computational fluid dynamics (CFD) (Bradshaw 1996). Several frameworks for compressor CFD exist and include: (1) the steady-state mixing-plane method (Denton 1990; Dawes 1991; (2) the unsteady sliding-plane method (Rai 1989); and (3) the average-passage method (Adamczyk 1985). Depending on which method is adopted, the action of deterministic stresses, Reynolds stresses, or their combined effect should be modeled properly to achieve physically meaningful results. However, it is widely acknowledged that contemporary modeling practice can compromise the reliability of CFD results (Denton 2010). Ultimately, modeling uncertainties will only be reduced through improved physical understanding.

To improve the understanding of momentum and energy exchange mechanisms in compressor annuli, it is imperative to acquire scale-resolved measurements in realistic, rotating, multi-bladed environments. Steady-state data are of limited value, since statistical moments of fluctuating quantities cannot be evaluated. Useful data must be highly resolved in both space and time, so that statistics of the unsteady, three-component flow can be examined in detail. Therefore, the development of practical experimental techniques which return a complete set of phase-averaged flow statistics in realistic compressor environments is of significant interest.

In the past, a number of experimental methods have been used to characterise compressor flow fields. These methods include: triple-element hot-wire probes (Lakshminarayana and Poncet 1974), fast-response total-pressure probes (Zierke and Okiishi 1982), laser doppler anemometry (Faure et al. 2001), and particle image velocimetry (Uzol et al. 2002). Some of these methods are described in further detail by Lakshminarayana (1981) and Sieverding et al. (2000). This work aims to complement past methods by acquiring phase-locked flow statistics using a a technique based on a rotated hot wire.

1.1 The rotated hot-wire technique

The rotated hot-wire technique was first mentioned by Dryden (1938). He described a procedure to measure the Reynolds shear stress in a turbulent boundary layer based on the difference between two measurements acquired at different sensor orientations. Following work done by Prandtl (1946), Hinze (1959), and Webster (1962), theoretical models of the sensor's angular response were established and more sophisticated rotated hot-wire techniques were developed. These methods have been used to characterise the turbulence structure in flow configurations that include: square ducts (Brundrett and Baines 1964); circular pipes (Fujita and Kovasznay 1968; Papavegros and Hedley 1979); circumferentially strained boundary layers (Bissonnette and Mellor 1974); curved rectangular ducts (De Grande and Kool 1981); and downstream of turbulence-generating grids (Sirivat 1989). In addition, a number of rotated hot-wire methods have been developed and employed in compressors.

Schmidt and Okiishi (1977) developed a rotated slant-wire method to measure periodic variations of velocity between the blade rows of a three-stage axial-flow compressor. A once-per-rotor-revolution trigger was used to phase-lock slanted hot-wire readings acquired at six different orientations. Following the solution of a set of non-linear equations, the mean velocity vector was recovered. Kuroumaru et al. (1982) devised a similar method whereby phase-locked, slant-wire measurements were collected downstream of an impeller at 12 different probe orientations. Based on a least squares solution of the ensemble-averaged data, three components of mean flow and six components of Reynolds stress were recovered. Inoue and Kuromaru (1984) applied a similar technique to study the dynamics of vortical structures downstream of a rotating blade row. A 12-orientation method was also used by Goto (1991), who studied the mean-flow field and Reynolds stresses across a stator blade row in a low-speed, single-stage compressor. Shin et al. (1994) examined rotor–rotor interactions in a counter-rotating unducted fan by traversing a rotated slant-wire upstream, downstream, and in-between the rotating blade rows. Hsu and Wo (1997) measured the three-dimensional wake structure downstream of a compressor rotor blade row using a rotated slant wire to quantify rotor–stator interference effects. Ma and Jiang (2004) examined radial profiles of mean velocity, turbulence intensity, and Reynolds shear stresses, based on rotated slant-wire measurements collected in the wake of a compressor rotor blade. In a recent study, Berdanier and Key (2016) devised a new data reduction technique for post-processing rotated slant-wire measurements. In this work, data were acquired in the tip region of a high-speed compressor at three different sensor orientations—effectively mimicking a triple-wire arrangement at each measurement point—and the time-mean pitch and yaw angles were recovered using a new look-up table method. Relative to the iterative solution of non-linear response equations, the look-up method required a fraction of the computational cost.

As an alternative to acquiring data with a rotated single wire, a multi-wire sensor could be used instead. Relative to a single wire, benefits of using a multi-wire sensor include: reduced total measurement time due to simultaneous wire sampling and direct evaluation of joint statistical properties in the amplitude, time, and frequency domains. Despite these advantages, multi-wire measurements in turbomachinery flows are known to suffer from the effects of prong interference (Díaz et al. 2016). As a result, only single-wire sensors were used throughout this work.

1.2 Motivation and objectives

Despite numerous measurements campaigns, details of phase-averaged deterministic stresses and phase-averaged Reynolds stresses between compressor blade rows remain scarce in the literature. To the best of the authors' knowledge, a rotated hot-wire method capable of resolving the time variation of deterministic and Reynolds stresses between compressor blade rows does not exist. The objective of this work is to develop such a method, with the aim of characterising the flow between compressor blade rows to a high level of statistical detail. This work establishes the theoretical, numerical, and experimental frameworks required to recover the phase-averaged velocity vector, phase-averaged deterministic stress tensor, and phase-averaged Reynolds stress tensor using a single rotated slant wire.

This paper is divided into six sections. Section 2 describes the derivation of the phase-averaged sensor response equations. Section 3 describes how the response equations are solved using a non-linear regression algorithm and the results from a series of a priori tests. Section 4 describes the experimental setup, including details of the compressor facility, data acquisition system, and calibration procedures. Section 5 details the experimental results, including phase-averaged velocities, deterministic velocities, phase-averaged deterministic stresses, and phase-averaged Reynolds stresses. Finally, in Sect. 6, the findings of this work are summarised.

2 PHASE-AVERAGED SENSOR RESPONSE EQUATIONS

This section describes the formulation of the phase-averaged sensor response equations and is divided into three parts. First, a relationship between the rotated, wire-fixed co-ordinate system, and the stationary compressor co-ordinate system is established. Second, a triple decomposition of the instantaneous velocity is invoked and equations for mean, deterministic, and stochastic components are derived. Finally, a set response equations which relate the flow statistics measured in the rotated, wire-fixed frame to the velocity statistics in the stationary compressor co-ordinate system are derived.

2.1 Co-ordinate systems

Throughout this work, the wire-fixed co-ordinate system of the slanted hot wire is designated y_i. With reference to Fig. 2, the y_1 axis lies tangential to the wire, the y_2 axis lies normal to the plane containing the sensor prongs, and the y_3 axis lies bi-normal to wire. The velocity components in the tangential (y_1), normal (y_2), and bi-normal (y_3) directions are v_1, v_2, and v_3, respectively.

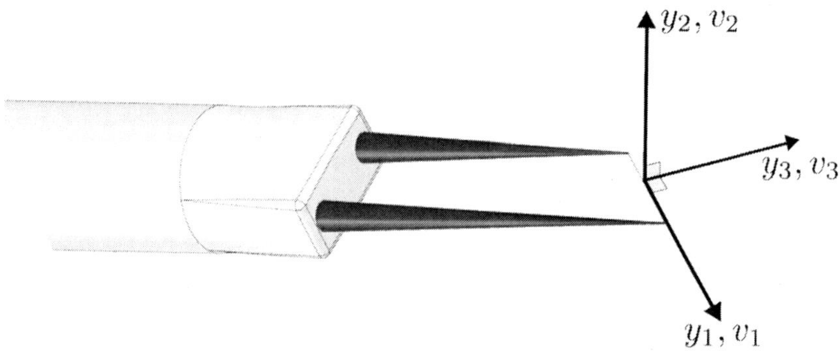

Figure 2 Schematic diagram illustrating the wire-fixed co-ordinate system y_i. The velocity components in tangential (y_1), normal (y_2), and bi-normal (y_3) directions are v_1, v_2, and v_3, respectively

The purpose of this rotated hot-wire method is to recover a set of velocity statistics defined in the stationary compressor co-ordinate system, x_i, based on a number of measurements acquired in the wire-fixed frame, y_i. With reference to Fig. 3, the probe stem is aligned with the x_3 axis and the probe setting angle, θ, is the angle formed between the rotated y_{12} plane and stationary x_{12} plane. Note that a positive change in θ corresponds to an anti-clockwise rotation around the probe stem axis, x_3. The probe setting angle is defined as zero when the x_1 axis, the sensor, and the prongs are coplanar and when the short prong lies upstream of the long prong. The wire slant angle, ψ, is the angle formed between the y_{13} and x_{13} planes and is held fixed at 45°.

A velocity vector u_i in the compressor co-ordinate system and its representation v_i in the wire-fixed co-ordinate system are related by the linear transformation

$$v_i = M_{ij} u_j \tag{1}$$

where

$$M_{ij} = \begin{bmatrix} \cos\psi\cos\theta & \cos\psi\sin\theta & \sin\psi \\ -\sin\theta & \cos\theta & 0 \\ -\sin\psi\cos\theta & -\sin\psi\sin\theta & \cos\psi \end{bmatrix} \tag{2}$$

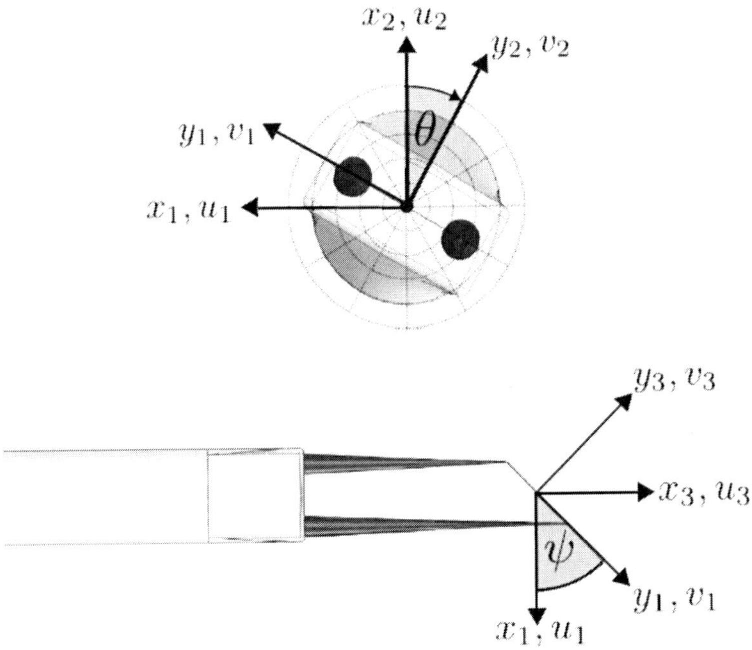

Figure 3 Schematic diagrams illustrating the wire-fixed co-ordinate system, y_i, stationary co-ordinate system, x_i, probe setting angle, θ, and wire slant angle, ψ

For a fixed probe orientation, the squared cooling velocity, q, can be evaluated using Jørgensen's equation (1971):

$$q = \delta_1^2 v_1^2 + \delta_2^2 v_2^2 + \delta_3^2 v_3^2 \tag{3}$$

where $\delta 1$, δ_2, and δ_3 are the tangential, normal, and bi-normal wire sensitivity coefficients, respectively. The velocity transformation (1) can be inserted into Jørgensen's Eq. (3) to obtain

$$q = G_{hk} u_h u_k \tag{4}$$

Where

Equation (4) relates the instantaneous cooling velocity to products of the Cartesian velocity for a known probe orientation. Next, a triple decomposition of the instantaneous cooling velocity (4) will be invoked to obtain expressions for the time-averaged, deterministic, and stochastic components.

2.2 Triple decomposition of instantaneous variables

For hot-wire measurements between compressor blade rows, it is useful to invoke a triple decomposition of the instantaneous cooling velocity:

$$\begin{aligned} q &= \overline{q}(\theta, \mathbf{x}) + \tilde{q}(\theta, \mathbf{x}, t) + q'(\theta, \mathbf{x}, t) \\ &= \langle q \rangle (\theta, \mathbf{x}, t) + q'(\theta, \mathbf{x}, t) \end{aligned} \tag{5}$$

where $\overline{q}$ is the time-averaged component, $\tilde{q}$ is the deterministic fluctuation, and q' is the true stochastic fluctuation. The phase-average $\langle q \rangle$ is obtained by evaluating

$$\langle q \rangle = \lim_{N \to \infty} \frac{1}{N} \sum_{n=0}^{N} q(\theta, \mathbf{x}, t + n\tilde{\tau})$$

where $\tilde{\tau}$ is the rotor passing period and N is the number of rotor revolutions. The deterministic fluctuation can be recovered by phase-averaging the triple decomposition (5) and rearranging to obtain

$$\tilde{q} = \langle q \rangle - \overline{q} \tag{6}$$

After inserting the triple decomposition (5) in to the instantaneous Eq. (4), and then applying the standard rules of phase-averaging outlined by Hussain and Reynolds (1970), the following equations are obtained:

$$\overline{q} = G_{hk}\left(\overline{u}_h \overline{u}_k + \overline{\tilde{u}_h \tilde{u}_k} + \overline{u'_h u'_k}\right) \tag{7}$$

$$\begin{aligned} \tilde{q} = G_{hk}(&\overline{u}_h \tilde{u}_k + \overline{u}_k \tilde{u}_h + \tilde{u}_h \tilde{u}_k - \overline{\tilde{u}_h \tilde{u}_k} \\ &+ \langle u'_h u'_k \rangle - \overline{u'_h u'_k}) \end{aligned} \tag{8}$$

$$q' = G_{hk}(\langle u_h \rangle u'_k + \langle u_k \rangle u'_h + u'_h u'_k - \langle u'_h u'_k \rangle) \tag{9}$$

In Eq. (8), the phase-averaged deterministic stresses are

$$\tilde{u}_h \tilde{u}_k = \left(\langle u_h \rangle - \overline{u}_h\right)\left(\langle u_k \rangle - \overline{u}_k\right) \tag{10}$$

and the phased-averaged Reynolds stresses are

$$\langle u'_h u'_k \rangle = \langle (u_h - \langle u_h \rangle)(u_k - \langle u_k \rangle) \rangle \tag{11}$$

Next, expressions that relate statistics of phase-averaged cooling velocity to the deterministic stresses (10) and Reynolds stresses (11) will be derived.

2.3 Phase-averaged sensor response equations

An equation for the phase-averaged cooling velocity, $\langle q \rangle$, can be obtained by summing time-averaged equation (7) and the deterministic equation (8) to give

$$\langle q \rangle = G_{hk}\left(\langle u_h \rangle \langle u_k \rangle + \langle u'_h u'_k \rangle\right) \tag{12}$$

To solve the phase-averaged equation (12), an additional equation that relates the phase-averaged Reynolds stresses (11) to statistics of the stochastic cooling velocity is required. A suitable expression can be derived by taking the outer product of Eq. (9) and then phase-averaging, to obtain

$$\begin{aligned}\langle Q \rangle = A_{ijhk}\Big[& \langle u'_h u'_i \rangle \langle u_k \rangle \langle u_j \rangle + \langle u'_k u'_i \rangle \langle u_h \rangle \langle u_j \rangle \\ & + \langle u'_h u'_j \rangle \langle u_k \rangle \langle u_i \rangle + \langle u'_k u'_j \rangle \langle u_h \rangle \langle u_i \rangle + C \Big]\end{aligned} \tag{13}$$

where

$$\langle Q \rangle = \lim_{N\to\infty} \frac{1}{N} \sum_{n=0}^{N} q'(t + n\tilde{\tau}, \theta) q'(t + n\tilde{\tau}, \theta) \tag{14}$$

and where A_{ijhk} is a rank four tensor defined here as

$$A_{ijhk} \equiv G_{ij} G_{hk}.$$

The last term on the right-hand side of Eq. (13) contains the third- and fourth-order velocity correlations and the product of the phase-averaged Reynolds stress tensor

$$\begin{aligned} C = {} & \langle u_h \rangle \langle u'_k u'_i u'_j \rangle + \langle u_k \rangle \langle u'_h u'_i u'_j \rangle + \langle u_i \rangle \langle u'_h u'_k u'_j \rangle \\ & + \langle u_j \rangle \langle u'_i u'_h u'_k \rangle + \langle u'_h u'_k u'_i u'_j \rangle - \langle u'_h u'_k \rangle \langle u'_i u'_j \rangle. \end{aligned}$$

To quantify the contribution of the term C towards the cooling velocity, a perturbation analysis of Eq. (13) was conducted. In the limit of a low turbulence (<10%) intensity flow, the instantaneous velocities can be written as

$$u_i = \langle u_i \rangle + \epsilon u'_i \quad \epsilon \to 0 \tag{15}$$

which, after some manipulation, can be used to derive the following set of sensor response equations:

$$\langle q \rangle = G_{hk} \langle u_h \rangle \langle u_k \rangle + \mathcal{O}(\epsilon^2) \tag{16}$$

$$\begin{aligned} \langle Q \rangle = \epsilon^2 A_{ijhk} \Big[& \langle u'_h u'_i \rangle \langle u_k \rangle \langle u_j \rangle + \langle u'_k u'_i \rangle \langle u_h \rangle \langle u_j \rangle \\ & + \langle u'_h u'_j \rangle \langle u_k \rangle \langle u_i \rangle + \langle u'_k u'_j \rangle \langle u_h \rangle \langle w_i \rangle \Big] \\ & + \mathcal{O}(\epsilon^3). \end{aligned} \tag{17}$$

Relative to the leading order terms, the term C is at least an order of magnitude smaller. As a result, the contribution of the term C towards the phase-averaged cooling velocity is herein considered negligible.

3 SOLUTION OF THE SENSOR RESPONSE EQUATIONS

This section describes the solution of the sensor response equations and is divided into three parts. First, the weighted non-linear regression algorithm employed to solve the phase-averaged response equations is described. Second, the accuracy of the non-linear regression model is demonstrated by conducting a series of a priori tests using synthetic data. Finally, the robustness of the non-linear regression output against small changes to the input parameters is verified by means of a sensitivity analysis.

3.1 Weighted non-linear regression method

If phase-averaged cooling velocities are available at several probe setting angles, then the sensor response equations (12) and (13) can be solved simultaneously using a weighted non-linear regression algorithm. For each probe setting angle, the residual between the modeled and measured data is

$$r_l = \langle q \rangle (\theta_l) - f_l(\theta_l) \tag{18}$$

$$s_l = \langle Q \rangle (\theta_l) - g_l(\theta_l) \tag{19}$$

where f and g correspond to the right-hand side of the phase-averaged equations (12) and (13), respectively. The sum-squared residual equations (18) and (19) define the cost function:

$$\mathcal{L} \equiv \frac{1}{2}\left(r_l r_l + \alpha s_l s_l\right) \tag{20}$$

where α is a weighting constant. The cost function L is minimised using the Newton–Raphson algorithm:

$$\mathbf{z}_{n+1} = \mathbf{z}_n + \gamma \mathcal{H}_n^{-1} \mathcal{J}_n \tag{21}$$

where subscript n denotes the iteration index, γ is a relaxation parameter, and z is the solution vector of phase-averaged velocity statistics:

$$z = \left[\langle u_1\rangle, \langle u_2\rangle, \langle u_3\rangle, \ldots \left\langle u_1' u_1'\right\rangle, \left\langle u_1' u_2'\right\rangle, \left\langle u_1' u_3'\right\rangle, \left\langle u_2' u_2'\right\rangle, \left\langle u_2' u_3'\right\rangle, \left\langle u_3' u_3'\right\rangle\right]^T.$$

The Jacobian J and the Hessian H are assembled at each Newton–Raphson update by evaluating the first- and second-order partial derivatives of the cost function (20) with respect to solution vector z_i:

$$\mathcal{J} = r_l \frac{\partial r_l}{\partial z_i} + \alpha s_l \frac{\partial s_l}{\partial z_i} \tag{22}$$

$$\mathcal{H} = \frac{\partial r_l}{\partial z_i}\frac{\partial r_l}{\partial z_i} + \frac{\partial r_l^2}{\partial z_i \partial z_i} + \alpha\left(\frac{\partial s_l}{\partial z_i}\frac{\partial s_l}{\partial z_i} + \frac{\partial^2 s_l}{\partial z_i \partial z_i}\right) \tag{23}$$

The Newton–Raphson algorithm (21) is initialised using values of $\langle u_i \rangle$ and $\langle R_{ij} \rangle$ obtained by solving the sensor response equations (16) and (17) separately. Equation (16) is solved using non-linear regression via Newton–Raphson algorithm, and Eq. (17) is solved using linear regression via QR factorisation. This two-step procedure returns values of $\langle u_h \rangle$ and $\langle R_{hk} \rangle$ which form the initial guess for the Newton–Raphson algorithm (21). Using this initialisation procedure, the cost function (20) is minimised to machine precision within ten iterations. The solution vector, z_i, is obtained by solving the linear system of equations via LU decomposition.

Next, the accuracy of the numerical solution to the sensor response equations is examined by performing a series of a priori tests using synthetic velocity statistics. As demonstrated by Buresti and Di Cocco

(1987), numerical tests are a convenient way to forecast the performance of rotated hot-wire methods before time-intensive measurement campaigns.

3.2 A priori tests

Velocity signals of the form

$$\begin{bmatrix} u_1(t) \\ u_2(t) \\ u_3(t) \end{bmatrix} = \begin{bmatrix} \overline{u}_1 \\ \overline{u}_2 \\ \overline{u}_3 \end{bmatrix} + \begin{bmatrix} \tilde{u}_1(t) \\ \tilde{u}_2(t) \\ \tilde{u}_3(t) \end{bmatrix} + \begin{bmatrix} u'_1(t) \\ u'_2(t) \\ u'_3(t) \end{bmatrix} \tag{24}$$

were synthesised to have known statistical properties. The magnitude of the mean axial velocity component was held fixed at $\overline{u}_1$=1.0, whereas the ratios $\overline{u}_2/\overline{u}_1$ and $\overline{u}_3/\overline{u}_1$ were varied systematically. The deterministic fluctuations, $\tilde{u}_i$, were modeled as a weighted sum of three sinusoids:

$$\tilde{u}_i(t) = A \sum_{m=0}^{2} \frac{\sin}{\sqrt{(m+1)}} \left[\frac{2(m+1)\pi}{\tilde{\tau}} \right] \tag{25}$$

where A denotes the amplitude. The stochastic velocity fluctuations, u'_i, were synthesised by correlating the output of a pseudo-random number generator with a Gaussian kernel. Reynolds shear stresses were simulated by taking linear combinations of the correlated sequences. The turbulence intensity was quantified using

$$\mathrm{Tu} \equiv \sqrt{\frac{1}{3}\overline{u'_i u'_i}} \times 100$$

A total of three test fields were synthesised—details are listed in Table 1.

Table 1 Statistical properties of the three synthetic velocity fields, including details of the time-averaged velocities, $\overline{u}_i$, time-averaged deterministic stresses, $\overline{\tilde{u}_i\tilde{u}_j}$, time-averaged Reynolds stresses, $\overline{u'_i u'_j}$, and turbulence intensity, Tu

Case	$\frac{\overline{u}_2}{\overline{u}_1}$	$\frac{\overline{u}_3}{\overline{u}_1}$	$\overline{\tilde{u}_1\tilde{u}_1}$, $\overline{u'_1u'_1}$ (×10^{-3})	$\overline{\tilde{u}_1\tilde{u}_2}$, $\overline{u'_1u'_2}$ (×10^{-3})	$\overline{\tilde{u}_1\tilde{u}_3}$, $\overline{u'_1u'_3}$ (×10^{-3})	$\overline{\tilde{u}_2\tilde{u}_2}$, $\overline{u'_2u'_2}$ (×10^{-3})	$\overline{\tilde{u}_2\tilde{u}_3}$, $\overline{u'_2u'_3}$ (×10^{-3})	$\overline{\tilde{u}_3\tilde{u}_3}$, $\overline{u'_3u'_3}$ (×10^{-3})	Tu (%)
A	0.1	0.2	1.58, 3.96	−0.54, −1.86	0.32, 0.17	0.19, 1.69	−0.11, −0.05	0.06, 1.86	5.00
B	0.2	0.3	3.32; 13.5	−0.99, −5.20	0.68, 2.71	0.29, 5.83	−0.20, −3.80	0.13, 10.3	10.0
C	0.3	0.4	5.72, 31.9	−1.66, −10.8	1.23, 3.80	0.52, 13.8	−0.38, −8.14	0.29, 21.4	15.0

Table 2 Absolute percentage errors on the time-averaged velocities, $\overline{u}_i$, time-averaged deterministic stresses, $\overline{\tilde{u}_i\tilde{u}_j}$, and time-averaged Reynolds stresses, $\overline{\tilde{u}_i\tilde{u}_j}$, relative to the target data listed in Table 1

Case	$\overline{u}_1$	$\overline{u}_2$	$\overline{u}_3$	$\overline{\tilde{u}_1\tilde{u}_1}$	$\overline{u'_1u'_1}$	$\overline{\tilde{u}_1\tilde{u}_2}$	$\overline{u'_1u'_2}$	$\overline{\tilde{u}_1\tilde{u}_3}$	$\overline{u'_1u'_3}$	$\overline{\tilde{u}_2\tilde{u}_2}$	$\overline{u'_2u'_2}$	$\overline{\tilde{u}_2\tilde{u}_3}$	$\overline{u'_2u'_3}$	$\overline{\tilde{u}_3\tilde{u}_3}$	$\overline{u'_3u'_3}$
A	0.14	0.17	0.12	0.24	0.16	0.45	0.22	0.12	0.31	0.43	0.27	0.29	0.12	0.33	0.25
B	0.18	0.20	0.18	0.13	0.12	0.21	0.28	0.17	0.13	0.18	0.23	0.28	0.21	0.31	0.33
C	0.21	0.22	0.12	0.17	0.16	0.12	0.33	0.22	0.42	0.37	0.27	0.31	0.25	0.34	0.45

For each test field, velocities were converted into a cooling velocity using Jørgensen's equation (4) which was then phase-averaged. The sequences generated for each probe setting angle are independent: this corresponds to the situation of consecutive measurements being taken at a fixed point in the flow and rotating the probe about its stem. Cooling velocities were synthesised at 15 probe setting angles distributed equally on the intervals $\theta \in [0,105]^\circ$ and $\theta \in [255,345]^\circ$ in 15° increments.

The accuracy of the non-linear regression algorithm was verified by comparing the values of the synthesised cooling velocity against the numerical prediction for each probe setting angle. Synthesised values of the phase-averaged cooling velocity, $\langle q \rangle$, are compared against the numerical prediction for case C in Fig. 4. The l_∞ norm, based on the difference between the synthesised and predicted data, was computed to be $O(10^{-2})$.

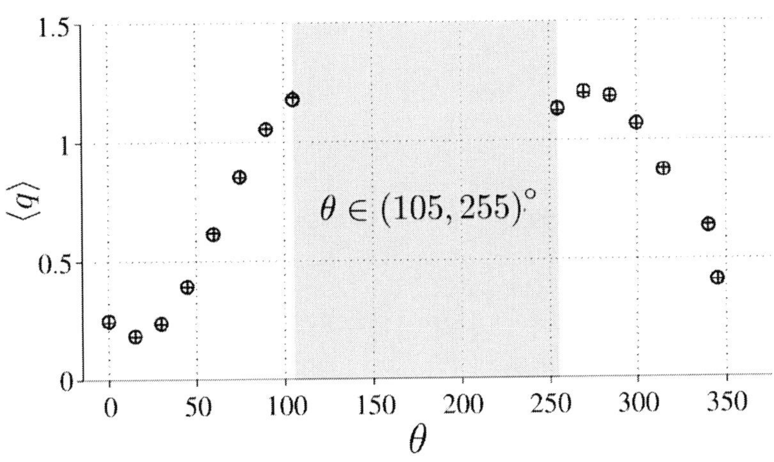

Figure 4 Phase-averaged cooling velocity $\langle q \rangle$ including the synthesised *open circle* and the predicted values (+) for the high turbulence intensity case C (see Table 1)

The absolute percentage error on the time–mean statistics, relative to the target data listed in Table 1, was quantified as

$$\text{Error} \equiv \frac{\left\| \overline{\mathbf{z}} - \overline{\mathbf{z}}_{\text{target}} \right\|}{\left\| \overline{\mathbf{z}}_{\text{target}} \right\|} \times 100$$

The absolute percentage error on each time-mean statistic was computed to be less than 1%1%—details are provided in Table 2.

3.3 Sensitivity analysis

In practice, the accuracy of the rotated hot-wire measurements will be susceptible to probe imperfections, probe positioning errors, and drift in the calibration coefficients. Therefore, it is important to quantify how small changes in the model input parameters

$$\lambda_m = \{\psi, \delta_1, \delta_2, \delta_3, \theta_l\} \tag{26}$$

affect the output of the non-linear regression model. ematically, the change in the non-linear regression output with respect to the model parameter set λ_m can be written as

$$\frac{\partial z_i}{\partial \lambda_m} = -\mathcal{J}^{-1}\left(\frac{\partial r_l}{\partial \lambda_m}\frac{\partial r_l}{\partial z_i} + r_l \frac{\partial^2 r_l}{\partial \lambda_m \partial z_i} + \alpha\left(\frac{\partial s_l}{\partial \lambda_m}\frac{\partial s_l}{\partial z_i} + s_l \frac{\partial s_l^2}{\partial \lambda_m \partial z_i} \right)\right) \tag{27}$$

After evaluating the total derivative of the left-hand side of equation (27), and then applying the Cauchy–Schwarz inequality, the sensitivity of the non-linear regression output can be written as

$$|dz_i| \le \left|\frac{\partial z_i}{\partial \psi}\right| |d\psi| + \left|\frac{\partial z_i}{\partial \delta_m}\right| |d\delta_m| + \left|\frac{\partial z_i}{\partial \theta_l}\right| |d\theta_l| \tag{28}$$

where $|d\psi|$, $|d\delta m|$, and $|d\theta_l|$ are the levels of uncertainty in the wire slant angle, directional sensitivity coefficients, and probe setting angles, respectively. For the purposes of this sensitivity analysis, the levels of uncertainty were held fixed at 5%. The non-linear regression output was found to be most sensitive to the wire sensitivity coefficient normal to the plane of the wire and the prongs, δ_2, followed by δ_3, δ_1, θ and ψ, as indicated by the relative magnitude of the corresponding terms on the right-hand side of Eq. (28). The sensitivity of each flow statistic to δ_2 can be written as

$$\left|\frac{\partial z_i}{\partial \delta_2}\right| = \underbrace{\left|\frac{\partial \overline{u_1}}{\partial \delta_2}\right|}_{S_1} + \underbrace{\left|\frac{\partial \overline{u_2}}{\partial \delta_2}\right|}_{S_2} + \underbrace{\left|\frac{\partial \overline{u_3}}{\partial \delta_2}\right|}_{S_3} + \underbrace{\left|\frac{\partial \overline{u'_1 u'_1}}{\partial \delta_2}\right|}_{S_4} + \underbrace{\left|\frac{\partial \overline{u'_1 u'_2}}{\partial \delta_2}\right|}_{S_5}$$

$$+ \underbrace{\left|\frac{\partial \overline{u'_1 u'_3}}{\partial \delta_2}\right|}_{S_6} + \underbrace{\left|\frac{\partial \overline{u'_2 u'_2}}{\partial \delta_2}\right|}_{S_7} + \underbrace{\left|\frac{\partial \overline{u'_2 u'_3}}{\partial \delta_2}\right|}_{S_8} + \underbrace{\left|\frac{\partial \overline{u'_3 u'_3}}{\partial \delta_2}\right|}_{S_9}.$$

Each of the above sensitivities, S_i, are shown in Fig. 5 for the highest turbulence intensity case C (see Table 1). At most, the sensitivities are of unit order and this confirms that the output of the non-linear regression model is robust against small changes to the input parameters.

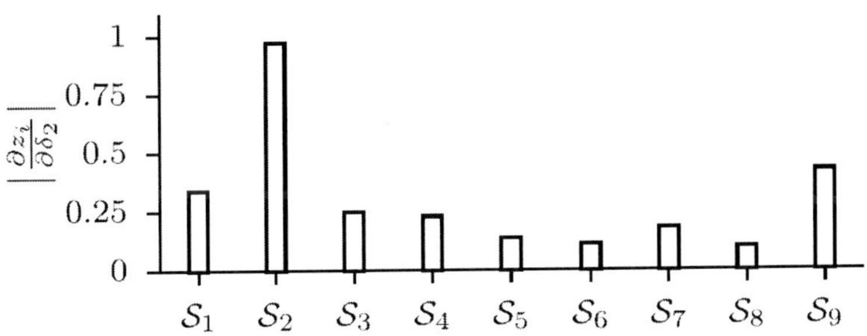

Figure 5 Sensitivities of the non-linear regression output, S_i, to small changes in the normal directional coefficient, δ_2, for the high turbulence intensity case C

4 EXPERIMENTAL SETUP

This section describes the experimental set up and is divided into three parts. First, descriptions of the research compressor facility and traversing mechanism are provided. Second, the hot-wire anemometry apparatus, sampling parameters, and calibration procedures are described. Finally, some preliminary hot-wire measurements are discussed and a validation case against an independent five-hole pressure probe measurement is presented.

4.1 Axial-flow compressor facility and traverse system

Measurements were acquired in a single-stage, axial-flow, low-speed compressor—a facility that has been used in previous studies (Goto 1991; Weichert and Day 2014). Ambient air is drawn into a 3400 mm diameter bell-mouth inlet through a thick gauze. The airstream passes through a honeycomb grid and two additional wire gauzes designed to straighten the flow. Following a 9:1 area contraction, the flow enters a parallel annulus with a mean radius of 762 mm and a hub-to-tip ratio of 0.8. The airstream then passes through a turbulence grid, a row a 49 inlet guide vanes (IGV), 51 rotor blades, and 49 stator blades. The stator blades have a chord length of 132.4 mm, a trailing edge diameter of 3 mm, and a pitch-to-chord ratio of 0.8. Downstream of the stator row trailing edge, the cylindrical annulus continues for an additional 20 stator chord lengths before the airstream exits into a dump diffuser. A conical throttle valve and an auxiliary fan positioned at the rear of the machine are used to set the flow rate through the compressor. The compressor was operated close to its design point, and the Reynolds number, based on the rotor speed at mid-height, U_m, and the stator chord, c, was held fixed at 2.4×10^5.

Before acquiring data, the temperature inside the compressor was allowed to settle into a statistically stationary state. By monitoring the output of a thermocouple positioned upstream of the rotor, a steady-state temperature was achieved after a settling period of around 30 min. After this, the time-averaged rotor inlet temperature was acquired simultaneously during each hot-wire measurement. A time history of the rotor inlet temperature and the probability density function (PDF) of the fluctuations are shown in Fig. 6. For all the results included in this paper, the temperature fluctuations were within ±0.25 °°C of the mean value.

A three-axis traverse gear secured to the outer casing of the compressor was used to control the position and orientation of the slanted hot-wire sensor. A circumferential slot in the casing permitted traverses on the plane perpendicular to the compressor axis to be performed. A pair of stepper motors were used to control the radial and tangential positions of the probe. The probe setting angle was also controlled using a stepper motor with an angular resolution of $\Delta\theta=\pm0.9^\circ$. The current traverse mechanism permits fully automated, rotated hot-wire measurements to be acquired from hub-to-casing across four stator passages. All rotated hot-wire measurements were acquired in the compressor stationary co-ordinate system, x_i.

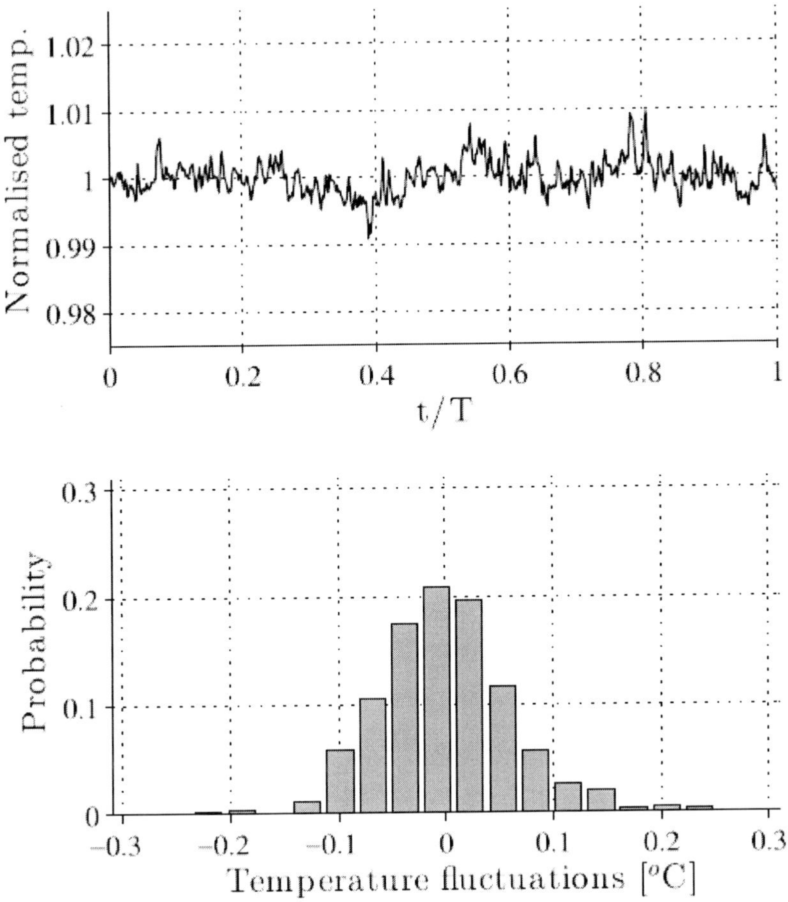

Figure 6 Temperature variations measured at the rotor inlet including a time-history of normalised temperature and the probability density function of temperature fluctuations

The blade-to-blade plane of the compressor stage is shown in Fig. 7. The velocity components in the axial (x_1), circumferential (x_2), and radial (x_3) directions are denoted u_1, u_2, and u_3, respectively. Measurements were acquired across two stator pitches at mid-span, at an axial distance x_1/c=0.2 from the stator trailing edge.

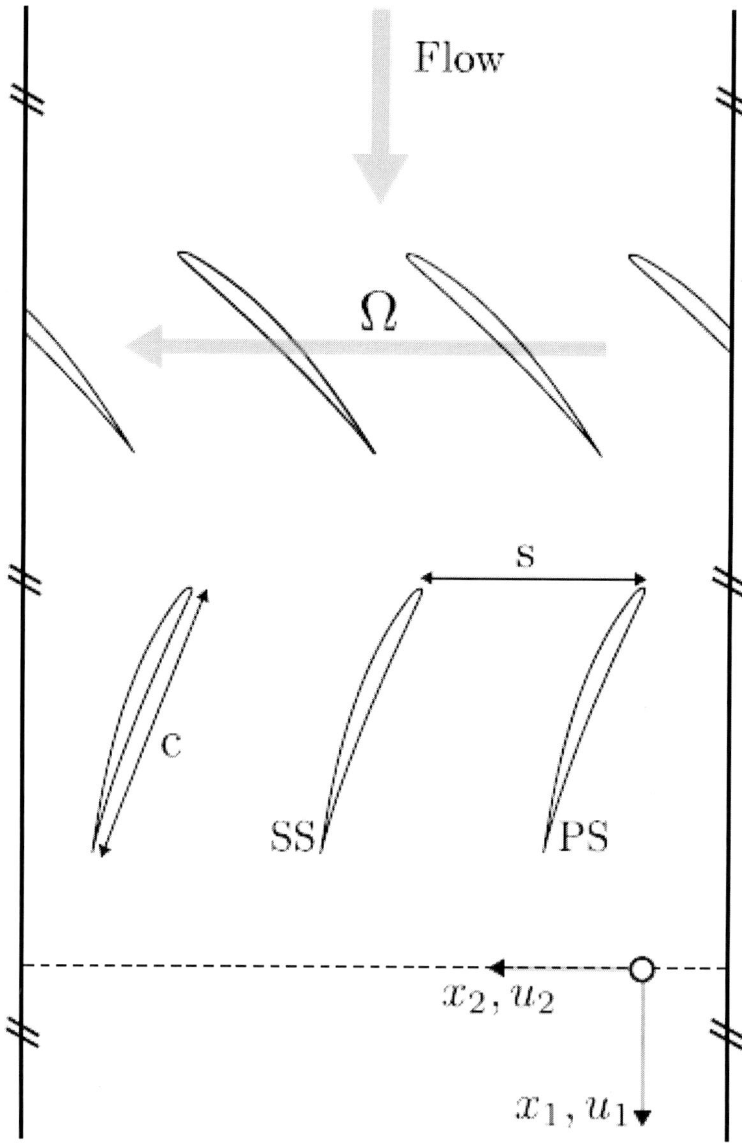

Figure. 7 Blade-to-blade plane at mid-span indicating the stator pitch, *s*, the stator chord, *c*, the suction-surface (SS), and the pressure-surface (PS) sides of the stator wake. All measurements were acquired across two stator pitches at an axial distance x_1/c=0.2 downstream from the stator row trailing edge (*dashed lines*)

4.2 Hot-wire anemometry and calibration procedure

Hot-wire measurements were acquired using a constant temperature anemometer (CTA). The bridge voltage was digitised using a 16-bit analogue-to-digital-converter interfaced to the output of a BNC multiplexer. The sensor was operated with an overheat ratio of 1.8. The ratio of the sensor sampling frequency, f_s, to the shaft frequency was held fixed at 1.2×10^4. The sensor cut-off frequency, f_c, was tuned to be half the value of f_s. The slanted hot-wire probe has two straight prongs which support a 3 mm platinum-plated tungsten wire inclined at a nominal angle of 45°, relative to the probe stem axis. The sensing element is etched halfway between the prongs to have a length-to-diameter ratio of 250.

Before and after each measurement, the hot-wire sensor was calibrated against a Pitot-static probe housed in a stand-alone, round jet facility. A detailed description of the calibration tunnel is provided by di Mare et al. (2017). Hot-wire measurements were acquired within the potential core of the jet, one nozzle diameter downstream of the nozzle exit, with the flow normal to the plane of the prongs. The squared sensor voltage, E^2, was correlated to the effective cooling velocity using the power law proposed by King (1914)

$$E^2 - E_0^2 = Bq^{\frac{n}{2}} \tag{30}$$

where E_0 is the voltage measured at zero velocity and where B and n are the King's Law coefficients. By recording the bridge output across a range of known jet velocities, the King's Law coefficients B and n were obtained using simple linear regression.

Throughout velocity calibration, the bridge voltage varied with both the velocity and temperature of the jet. Data from multiple velocity calibrations were collapsed on to a single master curve using a temperature correction of the form:

$$E^* = \left(E - E_0\right) + \alpha\Delta T \tag{31}$$

where E^* denotes the temperature-corrected bridge voltage, α is a user-prescribed coefficient, and $\Delta T\Delta T$ denotes the temperature difference, relative to a reference value. Four sets of temperature-corrected data are shown in Fig. 8 which collapse on to a master calibration curve using a value of α=0.008 for temperature changes within the range $\Delta T\pm2$°C.

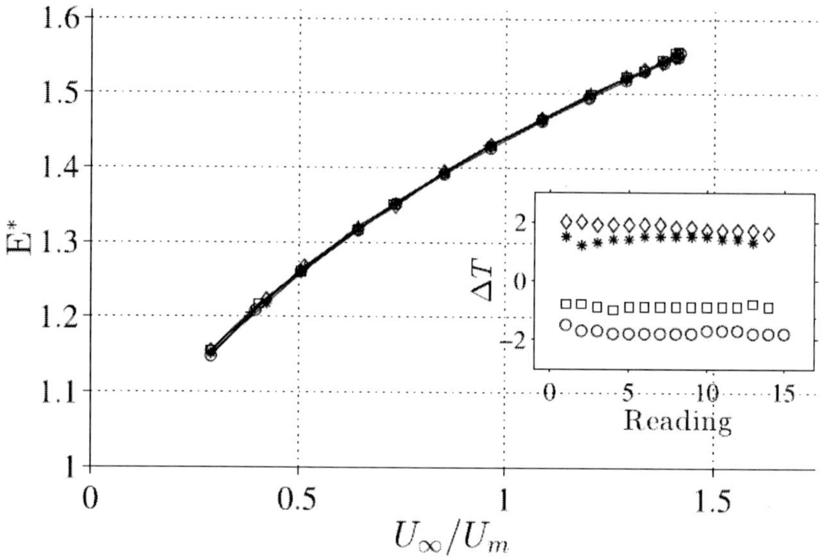

Figure 8 Master calibration curve showing temperature-corrected voltages. The inset figure shows the temperature difference throughout velocity calibration. Note that the data have been normalised using the rotor speed at mid-height, U_m

The wire sensitivity coefficients, δ_i, were determined via directional calibration. First, data were acquired at three fixed probe orientations, as shown in Fig. 9. The cooling velocity at each orientation was determined using the bridge voltage and King's Law coefficients. Using this data, the wire sensitivity coefficients were evaluated using the relation:

$$\delta_i^2 = \frac{\overline{q}_{\delta_i}}{\overline{q}_{\delta_2}} \tag{32}$$

where $\overline{q}_{\delta_1}, \overline{q}_{\delta_2}, \overline{q}_{\delta_3}$ denote the cooling velocity measured at tangential, normal, and bi-normal orientations, as shown in Fig. 9, respectively. Typical values of δ_1, δ_2 and δ_3 were found to be 0.1, 1.0, and 0.9, respectively. In addition to using Eq. (32), the values of δiδi were also estimated using a non-linear regression procedure. By acquiring data at different—but known—orientations of the wire and flow velocities in the calibration tunnel, the set of wire parameters

$$\mathbf{z} = \left[\psi, \delta_1^2, \delta_3^2\right]^\top \tag{33}$$

was determined by minimising the residual function

$$r_i = E_0^2 + B\overline{q}^{\frac{n}{2}}(\mathbf{z}, \mathbf{u}_i, \theta_i) - E_i^2 \quad (34)$$

using the Newton–Raphson algorithm described in Sect. 3.1. The values of δ_i estimated via non-linear regression were within 5% of the values predicted using Eq. (32). In addition, the wire slant angle ψ was predicted to within ± 0.2∘∘ of the measured value based on a magnified image of the sensor. With reference to the sensitivity analysis of the sensor response equations presented in Sect. 3.3, an uncertainty of $|d\delta_i|\approx 0.05$ will not significantly the alter the output of the non-linear regression algorithm. As a result, the values of δ_1, δ_2 and δ_3 used throughout this work were 0.1, 1.0, and 0.9, respectively.

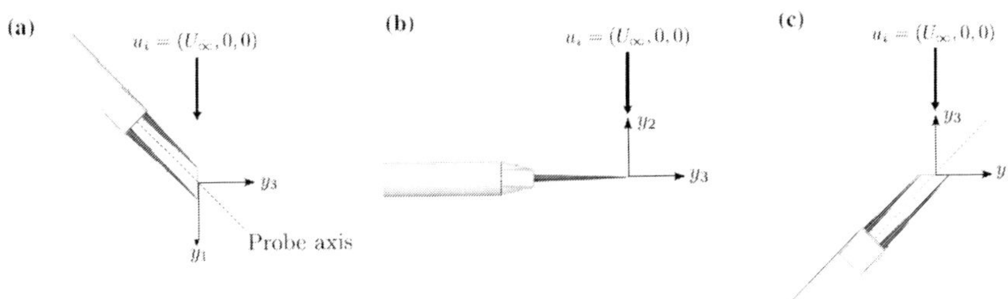

Figure 9 Schematic diagram showing the orientation of the calibration jet velocity vector, u_i=(U_∞,0,0), and the slanted hot-wire used to evaluate the **a** tangential (δ_1); **b** normal (δ_2) and **c** bi-normal (δ_3) wire sensitivity coefficients

The time-mean cooling velocity,, is compared against the velocity measured using the Pitot-static tube at 36 uniformly spaced probe setting angles (with the plane of the prongs perpendicular to the jet axis) in Fig. 10. The data sets match to within 3% for all probe setting angles outside the range $\theta \in (105,255)$∘. Within this range, the sensing element becomes immersed in the wake of the long prong, and as a result, significant levels of disagreement are observed.

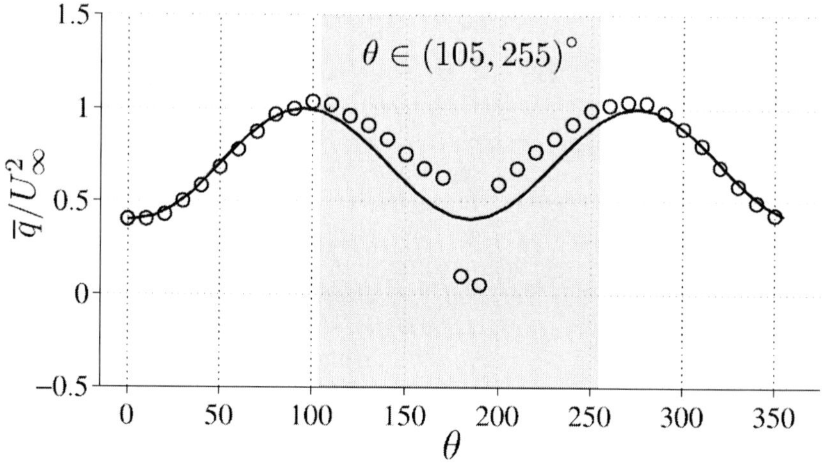

Figure 10 Validation of sensor calibration procedure showing measured (*open circle*) and modeled (*solid lines*) mean cooling velocity, $\bar{q}$, at 36 uniformly spaced probe setting angles. Note that the data have been normalised using the jet velocity, U_∞

4.3 Statistical convergence and preliminary checks

All hot-wire measurements acquired inside the compressor were recorded relative to a once-per-revolution rotor trigger, which was also used to determine the blade passing frequency:

$$\beta = K\Omega$$

where K and Ω denote the number of rotor blades and rotor frequency, respectively. Throughout the measurement campaign, the rotor frequency had a nominal value of 528 revolutions per minute. Small deviations in the rotor frequency were accounted for by linearly interpolating each once-per-revolution record to have 6528 points. The interpolated data were ensemble-averaged over N rotor revolutions to obtain phase-averaged statistics.

The convergence of time–mean statistics was verified using a hot-wire signal acquired at a fixed probe setting angle (θ=90°). The effect of increasing the number of the rotor revolutions, N, upon the mean-squared deterministic cooling velocity, $\overline{\tilde{q}^2}$, and the mean-squared stochastic cooling velocity, $\overline{q'^2}$, at three tangential positions is shown in Fig. 11.

Adequate convergence is achieved within a sampling period of 64 rotor revolutions at each tangential position.

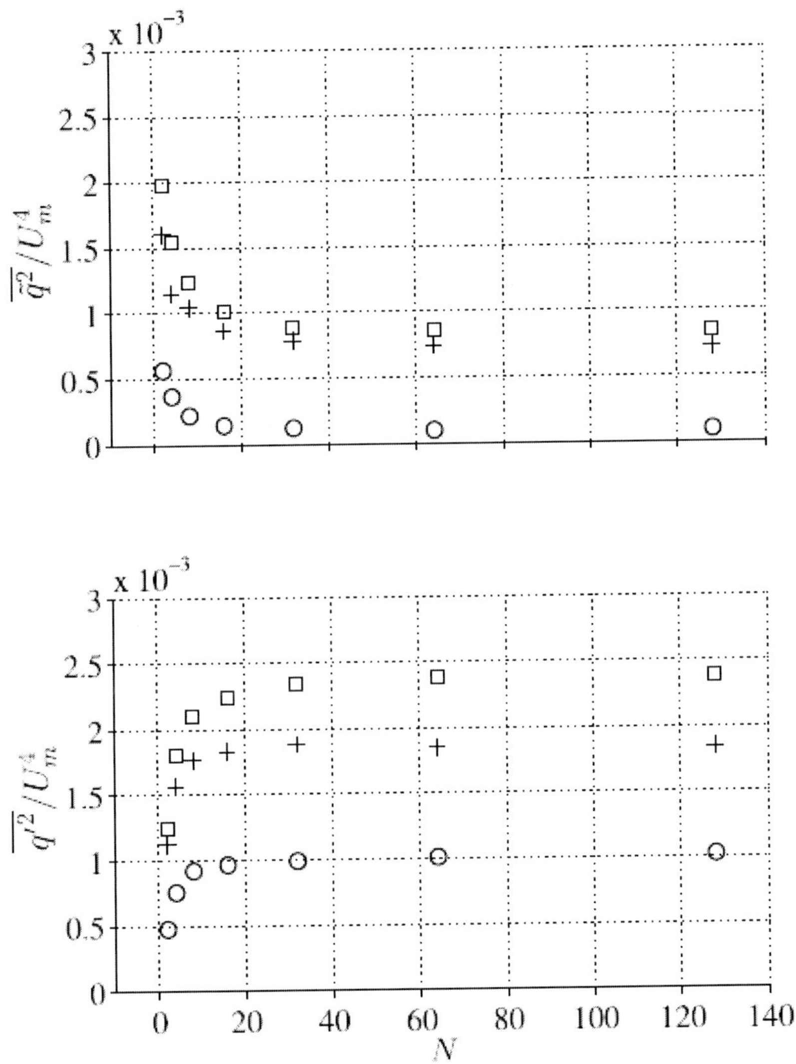

Figure 11 Convergence of the mean-squared deterministic cooling velocity, $\overline{\tilde{q}^2}$, and mean-squared stochastic cooling velocity, $\overline{q'^2}$, with increasing number of rotor revolutions, *N*. Data were acquired on the wake SS (*open square*), wake centre line (*open circle*), and wake PS (+)(+) and have been normalised using the rotor speed at mid-height, U_m

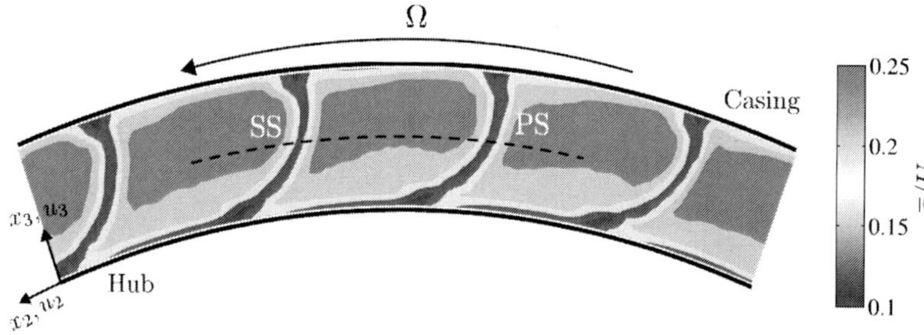

Figure 12 Four-passage area traverse showing contours of mean cooling velocity, $\overline{q}$. The *dashed line* indicates the position of a two-passage traverse at mid-span. Data were acquired at a fixed probe orientation (θ=90∘) and have been normalised using the blade speed at mid-height, UmUm

A four-passage area traverse was conducted to verify circumferential periodicity around the annulus. Data were acquired at a fixed probe orientation (θ=90∘) for a sampling period of 64 rotor revolutions across an equi-spaced grid of 16 radial points and 96 tangential points. Contours of the mean cooling velocity, $\overline{q}$, are shown in Fig. 12—excellent levels of passage-to-passage periodicity are observed.

To identify the probe orientations susceptible to prong interference, a single-passage line traverse at mid-span was conducted. For each measurement point, the probe was rotated through one revolution in 24 equal increments. Interference was detected in the range θ∈[120,240]∘ and these angles were omitted from all future measurements. After this, data were acquired at 15 different probe setting angles distributed equally on the intervals θ∈[0,105]∘ and θ∈[255,345]∘ in 15∘ degree increments for each measurement point. For each probe orientation, the phase-averaged cooling velocities were computed using an ensemble of 64 phase-locked time series.

The accuracy and reliability of the rotated hot-wire results were validated against an independent five-hole pressure probe (5HP) measurement. The 5HP was calibrated using the non-nulling procedure described by Treaster and Yocum (1978). The time-averaged velocity vector, $\overline{u}_i$, based on the 5HP and rotated hot-wire measurements is compared in Fig. 13, which also includes a 95% confidence interval on the hot-wire data. After using the time-averaged velocity vector, $\overline{u}_i$, to evaluate the total velocity

$$|u_i| = \sqrt{\overline{u}_1^2 + \overline{u}_2^2 + \overline{u}_3^2}$$

the maximum relative error on $|u_i|$ between the 5HP and hot-wire data was computed to be 9% which occurred on the wake suction side. An improved level of agreement may be possible by employing smaller, fast-response pressure probes. The steps required to fabricate, calibrate, and use miniaturised 5HP in turbomachinery flows have been discussed in detail by Georgiou and Milidonis (2014) and Grimshaw and Taylor (2016).

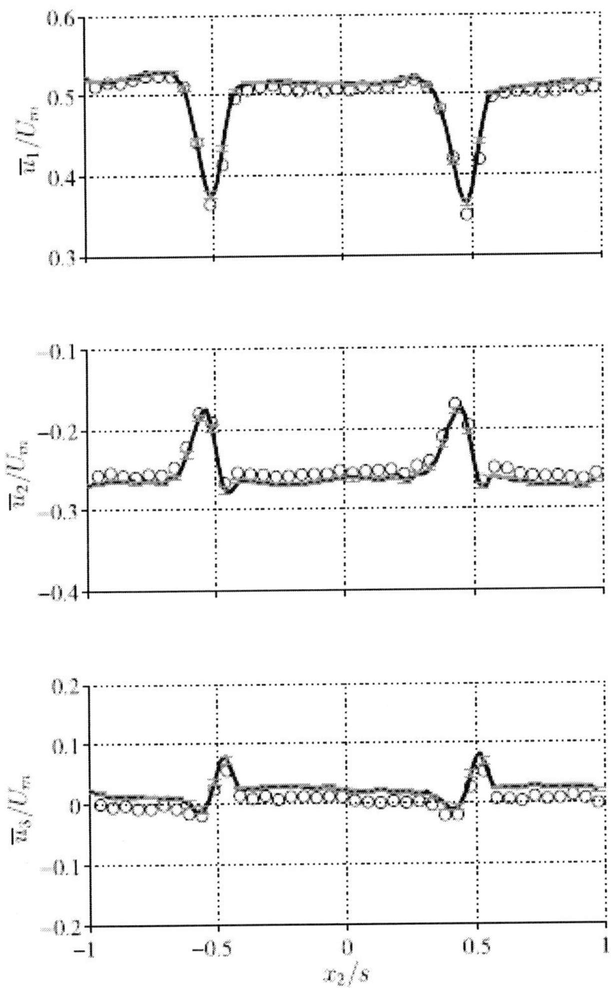

Figure 13 Comparison of time-averaged velocity vector, $\overline{u}_i$, based on the rotated hot-wire (*solid line*) and 5HP (*open square*) measurements. A 95% confidence interval on the hot-wire data (*solid line*) is also shown. Note that the data have been normalised using rotor speed at mid-height, U_m, and the stator pitch, s

5 PHASE-AVERAGED VELOCITY STATISTICS

This section presents the phase-averaged velocity statistics downstream of the compressor stator and is divided into three parts. First, the phase-averaged and deterministic velocity are presented. Second, the six components of the phase-averaged deterministic stress tensor are presented. Finally, the six components of the phase-averaged Reynolds stress tensor are examined and compared against their deterministic counterpart.

5.1 Phase-averaged and deterministic velocities

The phase-averaged flow field downstream of the compressor stator is shown in Fig. 14. Space–time contours of the axial, $\langle u_1 \rangle$, tangential, $\langle u_2 \rangle$, and radial, $\langle u_3 \rangle$, phase-averaged velocities are plotted for three blade passing periods. On either side of the stator wake centre line ($x_2/s \approx \pm 0.5$), the passing rotor wakes are visible in the "free-stream" region ($-0.4 < x_2/s < +0.4$). The rotor wakes interact with the edges of the stator wake and cause the stator wake to meander in both space and time. The rotor–stator interactions can be clarified by evaluating the deterministic velocity:

$$\tilde{u}_i(x_2,t) = \langle u_i \rangle (x_2,t) - \overline{u}_i(x_2).$$

The deterministic flow field downstream of the compressor stator is shown in Fig. 15. Space–time contours of the axial, $\tilde{u}_1$, tangential, $\tilde{u}_2$, and radial, $\tilde{u}_3$, deterministic velocities are plotted for three blade passing periods. Regions of negative deterministic velocity $(\tilde{u}_i < 0)$ correspond to individual rotor wakes, which intersect each stator wake and tangentially coalesce across the passage. The interaction of the rotor and stator wakes leads to vigorous deterministic fluctuations within the stator wake, where the axial, tangential, and radial components achieve a maximum absolute value of 0.08, 0.08, and 0.04 U_m, respectively. In general, the deterministic flow patterns shown in Fig. 15 are in-line with observations of Uzol et al. (2002), who interpreted compressor flow fields as a "lattice" of intertwined blade wakes.

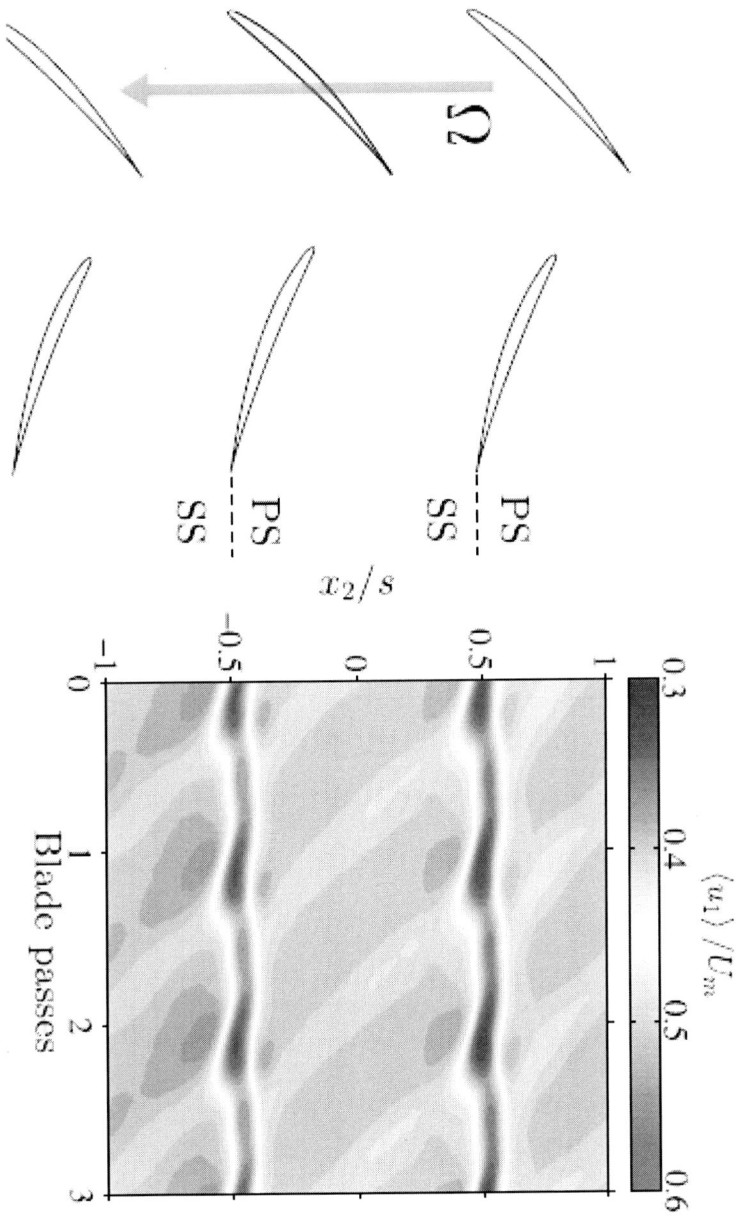
Ω
PS
SS
PS
SS
x_2/s
$\langle u_1 \rangle / U_m$
0.3
0.4
0.5
0.6
1
0.5
0
−0.5
−1
0
1
2
3
Blade passes

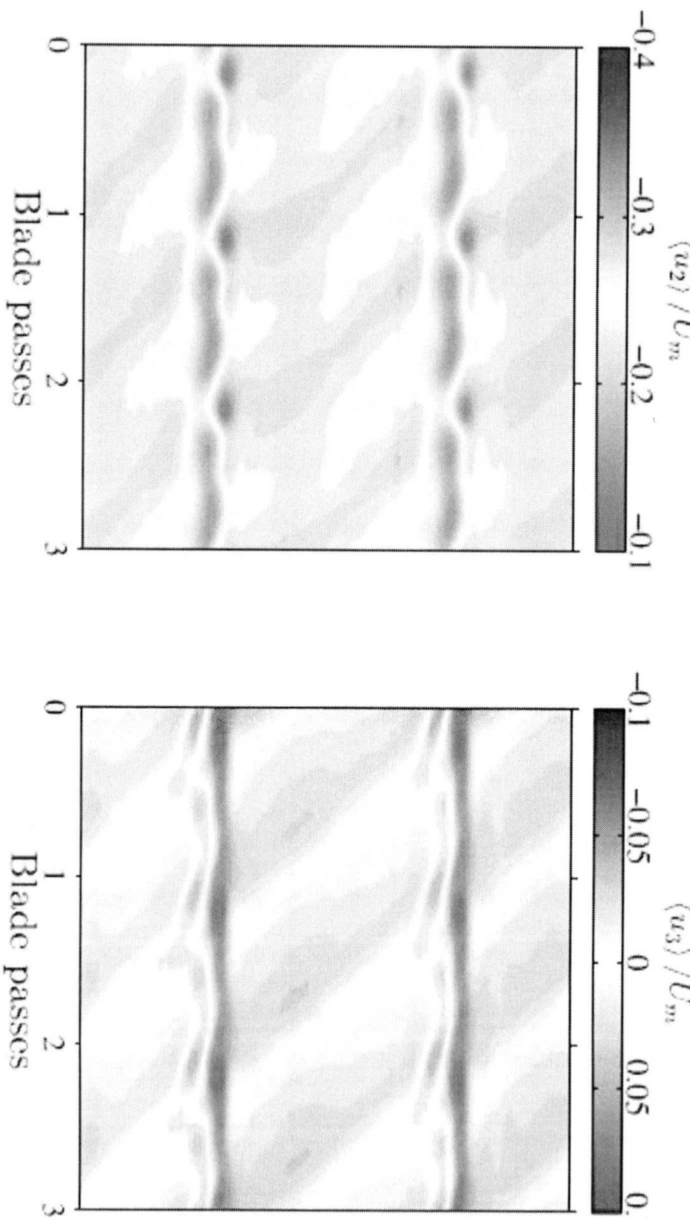

Figure 14 Phase-averaged velocities including the axial $\langle u_1 \rangle$, tangential $\langle u_2 \rangle$, and radial $\langle u_3 \rangle$ components. Note that data are shown for three blade passing periods and have been normalised using the rotor speed at mid-height, U_m, and the stator pitch, s

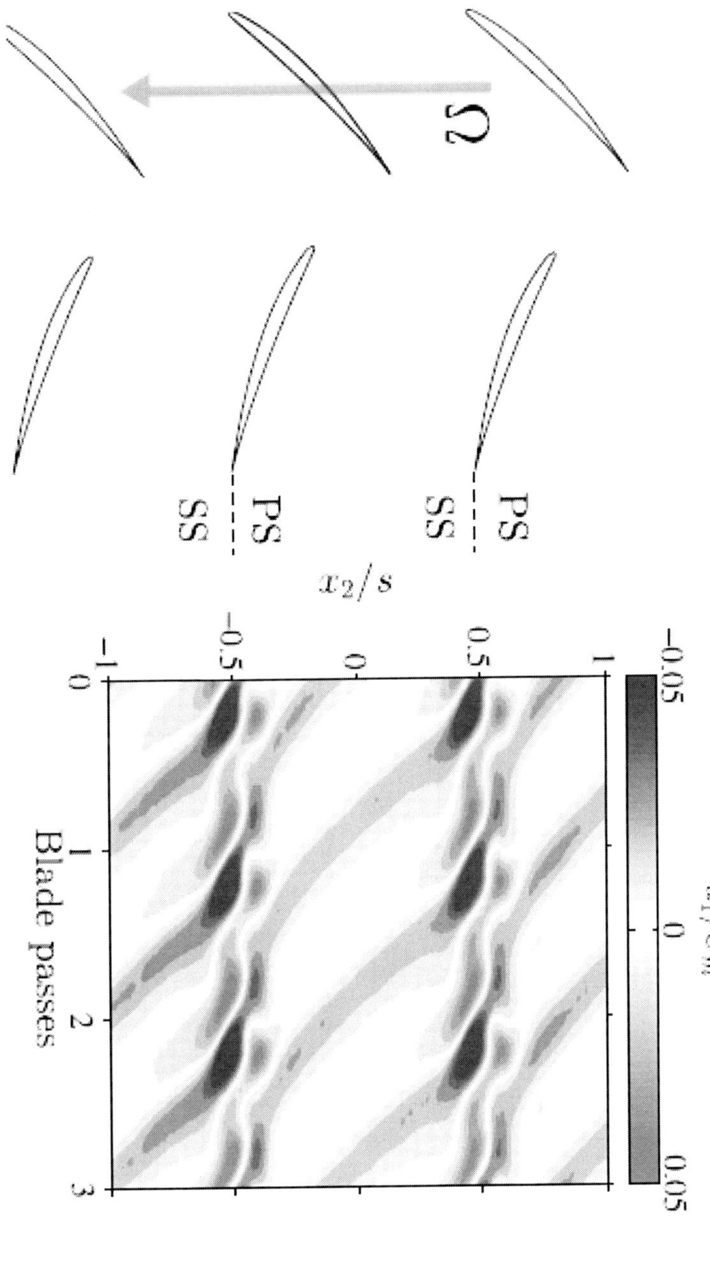
Ω
PS
SS
PS
SS
x_2/s
$\tilde{u}_1/U_m$
−0.05
0
0.05
1
0.5
0
−0.5
−1
0
1
2
3
Blade passes

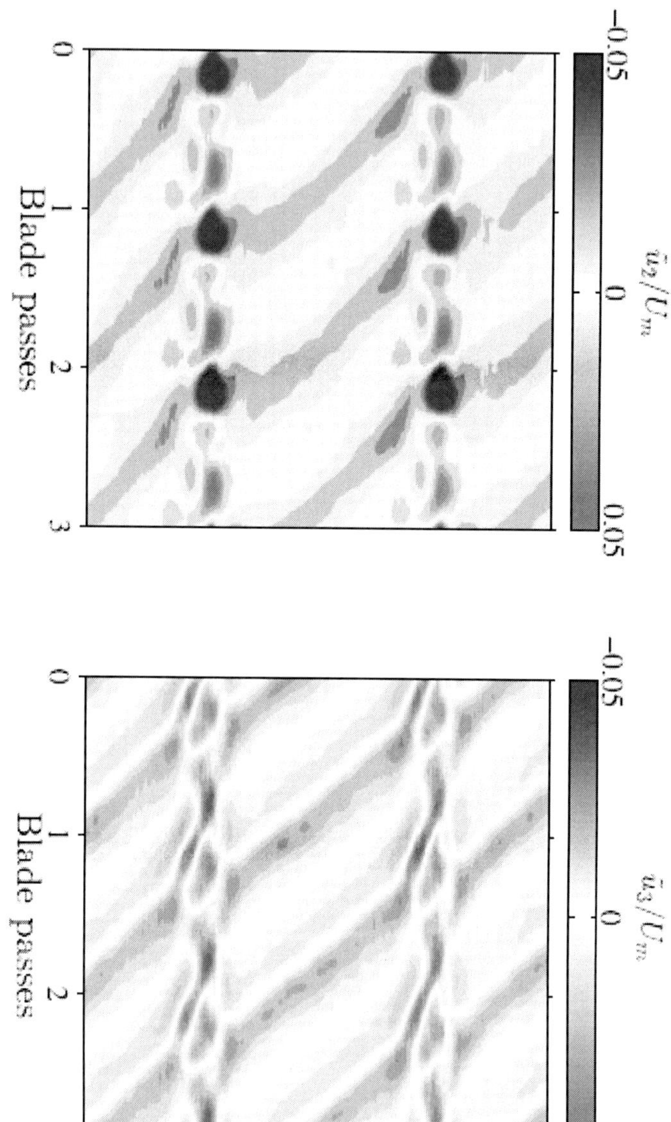

Figure 15 Deterministic velocities including the axial $\tilde{u}_1$, tangential $\tilde{u}_2$, and radial $\tilde{u}_3$ components. Note that data are shown for three blade passing periods and have been normalised using the rotor speed at mid-height, U_m, and the stator pitch, s

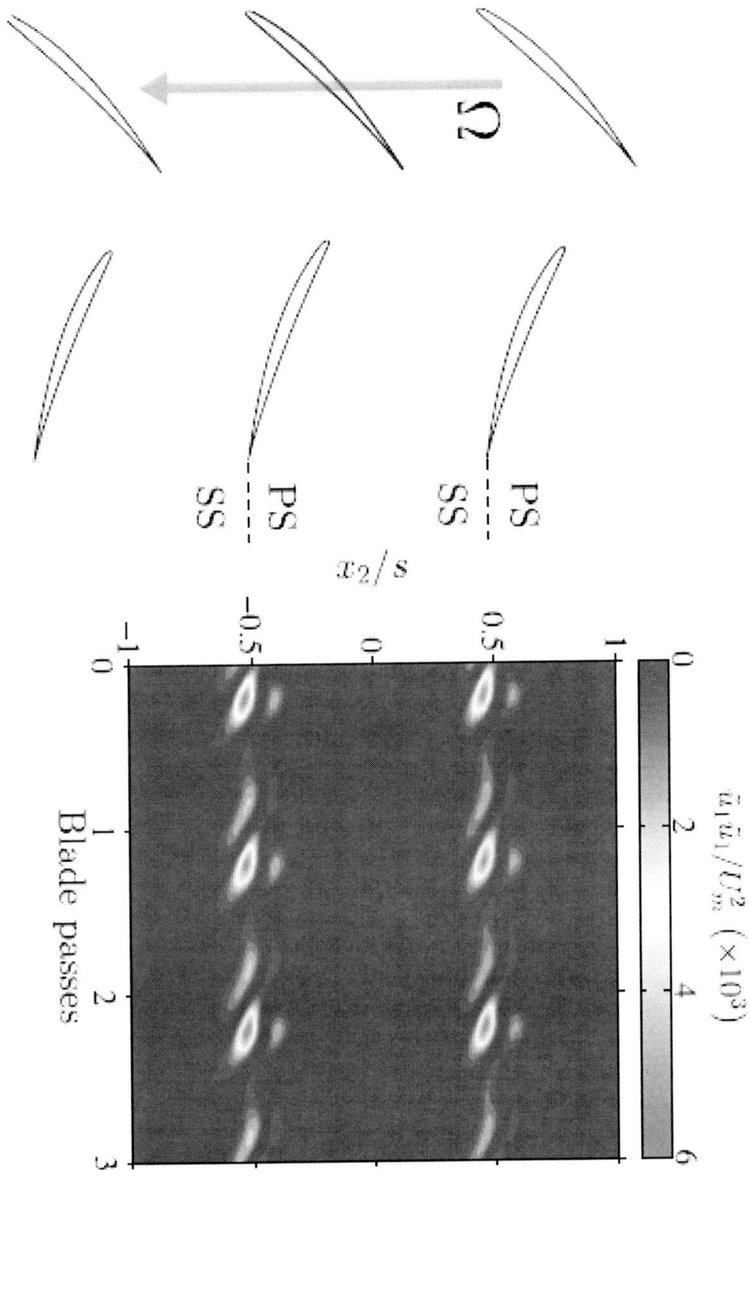
Ω
PS
SS
PS
SS
x_2/s
1
0.5
0
−0.5
−1
0
1
2
3
Blade passes
$\tilde{u}_1\tilde{u}_1/U_m^2$ (×10^3)
0
2
4
6

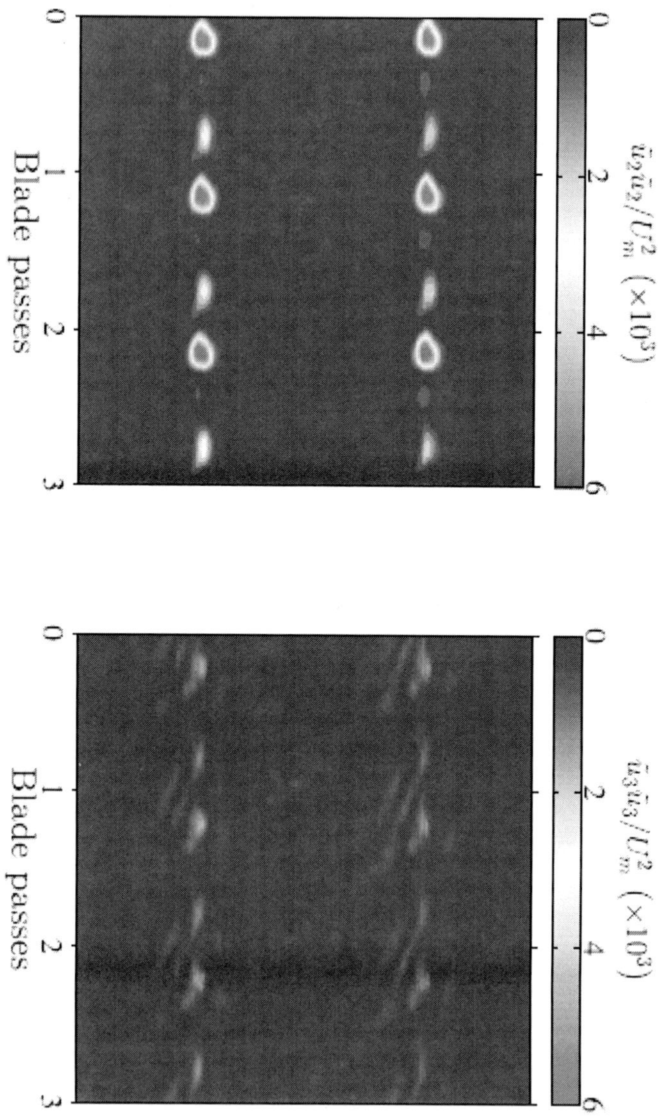

Figure 16 Phase-averaged deterministic normal stresses including the axial–axial $\tilde{u}_1\tilde{u}_1$, tangential–tangential $\tilde{u}_2\tilde{u}_2$, and radial–radial $\tilde{u}_3\tilde{u}_3$ components. Note that data are shown for three blade passing periods and have been normalised using the rotor speed at mid-height, Um,Um, and the stator pitch, *s*

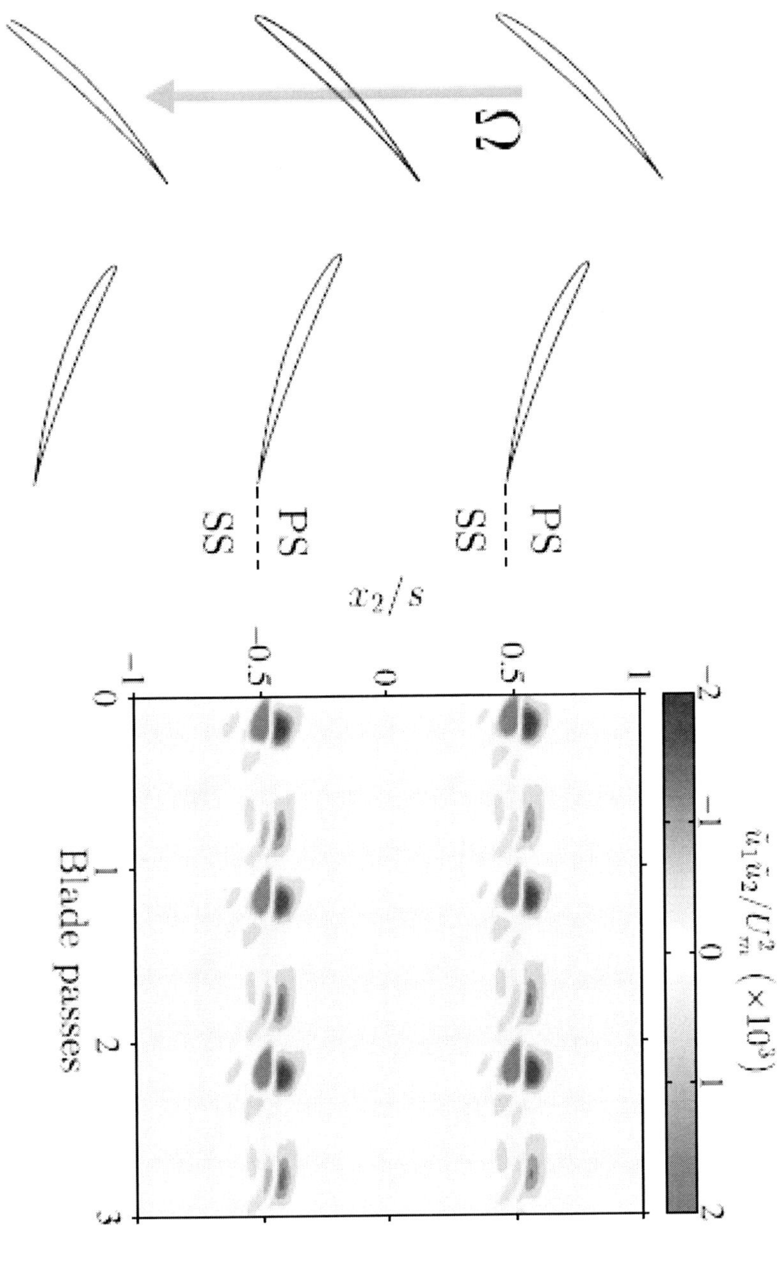
Ω
PS
SS
PS
SS
x_2/s
$\tilde{u}_1\tilde{u}_2/U_m^2$ $(\times 10^3)$
Blade passes

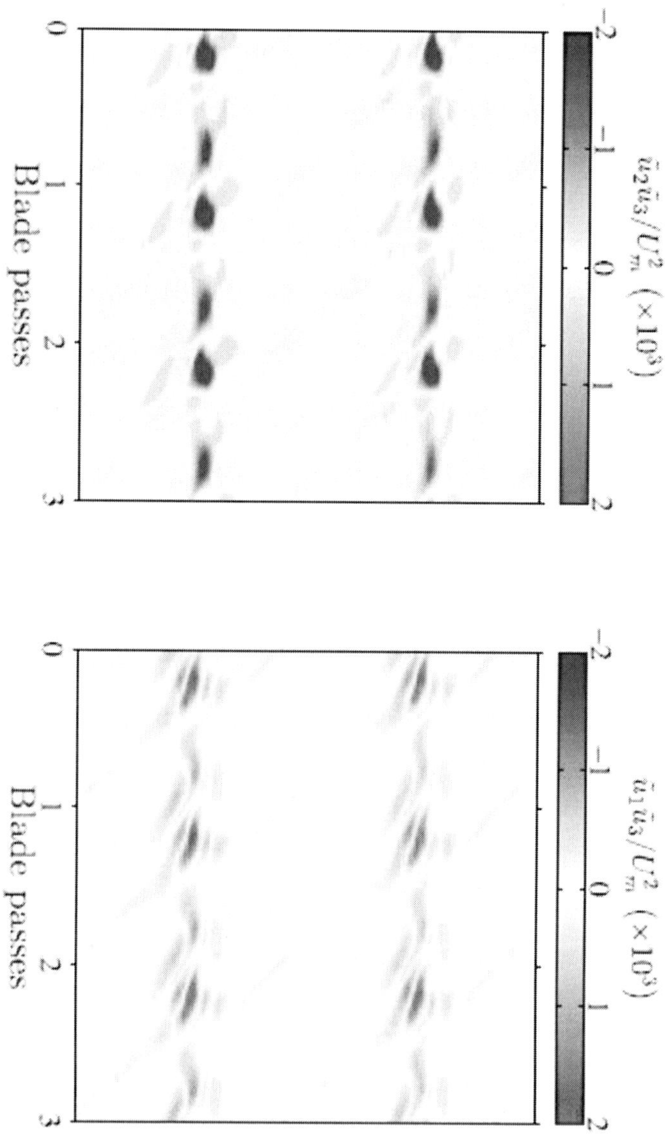

Figure 17 Phase-averaged deterministic shear stresses including the axial–tangential $\tilde{u}_1\tilde{u}_2$; tangential–radial $\tilde{u}_1\tilde{u}_3$, and axial–radial $\tilde{u}_2\tilde{u}_3$ components. Note that data are shown for three blade passing periods and have been normalised using the rotor speed at mid-height, U_m, and the stator pitch, s

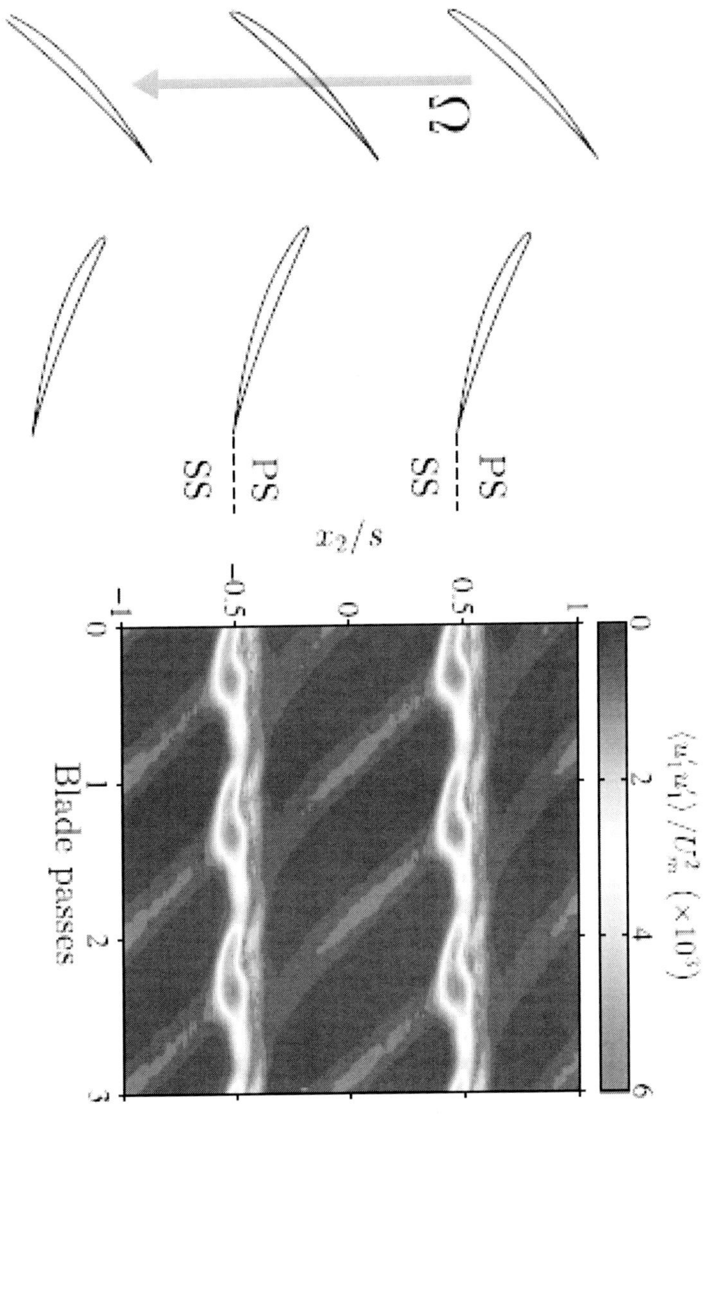
Ω
PS
SS
PS
SS
s/x2
Blade passes
0
1
2
3
−1
−0.5
0
0.5
1
0
2
4
6

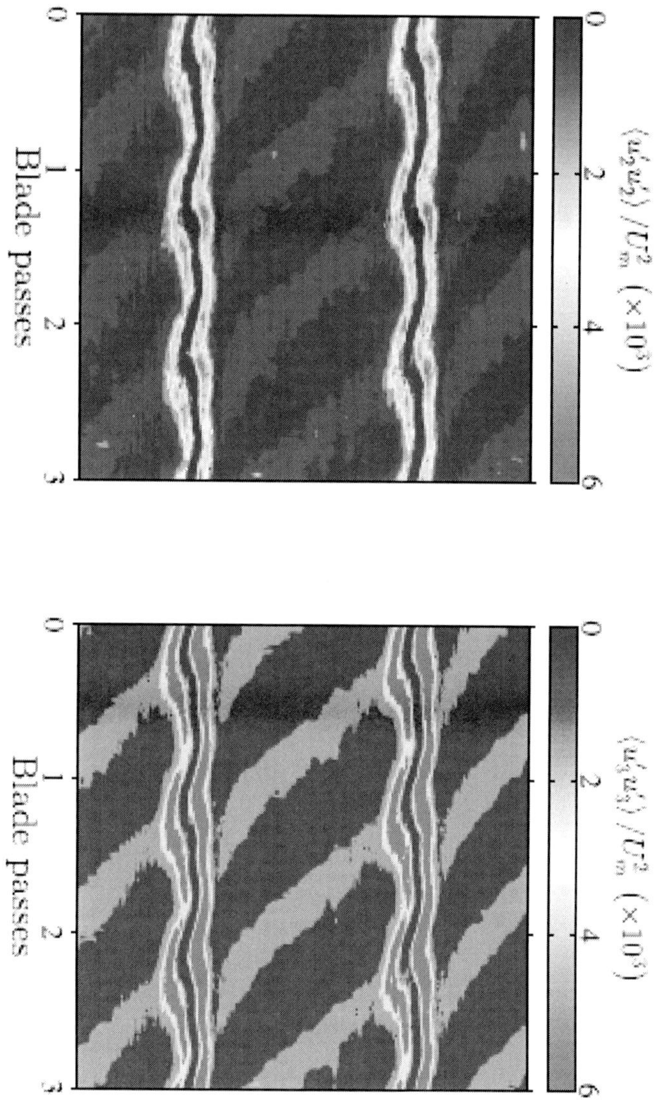

Figure 18 Phase-averaged Reynolds normal stresses including axial–axial $\langle u'_1u'_1\rangle$, tangential–tangential $\langle u'_2u'_2\rangle$, and radial–radial $\langle u'_3u'_3\rangle$ components. Note that data are shown for three blade passing periods and have been normalised using the rotor speed at mid-height, U_m, and the stator pitch, s

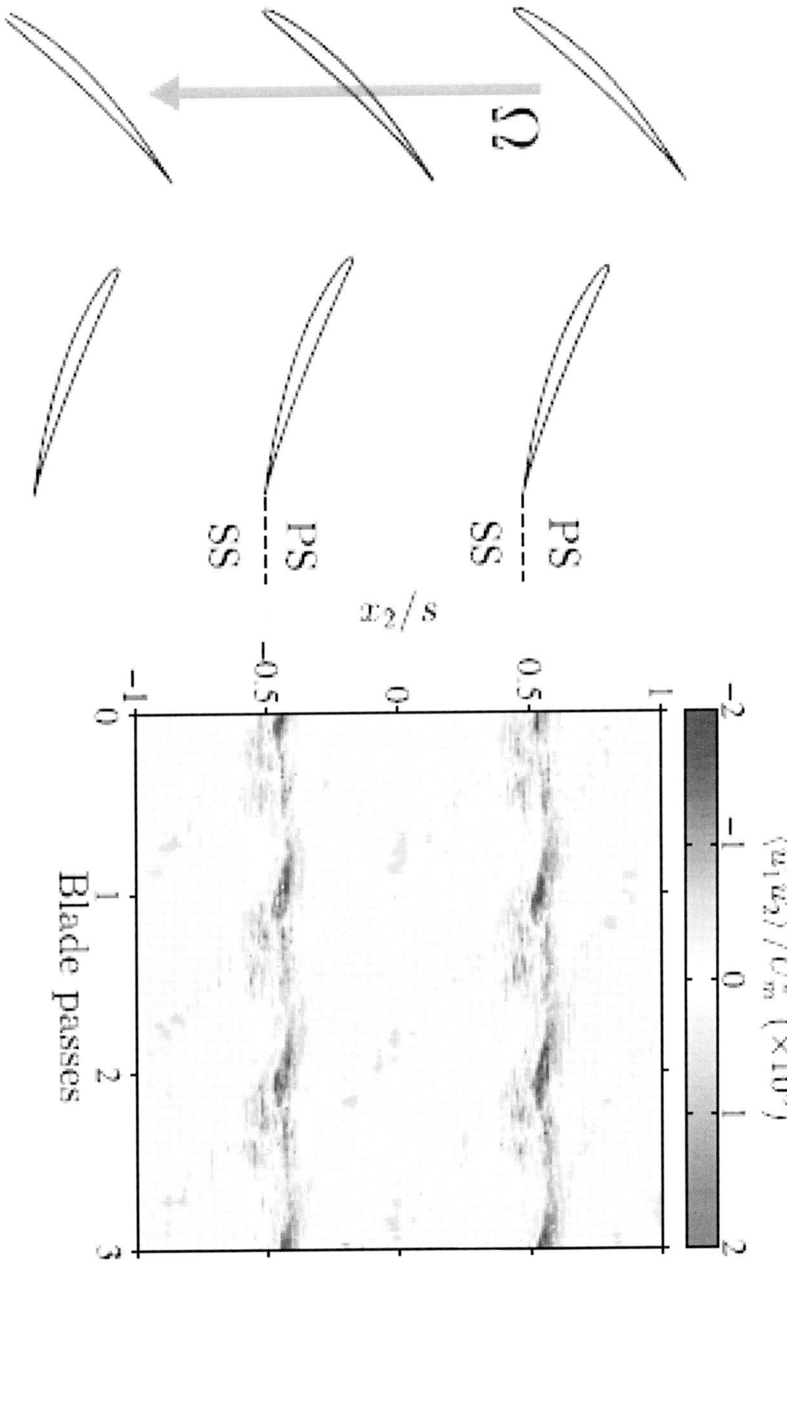

Ω
PS
SS
PS
SS
s/c_x
Blade passes

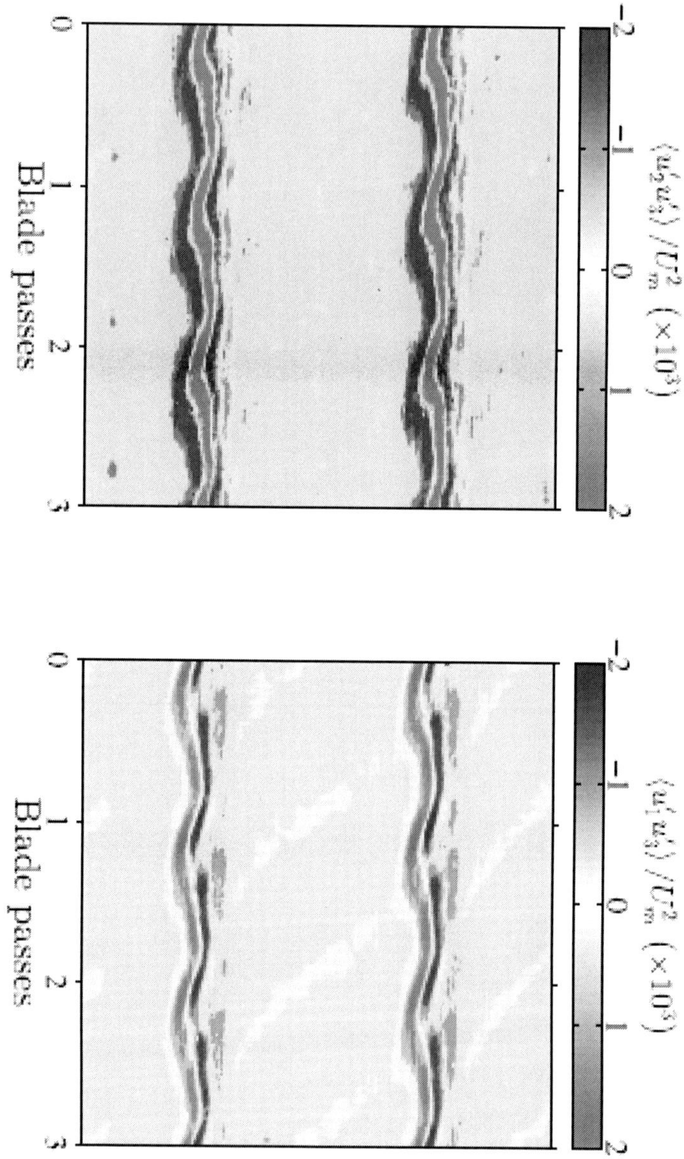

Figure 19 Phase-averaged Reynolds shear stresses including axial–tangential $\langle u'_1 u'_2 \rangle$, tangential–radial $\langle u'_2 u'_3 \rangle$, and axial–radial $\langle u'_1 u'_3 \rangle$ components. Note that data are shown for three blade passing periods and have been normalised using the rotor speed at mid-height, U_m, and the stator pitch, s

5.2 Phase-averaged deterministic stresses

In the preceding subsection, details of the phase-averaged velocity (see Fig. 14) and deterministic velocity (see Fig. 15) downstream of the compressor stator were provided. In this subsection, the six components of the phase-averaged deterministic stress tensor are presented and discussed.

The phase-averaged deterministic normal stresses downstream of the compressor stator are shown in Fig. 16. Space-time contours of the axial–axial, $\tilde{u}_1\tilde{u}_1$, tangential–tangential, $\tilde{u}2\tilde{u}_2$, and radial–radial, $\tilde{u}_3\tilde{u}_3$, components are plotted for three blade passing periods. In the free-stream region, the levels of deterministic normal stress can become very small ($\approx 0.01\ U_m$). On the other hand, considerable levels of time-varying deterministic normal stress occur within the stator wake—which agrees well with the deterministic velocity fields, as shown in Fig. 15. The time dependence of the normal stresses is related to the duty cycle of each rotor wake which supply a source of time-periodic shear. In addition to their intermittency, the components of normal stresses are distributed asymmetrically about the stator wake centre line. The axial–axial and tangential–tangential stresses show a biased towards the stator wake SS and PS, respectively. The maximum and minimum values of the deterministic turbulence intensity

$$\bar{u}_j \frac{\partial \bar{u}_i}{\partial x_j} = -\frac{\partial \bar{p}}{\partial x_i} + \frac{1}{Re}\frac{\partial^2 \bar{u}_i}{\partial x_j \partial x_j} - \underbrace{\frac{\partial\left(\overline{\tilde{u}_i \tilde{u}_j}\right)}{\partial x_j}}_{I} - \underbrace{\frac{\partial\left(\overline{u'_i u'_j}\right)}{\partial x_j}}_{II}$$

were computed to be 3 and 1% and occur within the stator wake and free-stream regions, respectively. The circumferential average of deterministic turbulence intensity at mid-span was computed to be 1.5%.

The phase-averaged deterministic shear stresses downstream of the compressor stator is shown in Fig. 17. Space–time contours of the axial–tangential, $\tilde{u}_1\tilde{u}_2$, tangential–radial, $\tilde{u}_2\tilde{u}_3$, and axial–radial, $\tilde{u}_1\tilde{u}_3$, components are plotted for three blade passing periods. The distribution of deterministic shear stresses is qualitatively similar to the normal stresses, as shown in Fig. 17. The strongest shear stresses are produced intermittently within the stator wake. In addition, each component of deterministic shear–stress exhibits a characteristic sign change about the stator wake centre line. In terms of magnitude, the dominant deterministic shear stress is the axial–tangential component, $\tilde{u}_1\tilde{u}_2$, followed by the axial–radial component, $\widetilde{u}_2\widetilde{u}_3$, and, finally, the tangential–radial component, $\widetilde{u}_1\widetilde{u}_3$.

5.3 Phase-averaged Reynolds stresses

In the preceding subsection, details of the phase-averaged deterministic normal stresses (see Fig. 16) and the phase-averaged deterministic shear stresses (see Fig. 17) were provided. In this subsection, the six components of the phase-averaged Reynolds stress tensor are presented and compared against their deterministic counterpart.

The phase-averaged Reynolds normal stresses downstream of the compressor stator are shown in Fig. 18. Space–time contours of the axial–axial, $\langle u'_1 u'_1\rangle$, tangential–tangential, $\langle u'_2 u'_2\rangle$, and radial–radial, $\langle u'_3 u'_3\rangle$, components are plotted for three blade passing periods. Relative to the deterministic normal stresses shown previously in Fig. 16, the Reynolds normal stresses exhibit some striking differences. First, the Reynolds stresses within the stator wake show a significantly lower level of intermittency and are more "fully-developed" in character. Second, the Reynolds normal stress exhibits a dual-peaked distribution about the stator wake centre line—a defining feature of normal stress statistics in the near-wake region (Hah and Lakshminarayana 1980). Finally, in the free-stream region, appreciable levels of Reynolds normal stress occur, which are in-phase with the passing rotor wakes. This observation implies that the deterministic flow structures support high levels of mean-squared stochastic unsteadiness in between adjacent stator wakes.

The maximum and minimum values of the turbulence intensity

$$I' = \frac{1}{U_m}\sqrt{\frac{1}{3}\langle u'_i u'_i\rangle} \times 100 \tag{35}$$

were computed to be 6.5 and 2.5% and occur within the stator wake and free-stream regions, respectively. The circumferential average of turbulence intensity was computed to be 3.7%, compared to 1.5% for the deterministic turbulence intensity.

The phase-averaged Reynolds shear stresses downstream of the compressor stator is shown in Fig. 19. Space–time contours of the axial–tangential, $\langle u'_1 u'_2\rangle$, tangential–radial, $\langle u'_2 u'_3\rangle$, and axial–radial, $\langle u'_1 u'_3\rangle$, components are plotted for three blade passing periods. Relative to the deterministic shear stresses shown previously in Fig. 17, the Reynolds shear stresses show a significant reduction of intermittency and exhibit low-frequency meandering motion about the stator wake centre line. In terms of magnitude, the dominant Reynolds shear stress is the tangential–radial correlation, $\langle u'_2 u'_3 \rangle$, followed by the axial–radial correlation, $\langle u'_1 u'_3 \rangle$, and finally, the axial–tangential component, $\langle u'_1 u'_2 \rangle$.

6 CONCLUSIONS

A rotated hot-wire method has been developed to characterise the unsteady, three-dimensional and multi-scale flow patterns which develop between compressor blade rows. Relative to past rotated hot-wire techniques, the current method benefits from superior levels of statistical fidelity and provides full details of the phase-averaged velocities, deterministic velocities, deterministic stresses, and Reynolds stresses. As a result, this method can be used to delineate the phase-averaged effect of rotor–stator interaction and non-linear turbulent convection. To improve the current understanding of momentum and energy exchange mechanisms in compressor annuli, the deterministic unsteadiness associated with the relative motion of blade rows must be separated from the genuine turbulence. This can be seen by considering the time-average of the phase-averaged Navier–Stokes equation:

where terms *I* and *II* show the phase-integrated effect of the deterministic and Reynolds stresses against the mean flow, respectively. The competing effects of deterministic and stochastic unsteadiness upon compressor performance have been a hotly debated topic ever since the works of Adkins and Smith (1982) and Gallimore and Cumpsty (1986). Furthermore, the phase-locked statistics reported in this paper will enable comparisons against other scale-resolving measurement techniques, such as three-component particle image velocimetry (Bhattacharya et al. 2016), and high fidelity numerical simulations (Brandvik and Pullan 2009) to be made.

Full details of the theoretical, numerical, and experimental aspects of a new rotated hot-wire method are provided in this paper and include: (1) a complete derivation of the phase-averaged sensor response equations following a triple decomposition of the instantaneous field variables; (2) the numerical solution of the sensor response equations using a weighted non-linear regression algorithm and (3) a demonstration of the technique in a low-speed, single-stage compressor research facility. Rotated hot-wire measurements were acquired at mid-span, 20% chord downstream from the stator trailing edge, whilst operating the compressor close to its design point. The phase-locked flow statistics reported in this paper reveal the space–time evolution of rotor–stator interactions (Figs. 14, 15), deterministic stresses (Figs. 16, 17), and Reynolds stresses (Figs. 18, 19) in the wake of a compressor stator. At mid-span, the relative magnitude of phase-integrated deterministic and Reynolds stress is comparable, whereas the relative magnitude of the phase-resolved fields can vary significantly. This variation is a due to the intermittency of the phase-averaged deterministic stresses—which become negligible for a large portion of each blade passing period. Unlike their deterministic counterpart, the Reynolds stresses exhibit a weaker phase dependence and remain more fully-developed within the stator wake.

Future measurement campaigns will focus on acquiring rotated hot-wire measurements in the compressor end-wall regions and downstream of the rotor. The relative magnitude of the deterministic and stochastic stress fields across the compressor operating map will also be studied. These data will be used to evaluate the appropriate terms in phase-averaged Reynolds stress transport equations which, ultimately, will lead to an improved understanding of transport phenomena in compressor flows.

ACKNOWLEDGEMENTS

The authors gratefully acknowledge Rolls-Royce plc and the UK TSB TUFT programme for funding this work and granting permission for its publication. All authors wish to thank Dr. J. J. Adamczyk, Dr. J. Bolger, Prof. N. A. Cumpsty, Dr. S. J. Gallimore, and Dr. J. S. S. Wong for their insightful comments and interest in this work. T.O.J. extends personal thanks to Mr. D. Barlow, Mrs. M. Folk, and Dr. H. C.-H. Ng for their assistance in the laboratory.

REFERENCES

1. Adamczyk JJ (1985) Model equation for simulating flows in multistage turbomachinery. ASME Paper No. 85-GT-226
2. Adkins GG, Smith LH (1982) Spanwise mixing in axial flow turbomachines. J Eng Gas Turbines Power 104:97–110
3. Berdanier RA, Key NL (2016) A novel data reduction technique for single slanted hot-wire measurements used to study incompressible compressor tip leakage flows. Exp Fluids 57:29. doi:10.1007/s00346-016-2114-z
4. Bhattacharya S, Berdanier RA, Vlachos PP, Key NL (2016) A new particle image velocimetry technique for turbomachinery applications. J Turbomach 138(12):124501
5. Bissonnette LR, Mellor GL (1974) Experiments on the behaviour of an axisymmetric turbulent boundary layer with a sudden circumferential strain. J Fluid Mech 63:369–413
6. Bradshaw P (1996) Turbulence modeling with application to turbomachinery. Prog Aerosp Sci 32:575–624
7. Bradshaw P, Cebeci T, Whitelaw JH (1981) Engineering calculation methods for turbulent flow. McGraw-Hill, London

8. Brandvik T, Pullan G (2009) An accelerated 3D Navier–Stokes solver for flows in turbomachines. ASME Turbo Expo 2009; Power for Land, Sea, and Air; GT2009-60052
9. Brundrett E, Baines WD (1964) The production and diffusion of vorticity in duct flow. J Fluid Mech 19:375–394
10. Buresti G, Di Cocco NR (1987) Hot-wire measurement procedures and their appraisal through a simulation technique. J Phys E Sci Instrum 20:87–99
11. Dawes WN (1991) Multi-blade row Navier–Stokes simulation of fan-bypass configurations. ASME paper 91-GT-148
12. De Grande G, Kool P (1981) An improved experimental method to determine the complete Reynolds stress tensor with a single rotating hot wire. J Phys E Sci Instrum 14:196–201
13. Denton JD (1990) The calculation of three dimensional viscous flow through multisage turbomachines. ASME paper 92-GT-364
14. Denton JD (2010) Some limitations of turbomachinery CFD. ASME Turbo Expo 2010; Power for Land, Sea, and Air; GT2010-22540
15. Díaz KA, Oro JF, Vega MG, Marigorta EB (2016) Effects of prong-wire interferences in dual hot-wire probes on the measurements of unsteady flows and turbulence in low-speed axial fans. Measurement 91:1–11
16. Dryden HL (1938) Turbulence investigations at the National Bureau of Standards. Appl Mech Int 5th Congr 5:362–368
17. Faure TM, Thierry M, Michon G, Miton H, Vassilieff N (2001) Laser Doppler anemometry measurements in an axial compressor stage. J Propuls Power 17(3):481–491
18. Fujita H, Kovasznay LSG (1968) Measurement of Reynolds stress by a single rotated hot wire anemometer. Phys Fluids 39:1351–1355
19. Gallimore SJ, Cumpsty NA (1986) Spanwise mixing in multi-stage axial flow compressors Part I. Experimental investigation. J Turbomach 108:2–9
20. Georgiou DP, Milidonis KF (2014) Fabrication and calibration of a sub-miniature 5-hole probe with embedded pressure sensors for use in extremely confined and complex flow areas in turbomachinery research facilities. Flow Meas Instrum 39:54–63
21. Goto A (1991) Three-dimensional flow and mixing in an axial flow compressor with different rotor tip clearances. ASME 1991 international gas turbine and aeroengine congress and exposition; 91-GT-89 1
22. Grimshaw S, Taylor J (2016) Fast settling millimetre-scale five-hole probes. In: ASME Turbo Expo 2016: Turbomachinery technical conference and exposition, American Society of Mechanical Engineers, pp V006T05A014–V006T05A014
23. Hah C, Lakshminarayana B (1980) Measurment and prediction of mean velocity and turbulence structure in the near wake of an airfoil. J Fluid Mech 115:251–282
24. Hinze JO (1959) Turbulence. McGraw-Hill, New York
25. Hsu S, Wo A (1997) Near-wake measurement in a rotor/stator axial compressor using slanted hot-wire technique. Exp Fluids 23(5):441–444
26. Hussain AKMF, Reynolds WC (1970) The mechanics of an organized wave in turbulent shear flow. J Fluid Mech 41(02):241–258
27. Inoue M, Kuromaru M (1984) Three-dimensional structure and decay of vortices behind an axial flow rotating blade row. ASME J Eng Gas Turbines Power 106:561–569
28. Jørgensen FE (1971) Directional sensitivity of wire and fibre-film probes. DISA information No. 11
29. King L (1914) Small cylinders in a stream of fluid: determination of the convection constants of small platinum wires with applications to hot-wire anemometry. Phil Trans R Soc (Lond) 214:373–432
30. Kuroumaru M, Inoue M, Higaki T, Abd-Elkhalek F, Ikui T (1982) Measurement of three dimensional flow field behind an impeller by means of periodic multi-sampling with a slanted hot wire. Bull JSME 39:1674–1681

31. Lakshminarayana B (1981) Techniques for aerodynamic and turbulence measurements in turbomachinery rotor. J Eng Power 103:374–392
32. Lakshminarayana B, Poncet A (1974) A method of measuring three-dimensional rotating wakes behind turbomachinery rotors. ASME J Fluids Eng 96:87–91
33. Ma H, Jiang H (2004) An experimental study of three-dimensional characteristics of turbulent wakes of axial compressor rotors. J Therm Sci 14:15–21
34. di Mare L, Jelly TO, Day IJ (2017) Angular response of hot wire probes. Meas Sci Technol 28(3):035303
35. Papavegros PG, Hedley AB (1979) A simple practical method for establishing turbulence characteristics by means of a single 45 degrees slant hot-wire probe in a field of known mean flow direction. J Phys E Sci Instrum 12:761–765
36. Prandtl L (1946) On boundary layers in three-dimensional flow. Reports and Transactions, No. 64, British M.A.P
37. Rai MM (1989) Three-dimensional Navier–Stokes simulations of turbine rotor-stator interaction. part i-methodology. J Propuls Power 5(3):305–311
38. Reynolds WC, Hussain AKMF (1972) The mechanics of an organized wave in turbulent shear flow. Part 3. Theoretical models and comparisons with experiments. J Fluid Mech 54:263–288
39. Schmidt DP, Okiishi TH (1977) Multistage axial flow turbomachinery wake production, transport and interaction. AIAA J 15:1138–1145
40. Shin H, Whitfield CE, Wisler DC (1994) Rotor–rotor interaction for counter-rotating fans, Part 1: Three-dimensional flowfield measurements. AIAA J 32:2224–2233
41. Sieverding HC, Arts T, Dénos R, Brouckaert JF (2000) Measurement techniques for unsteady flows in turbomachines. Exp Fluids 28(4):285–321
42. Sirivat A (1989) Measurement and interpretation of space–time correlation functions and derivative statistics from a rotating hot wire in grid turbulence. Exp Fluids 7:361–370
43. Treaster LA, Yocum AM (1978) The calibration and application of five-hole probes. DTIC Document
44. Uzol O, Chow Y, Meneveau C (2002) Experimental investigation of unsteady flow field within a two-stage axial turbomachine using particle image velocimetry. ASME J Turbomach 124:542–552
45. Webster CAG (1962) A note on the sensitivity to yaw of a hot-wire anemometer. J Fluid Mech 20:307–312
46. Weichert S, Day I (2014) Detailed measurements of spike formation in an axial compressor. J Turbomach 136(5):051006
47. Zierke WC, Okiishi TH (1982) Measurement and analysis of total-pressure unsteadiness data from an axial-flow compressor stage. J Eng Power 104(2):479–488

Index